Anionic Poly

Kinetics, Mechanism

Anionic Polymerization
Kinetics, Mechanisms, and Synthesis

James E. McGrath, EDITOR

*Virginia Polytechnic Institute
and State University*

Based on a symposium

sponsored by the Division

of Polymer Chemistry at

the 179th Meeting of the

American Chemical Society,

March 24–28, 1980.

A C S   S Y M P O S I U M   S E R I E S **166**

AMERICAN CHEMICAL SOCIETY
WASHINGTON, D. C.   1981

Library of Congress CIP Data

Anionic polymerization.
  (ACS symposium series; ISSN 0097-6156; 166)

  Includes index.

  1. Polymers and polymerization—Congresses.
  I. McGrath, James E. II. American Chemical Society.
Division of Polymer Chemistry. III. American Chemi-
cal Society. IV. Series.

QD380.A54   541.3'93   81–14911
ISBN 0–8412–0643–0 AACR2 ACSMC8 166 1–594
                                          1981

ACS Symposium Series

M. Joan Comstock, *Series Editor*

FOREWORD

The ACS SYMPOSIUM SERIES was founded in 1974 to provide a medium for publishing symposia quickly in book form. The format of the Series parallels that of the continuing ADVANCES IN CHEMISTRY SERIES except that in order to save time the papers are not typeset but are reproduced as they are submitted by the authors in camera-ready form. Papers are reviewed under the supervision of the Editors with the assistance of the Series Advisory Board and are selected to maintain the integrity of the symposia; however, verbatim reproductions of previously published papers are not accepted. Both reviews and reports of research are acceptable since symposia may embrace both types of presentation.

CONTENTS

PREFACE

This book is based on an international symposium on anionic polymerization. The need for such a symposium grew out of discussions between myself and other members of the Polymer Division Executive Committee, about two years earlier. At that time I pointed out that we had not had a symposium dealing with the subject since 1962! Clearly, much new scientific and important technological information had been generated since that time.

Anionic polymerization dates back at least to the early part of this century. Indeed, sodium-initiated butadiene polymers were investigated as potential synthetic rubbers many years ago. Unfortunately, the derived, high 1,2 microstructure shows a T_g of about 0°C. Electron transfer initiators also were studied by Scott in 1936.

The mid 1950's can be pinpointed perhaps as one of the golden eras of anionic polymerization (and, indeed, polymer science). Certainly, the discovery reported by Firestone that lithium-initiated polyisoprene had a structure quite similar to Hevea (natural rubber) should be noted. In addition, of course, one needs to comment on the important discovery by Professor Szwarc, M. Levy, and R. Milkovich that electron-transfer initiation by alkali metal polynuclear aromatic complexes could produce nonterminated living polymers of predictable molecular weights. These novel macromolecular carbanions were further shown to be capable of initiating other monomers to produce well-defined block polymers and to undergo reactions with reagents (e.g. carbon dioxide and ethylene oxide) that could provide functional end groups. In 1957 the Phillips Petroleum Company commercialized perhaps the first styrene–butadiene diblock copolymers. Many groups around the world began to intensively investigate kinetics, mechanisms, and synthesis possibilities of anionic polymerization. Here one should at least mention Professor Morton and his students at Akron (which include myself), Drs. Bywater and Worsfold in Canada, Professor Rempp and his many colleagues in France, and Professor Schulz in Germany. Many other scientists have contributed also to what we currently know about the subject but space prevents us from reviewing their efforts.

A few books and many reviews on anionic polymerization have appeared in the literature. Indeed, the contributors to the symposium and to this book also have written most of these! Briefly, one must cite the classic book of Professor Szwarc in 1968 and his later edited volumes on ion pairs. In addition, many of us have benefited from important reviews by Dr. Bywater and Professors Fetters and Rempp, among others. Currently, I am aware that Professor Morton has a new book that will be published soon.

One of the most important discoveries relating to synthesis and physical behavior was made by Dr. Milkovich while at the Shell Development Co. He and his colleagues showed that triblock copolymers containing polystyrene–polydiene–polystyrene blocks in appropriate sizes could behave as a physically cross-linked but linear thermoplastic elastomer. Thus Dr. Milkovich was involved with two very crucial discoveries in this field. Interestingly, he received his M. S. degree at Syracuse with Professor Szwarc and his Ph.D. at Akron with Professor Morton. I was pleased that Dr. Milkovich accepted my invitation to be a plenary speaker at the symposium, along with Professors Szwarc and Morton.

The symposium was, and this book is, truly an international contribution. There are papers from Australia, Belgium, Canada, France, Germany, Japan, Kuwait, Poland, the United Kingdom, and the United States. Invitations also were sent to U.S.S.R. scientists, but, unfortunately, none could attend. Nearly all of the papers presented at the symposium are published in this book. A few manuscripts were not received or were withdrawn for various reasons. The publication time was somewhat longer than initially perceived. I take responsibility for this and thank those authors who submitted manuscripts very promptly for their patience!

Those currently actively involved in anionic polymerization will recognize that despite the extensive progress that has been made over the past 25 years, many kinetic, mechanistic, and even synthetic aspects have not been elucidated fully. Thus, it should not be surprising that there are opposing points of view. In the past, controversies have occurred, and although I tried to minimize this factor, I was not completely successful in this regard. However, I was pleased that all of the speakers came to the same room and also contributed to this volume. A few of the papers may be a bit strong; however, I have decided to let the scientific community come to their own decisions on these matters.

This book contains very useful new work as well as critical reviews of specific areas. The scope is quite broad and ranges from ion pair structures to various features of the kinetics, mechanisms, and synthesis. Importantly, there are several fine industrial contributions that complement the academic studies. I hope that this book will be of interest and utility to students and scientists in academia, government, and industry.

I would like to thank all of the authors for their contributions to this book. I also would like to gratefully acknowledge the invaluable assistance of the ACS staff. Finally, the secretarial expertise provided by Debbie Farmer and Donna Perdue was indispensable.

The financial support of the symposium by The Firestone Tire and Rubber Co., Gulf Oil Chemicals Co., the 3M Co., the Shell Development Co., and the Polymer Materials and Interfaces Laboratory of our University was extremely helpful.

JAMES E. MCGRATH
Chemistry Department and
Polymer Materials and Interfaces Laboratory
Virginia Polytechnic Institute and State University
Blacksburg, Virginia

April 1981

1

Living and Dormant Polymers: A Critical Review

MICHAEL SZWARC

New York Polymer Center, New York State College of Environmental Science and
Forestry, Syracuse, NY 13210

A brief historical review of the concept of living
polymers and its ramifications is followed by consideration of various mechanisms of anionic polymerizations. The pertinent papers presented in this
meeting are surveyed. Special attention is devoted
to polymerizations involving Li counterions proceeding
in hydrocarbon solvents. It is stressed that living
and dormant polymers participate in such reactions
and the consequences of their presence are deduced.

Our studies of anionic polymerization that led us to the
concepts of living polymers and electron-transfer initiation resulted from a conversation I had with Professor Sam Weissman in
the Spring of 1955. He told me about his electron-transfer
investigations, such as naphthalene$^-$ + phenathrene $\rightleftarrows$ naphthalene
+ phenanthrene$^-$. As he talked, it occurred to me that electron-
transfer to styrene ought to convert its C=C bond into what
naively could be described as ·C-C$^-$. Then, the radical end should
initiate radical polymerization while anionic polymerization would
be simultaneously induced by the carbanionic end. Such a situation intrigued me, and I asked Sam whether he tried to transfer
electron to styrene. His answer was brief: "No use, it polymerizes." I asked then whether he would approve of our looking
into this problem and upon getting his consent, we started our
work that summer.

With the help of Moshe Levy and Ralph Milkovich, we developed
the all-glass, high-vacuum technique and established that electron
transfer to suitable monomers produces dimeric dianions by coupling the radical ends, while the carbanions induce anionic polymerization. Moreover, under the stringent purity conditions of
our experiments, the carbanions retain, virtually indefinitely,
their capacity of growth. Hence, the polymerization leading to
elongation of the previously formed chains may continue upon
"feeding" them with additional monomers. Our extremely simple
but highly convincing experiments were reported in 1956($\underline{1}$), and

0097–6156/81/0166–0001$05.00/0

the term "living polymers" was proposed to vividly convey the
idea that these interesting macromolecules resume their growth
whenever a suitable monomer is supplied. The ramifications of
this concept were clearly outlined: the feasibility of synthesizing polymers of narrow molecular-weight distribution,
tailoring block-polymers to desired patterns, the preparation of
polymers with functional end-groups, etc. These opportunities
equally apply to living polymers whether endowed with one or two
growth perpetuating ends. However, the latter provide a route
to synthesis of bifunctional polymers and easily yield the ABA
triblock polymers.

Our ideas rapidly caught the attention of polymer chemists
throughout the world and extensive studies of living polymers
began in many laboratories. These investigations progressed in
two directions: the synthetic one and the mechanistic. The
synthetic work led to refined preparations of mono-dispersed
polymers, allowing for more precise studies of the effects of
molecular weight distribution on the rheological properties of
polymer solutions and solid polymers. Studies of block-polymers
flourished, as is evident from the several symposia covering the
subject(2). The spectacular discovery by Ralph Milkovich(3) of
the triblock polymers yielding thermo-plastic rubber and the
unexpected difference in the properties of triblock and diblock
polymers added further impetus to those investigations. This
subject will be reviewed by Ralph Milkovich and by Teyssie later
during this symposium. Synthesis of functional and difunctional
polymers was greatly improved by Paul Rempp and his associates(4),
and we look forward to Schulz' discussion of this subject. Controlled preparation of various types of macromolecules, such as
star-shaped polymers, recently reviewed by Bywater(5), comb-
shaped polymers, etc., became feasible with the advent of living
polymers. In fact, the synthesis of various types of model
macromolecules by anionic polymerization will be reported here by
Rempp, Franta and Hertz.

Let me digress and ask two questions. Were we the first to
discover living polymers? The answer is no. Review of the literature showed that living polymers, although not named in that
way, were described by Ziegler(6) in 1929 and the terminationless
polymerization was considered by Mark and Dostal(7) and by
Flory(8). However, the potentiality of those systems was not
fully appreciated until our rediscovery of these interesting
molecules. Also, our work led to recognition of the necessary
conditions required for such studies, namely, the stringent
purity, vacuum techniques, etc.

Is anionic polymerization unique in providing living polymers? It is not. The cationically formed living polymers are
well known today and, in fact, Richards will describe here the
techniques of converting one kind of living polymers to another.

Ionic polymerization is much more colorful than radical polymerization. While only one species participates in radical

homo-polymerization, a variety of species may co-exist and
simultaneously contribute to ionic polymerization. Living polymers facilitate their characterization and provide ways of
studying their inter-conversion, mode of growth, and the degree
of their participation in the overall process. Indeed, these
subjects form the themes of numerous contributions to this
symposium.

Free ions and their respective ion-pairs participate in
numerous anionic and cationic polymerizations; their contribution
is affected by the nature of counter-ions, solvent, and solvating
agents whenever added to the reacting system. For example, the
effect of cryptates on some anionic polymerizations will be
discussed by Mademoiselle Boileau. Moreover, as revealed by the
elegant studies of Professor Johannes Smid(9), a variety of ion-
pairs may co-exist in polymerizing systems, each kind having its
own characteristic behavior.   The present status of this most
interesting field will be reviewed by Professor Smid, and the
dynamics of the interchanges taking place between those species
will be discussed by Dr. Andre Persoon, who developed the most
ingenious techniques allowing him to measure the rates of those
rapid reactions.

Triple ions form rather unusual species. They contribute
little, if at all, to the propagation but their formation may have
a buffering effect. The intramolecular triple ions' formation in
the cesium polystyrene system illustrates this phenomenon(10).
The mechanism is outlined in Scheme 1. The formation of triple
ions increases the concentration of Cs^+ ions and buffers, therefore, the dissociation of ion-pairs into free styryl ions which
are the main contributors to the propagation.

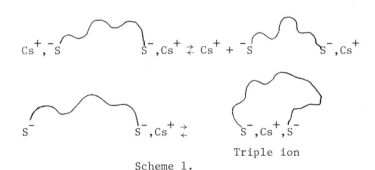

Triple ion

Scheme 1.

Generally, two distinct processes could be responsible for
the formation of triple ions, namely

$$2Cat^+,An^- \rightleftarrows Cat^+ + An^-,Cat^+,An^- \tag{a}$$

or

$$2Cat^+,An^- \rightleftarrows An^- + Cat^+,An^-,Cat^+ \tag{b}$$

Whenever equilibrium (a), which is concentration-<u>independent</u>,
is more important than the conventional dissociation of ion-pairs,
then its effect, in conjunction with the ion-pairs dissociation,
decreases the anticipated concentration of anions. The reverse
is true when equilibrium (b) prevails on (a).

A somewhat similar problem was encountered in the anionic
propagation involving the barium salt of living poly-styrene
possessing only one growth-propagating end per macromolecule(11).
The pseudo-first order rate constant of propagation was found to
be independent of the salt concentration. This result was
accounted for by a mechanism outlined in Scheme 2 with the
assumption that the free poly-styrene anions are the only species
contributing to the growth.

$$2Ba^{2+}, \; (S^- \rlap{\,}\text{\rotatebox{0}{$\wr\wr\wr$}})_2 \; \rightleftarrows \; Ba^{2+}, \; (S^- \text{\rotatebox{0}{$\wr\wr\wr$}}) + Ba^{2+}, \; (S^- \text{\rotatebox{0}{$\wr\wr\wr$}})_3$$

$$Ba^{2+}, \; (S^- \text{\rotatebox{0}{$\wr\wr\wr$}})_2 \; \rightleftarrows \; Ba^{2+}, \; (S^- \text{\rotatebox{0}{$\wr\wr\wr$}}) + S^- \text{\rotatebox{0}{$\wr\wr\wr$}}$$

Scheme 2.

Denoting by K_{triple} and K_{diss} the respective equilibrium constants

and by k_p the bimolecular rate constant of propagation by the free
poly-styrene anions, one finds the observed pseudo-first order
rate constant to be given by

$$k_p K_{diss}/K_{triple}^{1/2}$$

i.e., it is independent of the salt concentration because within
the investigated concentration range the relation

$$K_{triple} \;\; \gg \; K_{diss}/[Ba^{2+},(S^- \text{\rotatebox{0}{$\wr\wr\wr$}})_2]$$

was found to be valid.

A similar problem concerning the strontium salt will be discussed here by Professor Van Beylen, while the problem of the
barium salt of the poly-styrene endowed with two growth propagating ends will be dealt with by Dr. Francois.

The various kinds of growing species differ not only in their
propagation but also in their stereochemical preferences. Professor Hogen-Esch will review this subject in his talk on anionic
oligomerization of some vinyl monomer, and mechanisms of anionic,
stereospecific polymerization of 2-vinyl pyridine will be discussed by Dr. Fontanille. In this context, the interesting paper
of Schuerch et al.(12) deserves attention. Their work clearly
reveals the effect of cation solvation upon the mode of monomer's
approach to the growing centers.

The all cis polymerization of isoprene when carried out in
hydrocarbon solvents with Li counter ions provides the most
spectacular case of stereospecificity. It was reported by Stavely
and his colleagues(13) in 1956, and their discovery greatly

increased the interest in anionic polymerization. The profound
difference in stereospecificity of poly-isoprenyl lithium when
compared with the salts involving other alkali cations, and the
effect of solvation, fostered interest in the structure of polydienyl salts. The living character of these polymerizations
allows one to use the NMR as a tool for such studies, and indeed,
significant results were reported by various workers. The contributions of Dr. Worsfold, of Dr. Halasa and his colleagues, and
of Dr. Bywater cover some of these problems.

The understanding of the mechanism of propagation of polymers
involving lithium cations and proceeding in hydrocarbons stems
from the pioneering studies of Worsfold and Bywater(14). Not only
was the living character of these polymerizations unambiguously
established, but the kinetics of propagation of lithium polystyrene in benzene revealed the involvement in this process of
dormant, inactive dimeric polystyrenes which are in equilibrium
with a minute fraction of growing monomeric lithium polystyrenes.
This conclusion was deduced from the observed square-root dependence of the propagation rate on the concentration of the polymers, i.e.,

$$\text{the rate of propagation} = k_p (K_{diss}/2)^{1/2} [\text{Li-polyS}]^{1/2} [S]$$

where k_p is the propagation constant of the monomeric polymers
and K_{diss} is the dissociation contant of the dimers. Since this
relation applies even at polymer concentrations as low as 10^{-5}M,
Dr. Bywater pointed out in his talks that K_{diss} has to be of the
order of 10^{-6}M or less.

The above mechanism served as a prototype accounting for
other similar polymerizations. However, the kinetics of propagation of lithium polydienes revealed a dependence on the polymer
concentration lower than 1/2, apparently 1/4 or less order. This
prompted the assumption that the lithium polydienyls form higher
aggregates than dimers.

A useful quasi-quantitative technique allowing determination
of the degree of association of such polymers was reported by
Professor Morton(15). The viscosity of concentrated solutions
of high molecular weight polymers is proportional to a power of
their weight average molecular weight, M_w, viz.,

$$\eta \simeq M_w^\alpha$$

where α, although not precisely known, is within the range of 3.3
- 3.5 [see thorough review by Porter and Johnson(16)]. On
"killing" the aggregated polymers, say by adding a drop of methanol, the aggregates are converted into monomeric dead polymers,
and the viscosity of the solution drastically decreases. Thus,
the ratio of viscosities before and after methanol addition is
given by N^α, where N denotes the degree of association of the
living polymers.

In spite of some uncertainty in the precise value of α, this method gives N with reasonable accuracy. For example, if the ratio of viscosities is 10, then N = 2.01, for α = 3.3, and N = 1.93 for α = 3.5.

Using this approach, Professor Morton and his associates confirmed the dimeric nature of lithium polystryl in benzene, in agreement with the deduction of Worsfold and Bywater, but they claimed a dimeric nature also for lithium polydienyls in hydrocarbon media. The latter claim conflicted with the kinetic evidence (about 1/4 order) which, after initial denials(17), eventually was confirmed by the Akron group(18).

Controversy arose because the viscosity studies of other workers(19) led to different conclusions. The extension of these investigations, involving light-scatter studies, again led to conflicting results when performed in one laboratory or the other. Moreover, Dr. Fetters and Professor Morton were unable to account for the 1/4 order of propagation of the presumably dimeric polydienyls. Strangely enough, they stated(20) that "plausible mechanisms" accounting for the dimeric nature of the aggregates and 1/4 order of propagation were presented, without specifying a single one but quoting 12 irrelevant references in support of their claim.

Let me suggest a way by which the kinetic and viscosity results could be reconciled, although I have reservations about the validity of the mechanism I am proposing. Let us assume that the polymers with lithium cations are dimeric in hydrocarbon solvents. Their dissociation into free lithium cations is unlikely in hydrocarbon media, although I would not exclude the feasibility of forming the macro-anions. Hence, I reject the dissociation

$$(\text{Li}^+, \text{poly-dienyl}^-)_2 \rightleftharpoons 2\text{Li}^+ + 2\text{poly-dienyl}^-$$

However, an alternative mode of dissociation of the dimers is possible, namely

$$(\text{Li}^+, \text{poly-dienyl}^-)_2 \rightleftharpoons (2\text{Li}^+, \text{poly-dienyl}^-)$$
$$+ (\text{poly-dienyl}^-), \quad K_2$$

This equilibrium in conjunction with the equilibria

$$(\text{Li}^+, \text{poly-dienyl}^-)_2 \rightleftharpoons 2(\text{Li}^+, \text{poly-dienyl}^-), \quad K_1$$

and

$$(\text{Li}^+, \text{poly-dienyl}^-) + (\text{poly-dienyl}^-) \rightleftharpoons$$
$$\text{Li}^+, (\text{poly-dienyl}^-)_2, \quad K_3$$

leads to the following rate expression, provided that the free (poly-dienyl$^-$) ions are the only species contributing to the propagation. Thus,

$$\text{Rate of propagation} = k_p [(\text{poly-dienyl}^-)][M], \text{ i.e.,}$$

Rate of propagation/[M]   =

$$k_p K_2^{1/2}[(Li^+, P^-)_2]^{1/2}/(1 + K_1^{1/2}K_3[(Li^+,P^-)_2]^{\frac{1}{2}})^{\frac{1}{2}}$$

where Li^+,P^- is short for Li^+, polydienyl. For
$K_1^{1/2} \cdot K_3[(Li^+, \text{poly-dienyl})_2]^{1/2} \ll 1$, this expression yields
propagation rate proportional to the square-root of the living
polymers concentration; however, for

$$K_1^{1/2} \cdot K_3[(Li^+,\text{poly-dienyl}^-)_2]^{1/2} \gg 1$$

the rate becomes proportional to 1/4 power of living polymers
concentration. Since K_3 is large while K_1 is very small, both

extremes could be encountered.

The above-mentioned controversy led me to a critical review
of papers co-authored by Dr. Fetters(21). The results were perturbing, and reluctantly I had to conclude that at least two
examples illustrating the gravity of the problem should be
discussed.

The first example of questionable results published by
Morton and Fetters(22) is provided by their interesting attempt
to verify the dimeric nature of lithium salts of living polymers
in hydrocarbon media. Since these aggregates rapidly dissociate
into their unassociated forms which, in turn, undergo rapid re-
association, mixing of two types of homo-aggregates should lead
to the formation of mixed aggregates. Thus, the addition of a
lithium salt of a living oligomer, $(\text{oligomer},Li)_2$, to a lithium
salt of a living high-molecular weight polymer, $(\text{hmw},Li)_2$, should
convert some of the latter aggregates into mixed ones, i.e.,

(hmw,Li, oligomer,Li)

In the viscosity study the mixed dimers behave as if they were the
unassociated high-molecular weight polymers, i.e.,

(hmw,Li)

Hence, the viscosity of the solution of high-molecular weight
polymers substantially decreases on addition of the oligomer salt
and the viscometrically determined average degree of association,
N, should allow a reliable determination of the equilibrium
constant, K, of the athermal mixing process.

Accepting the dimeric nature of the aggregates, one concludes
that the mixing is represented by the following expression:

$$(\text{hmw},Li)_2 + (\text{oligomer},Li)_2 \rightleftarrows 2(\text{hmw},Li, \text{oligomer},Li)$$

$$K = [(\text{hmw},Li, \text{oligomer},Li)]^2/[(\text{hmw},Li)_2][\text{oligomer},Li)_2]$$

The lack of accounting for the symmetry factor of 2 in the
treatment of the heterodimers led the authors to a wrong derivation of K = 1, instead of the correct value of 4. Although the

viscometric technique should reliably give K $\approx$ 4, the reported
results were 0.85, 0.90, 1.1. for lithium polyisoprenyl in
hexane, and again, 0.97 and 1.1 for lithium polystyryl in benzene.
    The agreement of these results with the erroneous expecta-
tion of the authors is perturbing.
    The viscometric approach was adopted by Fetters and Morton
(17) for the determination of the dissociation constants and heat
of dissociation of the presumably dimeric lithium poly-isoprenyl
in hexane. The viscometer flow times were measured for the
living polymers and terminated ones, thus providing the $\underline{t}_1$ and $\underline{t}_2$
values. The 3.40 root of t_1/t_2, denoted by N, was calculated
from the experiments performed at 30°, 40° and 50°C, and its
derivation from the anticipated value of 2.00 was taken as the
fraction of dissociation. The data listed in Table V of Ref.
(17a) are partially reproduced here in Table I. For the sake of
brevity, only the first and last experiments of each series are
included in that table.
    Accepting the reasoning of the authors, one finds the
dissociation constant K to be

$$K = 2 C_o (2.000-N)^2/(N-1)$$

where C_o denotes the total concentration of living polymers. As
the factor 2.000-N is extremely small, K is enormously affected
by any minute errors in N.
    Two objections are raised to the validity of this approach:
        (1) It is technically unlikely to prepare and handle
high-molecular weight living polymers without inactivating at
least 2-3%. Hence, the small deviations of N from the expected
value of 2.00 (e.g., 1-3%) are meaningless.
        (2) The results are extremely sensitive to the numeri-
cal value of the exponent α. Even a minute deviation of α from
the claimed value of 3.4 dramatically affects the calculated
equilibrium constants. To demonstrate this fact, the K's were
calculated on the basis of α's slightly different from 3.40,
namely 3.37 and 3.42, both values reported by Morton and Fetters.
The results are given in Table I and are graphically shown in
Figure 1. The circles represent the K's claimed by the authors —
their precision is remarkable. Inspection of Figure 1 shows that
the uncertainties of the data are comparable to the range of log
K's investigated in that study. Hence, the claimed value of
$\Delta H = 37.4 \pm 0.8$ kcal/mole has no foundation. In fact, $\Delta H \sim$
50 kcal/mole is obtained for $\alpha = 3.42$ while $\Delta H \sim 25$ kcal/mole is
deduced for $\alpha = 3.37$.
    Apart from the objections raised above, the value ΔH of 37.4
kcal/mole is unacceptable for two reasons.
        (1) In conjunction with $K_{diss} \approx 10^{-6}$M that ΔH leads
to an impossibly large ΔS of +90 eu. For such a dissociation ΔS
may be at the most ≈ 20 eu, and this leads to ΔH of 14-15 kcal/
mole.

TABLE I.

$$N = \left\{ (\text{Time Flow of LP})/(\text{Time Flow Dead P}) \right\}^{1/\text{Exponent}}$$

10^3[LP]/M	N FOR DIFFERENT EXPONENTS			% UNASSOC. P FOR 3.40	RESPECTIVE K'S FOR		
	3.37	3.40	3.42		3.37	3.40	3.42
30°C					10^7K		
1.30	1.975	1.987	1.995	1.3%	16.7	4.4	0.65
3.50	1.979	1.991	1.999	0.9%	31.5	5.9	0.07
40°C					10^6K		
1.30	1.956	1.968	1.976	3.2%	5.2	2.8	1.5
3.50	1.961	1.973	1.981	2.6%	11.1	5.2*	2.6
					Erroneously reported as 3.8		
50°C					10^5K		
1.30	1.889	1.910	1.917	9%	3.6	2.3	1.95
3.50	1.932	1.943	1.951	5.7%	3.5	2.5	1.77

$$K = 2C_o \ (2.000-n)^2/N-1.000)$$

% Unassoc. = 100(2-N)

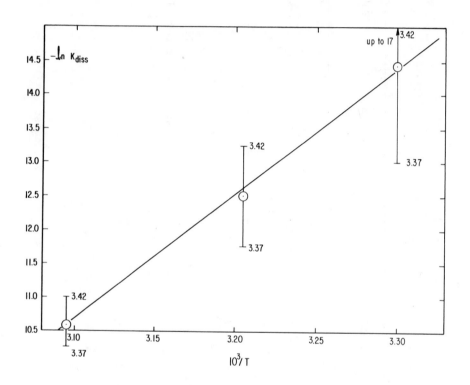

Figure 1.   The ln K$_{diss}$ for Li-polyisoprene in hexane.

(2)   Since the activation energy of dissociation must be at least as large as ΔH, the value of E_{diss} of 37.4 kcal/mole would lead to the rate constant of dissociation of 10^{-11} sec^{-1}, even for the largest plausible pre-exponential factor of 10^{15} sec^{-1}.   In such a case, hardly any dimer would dissociate in a time of an experiment.   For E_{diss} of 14-15 kcal/mole a reasonable k_{diss} of about 100 sec^{-1} is obtained.   These two examples, as well as the other discussed elsewhere (21), cast serious doubts on the reliability of the other claims of Dr. Fetters, including those reported at this meeting.*   It is interesting to quote in this context his recent statement printed on p. 740 of the Polymer Preprints, ACS Meeting, Spring 1979, in Honolulu, Hawaii:

"Parenthetically, we wish to note that the initial association state results obtained from viscometric technique were declared some time ago to be invalid as a result of underlined technical difficulties. Suffice it to mention that the author of this erroneous assessment (reference to my book CARBANIONS, LIVING POLYMERS, AND ELECTRON TRANSFER PROCESSES, Interscience Publishers, 1968, persumably pp. 499 and 500) has himself recently used this Technique; albeit without acknowledging the original or later association work, the development of the capping technique which he has exploited, nor his faulty commentary."

The underlining is mine.

Let me pass now to some problems of co-polymerization of lithium salts of living polymers. With Dr. Zdenek Laita we studied the rate of cross-over reaction converting lithium polystyryl into 1,1-diphenyl ethylene⁻, D⁻, end-groups in benzene (23). The stoichiometry of the reaction is

$$\text{\~\~\~\~} S^-, Li^+ + D \rightarrow \text{\~\~\~\~} SD^-, Li^+$$

S^- denoting the styryl unit. Since lithium poly-styryl in benzene is dimeric and the reaction is presumably carried out by the minute fraction of monomeric poly-styryl, we expected 1/2 order of the conversion, provided the excess of D is large. However, the conversion obeyed a first order law, i.e.,

$$d \ln(\text{\~\~\~\~} S^-, Li^+)/dt = k_i[D]; \quad [D] = [D]_o = \text{const.}$$

Moreover, the pseudo-first order rate constant, k_i, was found to

---

*In the paper submitted to this meeting, Fetters claims that the viscosity could not be measured if its value exceeds 10^3 poise. There is no inherent limit for viscosity measurements, even in conventional types of viscometers, provided that the tubes are sufficiently wide. His remark about the effect of 2.4% dilution is again erroneous. Such an effect is clearly shown by the data of Table III of Ref. (26).

be inversely proportional to the square-root of the initial concentration of $\sim\!\!\sim\!\!\sim S^-, Li^+$. The peculiar features were accounted
for by the following mechanism.

The polymers are present in the form of homo- and mixed
dimers, $(\sim\!\!\sim\!\!\sim S^-, Li^+)_2$, $(\sim\!\!\sim\!\!\sim SD^-, Li^+)_2$ and $(\sim\!\!\sim\!\!\sim S^-, Li^+, \sim\!\!\sim\!\!\sim SD^-,$
$Li^+)$. These are in equilibrium with minute concentrations of the
monomeric polymers, vis.,

$$(\sim\!\!\sim\!\!\sim S^-, Li^+) = u \text{ and } (\sim\!\!\sim\!\!\sim SD^-, Li^+) = v, \text{ with } v/u = f$$

Hence,

$$(\sim\!\!\sim\!\!\sim S^-, Li^+)_2 = (1/2)K_1 u^2, \quad (\sim\!\!\sim\!\!\sim SD^-, Li^+)_2 = (1/2)K_2 v^2$$

and

$$(\sim\!\!\sim\!\!\sim S^-, Li^+, \sim\!\!\sim\!\!\sim SD^-, Li^+) = K_{12} uv$$

The rate of conversion is given by $k_a Du$, $D = [CH_2 = CPh_2] = $ const.
Denoting the initial concentration of lithium polystyryl by C_o
and by g the fraction of lithium polystyryl present in any form
at time t, one finds

$$C_o = u^2(K_1 + 2K_{12}f + K_2 f^2), \quad g = (K_1 + K_{12}f)/(K_1 + 2K_{12}f$$

$$+ K_2 f^2) \text{ and hence } d\ln g/dt = (k_a D/C_o^{1/2})(K_1 + 2K_{12}f$$

$$+ K_2 f^2)^{1/2}/(K_1 + K_{12}f). \text{ For } K_{12} = (K_1 K_2)^{1/2} \text{ this}$$

$$\text{expression is reduced to } d\ln.g/dt = k_a D/K_1^{1/2} C_o^{1/2}$$

being in agreement with the experimental findings of pseudo-first
order conversion and the pseudo-first order rate constant being
inversely proportional to the square-root of the <u>initial</u> concentration of lithium polystyryl, i.e., $C_o^{-1/2}$.

For $k_{12} \neq (K_1 K_2)^{1/2}$ the plot of $\ln.g$ <u>versus</u> time becomes
curved, but its initial slope would still be inversely proportional to $[\text{lithium polystyryl}]_o^{1/2}$.

The interesting feature of this kind of reaction is their
memory. The observed rate at a chosen   concentration of
polystyryl depends on its initial concentration. This is evident
from inspection of Fig. 2; the effect arises from the influence
of products of the reaction on its rate. This is a general phenomenon. In the system discussed here, the formation of the
mixed dimer decreases the concentration of the reactive monomeric
lithium polystyryl and thus retards the reaction.

Mixed dimerization also affects the kinetics of anionic copolymerization of styrene and para-methyl styrene, a system
studied by O'Driscoll(24). The reaction was performed in benzene
with lithium counter-ions and the plots of ln.[styrene] or
ln(para-methyl styrene] <u>vs</u>. time were both linear. Their slopes,
denoted by λ_1 and λ_2, were, however, different from each other.

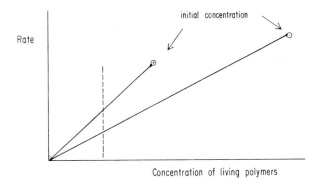

Figure 2.   Memory effect.

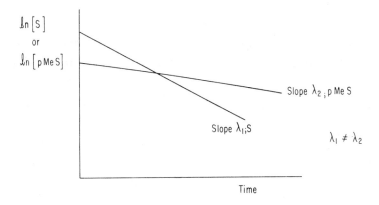

Figure 3.   Copolymerization of styrene and p-*methylstyrene initiated by RLi in benzene.*

This is seen in Fig. 3. These results were accounted for by
Yamagishi and myself(25) in the terms of the previous treatment.
Denoting again by u and v the concentration of the monomeric
living polymers terminated by styryl or para-methyl styryl unit,
respectively, and by C_o the concentration of all the polymers, we
find

$$K_1 u^2 + 2K_{12} uv + K_2 v^2 = C_o$$

$$k_{11} u + k_{21} v = \lambda_1 \qquad \text{and} \qquad k_{12} u + k_{22} v = \lambda_2,$$

with λ_1 and λ_2 being constant and the other symbols having their
usual meaning. Since u and v are variables, but not related to
each other through the equation,

$$k_{21} v [\text{styrene}] = k_{12} u [\text{para-methyl styrene}]$$

the first three equations have to be identical. This leads to
the conditions

$$K_{12} = (K_1 K_2)^{1/2}$$

and

$$k_{11}/k_{12} = k_{21}/k_{22} \quad \text{i.e.,} \quad r_1 r_2 = 1$$

Moreover

$$\lambda_1 = \gamma_1 C_o^{1/2} \qquad \text{and} \qquad \lambda_2 = \gamma_2 C_o^{1/2}$$

with γ_1 and γ_2 as two different constants, independent of the
concentrations or composition of the monomers and the concentration of the initiator, i.e., independent of C_o. Let it be
stressed again that all these conclusions are mathematical
consequences of the empirical findings revealed in Fig. 3, and the
assumptions of the existence of inert homo- and mixed dimers in
equilibrium with the active growing monomeric polymers.

In conclusion I thank the National Science Foundation for
the years of continuous support of our work.

Literature Cited

1. a) Szwarc, M.; Levy, M.; Milkovich, R. J. Amer. Chem. Soc.
      1956, 78, 2656.
   b) Szwarc, M. Nature 1956, 178, 1168.
2. a) Burke, J.J.; Weiss, V., Ed.; "Block and Graft Copolymers";
      Syracuse University Press: Syracuse, N.Y., 1973.
   b) Aggarwal, S. L., Ed.; "Block Polymers"; Plenum Press, 1970.
   c) LaFlair, R. T.; "Structure, Morphology and Properties of
      Block Polymers"; XXIII, IUPAC, 8, Butterworth, 1971.

3. a) Milkovich, R. (Shell), S. African Patent 642,271, 1964.
   b) Milkovich, R.; Holden, G.; Bishop, E. T.; Hendricks, W. R.
      Brit. Patent 1,035,873, 1966; U.S. Patent 3,231,635, 1966.
4. Rempp, P.; Loucheux, M. H. Bul. Soc. Chimique, France, 1958,
   1497.
5. Bywater, S. Adv. Polymer Sci. 1979, 30, 89.
6. a) Ziegler, K.; Bahr, K. Chem. Ber. 1928, 61, 253.
   b) Ziegler, K.; Colonius, H.; Schafer, O. Ann. Chem. 1929,
      473, 36; 1930, 479, 150.
   c) Ziegler, K.; Dersch, F.; Wolltham, H. Ann. Chem. 1934,
      511, 13.
7. Dostal, H.; Mark, H. Z. Phys. Chem. B. 1935, 29, 299.
8. Flory, P. J. "Principles of Polymer Chemistry"; Cornell Press,
   1953.
9. Hogen-Esch, T. E.; Smid, J. J. Amer. Chem. Soc. 1965, 87, 669;
   1966, 88, 307.
10. Bhattacharyaa, D. N.; Smid, J.; Szwarc, M. J. Amer. Chem. Soc.
    1964, 86, 5024.
11. DeGroof, B.; Van Beylen, M.; Szwarc, M. Macromolec. 1975, 8,
    396.
12. Schuerch, C.; et al. J. Amer. Chem. Soc. 1964, 86, 4481; 1967,
    89, 1396.
13. Stavely, F. W.; et al. Ind. Eng. Chem. 1956, 48, 778.
14. Worsfold, D. J.; Bywater, S. Can. J. Chem. 1960, 38, 1891.
15. Morton, M.; Bostock, E. E.; Livigni, R. Rubber Plastic Age
    1961, 42, 397.
16. Porter, R. S.; Johnson, R. Chem. Rev. 1966, 66, 1.
17.(a) Morton, M.; Fetters, L. J.; Bostock, E. E. J. Polymer Sci.
    1963, C1, 311.
   (b) Morton, M.; Fetters, L. J. J. Polymer Sci. 1964, A2, 3111.
18. a) Morton, M.; Pett, R. A.; Fetters, J. F.; IUPAC Symposium
       Tokyo, 1966, 1 69.
    b) Morton, M.; Fetters, L. J. Rubber Chem. Technol. 1975, 48,
       359.
19. e.g., Worsfold, D. J.; Bywater, S. Macromolec. 1972, 5, 393.
20. Fetters, L. J.; Morton, M. Macromolec. 1974, 7, 552.
21. Szwarc, M. Polym. Lett. 1980, 18, 493.
22. Morton, M.; Pett, R. A.; Fetters, L. J. Macromolec. 1970, 3,
    333.
23. Laita, Z.; Szwarc, M. Macromolec. 1969, 2, 412.
24. O'Driscoll, K. F.; Patsiga, R. J. Polymer Sci. 1965, A3, 1037.
25. Yamagishi, A.; Szwarc, M. Macromolec. 1978, 11, 1091.
26. Wang, H. C.; Szwarc, M. Macromolec. 1980, 13, 452.

RECEIVED April 7, 1981.

Current Status of Anionic Polymerization

MAURICE MORTON

Institute of Polymer Science, The University of Akron, Akron, OH 44325

Because of the strong interest in homogeneous anionic polymerization, especially during the past 20 years, there have been substantial advances, both in the understanding of the mechanisms involved as well as in the utilization of these systems for the synthesis of new materials. This overview will attempt to highlight the more important advances in both of the above areas, although admittedly with somewhat of a bias toward those particular endeavors pursued in our own laboratories.

Kinetics and Mechanism of Polymerization

Homogeneous anionic polymerization involving organoalkali initiators has been carried out in two classes of solvents, i.e., polar (ethers, amines), or non-polar (hydrocarbon) media. In the case of the polar solvents, it has been possible to use a wide range of organoalkali compounds, since they are soluble in these media; but this is not the case for hydrocarbon solvents which are mainly suitable for organolithium initiators. In all such studies, these anionic systems have a distinct advantage over other systems, since it is possible, under proper conditions, to avoid any chain termination reactions, so that the concentration of growing chains can correspond to the concentration of the initiator. Hence the kinetics of the polymerization reaction can thus represent the kinetics of the propagation reaction.

Evidence for the absence of termination or transfer reactions in the organolithium-initiated polymerization of styrene and isoprene is shown in Table I for representative examples of these polymers. It can be seen that these polymers exhibit the expected low M_W/M_n values, except in the case of the isoprene polymerized in the H_4furan, where a slow side reaction seems to occur between the solvent, on the one hand, and both the initiator (M_n vs M_S) and the growing chains (M_W vs M_n),on the other hand.

1. Kinetics in Polar Media. These polymerizations, initiated by organoalkali compounds mainly in ether solvents, have

0097–6156/81/0166–0017$06.00/0

TABLE I

Molecular Weight of Alkyllithium-Initiated Polymers[1,2]

Monomer	Solvent	Temp. (°C)	Molecular Weights ($\times 10^{-3}$)			
			M_S[a]	M_n[b]	M_w[c]	M_w/M_n
Styrene	Benzene	0	142	136	144	1.06
"	"	0	320	324	355	1.10
"	H_4-furan	-80	280	295	310	1.05
Isoprene	n-Hexane	29	53.8	53.2	61.2[d]	--
"	"	29	190	--	216	1.14
"	H_4-furan	0	130	172	217[d]	--

[a]Stoichiometric mol. wt.   [b]Osmotic   [c]Light scattering
[d]M_v values[3]

been intensively investigated, and there is now general agreement
that ionic dissociation actually occurs, especially in the case of
the more electropositive metals, i.e., sodium through cesium.
Thus the growing anionic chain can assume at least two identities:
the free anion and the anion-cation ion pair (several types of
solvated ion-pairs can also be considered). Furthermore, the
kinetics of these propagation reactions, which generally show a
fractional dependency on chain-end concentration ranging from one-
half to unity, can best be explained by assuming that the monomer
can react with both the free anion and the ion-pair (4, 5, 6), but
at different rates. Thus, for example, in the polymerization of
styrene by organosodium, the rate of polymerization (R_p) can be
expressed as

$$R_p/[S][RS_xNa]^{1/2} = k_p^- K_e^{1/2} + k_p^{\pm}[RS_xNa]^{1/2} \tag{1}$$

where S represents styrene monomer
    [RS_xNa] represents the total concentration of growing chains
             of both types, i.e., the concentration of initiator
             used.
    k_p^- = rate constant for propagation by free ions

    $k_p^{\pm}$ = rate constant for propagation by ion pairs

    K_e = ionization constant.

Equation (1) yields directly, from experimental data, the
value of the ion pair rate constant, $k_p^{\pm}$, while the free anion rate
constant, k_p^- , can be calculated if K_e is known. The latter has

actually been determined experimentally (4, 5, 6). Hence, in the above system, it has been found (4, 5, 6) that $\overline{k_p}$ is several orders greater than $k_p^{\pm}$ (ca. 10^4 vs. $\overline{10}^2 \overline{M}^{-1}S^{-1}$) and that K_e is of the order of 10^{-7}, when H_4furan is the solvent. The fact that much experimental data obey the kinetics of Equation (1) lends credence to the proposed mechanism.

    2. Kinetics in Non-Polar Media. Polymerization of vinyl monomers in non-polar solvents, i.e., hydrocarbon media, has been almost entirely restricted to the organolithium systems (7), since the latter yield homogeneous solutions. In addition, there has been a particularly strong interest in the polymerization of the 1,3-dienes, e.g., isoprene and butadiene, because these systems lead to high 1,4 chain structures, which yield rubbery polymers. In the case of isoprene, especially, it is possible to actually obtain a polymer with more than 90% of the cis-1,4 chain structure (7, 8, 9), closely resembling the microstructure of the natural rubber molecule.

    Early studies (1) of the kinetics of polymerization of styrene, isoprene and butadiene in hydrocarbon solvents indicated a half-order rate dependency on growing chain concentration, although there were conflicting data at that time (10, 11) which suggested even lower fractional orders for the dienes. Since the apparent half-order dependency could not be rationalized, as in the case of the polar media, by an ionic dissociation mechanism, some other form of association-dissociation phenomenon offered a possible answer. In view of the known tendency of organolithium compounds to undergo molecular association in non-polar media, the following scheme was proposed by us (1):

$$(RM_jLi)_2 \;\underset{k_p}{\overset{\longrightarrow}{\rightleftharpoons}}\; 2\ RM_jLi \tag{2}$$

$$RM_jLi + M \xrightarrow{k_p} RM_{j+1}Li \tag{3}$$

Furthermore, the proposed dimeric association of the polymer lithium species was actually confirmed (1) by viscosity measurements on the polymerized solutions, using the well-known relation (12) which applies to melts or concentrated solutions

$$\eta = KM^{3.4} \tag{4}$$

where η is the viscosity, M is the molecular weight, and K is a constant which includes the concentration of the polymer. Because of the large exponent in Equation (4), this method was found to be a sensitive measure of the state of association of the polymer lithium chain ends, and this was found to be very close to 2 for all three systems studied, i.e., polystyrene, polyisoprene and polybutadiene.

    Typical values of N, the "association number", or weight-average number of chains associated at their ends, are shown in Table II. The values of N were obtained from the relation

$$N = (t_a/t_t)^{1/3.4}$$

where t_a = flow time of "active" (non-terminated) chains.
t_t =   "   "   " terminated chains.

The somewhat higher values of N exhibited by the polybutadienyl lithium were ascribed to the occurrence of some crosslinking during the polymerization of this monomer, and this was
confirmed by use of a special "capping" technique, as shown later.

TABLE II

Association in Organolithium Polymer Solutions (1, 2)
(25°C)

Monomer	Solvent	RLi (x10³)	Flow Time (sec.)		N
			Active	Terminated	
Styrene	Benzene	1.23	825	81.1	1.98
Isoprene	n-Hexane	2.08	836	81.1	1.99
"	H₄-furan	1.95	3150	3130	1.00
Butadiene	n-Hexane	4.97	1770	112	2.89
"	n-Hexane	0.25	1030	75.5	2.16
"	H₄-furan	1.90	2580	2570	1.00

Unfortunately, we found later (7, 13) that the scheme proposed in Equations (2) and (3) was not valid for isoprene and
butadiene, since their kinetic order was considerably below one-
half, ranging from one-sixth to one-fourth, depending on the
solvent used. Rate plots for isoprene in 3 different solvents,
butadiene in n-hexane and butadiene in benzene, are shown in
Figures 1 to 4, respectively. There can be no doubt about the
low kinetic order exhibited by these reactions, much less than
the one-half order originally claimed. It is regrettable
that attempts were subsequently made by others to relate these
low fractional orders for isoprene (14) and butadiene (15) to
assumed higher states of association than 2, in order to make
these kinetics fit the mechanism proposed in Equations (2)
and (3). There was even some work (16) which purported
to prove the existence of such higher states of association
for these dienes. However, all such claims were shown to
be invalid by overwhelming experimental evidence (7, 17)
which definitely confirmed that not only polystyrene,
but polyisoprene and polybutadiene were associated in
pairs at their organolithium chain ends, in hydrocarbon media,
at lithium concentrations commonly used to prepare high polymers.

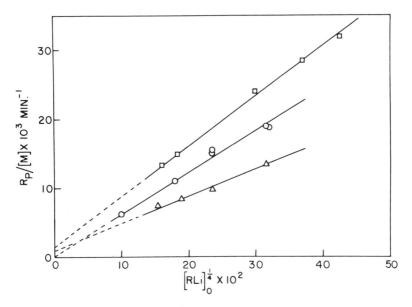

Figure 1. Dependence of rate (R_p) on organolithium concentration: isoprene in benzene ($\square$); cyclohexane ($\bigcirc$); n-hexane ($\triangle$); 30°C; $[M]_o = 2.0$ (7).

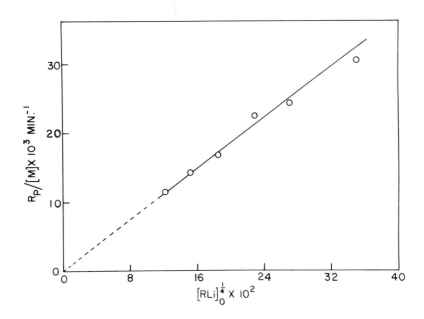

Figure 2. Dependence of R_p on organolithium concentration. Butadiene in n-hexane at 30°C; $[M]_o = 2.0$.

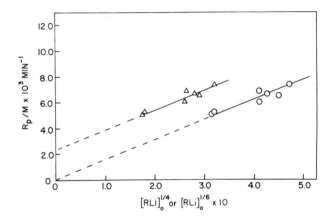

Figure 3.   Dependence of R_p on organolithium concentration.   Butadiene in benzene at $30°C$; $[M]_o = 2.0$; $[RLi]^{1/4}$ ($\triangle$); $[RLi]^{1/6}$ ($\bigcirc$).

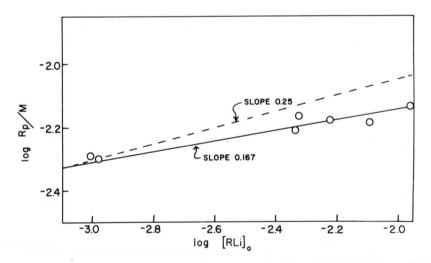

Figure 4.   Dependence of R_p on organolithium concentration.   Butadiene in benzene at $30°C$; $[M]_o = 2.0$.

In view of the evidence described above, and elaborated below, there is no reason to-day to relate the reaction order in the organolithium polymerization of the dienes to the state of association, as exemplified by the scheme described by Equations (2) and (3). This appears also to be true in the case of styrene, as indicated by some recent results (18) which also appear elsewhere in this publication. In other words, there seems to be no basis for assuming that organolithium polymerizations in non-polar media proceed by this type of mechanism, i.e., where only the unassociated species is active. In fact, it had been proposed (19) sometime ago that it would be more reasonable to assign such activity to the associated complex, and that the low fractional kinetic orders have no direct connection with the association-dissociation equilibrium.

    3. Molecular Association in Organolithium Polymerization. The data in Table II confirm the presence of an association between the organolithium chain ends of polystyrene, polyisoprene and polybutadiene in non-polar media (such association is apparently destroyed in the presence of solvating solvents, such as H_4-furan). Furthermore, the above polymer chains seem to be associated as dimers, although slightly higher association numbers appear to be observed for polybutadiene. Since it was suspected that some crosslinking was occurring in the case of the polybutadiene, a special "capping" technique was devised to overcome this difficulty. Solutions of polystyryl lithium were prepared and their flow times determined. These chains were then "capped" with a few units of butadiene or isoprene, and this could be done very efficiently in view of the fast "crossover" reaction between butadiene or isoprene and the polystyrene lithium chain ends (20). Hence the association behavior of the new chain ends should be the same as that of polybutadienyl lithium, without the risk of a crosslinking reaction interfering. Some typical data (21) are shown in Table III.
    It is apparent from these data that all of the polymers, including butadiene, exhibit an association as dimers, and that there is no reason to expect any higher states of association for polyisoprene or polybutadiene. This is confirmed not only by the viscosity data on the active vs. terminated "capped" polymers, but also by the fact that there was no significant increase in viscosity when the polystyryl lithium was "capped" by butadiene or isoprene, i.e., all three types of chain ends are associated in the same way, as dimers.
    The change in flow time that should occur when a dimeric association of chain ends might change to a higher state, e.g., tetrameric, was well illustrated by data based on linking poly-isoprenyl lithium by means of silicon chloride linking agents ('17) ranging from dimethyldichlorosilane to tetrachlorosilane (silicon tetrachloride). Such data are shown in Table IV. It is obvious

TABLE III

Association of Polystyryl Lithium "Capped" with
Butadiene and Isoprene (30°C) (21)

Solvent	[RLi] ($\times 10^3$)	[M]	Diene*	Flow Time (sec.)			N
				Polystyryl Lithium	"Capped" Polymer		
					Active	Terminated	
Cyclohexane	2.40	3.0	Isoprene	1567	1591	158.4	1.97
Benzene	2.40	3.2	"	1899	1974	187.5	2.00
"	0.63	1.5	Butadiene	--	268.9	26.2	1.98
"	0.60	1.5	"	--	398.8	38.4	1.99

*The added diene increased the molecular weight by 2000, thus causing the observed slight increase in the viscosity.

TABLE IV

Viscosities of Associated vs. Linked Polyisoprene (17)

(Cyclohexane, 25°C $v_2 \sim 0.20$)

Linking Agent	Flow Time (sec.)			Number Ave. Mol. Wt. ($\times 10^{-5}$)		
				Primary		Linked
	Active	Linked	Terminated	M_s	M_n	M_n
$SiCl_4$	1320	2340	2280	1.2	1.2	3.35
$SiCl_4$*	2480	4800	4770	1.5	1.6	6.0
CH_3SiCl_3	500	825	820	1.3	1.35	3.6
$(CH_3)_2SiCl_2$	235	229	229	0.81	0.85	1.6

M_s = stoichiometric mol. wt. M_n = osmotic mol. wt.

*Polyisoprene "capped" with butadiene.

at once that the tri- and tetrachlorosilanes cause a sharp
increase in the flow times of the active polymers, presumably by
linking the chain ends together into tri- and tetra-branched
species. The actual extent of linking is, of course, shown by a
comparison of the M_n values of the primary end linked polymer
chains. These values also demonstrate the well-known fact (17)
that the silicon tetrachloride cannot efficiently link polyisoprenyl lithium into a "tetra-star" but does so in the case of
polybutadienyl lithium.

The fact that the difunctional dimethyldichlorosilane had no
effect on the flow time of the polyisoprenyl lithium offers perhaps the most convincing evidence that these chain ends must have
been associated in pairs prior to the linking reaction. The
corresponding M_n values provide the necessary confirmation that
the linking reaction has actually taken place. It should also be
noted that flow times of the terminated linked polymers can also
be used to detect the presence of some unlinked, but associated,
chains.

One additional item of experimental evidence for the dimeric
association of polyisoprenyl lithium was provided by a light
scattering study (21), in n-hexane at 25°C., where it was found
that the molecular weight of the terminated polymer was very
close to one-half that of the active polymer. All of these data
seem to leave no doubt that the active chain ends in the organolithium polymerization of styrene, isoprene and butadiene, in
non-polar solvents, are associated as pairs, at least at chain-
end concentrations of 10^{-2} M or less. This conclusion has also
been supported by data obtained in four other laboratories (22,
23, 24, 25), based on three different experimental techniques,
viz. cryoscopy (22), viscosity (23, 24), and low angle laser
light scattering (25).

It must be concluded, therefore, that the kinetic scheme
proposed in Equations (2) and (3) cannot be valid for these
systems. Although the seeming correspondence between the half-
order kinetics and the dimeric association in the case of styrene
might validate the above kinetic scheme, even this hypothesis
has been recently contradicted (18). Thus Fetters and Young
(18) showed that even though the organolithium polymerization of
styrene in aromatic ethers, such as anisole and diphenyl ether,
had previously been found to obey half-order kinetics, these
solvents did have a marked effect in disrupting the dimeric
association of the chain ends. Hence it is doubtful that
Equations (2) and (3) can be applied even in the case of styrene.

Another complication introduced by the associative properties
of organolithium solutions in non-polar solvents is the fact that
the alkyllithium initiators are themselves associated and can be
expected to "cross associate" with the active polymer chain ends.
Thus some of our studies (26) on the effect of added ethyl
lithium on the viscosity of polyisoprenyl lithium solutions in
n-hexane support the following association equilibrium

$$(C_2H_5Li)_6 + (PLi)_2 \xrightarrow{K_e} 2(PLi \cdot 3C_2H_5Li) \qquad (3)$$

where PLi represents a polyisoprenyl lithium, and the equilibrium highly favors the cross-complex ($K_e \sim 6$). Hence it is obvious that such a cross-complexation would affect the kinetics of the initiation reaction, rendering any such studies of dubious value.

4. Stereospecificity in Diene Polymerization

A. Chain Microstructure. In addition to their non-terminating character, the homogeneous anionic polymerizations discussed herein have one additional feature, viz., control of chain structure of dienes. Although this does not seem to apply to the tacticity of these polymers, it does apparently affect the geometrical isomerism of the polydiene units. The most striking example of such an effect is, of course, the synthesis of a high cis-1,4 polyisoprene by lithium catalysis (8). It has already been shown by many investigators (27, 28, 29, 30) that both the nature of the alkali metal counter-ion and the type of solvent affect the chain unit structure of the polydienes, the more electropositive metals and/or the more solvating (polar) solvents favoring the side-vinyl chain unit structures (1,2 or 3,4). The effect of such solvents in increasing the ionic character of the carbon-metal bond has also been correlated with the ability of the dienes, e.g., isoprene, to copolymerize with styrene (31), as illustrated in Table V. These data are consistent with the previously discussed hypothesis that, in polar

TABLE V (31)

Effect of Alkali Metal Intiators and Solvents on
Copolymerization of Styrene with Isoprene at 25°C

Initiator	Solvent	Wt. % Styrene[a]	% 1,4
n-C$_4$H$_9$Li or Li	H$_4$-furan	80	30
Na	H$_4$-furan	80	32
Na	Dimethyl ether	75	36
Li	Diethyl ether	68	51
Na	Hydrocarbons	66	45
Li	Triethylamine	60	52
n-C$_4$H$_9$Li	Diphenyl ether	30	82
Li or n-C$_4$H$_9$Li	Hydrocarbons	16	95

[a]Composition of initial copolymer at equimolar charge ratio.

solvents, such as ethers, the propagating chain ends are ion pairs in equilibrium with free anions.

In the case of the much-studied organolithium polymerization of butadiene and isoprene, the effect of ethers and other polar solvents is to change the chain structure from one containing

>90% 1,4 to one having predominantly 1,2 (or 3,4) units. As
expected for these homogeneous systems, the placement of the
chain units is random (32). However, even in the case of non-
polar media, the chain unit structure may be affected by such
factors as the concentration of the lithium. A recent investigation (33), using high resolution proton magnetic resonance, has
shown that the cis-trans ratio in both polyisoprene and polybutadiene is affected both by the concentration of lithium and by the
presence of different hydrocarbon solvents, the highest cis content being obtained at the lowest lithium concentration and in the
absence of any solvents. Typical results for polyisoprene are
shown in Figure 5. It was also noted that the proportion of side
vinyl units in both polyisoprene and polybutadiene was not
noticably affected by these variables.

      B. Structure Studies of the Propagating Chain End in
Organolithium Polymerization of Dienes. The remarkable effects
of solvents on the chain microstructure of the lithium-
polymerized conjugated dienes made it of great interest to study
the structure of the propagating chain ends in these systems.
This has been done, mainly by proton magnetic resonance spectroscopy (22, 34, 35) on butadiene (22, 35), isoprene (34, 35, 36),
2,3-dimethylbutadiene (35, 36), several pentadienes (37, 38) and
2,4-hexadiene (37, 38). Typical spectra for polybutadienyl
lithium are shown in Figure 6. As can be seen, the "transparent"
perdeuterobutadiene was used to build chains of about 20 units.
This was done in order to avoid any effects of initiator on the
first few chain units (25). The effect of the polar medium
(tetrahydrofuran) on the Y hydrogen of the chain-end unit is to
move the peak upfield, from 4.7 ppm to 3.3 ppm, and to indicate a
more delocalized carbon-lithium bond. This, of course, ties in
well with the high side-vinyl content of the polymers made in
such media.
      On the basis of such spectra obtained in polar and non-polar
media, a mechanism was proposed (35) in which the carbon-
lithium bond participated in an equilibrium between a localized
and delocalized structure, the former being responsible for the
high 1,4 chain structure while the latter, highly favored in polar
media, accounted for the high side vinyl chain structure. A
schematic of this hypothesis is shown in Figure 7. It should be
noted that σ-bonding is restricted to the α carbon in non-polar
media, but is proposed for both the α and the Y carbon in polar
media. This means that the σ-π bond equilibrium illustrated in
Figure 7 should lead to a cis-trans equilibrium in the 1,4
propagating chain-end units only in polar media. This was
actually observed (35), together with the absence of any such
isomerization in non-polar media (38), although there is some
controversy (34) on this point. In fact, a cis-trans isomerization of the 4,1 propagating chain-end units of polyisoprenyl

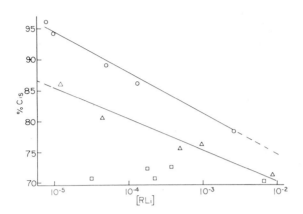

Figure 5.   Effect of organolithium concentration on polyisoprene chain structure: undiluted (○); 0.5M in n-hexane (△); 0.5M in benzene (□).

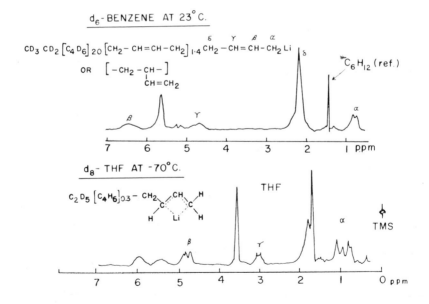

Figure 6.   100-MHz H-1 NMR spectrum of polybutadienyllithium

lithium has been advanced (39, 40) as a kinetic mechanism to account for the effect of lithium concentration on the cis-trans ratio of the chain microstructure formed in non-polar media.

An interesting phenomenon concerning the carbon-lithium bond of polydienyl lithium was observed in the case of 2,4-hexadiene (37, 38). Figure 8 shows the H[1] magnetic resonance spectra of an oligomer of 2,4-hexadiene initiated by s-butyl lithium (the term "pseudoterminated" refers to the process of adding butadiene-d_6 in order to remove the lithium atom from the chain end without introducing the complexities possibly caused by addition of a polar terminating agent, e.g., water, methanol, etc.). It can be clearly seen that the γ-hydrogens resonate principally at 3.3 ppm, which corresponds to their usual chemical shift in polar solvents, as observed in the case of the other polydienes. Hence it appears that the carbon-lithium bond in this case is largely delocalized even in a non-polar medium. This is also confirmed by the fact that these solutions are strongly colored (red) and show only partial chain-end association by viscosity measurements (ca. 60% dissociated)(37). The explanation probably lies in the fact that the lithium in this case is bonded to a secondary carbanion, which can be considered as much less stable than a primary carbanion, so that a delocalization of the bonding electrons occurs as a stabilizing mechanism.

Advances in Anionic Polymer Synthesis

Among the main accomplishments in the use of these non-terminating anionic polymerizations have been the synthesis of the following polymer structures:
   a) Block copolymers
   b) Star-branched polymers
   c) α,ω-Difunctional chains for network formation

Some of our work in these areas is described below.

Block Copolymers

The most striking technological development based on anionic block copolymers was probably the discovery (41) of "thermoplastic elastomers", based on the ABA triblock copolymers, where A is a polystyrene block and B refers to a polybutadiene (or polyisoprene) block. The result is a well-defined and uniform two-phase morphology, consisting of very small (∿20 nm) polystyrene domains, which are aggregates of the polystyrene end blocks and are therefore chemically bonded and regularly dispersed within the elastic polybutadiene matrix. A transmission electron microphotograph of a thin film of a styrene-isoprene-styrene triblock copolymer is shown in Figure 9. Such a "network" therefore is permanently thermoplastic, yet has the properties of an elastic network at room temperature and lower. In these

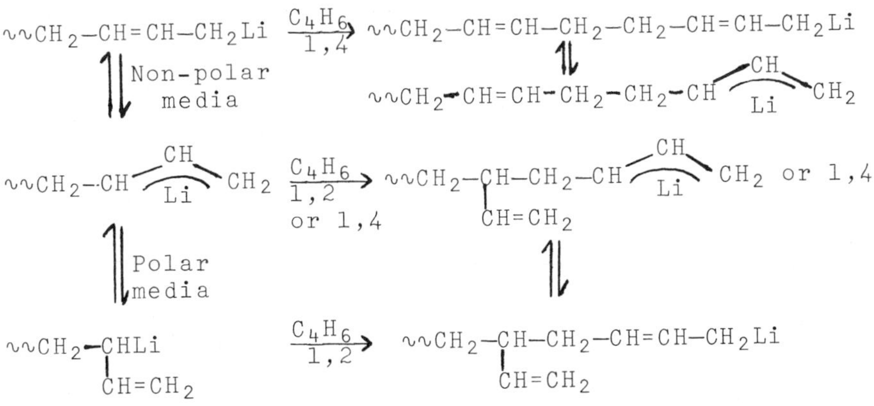

∿∿CH2—CH=CH—CH2Li
C4H6
1,4
∿∿CH2—CH=CH—CH2—CH2—CH=CH—CH2Li
Non-polar
media
∿∿CH2—CH=CH—CH2—CH2—CH
CH
Li
CH2
∿∿CH2—CH
CH
Li
CH2
C4H6
1,2
or 1,4
∿∿CH2—CH—CH2—CH
CH=CH2
CH
Li
CH2 or 1,4
Polar
media
∿∿CH2—CHLi
CH=CH2
C4H6
1,2
∿∿CH2—CH—CH2—CH=CH—CH2Li
CH=CH2

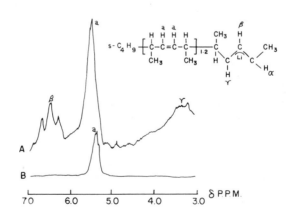

Figure 7. Proposed mechanism for butadiene polymerization.

a
β
γ
A
B
70
6.0
5.0
4.0
3.0
δ P.P.M.
s-C4H9
1.2
CH3
Li
Hα
Macromolecules

Figure 8. 60-MHz H-1 NMR spectrum of poly-2,4-hexadienyllithium in d_6-benzene (38): living (A) and pseudoterminated (B).

systems, the end-blocks typically have a molecular weight of
about 15,000, while the center elastic block has a molecular
weight of about 70,000.

Work in our laboratories over the past decade (42, 43) on
the above triblock copolymers, as well as on other chemical
variations, has thrown a great deal of light on their structure-
property relations. These studies have demonstrated the role of
such factors as block incompatibility, block size distribution,
glass transition temperature of the end blocks, and interfacial
energy of the two phases. Thus, for example, a greater incompatibility between the two phases leads to a better phase separation and hence to less softening of the glassy domains, by
solubilization of the elastic matrix, resulting in higher fracture
strength. On the other hand, too great a difference in solubility parameters of the two phases leads to a higher viscosity
and concomitant processing difficulties. On top of all this, one
must take into account the interfacial adhesion between the
matrix and the glassy domains, which also exerts a controlling
influence on the mechanical failure of these materials. This
has been strikingly demonstrated in the case of triblock copolymers based on polysiloxanes (43).

Among the variations in chemical structure of these triblock
copolymers developed in our laboratories were the use of
poly(α-methyl styrene)(43, 44) as end blocks, and poly(alkylene
sulfides) (43, 45) and polydimethylsiloxanes (43, 46) as center
blocks. The reactions of cyclic sulfides with organolithium is
illustrated in Figure 10. Both the propylene sulfide and the
methyl thietane can be used for the center block with styrene or
α-methylstyrene end blocks, but the chemistry shown in Figure 10
indicates that the propylene sulfide leads to formation of a
lithium thiolate, which is incapable of initiating a vinyl
polymerization. Hence this system requires a chemical linkage
of a styrene-(propylene sulfide) diblock into a triblock. With
the 4-membered cyclic sulfides, however, it is possible to carry
out a sequential polymerization of the three blocks, since the
opening of the cyclic sulfide in this case leads to a carbanion,
capable of initiating the polymerization of vinyl monomers.

The formation of triblocks based on polydimethylsiloxane (46)
is illustrated in Figure 11. It should be noted that the poly-α-
methylstyrene blocks were "capped" with butadiene in order to
circumvent the relatively slower initiation of the hexamethylcyclotrisiloxane by the poly-α-methylstyrene lithium. The high
degree of incompatibility of these two types of blocks led to the
desired phase separation, as shown by the transmission electron
microphotograph in Figure 12. However, these polymers also
exhibited a very high viscosity for this reason (47). In addition, they showed a rather low tensile strength, presumably due
to the poor adhesion between the glassy domains and the poly-
siloxane matrix (43).

Figure 9. TEM of a styrene–isoprene styrene triblock copolymer

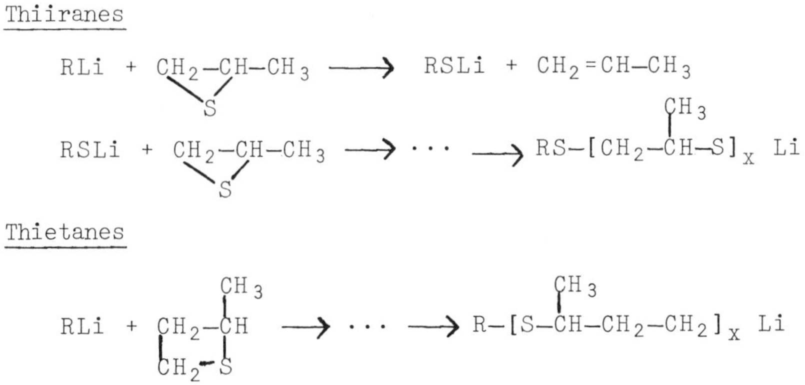

Thiiranes

$$RLi + CH_2-CH-CH_3 \longrightarrow RSLi + CH_2=CH-CH_3$$
$$\overset{\diagdown}{}\underset{S}{\diagup}$$

$$RSLi + CH_2-CH-CH_3 \longrightarrow \cdots \longrightarrow RS-[CH_2-\overset{\overset{\displaystyle CH_3}{|}}{C}H-S]_X Li$$
$$\overset{\diagdown}{}\underset{S}{\diagup}$$

Thietanes

$$RLi + \overset{\overset{\displaystyle CH_3}{|}}{\underset{\underset{\displaystyle CH_2-S}{|}}{CH_2-CH}} \longrightarrow \cdots \longrightarrow R-[S-\overset{\overset{\displaystyle CH_3}{|}}{C}H-CH_2-CH_2]_X Li$$

Figure 10. Organolithium polymerization of cyclic sulfides.

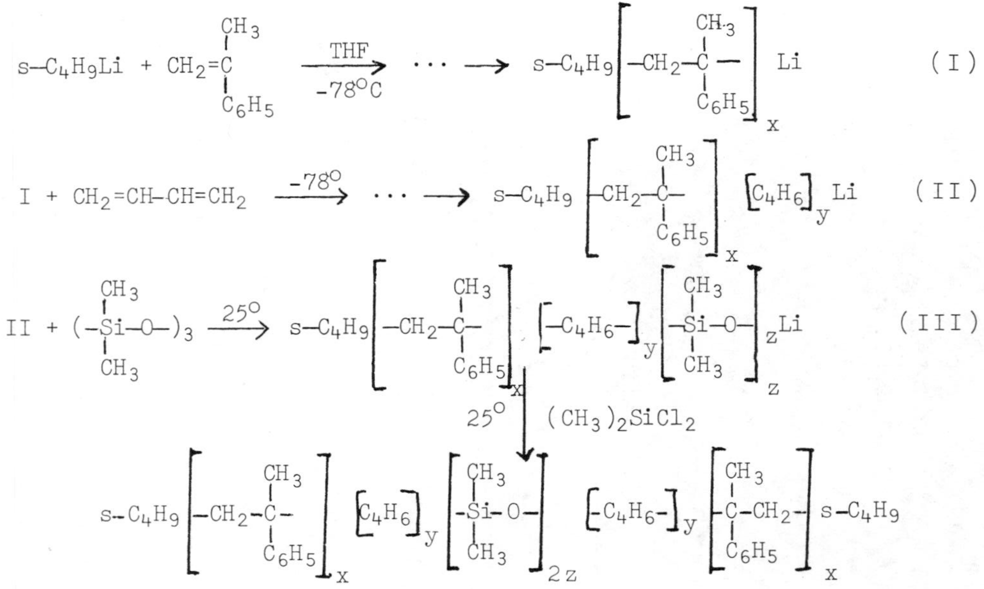

Figure 11.   Synthesis of an α-methylstyrene–siloxane–α-methylstyrene triblock copolymer: x = ~ 100, y = ~ 10, and z = ~ 300–900.

Applied Polymer Symposia

Figure 12.   TEM of an α-methylstyrene–dimethylsiloxane–α-methylstyrene triblock copolymer (46).

$$\underline{s}-C_4H_9Li + CH_2{=}\overset{\overset{\textstyle CH_3}{|}}{C}{-}CH{=}CH_2 \longrightarrow \cdots \longrightarrow \underline{s}-C_4H_9 \, [C_3H_8]_x Li \qquad (I)$$

$$Cl_3SiCH_2CH_2SiCl_3 + 6CH_2{=}CHMgBr \longrightarrow$$
$$6MgBrCl + (CH_2{=}CH)_3SiCH_2CH_2Si(CH{=}CH_2)_3 \qquad (II)$$

$$II + 6SiHCl_3 \longrightarrow (Cl_3SiCH_2CH_2)_3SiCH_2CH_2Si(CH_2CH_2SiCl_3)_3 \qquad (III)$$

$$I + III \longrightarrow$$

$$(\underline{s}-C_4H_9[C_3H_8]_x)_9(SiCH_2CH_2)_3SiCH_2CH_2Si(CH_2CH_2Si)_3([C_5H_8]_x\underline{s}-C_4H_9)_9$$

Figure 13.   Synthesis of an 18-arm, star-branched polyisoprene (49).

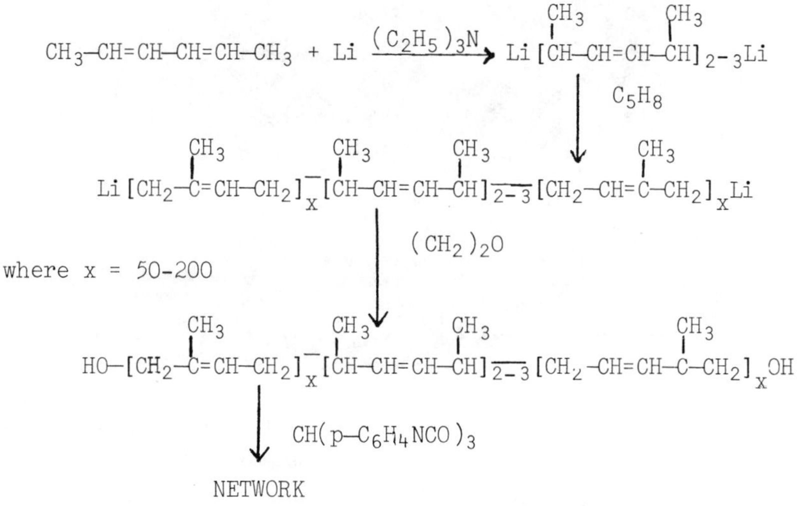

$$CH_3-CH=CH-CH=CH-CH_3 \; + \; Li \; \xrightarrow{(C_2H_5)_3N} \; Li \, [\underset{CH_3}{CH}-CH=CH-\underset{CH_3}{CH}]_{2-3} Li$$

$$\downarrow C_5H_8$$

$$Li \, [CH_2-\underset{CH_3}{C}=CH-CH_2]_x^- \, [\underset{CH_3}{CH}-CH=CH-CH]_{\overline{2-3}} \, [CH_2-CH=\underset{CH_3}{C}-CH_2]_x Li$$

where x = 50-200

$$\downarrow (CH_2)_2O$$

$$HO-[CH_2-\underset{CH_3}{C}=CH-CH_2]_x^- \, [\underset{CH_3}{CH}-CH=CH-CH]_{\overline{2-3}} \, [CH_2-CH=CH-\underset{CH_3}{C}-CH_2]_x OH$$

$$\downarrow CH(p-C_6H_4NCO)_3$$

NETWORK

Figure 14.   Synthesis of uniform polyisoprene networks.

Star-Branched Polymers

In recent years a variety of star-branched polystyrenes, polyisoprenes and styrene-isoprene block copolymers have been prepared by Fetters and coworkers (48). The synthetic route involved organolithium polymerization followed by linking of the lithium chain ends, either by a polyfunctional reagent such as a chlorosilane or by divinyl benzene. In this way it was found possible to prepare star-branched polymers of precise structure, with up to 15 or more monodisperse branches, and to examine their properties. A typical scheme is shown in Figure 13 for the synthesis of an 18-arm star polyisoprene. The efficiency of the linking reaction is demonstrated by the narrow, single peak of the distribution curve in the gel permeation chromatogram (49). Such precise polymers have helped to elucidate the solution behavior, chain dimensions, and rheology of branched polymers.

Uniform Polyisoprene Networks

It has long been considered highly desirable to be able to construct an elastic network having a relatively uniform size of network chains. This is because of the random manner in which crosslinks are introduced during the conventional process of rubber vulcanization, and because there is no theoretical relation which can predict the effect of a distribution of network chain size on the mechanical properties of the network. One way to achieve such a network would be through the anionic polymerization of a suitable monomer, e.g., isoprene, by a difunctional initiator, which could lead to a relatively monodisperse α,ω-difunctional polyisoprene suitable for linking into a uniform network by a polyfunctional linking agent.

Such a network has recently been accomplished (50, 51), using the scheme shown in Figure 14. Since the polyisoprene diols showed a high degree of monodispersity in chain length as well as difunctionality, and since sol-gel studies showed a high extent of reaction with the tri-isocyanate, it could be assumed that the resulting networks were of a high degree of uniformity. Measurements of their mechanical properties indicated that these uniform elastic networks exhibited both a more ideal elastic behavior as well as a higher strength than comparable random networks.

ABSTRACT

The current status of homogeneous anionic polymerization by alkali metal derivatives is reviewed along two lines:

a) Mechanism and kinetics. b) Synthesis of unique polymers.

In the case of the mechanism, a distinction is made between polymerization in polar and non-polar media. In the former type of solvents (ethers, amines, etc.) it has been proposed that the propagating chain end consists of an ion pair in equilibrium with the free ionic species, both types of chain end being capable of chain propagation, although the free carbanion is the more reactive, by several orders of magnitude. In the case of non-polar solvents (mainly hydrocarbons), the polymerizations are largely restricted to the soluble organolithium systems, and there is no evidence of the ion solvation necessary for ionic dissociation. Instead, there is convincing evidence that the propagating chain ends, in such cases as styrene, isoprene and butadiene, are associated in pairs, as might be expected for such polar species in non-polar media. There is no clear-cut evidence whether the active chains are the associated or dissociated species, or both, since there is no simple relation between the state of association and the kinetic order, which can vary from one-half to one-sixth, depending on the monomer and solvent used.

A brief review is presented of the unique synthetic capabilities inherent in these homogeneous anionic systems, including the following macromolecular structures: 1) block copolymers, 2) star-branched polymers, 3) α,ω-difunctional polymers linked into networks by polyfunctional linking agents.

Literature Cited

1. Morton, M.; Bostick, E.E.; Livigni, R. A.
     Rubber Plast. Age, 1961, 42, 397.
2. Morton, M.; Fetters, L. J.; Bostick, E. E.
     J. Polym. Sci., 1963, C1, 311.
3. Morton, M.; Bostick, E. E.; Clarke, R. G.
     J. Polym. Sci., 1963, A1, 475.
4. Hostalka, H.; Figini, R. V.; Schulz, G. V.
     Makromol. Chem., 1964,71, 198.
5. Hostalka, H.; Schulz, G. V. Z. Physik. Chem.
     (Frankfurt), 1965, 45, 286.
6. Battacharyya, D. N.; Lee, C. I.; Smid, J.;
     Szwarc, M. Polymer, 1964, 5, 54; J. Phys.
     Chem., 1965, 69, 612.
7. Morton, M.; Fetters, L. J. Rubber Chem. Technol.,
     1975, 48, 359.
8. Stavely, F. W. and coworkers. Ind. Eng. Chem.,
     1956, 48, 778.
9. Schue, F.; Worsfold, D. J.; Bywater, S. Can. J.
     Chem., 1964, 42, 2884.
10. Lundborg, C.; Sinn, H. Makromol. Chem., 1960, 41,
     242; Sinn, H.; Lundborg, C., ibid., 1961, 47,
     86.
11. Spirin, Yu. L.; Polyakov, D. K.; Gantmakher, A.R.;
     Medvedev, S.S. J. Polym. Sci., 1961, 53, 233.
12. Bueche, F. J. Polym. Sci., 1960, 45, 267.
13. Morton, M.; Pett, R. A.; Fellers, J. F. Preprints
     IUPAC Macromol. Symp. Tokyo, 1966, 1, 69.

14.   Worsfold, D. J.; Bywater, S. Can. J. Chem., 1964,
      42, 2884.
15.   Johnson, A. F.; Worsfold, D. J. J. Polym. Sci.,
      1965, A3, 449.
16.   Worsfold, D. J.; Bywater, S. Macromolecules, 1972,
      5, 393.
17.   Fetters, L. J.; Morton, M. Macromolecules, 1974,
      7, 552.
18.   Fetters, L. J.; Young, R. N. Polymer Preprints,
      1980, 21(1), 34.
19.   Brown, T. L. J. Organomet. Chem., 1966, 5, 191;
      Adv. Organomet. Chem., 1965, 3, 365.
20.   Morton, M.; Ells, F. R. J. Polym. Sci., 1962, 61,
      25.
21.   Morton, M.; Fetters, L. J.; Pett, R. A.; Meier,
      J. F. Macromolecules, 1970, 3, 327.
22.   Glaze, W. H.; Hanicak, J. E.; Moore, M. L.;
      Chadhuri, J. J. Organomet. Chem., 1972, 44,
      39.
23.   Al Jarrah, M. M.; Young, R. N. Polymer, 1980, 21,
      119.
24.   Makowski, H. S.; Lynn, M. J. Macromol. Chem.,
      1966, 1, 443.
25.   Hernandez, A.; Semel, J.; Broecker,H. C.;
      Zachmann, H. G.; Sinn, H. Makromol. Chem.,
      Rapid, Comm., 1980, 1, 75.
26.   Morton, M.; Pett, R. A.; Fellers, J. F.
      Macromolecules, 1970, 3, 333.
27.   Foster, F. C.; Binder, J. L. Adv. Chem. Ser.,
      1957, 17, 7.
28.   Morita, H.; Tobolsky, A. V. J. Am. Chem. Soc.,
      1957, 79, 5853.
29.   Tobolsky, A. V.; Rogers, C. E. J. Polym. Sci.,
      1959, 40, 73.
30.   Stearns, R. S.; Forman, L. E. J. Polym. Sci.,
      1959, 41, 381.
31.   Tobolsky, A. V.; Rogers, C. E. J. Polym. Sci.,
      1959, 38, 205.
32.   Santee, E. R.; Malotky, L. O.; Morton, M. Rubber
      Chem. Technol., 1973, 46, 1156.
33.   Rupert, J. Ph.D. Dissertation, University of
      Akron, 1975.
34.   Schue, F.; Worsfold, D. J.; Bywater, S.
      J. Polym. Sci., 1969, B7, 821; Macromolecules,
      1970, 3, 509.
35.   Morton, M.; Sanderson, R. D.; Sakata, R.
      J. Polym. Sci., 1971, B9, 61; Macromolecules,
      1971, 6, 181.
36.   Morton, M.; Sanderson, R. D.; Sakata, R.; Falvo,
      L. A. Macromolecules, 1973, 6, 186.

37. Morton, M.; Falvo, L. A.; Fetters, L. J.
         J. Polym. Sci., 1972, B10, 561.
38. Morton, M.; Falvo, L. A. Macromolecules, 1973, 6,
         190.
39. Brownstein, S.; Bywater, S.; Worsfold, D. J.
         Macromolecules, 1973, 6, 715.
40. Lachance, P.; Worsfold, D. J. J. Polym. Sci.,
         Polym. Chem. Ed., 1973, 11, 2295.
41. Holden, G.; Milkovich, R. U. S. Pat. 3,231,635;
         Belg. 627,652 (Shell), appl. Jan. 1962.
42. Morton, M. Encyclopedia of Polymer Science and
         Technology, Vol. 15, John Wiley and Sons,
         New York, 1971, p. 508.
43. Morton, M. J. Polym. Sci., Polym. Symp. No. 60,
         1977, 1.
44. Fetters, L. J.; Morton, M. Macromolecules, 1969,
         2, 453.
45. Morton, Maurice; Kammereck, Rudolf F.; Fetters,
         L. J. Macromolecules, 1971, 4, 11.
46. Morton, Maurice; Kesten, Y.; Fetters, L. J.
         J. Polym. Sci., Appl. Polym. Symp. No. 26,
         113 (1975).

RECEIVED March 5, 1981.

Synthesis of Controlled Polymer Structures

RALPH MILKOVICH

ARCO Polymers, Inc., Newtown Square, PA 19073

The development of anionic chemistry over the past 30 years has led to the emergence of new processes and products of industrial importance, the most significant being a family of thermoplastic elastomers. These unique elastomers are presently commercialized by Shell Chemical Company as Kratons and by Phillips Chemical Company as Solprenes. Their uniqueness is the result of deliberate design of the polymeric structure and composition.

The evolution of anionically synthesized block polymers started with an exciting academic discovery and culminated as an industrial economic reality. The recognition of the significance of the electron transfer reaction from sodium to naphthalene in an ether solution, as shown in Figure 1, led to utilization of the electron transfer to a polymerizable monomer. Figure 2 shows a key step toward the realization that quantitative polymerization reactions could be carried out leading to the synthesis of pure block polymers of controlled structure (1, 2). The design and synthesis of two-phase block polymers utilizing the "living" polymer concept, will be discussed.

A more recent development, MACROMER®, a new tool that expands our capability for synthesizing controlled polymer structures, will also be discussed.

Anionic Two-Phase Block Polymers

Styrene-butadiene elastomers were developed because of this country's need to have an independent reliable source of rubber during World War II. This rubber was developed as a random copolymer containing about 25% styrene, and is still prepared by emulsion technology. Today it is also being prepared by solution anionic chemistry. The random copolymer has no useful properties until it is compounded and vulcanized as in the production of automobile tires, for example. If the basic ingredients of this copolymer (the styrene and butadiene monomeric units) are segregated into segments of homopolymers, several interesting structures can be realized, i.e.:

a b b a b a a b b b a b a a a b b

RANDOM COPOLYMER

a a a a a a a a b b b b b b b b b

BLOCK COPOLYMER

This A-B block polymer can be synthesized anionically. The "A" or polystyrene segment in this case is prepared by the reaction of styrene monomer with butyllithium in an appropriate solvent. The ratio of styrene to initiator used will result in a polystyrene segment of predetermined molecular weight and a very narrow molecular weight distribution. Once all the styrene monomer is quantitatively consumed, a styryl anion will be present at the end of the polymer chain if great care is taken to avoid the presence of any other species reactive to anions. Now the "B" monomer units, or the butadiene, are added to the "living" polystyrene solution and the polymer chains continue to grow in size as the polydiene polymer segment is formed. Again, the ratio of monomer added to active species present will control the size of the diene polymer segment. Also, once the monomer is consumed, a "living" polymer species is retained - this time it is a dienyl anion. This anion can be further reacted with a large variety of chemicals as shown in Figure 3.

In this example styrene-isoprene is being used instead of styrene-butadiene. The first reaction is a direct termination of the styrene-isoprene terminal group with a protonating reagent to form a diblock (3). In the second reaction this diblock can be functionalized by reacting it with a variety of materials. Simple functional groups such as carboxyl or hydroxyl groups can be introduced. This anion can also be used as the basis for obtaining typical vinyl-type activity. Macromonomers of this type are trademarked MACROMER® by CPC International (4, 5).

Another reaction that can be carried out is a linking reaction. It is a terminating reaction where one uses a di- or trihalide and forms a tri- or tetra-block polymer (3).

The most obvious reaction is simply to add more styrene and convert the isoprene anion to a styryl anion and grow it to a desired size and form the styrene-isoprene-styrene triblock of which we are more familiar in all the theoretical work that has been reported (6, 7). The last example employs a bifunctional monomer, divinylbenzene, to form miniblocks of divinylbenzene as it reacts with a number of diblocks into what is called a starblock configuration (8). The active anionic sites are now on the divinylbenzene. These materials are then terminated by protonating agents to obtain the final product. The number of arms or diblocks that united into a starblock of this type is controlled by the relative amounts of the diblock to the divinylbenzene.

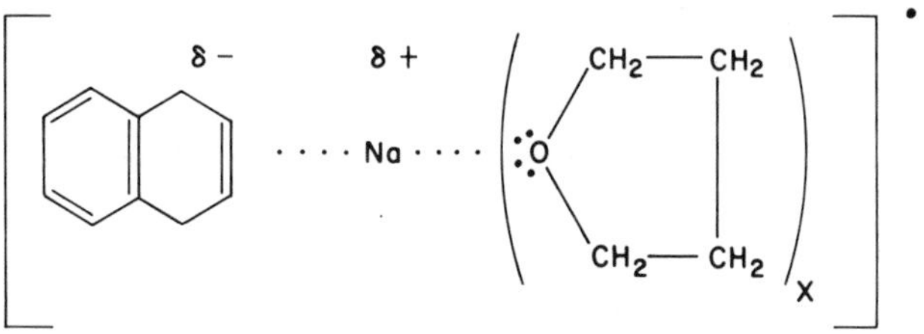

Figure 1. *Formation of sodium naphthalene complex.*

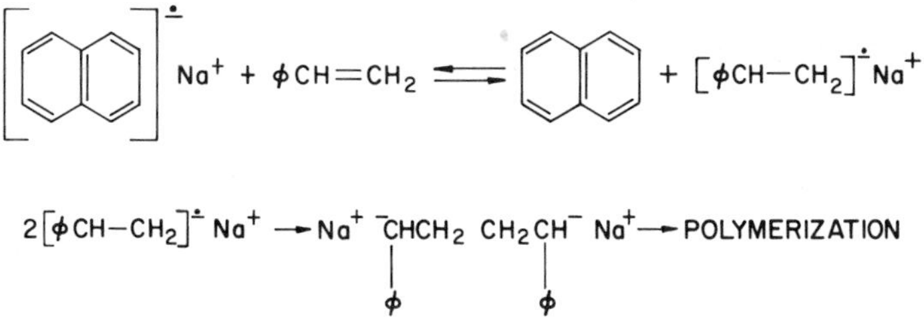

Figure 2. *Initiation of polymerization of styrene by electron transfer.*

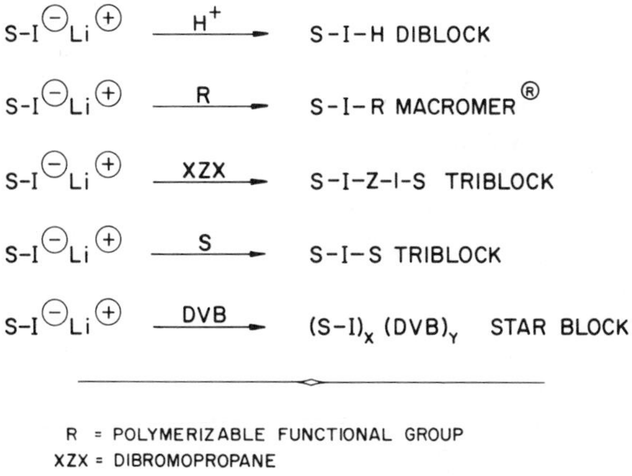

R = POLYMERIZABLE FUNCTIONAL GROUP
XZX = DIBROMOPROPANE

Figure 3. *Chemical modification of styrene–diene living block polymer.*

Professor Fetters at the University of Akron has indicated
that starblocks having over 15 arms can be synthesized. He has
reported studies contrasting differences in properties of the
starblock configuration over other types of polymer configurations (9).

The course and some effects of these various synthetic steps
in the synthesis of block polymers have been investigated via GPC
analysis. The very small hump on the left in Figure 4 represents
the polystyrene residue or block segment. It is there because the
added diene may have contained impurities that terminated a small
amount of the polystyrene. It forms a good detecting area to
indicate the possible size of that segment. The next hump on the
shoulder is the diblock and the larger peak is the triblock or
linked material. If the triblock were prepared by addition of
styrene monomer into an "A-B" block rather than a linking reaction, the hump would not be as readily detected because of the
small change in molecular weight. The result would be the smooth,
symmetrical curve that we are more accustomed to seeing in the
literature. However, the presence of this diblock peak could be
desirable and very beneficial. Some triblocks are difficult to
thermally process. Perhaps the presence of some of the diblock
material will be a benefit in aiding the processing of the material depending on its composition, molecular weight, and intended
application.

The GPC analysis in Figure 5 shows the result of a synthesis
of a starblock polymer. In this case the various steps and components of a starblock synthesis are seen more clearly. The first
peak on the left is the polystyrene peak (some of it was terminated when the isoprene was added to it) and the remainder grew to
a peak the size of the second one, the diblock. The presence of
the diblock is again due to impurities when the divinylbenzene was
added, or the technique used was not highly refined. The final
peak shows that it is substantially higher in molecular weight
than the diblock, and that it is certainly more than double its
molecular weight. It has been estimated to have about 10 or 12
arms and is an example of how one can utilize chemistry to develop different configurations of block polymers.

The compositional and two-phase morphological relationships
of "A-B" blocks, the "A-B-A" and starblocks have been studied
intensively. It has been demonstrated that there is a substantial
difference between random copolymers and block polymers, and this
difference is based solely on the architectural arrangement of the
monomeric units. One of the most important differences is that
one Tg is observed in the random copolymer, which is related to
the overall composition of the polymer. The block polymer has
been shown to have two Tg's - one for polystyrene and one for the
polydiene segment, and that these Tg's are not affected by the
composition of the block copolymer. Since we can now synthesize
large quantities of these pure block polymers, more detailed
physical studies can be carried out. The two Tg's observed in

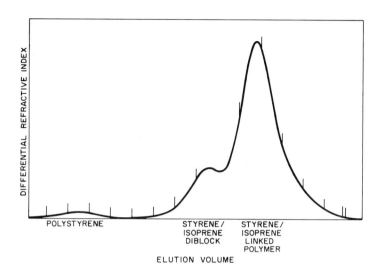

Figure 4.   GPC of block polymer formed by dihalide coupling reaction.

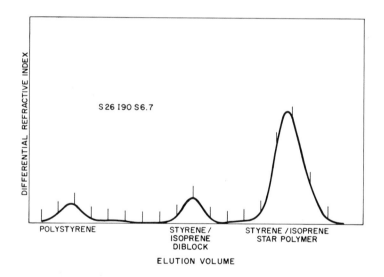

Figure 5.   GPC of a starblock: Shell Dutch Patent 646,835, Table III, Example B;
Shell Canadian Patent 716,645, Table III, Example B.

block polymers indicate the presence of two distinct incompatible entities. We know that when polystyrene and a polydiene are blended together, two Tg's are observed. However, the mixture is also opaque because the polymers are thermodynamically incompatible. The block polymers, on the other hand, are optically clear, indicating that if two distinct polymers are present as one molecule, then one polymer segment must be dispersed very efficiently in the other to the extent that it does not scatter light. After much discussion with my collaborator, Dr. Geoffrey Holden, we decided that the morphological concept as represented in Figure 6 explains the physical phenomena observed. The polystyrene segments are aggregates as spheres tied together by the elastic polydiene segments. Dr. Dale Meier, when he first saw some of the data generated on the block polymers at the Shell Rubber Division Laboratories in Torrence, California, was able to calculate on what one would say the back of an envelope the number of domains, the sizes of the domains, and the number of chains per domain. It has been proven over the years that we were correct in our assumptions for the composition studied at that time. We now know that the composition of block polymers can grossly alter their morphology.

Although the "A-B" or styrene-diene diblocks have certain advantages over random copolymers, it is after the second plastic phase is added to form at least an "A-B-A" or a star-type block polymer that the physical characteristics of block polymers as thermoplastic elastomers are realized as is indicated in Figure 7. The polystyrene molecular weights are all about 26,000 and the polyisoprene molecular weights are about 94,000. These examples were selected from patents (3, 7, 8) to provide some continuity in showing these physical effects. Although the diblock has essentially no tensile strength (about 100 psi), the triblock configuration (formed by adding styrene to a "living" diblock then terminating it) gives a polymer structure showing a dramatic increase in tensile strength to about 3000 psi. The triblock prepared by a coupling reaction using dibromopropane in this case shows a lower tensile strength which is probably due to the presence of diblock material. On the other hand, the starblock shows the highest tensile strength. This could be due to the fact that there are a multitude of thermal crosslinks for the same polymer molecule because of the number of blocks linked, whereas the triblock has only one polystyrene segment at each end of the molecule.

For these physical effects to occur in the most efficient manner, several criteria must be satisfied:
- The polymer segments, first of all, must be thermodynamically incompatible.
- The polymer segments must be linked together chemically.
- The Tg's of the two polymer segments must be substantially different.

- The molecular weight distribution of the plastic
  or high Tg polymer segment must be very narrow
  to achieve the maximum phase separation.
- The molecular weight of the plastic dispersed
  phase must be above a critical molecular weight.
- To obtain high strength, more than one high Tg
  polymer segment per molecule must be in a dispersed phase.

Graft Polymers

MACROMER® (10) is a trademark by CPC International of a new
family of monomers. Because they are synthesized via anionic
chemistry, their molecular weight is controlled by the ratio of
monomer to initiator and they also have very narrow molecular
weight distributions. The typical polymeric portions of MACROMER®
that have been investigated are polystyrene, polydiene, and blocks
of the two (5, 10). Some of the typical MACROMER® functional
groups that were examined are shown in Figure 8. These are shown
to indicate the wide variety of functional groups that are useful
for various polymerization mechanisms (4).
   There are some problems associated with the synthesis of
MACROMERS® when the styryl anion is reacted with certain functional monomers such as:

Maleic Anhydride   or

Vinyl
Chloroacetate

With maleic anhydride or vinyl chloroacetate the various points of
anionic attack are indicated by the arrows - double bonds,
α-hydrogen, carbonyls, and of course, the halide. It was found
that some of these side reactions could be avoided by simply reacting the very basic styryl anion with a chemical which would
result in a lower base strength anion. This weaker anion will not
react with the carbonyl and α-hydrogens and other possible sites
of the functionalizing reaction and limit itself to simple halide
reactions. This was done by using ethylene or propylene oxide as
shown in Figure 9 for the synthesis of a polystyrene MACROMER®
with a methacrylate functional group (8). This intermediate
reaction step now opens up a whole new opportunity and challenge

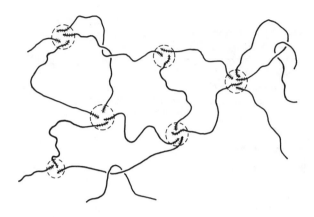

Figure 6.   Conceptual morphological structure of two-phase SIS block polymers.

	TENSILE STRENGTH AT BREAK (PSI)	ELONGATION AT BREAK (PSI)
DIBLOCK	95	1300
TRIBLOCK (S)	3000	1100
TRIBLOCK (XZX)	2100	1300
STAR BLOCK	3800	1100

POLYSTYRENE MOLECULAR WEIGHT $\sim$ 26,000 g/MOLE

POLYISOPRENE MOLECULAR WEIGHT $\sim$ 94,000 g/MOLE

Figure 7.   Physical properties of styrene–isoprene block polymers.

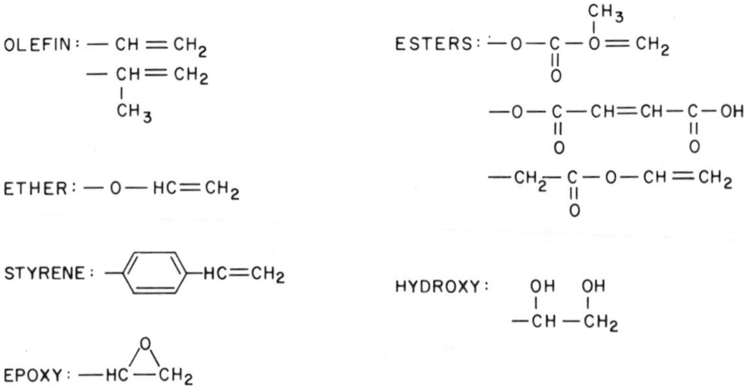

OLEFIN : $— CH = CH_2$
$— CH = CH_2$
$\underset{CH_3}{|}$

ETHER : $— O — HC = CH_2$

STYRENE : $—\langle\!\!\bigcirc\!\!\rangle— HC = CH_2$

EPOXY : $— HC \overset{O}{\diagdown\!\!\diagup} CH_2$

ESTERS : $— O — \underset{\underset{O}{\|}}{\overset{\overset{CH_3}{|}}{C}} — O = CH_2$

$— O — \underset{\underset{O}{\|}}{C} — CH = CH — \underset{\underset{O}{\|}}{C} — OH$

$— CH_2 — \underset{\underset{O}{\|}}{C} — O — CH = CH_2$

HYDROXY : $\underset{— CH — CH_2}{\overset{OH \quad OH}{| \quad\quad |}}$

Figure 8.   Typical MACROMER functional groups.

to the polymer scientist. It offers the means to prepare functionalized polymers that can be used in a variety of polymerization mechanisms, i.e. free-radical, ionic, condensation, or
coordination systems to prepare controlled polymer structures,
rather than being limited in the case of the styrene and dienes
to anionic chemistry.

Since this work was directed toward potential commercializa-
tion, it utilized the various modes of copolymerization or polymerization systems – solution, suspension, and latex.

Some of the primary comonomers that were investigated in the
MACROMER® studies were the acrylates, vinyl chloride, styrenes,
ethylene, ethylene/propylene, acrylonitrile, and N,N-dimethyl-
acrylamide.

A concept has been presented for preparing a large macromonomer – the molecular weight can be 10,000 or 20-30,000 grams/
mole – with a single unsaturated group at the end of the polymer
chain. How can it be proved that a molecule this big, if indeed
it has functionality, will react and how will it react with typical commercially available low molecular weight monomeric species?

The best test for functionality would be in a copolymerization study. A polystyrene with a methacrylate terminal functional
group was prepared. A review of relative reactivity ratios indicated that vinyl chloride reacts very rapidly with methacrylates.
Therefore, a copolymerization of the polystyrene terminated with
a methacrylate functional group in vinyl chloride would be a good
test case, and one should observe the disappearance of the
MACROMER® if the reaction is followed by using GPC analysis.

The GPC traces that were taken during the course of a copolymerization reaction of vinyl chloride and an 11,000 MW polystyrene methacrylate-terminated macromonomer are shown in Figure
10. The very sharp peak on the left is the MACROMER® and the
curves on the right are the growing copolymer peaks. One sees
that by the time 55% of the vinyl chloride conversion is completed
that the MACROMER® is virtually completely copolymerized and that
its peak has almost disappeared. This at least indicates the
existence of a functional MACROMER®.

A second test was done by using butyl acrylate as the comonomer as shown in Figure 11. The reactivity ratios in this case
are such that the methacrylate functionality would react slower
with acrylates than with vinyl chloride. As predicted the butyl
acrylate is at 62% conversion before the MACROMER® peak is significantly diminished. These data add validity to the hypothesis
that the placement of side chains in the backbone is dependent on
the terminal group of the macromonomer and the relative reactivity
of its comonomer.

An additional observation concerning MACROMER® is that it
will not homopolymerize, and thus did not have to be stabilized.

It was demonstrated that MACROMER® will copolymerize with
conventional monomers in a predictable manner as determined by the
relative reactivity ratios. The copolymer equation:

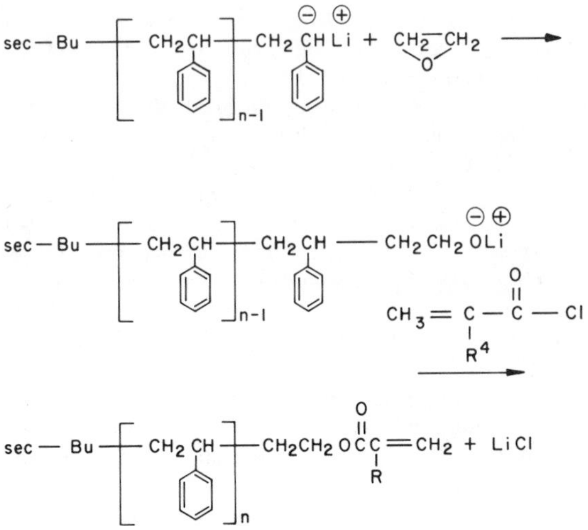

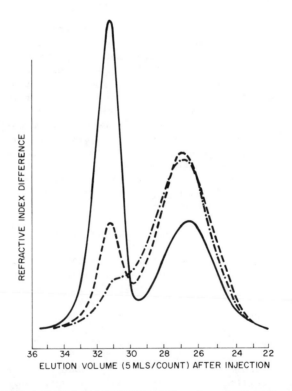

Figure 9.   *MACROMER synthesis.*

Figure 10.   *GPC of S11MA MACROMER/VCl product: 4.6% VCl conversion
(—); 9.8% VCl conversion (– – –); 55.4% VCl conversion (— · —).*

$$\frac{dM_1}{dM_2} = \left[\frac{M_1}{M_2}\right] \left[\frac{r_1 M_1/M_2+1}{\dfrac{M_1}{M_2} + r_2}\right]$$

can be reduced to the approximation

$$\frac{dM_1}{dM_2} \sim \frac{M_1}{r_2 M_2}$$

when M_1 is very low in molar concentration. Thus MACROMER® (M_1)
copolymerization with other monomers (M_2) can be described by r_2
values and the monomer feed compositions to give the equation

$$r_2 = \frac{dM_2/M_2}{dM_1/M_1} = \frac{\text{Percent Conversion } M_2}{\text{Percent Conversion } M_1}$$

In a monofunctional MACROMER® having a 15,000 molecular
weight, the functionality can be about one-half of a weight per
cent of the total monomer itself. Therefore, considering a copolymer charge on a weight basis of 50/50 wt %, the feed will only
contain about 3×10^{-3} moles of MACROMER®. This is unique in
itself; the opportunity to carry out copolymerization studies
where one monomer is of such a low relative concentration to the
other. This presents an opportunity to study areas of copolymerization that may have been questioned in the past, i.e.: "Will
a functional group attached to a very large polymer chain react
with the same predictability as a typical low molecular weight
monomer?" and "Are conventional relative reactivity ratios still
valid when one of the monomers is in a very low molar concentration?"

Some experimental data that were obtained through a series of
polymerization studies with a methacrylate-terminated MACROMER®
with vinyl chloride are shown in Figure 12. The Alfrey-Goldfinger
equation was used to calculate the copolymer composition for comparison to the actual copolymer composition as estimated from GPC
analysis. A reasonably close agreement was achieved of the actual
and the theoretical copolymer compositions, which indicates that
the r values are in the region of $r_1 = 10$ and $r_2 = 0.1$.

Thus far, this phase of this research area has shown:
• That very large macromonomers having a selected
    functional group can be quantitatively synthesized.
• That these macromonomers do indeed copolymerize
    with conventional monomers.
• That the reactivity of the MACROMER® is predictable.

A polystyrene with a functionality such as a methacrylate
group copolymerized with a mixture of ethyl and butyl acrylate
should yield a graft structure meeting the criteria of a thermoplastic elastomer as shown in Figure 13. The data in this figure
show that as the MACROMER® content is increased, the tensile

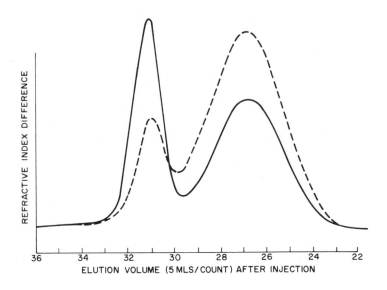

Figure 11.   GPC of S11MA MACROMER/BA product: 36.2% BA conversion
(—); 62.3% BA conversion (– – –).

M_1 FEED (%)	CONVERSION		COPOLYMER COMP. M_1	
	M_2 (%)	M_1 (%)	ACTUAL (a) (%)	THEORY (b) (%)
23.7	12.2	85.9	68.7	75.6
35	8.3	77.9	83.5	84.4
50	4.6	31.9	87.4	90.9
	9.8	83.0	89.4	

(a), CALC. FROM GPC ANALYSIS

(b), CALC. FROM THEORY FOR $r_1 = 10$, $r_2 = 0.1$

ALFREY–GOLDFINGER Eq.

$$\frac{d(M_1)}{d(M_2)} = \frac{(M_1)}{(M_2)} \cdot \frac{r_1 \dfrac{(M_1)}{(M_2)} + 1}{\dfrac{(M_1)}{(M_2)} + r_2}$$

Figure 12.   MACROMER $[M_1]$/vinyl chloride $[M_2]$ copolymerization.

strength is increased. A yield strength is observed at 30%
MACROMER® content and increases with increasing MACROMER® content.
Of particular interest is the fact that at 30 or 40% MACROMER®
contents, tensile strengths of 1700 and 2200 psi at a reasonable
elongation percentage are obtained. These are equivalent to tensile strengths obtained for typical vulcanized acrylic rubbers.
These rubbers were prepared via a suspension polymerization process. We have achieved this thermoplastic effect here by using
the phase separation of a polystyrene MACROMER® and the acrylic
rubber moieties. Again, it should be stressed that these materials are completely optically transparent, being clear and colorless. Figure 14 shows the typical stress-strain behavior and
yield points of the acrylic based thermoplastic elastomer that one
observes in the styrene-diene-styrene triblock polymers. The use
of MACROMER® has enabled the synthesis of new, interesting materials that can be carried out on a large scale using suspension
polymerization technology. In this case a controlled graft structure is synthesized having parallel physical effects to block
polymer structures.

```
a-a-a-a-a-a-a-a-a-a-a
b                   b
b                   b
b                   b
b                   b
```

The side chains are of a predetermined size and composition via
anionic chemistry and based on the reactivity placed at the ter-
minus of the MACROMER® and the comonomer with which it will be
reacted. The number of side chains per backbone and a distance
apart can be reasonably estimated. This distance between side
chains must be sufficient that the backbone can manifest its Tg.
Considering the earlier comments of molar concentration of the
MACROMER® in a typical copolymerization recipe, there will not be
many MACROMERS® per backbone on a statistical basis.

Since these materials are incompatible and are chemically
linked, one would expect to have a domain type structure as con-
ceptualized in Figure 15. Physical blends of these polymers were
prepared and again, as in the case of the styrene-dienes, were
opaque and cheesy. The copolymer graft structures were optically
clear and tough, indicating that one of the phases is efficiently
dispersed in the matrix of the other and that the dispersed phase
is below the wavelength of light. The same physical phenomenon
associated with block polymers is shown to occur in a graft
polymer.

The use of MACROMER® technology is another synthetic tool
that the polymer scientist has in being able to prepare a wide
variety of new types of polymer structures which could be of some
potential commercial interest, such as:

MACROMER WT %	YIELD STR (PSI)	TS (PSI)	ELONG (%)
25		1300	730
30	270	1700	550
40	1200	2200	400
45	1800	2500	350
50	2700	3000	240

S16MA 1:1 EA:BA

Figure 13.   MACROMER/acrylic copolymers.

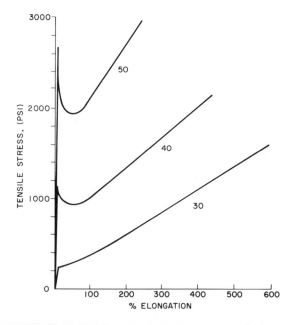

Figure 14.   Stress–strain curves for 1:1 ethyl acrylate:butyl acrylate copolymers with S16MA MACROMER at 30%, 40%, and 50% MACROMER contents.

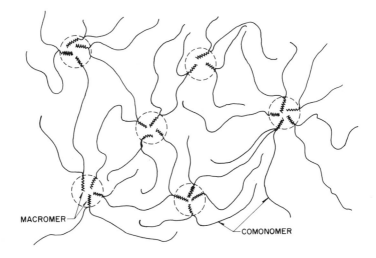

Figure 15.   Morphological structure of controlled graft polymers.

- Low Tg dispersed phase in high Tg matrix
  (impact plastics),
- High Tg dispersed phase in low Tg matrix
  (thermoplastic elastomers),
- High Tg dispersed phase in crystalline
  polymer matrix,
- Low Tg dispersed phase in crystalline
  polymer matrix,
- High Tg dispersed phase in high Tg matrix,
- Hydrophilic-hydrophobic, and
- Controlled branch homopolymer.

The high Tg dispersed phase in a high Tg matrix is quite interesting and would be appealing to rheologists. Copolymers with
vinyl chloride or acrylonitrile were found to have enhanced
ability to be pressed into thin sheets. In fact, the acrylonitrile styrene MACROMER® copolymer was successfully processed
through an Instron capillary rheometer. The hydrophilic-
hydrophobic material was developed by reacting the polystyrene
MACROMER® with N,N-dimethylacrylamide which developed into a very
interesting hydrogel (11). The final example is perhaps of more
of a theoretical interest. Thus, it is now possible with this
technique to prepare controlled branched homopolymers and compare
their solution, physical, and melt properties to straight linear
homopolymers.

Conclusion

The development of anionic chemistry has placed a number of
powerful tools in the hands of the polymer chemist. Polymer
molecules of predetermined molecular weight, molecular weight
distribution, composition, and configuration can now be synthesized nearing the purity of simple organic molecules. Controlled
polymeric structures have been realized that are highly desirable
as models to advance theoretical studies and indeed have vast
economic values to industry as profitable consumer items.

Literature Cited

1. R. Milkovich, M. Szwarc and M. Levy, J.A.C.S. $\underline{78}$, 2656
    (1956).

2. R. Milkovich, Polymerization Initiated by Electron Transfer
    to Monomer, M.S. Thesis, S.U.N.Y., Syracuse, NY (1957).

3. R. Milkovich, Process for the Preparation of Elastomeric
    Block Copolymer, Republic of S. Africa 227,164 (1964).

4. R. Milkovich, M. T. Chiang, U.S. Patents claiming functional
    groups and methods of preparation:
    3,842,050 (1974); 3,842,057 (1974); 3,842,058 (1974);
    3,842,059 (1974); 3,846,393 (1974); 3,862,098 (1975);
    3,862,101 (1975); 3,862,102 (1975).

5. R. Milkovich, M. T. Chiang, Polymerizable Diblock, etc.
    U.S. Patent 3,842,146 (1974); also 3,862,267 (1975).

6. G. Holden and R. Milkovich, U.S. Patent 3,231,635; Process
    for the Polymerization of Block Polymers (1966).

7. R. Milkovich, G. Holden, E. T. Bishop and W. R. Heindricks,
    Block Copolymers and Compositions Containing Them, Republic
    of S. Africa 284,464 (1964).

8. R. Milkovich, Block Polymers and Process for Preparing Them,
    Canadian Patent 716,645 (1965).

9. B. J. Bauer and L. J. Fetters, Rubber Chem. and Technol.,
    51, 406 (1978).

10. R. Milkovich, M. T. Chiang, Chemically Joined Phase Separated
    Thermoplastic Graft Copolymers, U.S. Patent 3,786,116 (1974).

11. R. Milkovich, M. T. Chiang, Chemically Joined, Phase Sepa-
    rated, Self Cured Hydrophilic Thermoplastic Graft Copolymers
    and Their Preparation, U.S. Patents 3,928,255 (1975) and
    4,085,168 (1978).

RECEIVED March 5, 1981.

Synthesis of Model Macromolecules of Various Types via Anionic Polymerization

PAUL REMPP, EMILE FRANTA, and JEAN HERZ

Universite Louis Pasteur, Centre de Recherches sur les Macromolecules, 6 rue Boussingault, 67038 Strasbourg Cedex, France

The need for well defined polymer species of low polydispersity and of known structure arises from the increasing interest in structure-properties relationship in dilute solution as well as in the bulk. A great variety of methods have been attempted, to synthesize so-called model macromolecules or tailor made polymers-over the past 20 years. The techniques based on anionic polymerization, when carried out in aprotic solvents, have proved best suited for such synthesis, because of the absence of spontaneous transfer and termination reactions that characterize such systems. The "living" polymers obtained are fitted at chain end with carbanionic sites[1], which can either initiate further polymerization, or react with various electrophilic compounds, intentionally added to achieve functionalizations. Another advantage of anionic polymerizations is that difunctional initiators are available, yielding linear polymers fitted at both chain ends with carbanionic sites. In this paper we shall review the various utility of anionic polymerization to the synthesis of tailor made well defined macromolecules of various types.

I - HOMOPOLYMERS

    I -<u>General</u>

       When a suitable monomer is polymerized by utilizing an efficient anionic initiator where the rate constant of initiation is higher than the rate constant of propagation, a linear polymer exhibiting a sharp Poisson-type molecular weight distribution is obtained.[1,2] Furthermore, the number average molecular weight of the polymer is determined by the molar ratio of monomer-to-initiator used. The range of molecular weights easily accessible by anionic polymerization techniques extends from 1000 to 500,000 approximately. For <u>very</u> <u>low</u> molecular weights the initiation reaction can no more be considered instantaneous and broadening of the molecular weight distribution results. For <u>very</u> <u>high</u> molecular weights the amount of initiator involved is very small, and a precise control of the molecular weight of the polymer becomes difficult. This is true even when drastic measures are taken to

0097–6156/81/0166–0059$05.00/0

avoid all kinds of "killing" impurities in the reactor as well as
in the monomer and solvent used.

Some anionically polymerizable monomers are listed below,
roughly in the order of increasing electroaffinity, i.e. in the
order of increasing tendency to polymerize anionically.

Table I

Anionically Polymerizable Monomers Listed in the Order of
Increasing Electroaffinity

- Dimethylaminostyrene
- Methoxystyrenes
- Methylstyrenes (p, o, m, and α-methyl)
- Vinylmesitylene
- Styrene
- p-chlorostyrene
- Vinylnapthalenes
- Isoprene (piperylene . . .)
- Butadiene
- Vinylpyridines (2 or 4)
- Alkylmethacrylate (alkyl = C_1 to C_{18}
- Acrylonitrile
- Oxiranes
- Lactones
- Cyanoacrylates
- Thiiranes

The choice of the initiator best suited for a given polymerization is of great importance. Fast initiation requires that
the nucleophilicity of the initiator be high enough for the monomer to be polymerized. However too high a nucleophilicity sometimes increases the probability of side reactions. For example,
cumylpotassium is well suited to initiate the polymerization of
styrene; if the same initiator is used for methylmethacrylate
polymerization, a side reaction on the ester carbonyl is likely
to occur. Diphenylmethylpotassium, a much weaker nucleophile, is
still an efficient initiator for methylmethacrylate, without side
reaction, but it is unable to initiate rapidly and quantitatively
the polymerization of styrene.

Monomers devoid of polar groups generally undergo anionic
polymerization in a predictable manner. With polar monomers
sometimes side reactions occur during the process; transfer reactions in the case of acrylonitrile, or propylene oxide, and
even more so with alkylacrylates; deactivations (or "killing")
reactions in the case of halogen substituted styrene or dienes.

Anionic "living" polymerization can be carried out in polar
as well as in non-polar solvents, provided they are aprotic, and
contain no strongly electrophilic functions. The rate of chain
growth is much faster in polar than in non-polar solvents.

However, for the purpose of synthesis, they are almost equivalent. Only for diene polymerizations, when high 1,4 unit contents are usually desired, it is of importance to choose non-
polar solvents (and Li as a counter ion) as the polymerization
media.

   2 - Functional Polymers
      The terminal carbanionic sites of "living" polymers can
be reacted with various electrophilic compounds of yield (α)-
functional polymers. Esters, nitriles, acid chlorides, anhydrides, lactones, epoxides, benzyl or allyl halogenides have been
used for their high reactivity with metal organic sites, to yield
appropriate functions.[3] Carbon dioxide is also an efficient reagent to yield terminal carboxylic functions.

      More sophisticated functions, such as acyllactam [4],[5]
vinyl silane [6], isocyanate [4], phosphonic ester [7], etc. have been
attached at the end of polymer chains, in nearly quantitative
yields. Also polymers fitted with dyes or fluorescent groups at
chain ends have been synthesized.[8]

      In all these functionalization processes attention must
be paid to avoid side reactions in order to attain high yields.
The reagents chosen should not contain any acidic hydrogen, nor
any group that could lead to a competing reaction.

      Special interest has been devoted to the reaction of
oxygen on carbanionic sites.[9] It has been established that triplet oxygen acts as an electron acceptor, and that the reaction
may yield either terminal hydroperoxy or alcohol functions.
Depending on the reaction conditions, the latter group which
originates from a disproportionation process between a hydro-
peroxidate and a carbanion may be observed. By choosing proper
conditions one can get polymers terminated quantitatively with
hydroperoxide groups. The latter can be used in a later step as
a radical initiator to prepare new block copolymers.

      Functional polymers have been of increasing interest in
recent years. The derived end groups can be used for spectrometric (UV or NMR) determinations of the molecular weight M_n, or
for excimer investigations. Telechelic oligomers or polymers are
accessible via "living" anionic polymerizations [10], and they can
undergo chain extension reactions by means of difunctional reagents, as will be seen below. Functional polymers can also be
used as starting points for block copolymer synthesis or for
model-network preparations. It also should be mentioned that
chemical functions are responsible for some new desirable polymer
modifications (e.g. flame retardant polymers).[7]

   3- Cyclic (Ring Shaped) Macromolecules
      The synthesis of well defined, ring shaped macromole-
cules has been independently achieved reacently in three different laboratories.[11-13] A linear "living" dicarbanionic precursor
polymer is reacted with a stoichiometric amount of an efficient
electrophilic coupling agent, such as dibromo-p-xylene or dimethyldichlorosilane. The reaction has to be carried out at

very high dilution in order to favor the intramolecular reaction
with respect to the intermolecular one. The cyclic macromolecules formed are mixed with a high molecular weight, linear
"polycondensate", but careful fractional precipitation allows a
close to quantitative separation of the ring shaped macromolecules. The yields of cyclization depend upon the concentration
used, the molecular weight of the precursor, and the nature of
the solvent used. They generally range from 15 to 45%, for molecular weights of the order of 5000 to 50,000, and concentrations
chosen far below 1% by weight. This is, to our knowledge, the
first time that ring shaped macromolecules were not obtained that
were a result of a ring-chain equilibrium.[14] The anionic cyclization is an irreversible reaction and it yields cyclic macromolecules exhibiting a molecular weight chosen at will, and a
narrow molecular weight distribution.[11-13]

   4 - Segmented Polymers
      The reaction described above can also be carried out at
higher concentration whereby the probability of intramolecular
reaction (cyclization) vanishes. So called chain extension processes result from the stoichiometric reaction of a "living" bifunctional precursors with an efficient bifunctional electrophilic deactivator. This polycondensation reaction induces a
very large increase of the molecular weight, but is also results
in an enhanced polydispersity.[15-16] Fractionation is necessary
if well defined substances are required. However the average
distance between successive hinges along the chain fluctuates
only very little.
      Functional hinges have been used recently, in order to
provide for ionizable sites at constant distances along the
chain.[17] Low charge density polyelectrolyte chains were obtained
according to this method. Also graft copolymers with regular
placement of grafts have been synthesized using a similar
pathway, as   will be seen below.[16]

   5 - Branched Polymers
      Special interest has been devoted in recent years to
comb shaped polymers. This architecture consists of a backbone
carrying p randomly distributed grafts of known length.
Moreover, star shaped polymers, which have p branches of known
length connected by one of their ends to a central nodule are
also of interest. The size of the nodule should be kept small
with respect to the whole star molecule. The methods developed
to synthesize these tailor made polymers have been reviewed
recently.[18,19]
      The synthesis of comb polymers derives from that of
graft copolymers which is described below.[20,21]
      For the synthesis of star shaped macromolecules, two
approached are possible:
      - One can react a monofunctional "living" precursor polymer
        stoichiometrically with a multi-functional deactivator,
        whereby the number of branches p should be given by the

functionality of the deactivator.[22] Star molecules with
up to 12 branches were obtained by this method.[23],[24]
similarly star molecules with 4 branches linked to a
central lead atom were made by reacting the precursor
with tetra(diphenylethylene) lead.[25] The star polymer
obtained using these methods are well defined, provided
the efficiency of the antagonist functions is
satisfactory.

$$Pb(C_6H_4-\overset{"}{\underset{CH_2}{C}}-C_6H_5)_4$$

- Star-shaped macromolecules have also been synthesized by
  using the monofunctional "living" precursor as an initiator for the polymerization of a small amount of a divinyl monomer. A small crosslinked nodule is formed,
  which is connected with the p chains that have contributed to its initiation.[26] It turns out that fluctuations on the value of p within a sample remain rather
  small, and consequently the star polymers obtained by
  this method can also be considered as tailor-made polymers. Recently star molecules with deuterium labeled
  central nodule have been synthesized according to the
  same method.[27]

    The synthesis of extremely high molecular weight star
polymers has been achieved by another method.[28] First divinylbenzene is reacted at very low concentration with an anionic initiator (sec. BuLi) to yield a suspension of poly-DVB nodules fitted with numerous initiating sites. Then styrene is added, and
each initiating site should give yield to a branch. However, the
polydispersity of the samples obtained is very high.

   6 - Model Networks

    Model networks are tridimensional crosslinked polymers
whose elastically effective network chains are of known length
and of narrow molecular weight distribution. The techniques used
to synthesize such networks are derived from those developed for
the synthesis of star shaped macromolecules, whereby the initiator used must be bifunctional instead of monofunctional.[29],[30]
The first step involves preparation of a bifunctional "living"
precursor, of known molecular weight and low polydispersity.
Crosslinking can be achieved either by the addition of stoichiometric amounts of a multifunctional electrophilic deactivator, or
by adding small a small amount of a bifunctional monomer (such as
DVB or ethylene glycol dimethacrylate), the polymerization of
which will be initiated by the carbanionic sites of the precursor. In either case, the precursor chains become the elastically
effective chains of the networks. The experimental conditions

can be chosen as to keep the proportion of defects (pendant
chains, loops. . .) very low, and to prevent syneresis from occuring upon crosslinking.

The latter method was also used to synthesize model networks with labeled branch points[30],[31] for structure determination
by means of X-ray scattering or neutron scattering.

Anionic polymerization is also well suited for the synthesis of networks exhibiting a known proportion of defects distributed at random in the network.[32] By using a given proportion
of a monofunctional initiator, together with the bifunctional
one, it is possible to introduce a known amount of pendent
chains, the average length of which is half that of the elastic
chains. By crosslinking a mixture of two (or more) "living" precursors of different length, one can introduce at will additional
polydispersity among the elastic chains. This is of great interest for a detailed investigation of the structure properties relationships in model-networks.

II - BLOCK AND GRAFT COPOLYMERS

I - Block Copolymers

Fast growing interest in block copolymers originates
from the intramolecular phase separation, which is responsible
for some unique properties involving many potential applications.
In this field again major progress has been accomplished due to
the availability of the "living" anionic polymerization techniques.

A rather obvious method to synthesize block copolymers
is to use a "living" precursor as the anionic initiator for the
polymerization of second monomer.[33],[34] However, this method
requires that the initiation reaction be fast, quantitative and
free of side reactions. This means that the nucleophilicity of
the carbanionic sites should be sufficient to attack the second
monomer added; in other words, the monomers have to be added in
the order of increasing electroaffinity.

On the other hand, it sometimes happens that the nucleophilicity of the end standing precursor carbanion has to be decreased to prevent side reactions. An easy way to achieve that
is an intermediate addition of 1,1-diphenylethylene.[35] This procedure is used especially when the second monomer to be added is
a methacrylic ester, to prevent attack of the ester carbonyl.

If the experimental conditions have been chosen properly the molecular weight and the composition of the sample can be chosen at will, and the fluctuations in composition as well as the polydispersity in molecular weight remain small.

This method applies to a great number of systems some of which are listed below.

Table II
Some Block Copolymers Obtained by
Anionic Polymerization Techniques

- Two amorphous glassy blocks
    - polystyrene - polydimethylaminostyrene
    - polystyrene - polyvinylpyridine (2 or 4)
    - polystyrene - polymethylmethacrylate
    - poly-2-vinylpyridine - poly-4vinylpyridine

- One glassy and one elastomeric block (or poly-α-methyl-styrene)
    - polystyrene (or poly-α-methylstyrene) - polyisoprene
    - polystyrene - polybutadiene
    - polystyrene -polybutylmethacrylate
    - polystyrene (or polyvinylmesitylene)-
                   polydimethylsiloxanee
    - polystyrene - polypropylene sulfide
    - polymethylmethacrylate - polybutylmethacrylate
    - polyvinylpyridine - polypropylene sulfide

- One glassy and one crystalline block
    - polystyrene - polyethylene oxide
    - polystyrene - polycaprolactone (polypropiolactone)
    - polymethylmethacrylate - polycaprolactone

- One crystalline and one elastometric block
    - polyisoprene (or polybutadiene) - polyethylene oxide
    - polycaprolactone - polydimethylsiloxane

- Two crystalline blocks
    - polyethylene oxide - polycaprolactone

The synthesis of triblock copolymers B-A-B can be achieved by means of a bifunctional initiator: a bifunctional α,ω-dicarbanionic poly-A precursor is formed, and is used in a second step as the initiator for the polymerization of monomer B, with the same conditions to be observed as above. A number of efficient bifunctional initiators are commonly used, in polar solvent media. Recent work, carried out in several laboratories[36-38], aimed at the preparation of efficient bifunctional initiators, soluble in non polar solvents. Such systems are especially useful when a polydiene (elastomeric) center block is wanted, as in Kraton-type thermoplastic elastomers.

Another way of synthesizing B-A-B triblock copolymers is to use a coupling reaction.[34] Monocarbanionic poly-B precursor is used to initiate the polymerization of A. The living two block copolymer is then reacted stoichiometrically with an efficient bifunctional coupling agent, such as dibromo-p-xylene or dimethyldichlorosilane, or even phosgene. This coupling reaction yields the triblock copolymers.

Another way of synthesizing block copolyers it to have two polymers which possess mutually reacting chain ends. A picturesque example is the mutual deactivation of "living" cationic polytetrahydrofuran and of "living" anionic polystyrene.[39]

There are however many other examples of reactions between ω-functional polymers, yielding two-block, triblock-or multiblock copolymers, depending upon the functionality of both precursors. Functionalization can be achieved anionically, or by polycondensation, or by using transfer reactions. However, the anionic techniques are the only method to achieve an adequate molecular weight control and a low polydispersity.

Chemically unlike polymers are incompatible, and it sometimes happens that the reaction medium is heterogeneous at the beginning. However, once some block copolymer is formed it acts as a "compatibilizer" and the reaction medium gradually becomes homogeneous. Many examples of such reactions could be quoted. A recent one is the hydrosilylation reaction carried out between a polystyrene fitted at a chain end with vinylsilane groups, and an α,ω-dihydrogenopolydimethylsiloxane. This process is carried out at high concentration and it yields polystyrene-polydimethylsiloxane-polystyrene block copolymers.[40]

$$
\text{Polystyrene} -\underset{\underset{\emptyset}{|}}{\overset{\overset{\text{H}}{|}}{\text{C}}}-\underset{|}{\overset{|}{\text{Si}}}-\text{CH=CH}_2 \quad + \quad \text{H}\underset{|}{\overset{|}{\text{Si}}}-\text{OPDMS-O}\underset{|}{\overset{|}{\text{Si}}}-\text{H}
$$

catalyst, e.g. H_2PtCl_6

Polystyrene $\underset{\underset{\emptyset}{|}}{\overset{|}{\text{CH}}}-\underset{|}{\overset{|}{\text{Si}}}-\text{CH}_2-\text{CH}_2-\underset{|}{\overset{|}{\text{Si}}}-\text{O}-$ PDMS $-\text{O}\underset{|}{\overset{|}{\text{Si}}}-\text{CH}_2-\text{CH}_2-\underset{|}{\overset{|}{\text{Si}}}-$ PStyrene

There are also cases where a functional polymer is required to initiate the polymerization of a second monomer. An

obvious example of this procedure is given by the polymerization
of a lactam (pyrrolidone) in the presence of a polymer fitted at
a chain end acyllactam functions.[4,5] Lactams as well as Leuchs
anhydrides (oxazolidine diones) polymerize according to a so-
called "activated monomer" process onto a "promotor" function.

In recent years several attempts have been made to prepare polymers possessing chain end functions capable of giving
rise to free radical or to cationic sites.[41] This research has
been mostly aimed at extending the possibilities of synthesis of
block copolymers, in which only one of the blocks is obtained
anionically. The synthesis of ω-hydroperoxy polymers[9] has already been mentioned. Peroxy-or peranhydride functions have also
been introduced into polymer chains.[42] Subsequent radical polymerization of a second monomer results in block copolymers.

Reaction of living polystyrene with excess phosgene
yields a polymer chain fitted with functional acyl chloride end
groups. Upon addition of silver hexafluoroantimonate an end-
standing oxocarbenium site is formed, that can be used to initiate the cationic polymerization of THF.[43]

2 - Graft Copolymers

The synthesis of graft copolymers can be achieved either
by "grafting from" or by "grafting onto" processes.[33] The former
method seems more versatile, but it does not allow an adequate
structure control, and it often yields rather polydisperse
samples. On the contrary "grafting onto" methods allow a precise
control of the size and of the number of grafts, but it is only
applicable to a limited number of systems.

"Grafting from" reactions

This method requires fitting a backbone chain with metal
organic sites, capable of initiating the polymerization of
another monomer, to yield the grafts. Metalation of a
polymer[33,44,45] backbone has been performed by various
procedures: direct metalation, metal halogen interchange, addition of organometallics substitution reactions involving
lithium-organic compounds activated with TMEDA. Once the backbone is metalated it often aggregates and becomes insoluble.
Therefore when a monomer is added to form grafts, the initiation
process is generally slow and it does not occur quantitatively.
This induces large fluctuations on the length of the grafts, and
it prevents any estimation of the number of grafts actually formed.

A special case of "grafting from" reaction is the chemical modification that can be induced on a metalated backbone
chain by micromolecular reagents. A recent example[46] of such reactions is the oxidation of metalated sites on a polystyrene
chain, yielding vinylphenol units. It has been shown that in
this case metalation (using BuLi, TMEDA) is located on the phenyl
rings.

"Grafting onto" reactions
        This grafting method is based upon the reactivity of the
carbanionic sites of a "living" polymer onto electrophilic functions distributed along the backbone chain.[47] Living polystyrene
or polydienes could thus be grafted onto backbones bearing ester
functions, or anhydride functions, or epoxide groups.[48] Also
benzylhalide functions[21],[49] are well suited for such reactions;
they can be obtained by cationic chloromethylation of polystyrene
under mild conditions.[50] Nitrile and pyridine groups were also
shown to react under proper conditions with "living" polymer
carbanions of high nucleophilicity.[33]
        Backbone and grafts can be characterized independently,
and the graft copolymers obtained can be quoted therefore as
model macromolecules. It was shown also that grafting occurs at
random[47], and that fluctuations in composition are negligible,
the molecular weight distribution of the graft copolymer being
homothetic to that of the backbone chain. Very high molecular
weights can be attained easily by such "grafting onto" reactions,
even though the extent of grafting is limited by steric factors.
        Recently this grafting methods has been used to synthe-
size amphiphilic graft copolymers in which hydrophilic grafts are
linked to a hydrophobic backbone. Partly chloromethylated polystyrene is used to deactivate either monofunctional "living"
polyethylene oxide[51] or monofunctional "living" polyvinylpyri-
dine. In the latter case subsequent quaternization yields polyelectrolyte grafts.[52]
        As already mentioned, well defined comb-like homopoly-
mers can be made by the same technique[20],[21]; here a living polystyrene is reacted on a partly chloromethylated polystyrene backbone, whereby random distribution of the grafts along the chain
occurs. It was also successfully demonstrated that one can synthesize comb polymers with grafts distributed at regular intervals along the chain.[16] Here the backbone is a "segmented" polystyrene with electrophilic functions remaining on the hinges, obtained by a chain extension reaction; monofunctional "living"
polystyrene is then reacted with the functions remaining on the
hinges, to produce the grafts. It has been established that the
placement of the grafts (regular or at random) does not influence
significantly the morphology of the comb polymer.

                            CONCLUSION

        Anionic polymerization techniques have contributed to a
very large extent to the development of tailor-made molecules of
various types. The long life time of the active sites is a
factor of decisive importance for such synthesis: it enables
one to choose at will the molecular weights of the polymers to be
made, and it ensures narrow molecular weight distribution of the
samples. It involves the possibility of functionalizations at
one or more chain ends, and of coupling reactions with bifunc-

tional electrophilic deactivators. It permits subsequent initiation of a second monomer to yield block copolymers. The occurence of difunctional initiators gives rise to further applications, expecially for the synthesis of triblock copolymers and of model networks.

The main limitation of these methods is the rather small number of monomers that undergo anionic polymerizations without transfer or termination steps. Nevertheless, research on the various applications of anionic polymerization as an efficient tool for the synthesis of tailor made polymers has been pursued very actively over the past 20 years, and it has yielded a great number of results of significance which has also opened a broad field of potential applications.

LITERATURE CITED

1. M. Szwarc, Advances in Polym. Sci. 12, 127 (1966).
2. S. Bywater, Progress in Poly. Sci. 4, 27 (1974).
3. M. Morton, L. J. Fetters, Macromol. Rev. 2, 71 (1967).
4. Y. Nitadori, E. Franta, P. Rempp, Makromol. Chem. 179, 927 (1978).
5. M. Schmitt, E. Franta, D. Froelich, P. Rempp, to be published (Makromol. Chem.).
6. P. Chaumont, J. Herz, P. Rempp, Europ. Polym. J. 15, 537 (1979).
7. J. Brossas, G. Clouet, C. R. Acad. Sci. C-280, 1459 (1975).
8. G. Beinert, G. Weill, to be published.
   M. Winnik, D. H. Richards, to be published.
9. J. M. Catala, G. Reiss, J. Brossas, Makromol. Chem. 178, 1249 (1977).
10. M. Morton, L. J. Fetters, J. Inomata, D. Rubio, R. Young, Rubber Chem. Techn. 49, 303 (1976).
11. B. Vollmert, J. X. Huang, Makromol. Chemie, 182 (1981).
12. D. Geiser, H. Hocker, Polymer Bulltein, 2, 591 (1980).
13. G. Hild, A. Kohler, P. Rempp, Eur. Pol. J. , 16, 525 (1980).
14. J. S. Higgins, K. Dogson, J. S. Semlyen, Polym. 20, 553 (1979).
15. J. C. Galin, M. Galin, Makromol. Chem. 160, 321 (1972).
16. C. Strazielle, J. Herz, Europ. Polym. J. 13, 223 (1977).
17. J. C. Galin, Martenot, to be published.
18. S. Bywater, Adv. Polym. Sci. 30, 90 (1979).
19. B. J. Bauer, L. J. Fetters, Rubber Chem. Techn. 51, 406 (1978).
20. J. Pannel, Polym. 12, 558 (1971), 13, 2 (1972).
21. F. Candau, E. Franta, Makromol. Chem., 149, 41 (1971).
22. M. Hert, J. Herz, C. Strazielle, Makromol. Chem., 160 (1972).

23. J. E. L. Roovers, S. Bywater, Macromol. 5, 384 (1972),
       7, 443 (1974).
24. N. Hadjichristidis, A. Guyot, L. J. Fetters, Macromol.
       11, 668 (1978).
25. G. Beinert, J. Herz, Makromol. Chem. 181, 59 (1980).
26. D. J. Worsfold, J. G. Zilliox, P. Remmp, Can. J. Chem. 47,
       3379 (1969).
27. A. Kohler, J. G. Zilliox, P. Rempp, J. Polacek, I. Kossler,
       Europ. Polym. J. 8, 627 (1972).
28. H. Eschweg, W. Burchard, Polym. 16, 180 (1975), Makromol.
       Chem. 173, 235 (1973).
29. J. Herz, P. Rempp, W. Borchard, Adv. Polym. Sci., 26,
       105 (1978).
30. P. Lutz, C. Ipcot, G. Hild, P. Rempp, Brit. Polym. J.,
       151 (1977).
31. A. Belkebir-Mrani, G. Beinert, J. JErz, A. Mathis,
       Europ. Polym. J. 12, 243 (1975).
32. J. Bastide, C. Picot, S. Candau, J. Polym. Sci., A-2,
       17, 1441 (1979).
33. P. Rempp, E. Franta, Pure and Appl. Chem. 30, 229 (1972).
34. L. J. Fetters, J. Polym. Sci., C-26, 1 (1969).
35. D. Freyss, P. Rempp, H. Benoit, Polym. Letters, 2, 217
       (1964).
36. R. P. Ross, H. W. Jacobson, W. H. Sharkey, Macromol. 10,
       287 (1977).
37. E. A. Mushina, L. Muraviova, T. Samedova, B. A. Krentsel,
       Europ. Poly. J., 15, 99 (1979).
38. G. Beinert, P. Lutz, E. Franta, P. Rempp, Makromol. Chem.,
       179, 551 (1978).
39. Y. Yamishita, et al. J. Polym. Sci. B-8, 481 (1979).
40. P. Chaumont, B. Beinert, J. Herz, P. Rempp, Polymer, 22,
       663 (1981).
41. T. Souel, F. Schue, M. Abadie, D. H. Richards, Polym.,
       18, 1292 (1978).
42. G. Riess, F. Palacin, IUPAC Symp. Preprints - Helsinki,
       1, 123 (1970).
43. E. Franta, J. Lehmann, L. Reibel, S. Penczek, J. Polym.
       Sci., C-56, 139 (1976).
44. G. Clouet, J. Brossas, Makromol. Chem. 180, 867 (1979).
45. A. J. Chalk, A. S. Hay, J. Polym. Sci., A-1(7), 691,
       1359 (1969).
46. J. M. Catala, J. Brossas, Polym. Bull. 2, 137 (1980).
47. Y. Gallot, P. Rempp, J. Parrod, Polym. Letters, 1, 329
       (1963).
48. P. Takaki, R. Asami, M. Pizuno, Macromol., 10, 604 (1977).
49. D. Rahlves, J. Roovers, S. Bywater, Macromol. 10,
       604 (1977).
50. F. Candau, P. Rempp, Makromol. Chem., 122, 15 (1969).
51. F. Candau, F. Afshar-Taromi, P. Rempp, Polym., 18,
       1253 (1977).
52. J. Selb, Y. Gallot, Polym. 20, 1268, 1273 (1979).

RECEIVED June 5, 1981.

Relationship of Anion Pair Structure to Stereospecificity of Polymerization

S. BYWATER

Division of Chemistry, National Research Council of Canada,
Ottawa, Canada K1A OR9

Spectroscopic measurements can be made on the active centers in anionic polymerization. Oligomeric model compounds, for example, of butadiene or isoprene show the existence of variable proportions of cis and trans forms at equilibrium dependent on counter-ion and solvent. The new active center formed immediately on monomer addition may not, however, be in its equilibrium configuration. In this case, the rate of isomerization to the stable form becomes important in microstructure determination if it is assumed that the structure is "frozen-in" at the next monomer addition. Charge distribution and counter-ion position also change with reaction conditions, which must have some influence on the proportion of 1,4 and vinyl units in the polymer. Particular examples of correlations between ion-pair properties and polymer microstructure are discussed.

One of the interesting features of anionic polymerization is the variation in microstructure of the polymers formed under different conditions. Both vinyl and diene monomers show these effects, most markedly with acrylates, butadiene and isoprene and particularly with lithium as counter-ion. Early models for stereospecificity are essentially static ones, for example with the lithium/hydrocarbon/isoprene system, the formation of a six-membered intermediate with coordination of the C-Li bond to the incoming monomer in its cis form (1). The configuration of each terminal unit would then be determined in the reaction leading to its formation with no subsequent changes possible. Similarly with acrylates, complexation of the lithium with carbonyl groups on incoming monomer and units of the polymer chain is the dominant feature of mechanisms for isotactic polymer formation (2). Probably the first departure from these static models was

0097–6156/81/0166–0071$05.00/0
Published 1981 American Chemical Society

suggested by Bovey et al (3) in acrylate polymerization. They
were able to show using specifically deuterated monomers that
although in hydrocarbon solvents monomer attack did occur so as to
maintain aligned ester-groups in monomer and polymer, with traces
of ethers present, the monomer could approach with ester groups
opposed. This would normally produce a racemic diad. Subsequent
rotation of the terminal group to maximize electrostatic interactions and hence bring back a meso configuration was found to
occur. In isoprene polymerization too, there were indications
that the simple picture was not adequate, for the cis-1,4 content
was found to be dependent on reaction conditions, particularly on
the initiator concentration (4). The fact that butadiene under
similar conditions gave a mixed cis/trans polymer also suggested
the mechanism was more complex. Both these observations suggest
that kinetic as well as thermodynamic factors are important in
stereostructure determination.

In order to increase the understanding of these processes, a
valuable tool is the study of configurations and isomerization
rates of low molecular weight models of the polymer chain. The
ability to study such compounds is one of the major advantages of
studies in anionic polymerization. The active centers are stable
under many conditions and can be studied at leisure by a number of
techniques, particularly NMR and optical spectroscopy. The dienes
provide a good example. Addition of sec. or t-butyllithium to
isoprene or butadiene under controlled conditions yields a model
one unit active chain usually with small amounts of two unit
material (5,6). The latter can be removed if necessary by
converting to the mercury compound which can be distilled to
remove dimer. The mercury compound can subsequently be
reconverted to the lithium (or other alkali metal) derivative by
simple treatment with an alkali metal (7). With the acrylates,
due to extensive attack of lithium alkyls on the ester group, this
approach cannot be used, but models can be made by metallation of
suitable compounds such as methylisobutyrate (8).

N.M.R. studies on diene models show that the preferred
configuration of the lithium compound in hydrocarbons is trans
although some cis structure does exist in equilibrium i.e. the
active center itself does exist in two forms (9,10).

In polar solvents such as THF, generally the cis form is more

stable for the whole series of alkali metals (11) (the lithium
derivative of butadiene in diethylether is an exception). If the
lithium isoprene model is transferred from hydrocarbon solvents to
THF at low temperature, its configuration is frozen in the form
stable in hydrocarbons. On warming however, at -40° a slow
isomerization occurs to the form stable in THF (12). This gives
an estimate of the isomerization rate under a specific set of
conditions. More important are the rates in hydrocarbon solvents.
These can be determined in a different way. Di-isoprenyl mercury
as prepared by the technique described above is mainly in its cis
form and when reacted with lithium suspension in hydrocarbon
solvents at low temperatures gives cis-isoprenyl lithium which
slowly isomerizes to the equilibrium trans-rich mixture. Rates
can be measured between -20° and 0° (13). Estimates can also be
made of the isomerization rates of a two unit active chain which
turn out to be somewhat different. These latter rates are
probably closer to that of high polymer.

The microstructure of polyisoprene prepared by lithium
initiation in hydrocarbons is 95% 1,4 under all conditions. The
trans 1,4 content however falls from about 20% to zero as the
monomer/initiator ratio increases leading finally to a 95% cis 1,4
polymer. This variation can be explained with the following
scheme.

$$\sim\!\!\sim\!\!cis^* + M \longrightarrow \sim\!\!\sim\!\!cis, cis^*$$

$$\downarrow\uparrow$$

$$\sim\!\!\sim\!\!trans^* + M \longrightarrow \sim\!\!\sim\!\!trans, cis^*$$

where $\sim\!\!\sim\!\!cis^*$ represents a cis active center, an internal unit
appearing unstarred. The newly formed active center is entirely
in the cis form i.e the reaction is stereospecific. This point
can actually be proved within experimental error. When new
monomer adds, a cis center is converted to a cis chain unit and of
course a trans one to a trans unit. It is unlikely that monomer
addition would change its configuration. Now if monomer addition
is slow, the initially formed cis active center may have time to
rearrange to the more thermodynamically stable trans form and in
the limit of very slow polymerization rate, the population
distribution will be the equilibrium one. If however the rate of
monomer addition is fast compared to isomerization the whole
active center population will remain cis and with it the polymer
configuration. On this simple picture the low rate plateau would
give a 66% trans content in the polymer and the high rate limit
zero percent if rates of addition of monomer to the two types of
center are equal. In fact it can be shown that they are not, the
cis active centers adding monomer eight times faster than the
trans centers. Knowing the relative monomer addition rates,
isomerization rates and equilibrium populations it is possible to
calculate the cis-trans ratio in the polymers and compare it with

experiment. Good agreement is found; the importance of kinetic
and thermodynamic factors in this case is well established.
Attempts to repeat these measurements on butadiene polymerization
led to the result that the isomerization rates were faster - too
fast to measure by the techniques in use. No detailed examination
was therefore possible but qualitatively it is clear that this
will result in a microstructure much closer to the equilibrium
population of active center configurations i.e. less cis 1,4 units
in the polymer than with isoprene although this again is probably
a cis-stereospecific reaction.

Polymerization of dienes in polar solvents is a more complex
problem for large amounts of vinyl unsaturation occur in the
polymer. Some 1,4 structures do persist particularly in the
ethers of lower dielectric constant so it is of interest to
inquire if cis/trans isomerization of the active centers is again
of importance. This can be shown to be the case in butadiene
polymerization in THF. Optical spectroscopy provides the needed
tool. Cis and trans active centers both show the near-UV
absorptions characteristic of delocalized allylic ions and their
pairs. The absorption maxima are however shifted, the trans
centers absorbing at longer wavelength (14,15). If at -40° a
relatively large amount of monomer is added to a polybutadienyl-
sodium solution at equilibrium, an almost instantaneous shift of
the absorption maximum occurs to longer wavelength, which slowly
returns to its original position when polymerization is over. It
appears that in this solvent trans centers are preferentially
formed on monomer addition which cannot relax to the more stable
cis form at this temperature. The situation is exactly opposite
to that observed in non-polar solvents. At higher temperatures
however equilibrium is maintained. Again we are dealing with a
reaction which has a preferential product which can be different
from the the equilibrium form, in this case indicated by lowering
the temperature rather than increasing the monomer concentration.
Parallel measurements on polymer microstructure show that these
changes are reflected in the product. Although the total 1,4
content is low, it is almost entirely trans at low temperatures
(16). Differing rates of monomer addition to cis and trans
centers are also indicated by a sharp change of the temperature
coefficient of polymerization rate at -30° where "freezing-in" of
structure first becomes important - the trans centers add monomer
faster in this solvent. In view of these observations care should
be taken in interpretation of literature data on microstructure,
often determined at one particular temperature and monomer/
initiator ratio. The results may not be typical of the reaction
under all conditions.

In polar solvents the other important variable is the 1,4 to
vinyl ratio. Vinyl unsaturation is always an important part of
the polymer structure but its amount does depend on counter-ion.
It is tempting to correlate this with charge distribution at α and
γ positions of the delocalized allylic active centers. In ^{13}C NMR

spectra of the 1:1 adduct of t-butyllithium and butadiene for example the resonance of the γ-carbon moves upfield and that of the α-carbon moves downfield on moving from a hydrocarbon to an ether solvent. In THF further shifts of the same type occur on changing the counter-ion from Li^+ to K^+ (11). These chemical shift changes can be interpreted as being caused by redistribution of charge from α to γ positions in the ion pairs (Table 1).

Table I

Charge distribution on butadiene one unit model and % 1,2 structure in polybutadiene.

	% of charge[a] at γ	% 1,2		
		diethylether	dioxane	THF
Li (benzene)	22	–	–	–
Li (THF)	37	73	87	96
Na (THF)	43	–	85	91
K (THF)	47	58	55	83
Cs (THF)	50	44	41	74

a) as percentage of total charge at $\alpha + \gamma$

It will be seen that there is an almost equal distribution of the charge between α and γ positions in THF for the heavier alkali metal counter-ions. If we suppose that increased charge produces an increased reactivity at a given position, then more vinyl unsaturation will be produced in THF than in hydrocarbon solvents and the highest vinyl content with heavier alkali metal counter-ions. The order in THF is however reversed, i.e. the highest vinyl structures are produced by lithium catalysis (17) although microstructure determinations in this solvent normally apply to reactions with an appreciable free anion contribution and hence cannot be simply interpreted. In dioxane (18) and diethylether (19) however this complication should be absent and charge distribution surely must follow the same pattern with counter-ion. Again the highest 1,2 content in the polymer occurs with lithium as counter-ion. The large counter-ions produce a fairly evenly balanced 1,4/1,2 ratio as might be expected from an even α/γ charge distribution and a large counter ion tending to block both positions. It seems plausible to suppose that a highly solvated lithium cation and moderately solvated sodium cation in ether solvents situated closer to the α-position effectively block the terminal position leaving preferential attack at the γ position.

Butadiene is a relatively simple case being a symmetrical monomer. With isoprene on the other hand a further complication arises because the monomer can add at its 1 or 4 position giving a 4,1 active center ($\sim CH_2-CH \overset{\text{CH}_3}{=} \overset{-}{C} = CH_2 X^+$) a) or a 1,4 center

$(\sim CH_2-\overset{\overset{\displaystyle CH_3}{|}}{C}{=}CH{=}\overline{CH}_2 X^+)$ b). As the addition is in practice not
reversible the mode of monomer addition decides immediately if the
active centre will produce a) 1,4 or 3,4 b) 4,1 or 1,2 units, the
final internal choice in each case only being decided at the next
monomer addition. All evidence on configurational preference so
far obtained has been on type a) centres normally formed almost
entirely by reaction of butyl lithium with monomer in hydrocarbon
solvents (20). The presence of both 1,2 and 3,4 units in polymers
formed in polar solvents shows that here both modes of addition
occur (Table II). Between 20 and 30% of the vinyl units have a

Table II

Total vinyl unsaturation and proportion of 1,2 structure in
polyisoprenes.

Counter-ion	diethyl ether		dioxane	
	total vinyl	% 1,2[a]	total vinyl	% 1,2
Li	65%	20	93%	18
Na	83%	26	91%	13
K	62%	32	62%	19
Cs	47%	33	53%	23

a) Percentage of total vinyl

1,2 structure in diethyl ether and dioxane. With the free anion
the proportion rises to ~40% according to results obtained in
dimethoxyethane where it appears to be responsible for most
polymer formation. It should be noted that these figures cannot
be used to calculate the proportion of the two types of active
centres present because again their reactivities with monomer may
be different. Their reactivities at α and γ positions may be also
different leading to different 1,4 (4,1) to vinyl ratios.
Structures of type b) form have not so far been prepared although
there is some evidence that they prefer to be in the trans form
even in THF in contrast to type a) centres (19). Isolation of
models of type b) is required to confirm their configurational
preference and to measure the different rates of monomer addition
of the two before a good understanding of the mechanism of
isoprene polymerization can be obtained.
    While for the dienes we have seen much information can be
obtained from simple one unit active polymer chain models, this is
less helpful in the case of vinyl monomers where microstructure
depends on the relationship between two or more monomer units in
the chain and orientation of the incoming monomer. Little work
has been done in this field (18). A mechanism has, however, been
proposed to explain the changes in microstructure observed in poly
α-methylstyrene formed by lithium catalysis in THF which does

attempt to explain these changes in terms of isomerization of the terminal unit with respect to the penultimate one (21). At low monomer concentration and high temperature, equilibrium between terminal meso and racemic diads would be maintained and microstructure determined by the thermodynamic stability of the meso and racemic units. The racemic form is supposed to be more stable by 700 cal/mole, its concentration increasing from 72% at 0° to 87% at -100° at equilibrium. At higher monomer concentrations and lower temperatures kinetic factors were claimed to dominate with the meso form being less easily formed with a higher activation energy. Once again the importance of both kinetic and thermodynamic effects is emphasized, illustrating the widespread importance of both of them in stereospecific polymerization models.

Literature Cited

1. Stearns, R.S. and Forman, L.E. J. Polym. Sci. (1959) 41, 381.
2. Glusker, D.L., Lysloff, I. and Stiles, E. J. Polym. Sci., (1961) 49, 315.
3. Fowells, W., Schuerch, C., Bovey, F.A. and Hood, F.P. J. Am. Chem. Soc. (1967) 89, 1396.
4. Gebert, W., Hinz, J. and Sinn, H. Makromol. Chem. (1971) 144, 97.
5. Schué, F., Worsfold, D.J. and Bywater, S. J. Polym. Sci. (Pol. Lett. Ed.) (1969) 7, 821.
6. Glaze, W.H. and Jones, P.C. Chem. Commun. (1969) 1434.
7. Bywater, S., Lachance, P. and Worsfold, D.J. J. Phys. Chem. (1975) 79, 2148.
8. Lochmann, L. and Lim, D. J. Organometal. Chem. (1973) 50, 9.
9. Brownstein, S., Bywater, S., and Worsfold, D.J. Macromolecules (1973) 6, 715.
10. Glaze, W.H., Hanicak, J.E., Moore, M.L. and Chaudhuri, J. J. Organometal Chem. (1977) 44, 39.
11. Bywater, S. and Worsfold, D.J. J. Organometal Chem. (1978) 159, 229.
12. Schué, F., Worsfold, D.J. and Bywater, S. Macromolecules (1970) 3, 509.
13. Worsfold, D.J. and Bywater, S. Macromolecules (1978) 11, 582.
14. Garton, A. and Bywater, S. Macromolecules (1975) 8, 694.
15. Garton, A., Chaplin, R.P. and Bywater, S. Eur. Polym. J. (1976), 12, 697.
16. Garton, A. and Bywater, S. Macromolecules (1975) 8, 697.
17. Rembaum, A., Ells, F.R., Morrow, R.C. and Tobolsky, A.V. J. Polym. Sci. (1962) 61, 166.
18. Salle, R. and Pham, Q.T. J. Polymer Sci. (Chem. Ed.) (1977) 15, 1799.
19. Dyball, D.J., Worsfold, D.J. and Bywater, S. Macromolecules (1979) 12, 819.
20. Schué, F. and Bywater, S. Bull. Soc. Chem. Fr. (1970) 271.
21. Wicke, R. and Elgert, K.F. Makromol. Chem. (1977) 178, 3085.

RECEIVED February 12, 1981.

Some Aspects of Ion Pair–Ligand Interactions

JOHANNES SMID

Chemistry Department, College of Environmental Science and Forestry,
State University of New York, Syracuse, NY 13210

Changes in the inter- and intramolecular interaction
and in the structure of carbanion pairs on adding
cation-binding ligands (ethers, crown ethers, polyamines) is briefly reviewed.

The complexity of anionic polymerization arises from the
existence of a variety of species in the type of solvents often
used in these reactions: free ions, different kinds of ion
pair, triple ions, ion pair dimers, etc. Cation-binding ligands,
added to catalyze the polymerization or to modify the product
structure, further increase the number of species:

$$A^- + M^+ \rightleftharpoons A^-||M^+ \rightleftharpoons A^-,M^+ \qquad \text{Ion Pairs}$$

$$A^-,M^+ + A^- \rightleftharpoons A^-,M^+,A^- \qquad \text{Triple Ions}$$

$$A^-,M^+ + M^+ \rightleftharpoons M^+,A^-,M^+$$

$$n\ A^-,M^+ \rightleftharpoons (A^-,M^+)_n \qquad \begin{array}{l}\text{Ion Pair}\\\text{Aggregates}\end{array}$$

$$A^-,M^+ + L \rightleftharpoons A^-,M^+,L \qquad \begin{array}{l}\text{Externally Complexed}\\\text{Tight Ion Pairs}\end{array}$$

$$A^-,L,M^+ \rightleftharpoons A^- + M^+,L \qquad \begin{array}{l}\text{Solvent- or Ligand}\\\text{Separated Ion Pairs}\\\text{and Free Ions}\end{array}$$

The ionic equilibria are influenced by many factors: concentration of ionic species, nature of cation and anion (charge
delocalization, bulky substituents), dielectric constant, donicity
number and specific structural features of the solvent molecules,

0097–6156/81/0166–0079$05.00/0

the presence of salts or cation binding ligands (e.g., glymes,
crown ethers, polyamines, cryptands, podands), temperature,
pressure, etc. Any of these variables can alter the relative
amounts of the ionic species, each of which can propagate the
chain with its own characteristic rate constant and stereospecificity.

Modification of ion pair structures by ion-binding ligands
have been studied by several techniques, e.g., optical, infrared
and raman spectroscopy, conductivity, electron spin resonance-
and ^1H, ^{13}C and alkali nuclear magnetic resonance spectroscopy,
dipole moment measurements, viscometry and mechanistic studies.
Much of the information pertinent to anionic polymerization has
been derived from non-polymerizable ionic systems, e.g., carbanion salts derived from fluorene, diphenylmethane, triphenyl-
methane and 4,5-methylenephenanthrene as well as salts of other
charge delocalized carbanions, radical anions, enolates,
phenoxides, carbazyl, etc. These and other investigations have
immensely contributed to a deeper understanding of the mechanism
of anionic polymerization and have been extensively reviewed
elsewhere (1-7).

Some aspects of ligand interactions with carbanion pairs
leading to different types of ion pair complexes will be briefly
reviewed. The discussion will focus chiefly on crown ethers and
polyamines as ligands interacting with alkali and alkaline earth
cations. The polyamines have been used extensively in reactions
with organolithium reagents (8), while crown ethers, together
with the linear glymes, have been employed in several studies of
salts of carbanions, oxyanions, sulfides, carboxylates, etc.
Optical spectroscopy has been one of the important tools in
obtaining information on complex formation constants and on the
nature of the ion pair-solvation and ligand complexes.

Crown ether complexes of fluorenyl carbanion salts

Solvent or ligand interactions with tight ion pairs produce
externally complexed tight ion pairs and/or ligand separated ion
pairs. The stability of the complexes depends on solvent,
temperature, type of crown and the nature of the cation. For
example, in ethereal solvents benzo-15-crown-5 and fluorenyl
sodium (Fl^-,Na^+) form the two isomeric complexes I and II depicted in reaction 1, but the ratio I/II is highly solvent
sensitive (9) (if the bound solvent in II is included in the
structure of II, the two complexes of course can actually not be
considered isomeric).

In these equations, Cr refers to a crown ether ligand while
S denotes a solvent molecule. The notation S_n following Fl^-,Na^+
signifies that in solvents such as dioxane, ethyl ether, tetrahydropyran (THP) or tetrahydrofuran (THF) the tight ion pairs,
especially when small alkali ions are involved, contain externally complexed solvent molecules. These must be removed before

$$\text{Fl}^-,\text{Na}^+,\text{S}_n + \text{Cr} \underset{}{\overset{K_1}{\rightleftharpoons}} \text{Fl}^-,\text{Na}^+,\text{Cr} + n\text{S}$$

$$\overset{K_2}{\searrow} \qquad \overset{K_3}{\nearrow} \qquad \text{(I)} \tag{1}$$

$$\text{Fl}^-,\text{Cr},\text{Na}^+,\text{S}_m + (n-m)\text{S}$$

(II)

a crown ligand can form the external tight ion pair complex I. The energy of desolvation is expected to increase in the order ethyl ether $<$ THP $<$ THF $<$ DME (1,2 dimethoxyethane) (2). The <u>loose</u> ion pair complex II is likely to retain one or two solvent molecules since the cation may slightly protrude from the other side of the crown cavity to allow interaction with solvent molecules. Hence, a better cation-solvating solvent is likely to shift equilibrium 2

$$\text{Fl}^-,\text{Cr},\text{Na}^+,\text{S}_m \rightleftharpoons \text{Fl}^-,\text{Na}^+,\text{Cr} + m\text{S} \tag{2}$$

II   I

in favor of the loose ion pair complex II. This explains the observed values (9) for the ratios K_3 = II/I with benzo-15-crown-5 and Fl^-, Na^+, viz., 0 (ethyl ether), 0.52 (THP) and 1.8 (THF).

The effect of <u>dielectric constant</u>, D, on the formation constant of a loose ion pair complex is probably not large in spite of the increase in the interionic ion pair distance. The microscopic rather than the macroscopic dielectric constant determines to a large extent the difference between the coulomb attraction energy of two ions in a tight and loose ion pair. For example, the formation constant of the loose ion pair complex III between fluorenyl sodium and triisopropanol amine borate at 25°C is nearly identical in THF (K = 102 M^{-1}, D = 7.4), THP (K = 104 M^{-1}, D = 5.6) and dioxane (K = 74 M^{-1}, D = 2.4) (10).

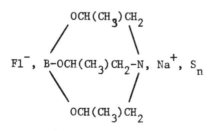

III

For complex III, the Na^+ is probably as accessible to solvation by solvent molecules as is the Na in the tight Fl^-,Na^+ ion pair. Hence, no externally bound solvent molecules need to be removed. This may be different in other systems. For example, the formation constant of a loose ion pair complex between Fl^-, Na^+ and tetraglyme (tetraethylene glycol dimethyl ether) is nearly four times lower in dioxane than in THF (10). This may be caused by specific solvent effects rather than by the difference in solvent dielectric constant. The flexible glyme ligand wraps itself around the Na^+ ion, and this may make it more difficult for solvent molecules to remain bound to Na^+ in the glyme-separated ion pair.

The specific structure of the cation binding ligand is critical in determining the type of ion pair complex. For example, in comparison to benzo-15-crown-5 (see above), benzo-18-crown-6 and Fl^-,Na^+ form exclusively a loose ion pair complex, at least in THF and THP, although with Fl^-,K^+ both ion pair complexes are formed (9,11). Substituents close to the crown cavity can also affect the stability of the ion pair-ligand complex. An interesting example is the crown ether IV, an excellent chelating agent for Li^+ (12). It was observed (13) that in chloroform the tight lithium picrate ion pair converts to the crown -

IV

separated ion pair complex on addition of IV. However, in THP a crown complexed tight ion pair is formed in spite of the better cation-solvating properties of THP which normally favor a loose ion pair complex. Apparently, the bulky methyl groups prevent the Li^+ ion in the loose ion pair from complexing THP molecules. Hence, while in chloroform the formation of the loose ion pair - IV complex requires energy to overcome the coulomb forces on enlarging the interionic ion pair distance, in THP it would also require the removal of THP molecules bound to the tight lithium picrate ion pair. Apparently, complexation to IV does not provide enough energy to accomplish both, and in THP only the crown-complexed tight ion pair is stable. Examples of

this kind show the difficulty in predicting the nature of the ion
pair-complex.

An important observation is the concentration dependency of
the ratio II/I (see equations 1 and 2). In a study of the complex formation between benzo-15-crown-5 and fluorenylsodium in
THF, 2-methyl-THF and THP the ratio of the two complexes I and
II was found to depend not only on solvent but also on the
total ion pair concentration (14). The results were rationalized
by assuming the formation of dimers of II on increasing the ion
pair concentration as shown in reaction 3. Aggregation of complex I is

$$2 \text{ Fl}^-,\text{Cr},\text{Na}^+ \underset{}{\overset{K_a}{\rightleftharpoons}} (\text{Fl}^-,\text{Cr},\text{Na}^+)_2 \qquad (3)$$

hindered by the externally complexed crown molecule. The
aggregation constants, K_a, depend on the solvent dielectric
constant, their values at 25°C being 62 M^{-1} in THF, 87 M^{-1} in
Me-THF (D = 6.24) and 168 M^{-1} in THP.

The tendency of ligand-complexed loose ion pairs to
aggregate is often reflected in the solubility behavior of these
complexes. For example, while Fl^-,Li^+ has a high solubility in
THF, its loose ion pair complex with dibenzo-14-crown-4 has a
solublility less than 10^{-4}M (14). Difluorenylbarium is soluble
in THF up to 0.05 M, but the solubilities of the mixed tight-
loose ion pair complexes with glymes and crown ethers (e.g., Fl^-,
Ba^{++},G,Fl^-) are only in the order of 0.001-0.01 M (15). Similar
findings have been reported for polyamine complexes with organolithium complexes (see next section). Hence, while external
ligand complexation to a tight ion pair can dramatically enhance
salt solubility in apolar media (16), solubility may decrease
when interaction with ligands produces loose ion pairs.

Differences in aggregation constants between tight and
loose ion pair solvation or ligand complexes make it more
difficult to compare results of investigators working on the
same systems but in different concentration ranges. For example,
U.V-visible spectrophotometric data can easily by obtained at
ion pair concentration between 10^{-3}-10^{-5} M, but infrared or nmr
measurements on ion pairing are frequently carried out at 0.1 M.
Conclusions drawn from these studies as far as ion pair
structures are concerned may differ because aggregation at higher
concentration can lead to shifts in equilibria between different
ion pair species.

The structure of the ion pair complex is also sensitive to
the anion, especially the extent of charge delocalization and
the presence of substituents close to the anionic site. Loose
ion pair formation requires a considerable increase in the interionic ion pair distance, and a charge localized anion, therefore,

will favor external ligand complexation since this requires less
stretching of the ionic bond. For complexes of benzo-15-crown-
5 with potassium fluorenyl and potassium picrate, the complex
formation constants leading to the 1:1 externally complexed tight
ion pairs are nearly identical for the two salts, i.e., 8000 M^{-1}
for potassium fluorenyl (9) and 6000 M^{-1} for potassium picrate
(17). Addition of a second crown produces the loose ion pair
A^-,Cr,K^+,Cr. However, the complexation constant for adding the
second crown is 1800 M^{-1} for the fluorenyl carbanion and only
200 M^{-1} for the picrate salt. The lower value for picrate may
in part be due to less charge delocalization, e.g., the free ion
dissociation constant for potassium fluorenyl in THF is 1.6 x
$10^{-7}M$ (18) as compared to 9.2 x $10^{-8}M$ for potassium picrate (17).
The two NO_2 substituents close to the O^-,M^+ bond in picrate may
also hinder the enlargement of this ionic bond and the insertion óf a crown ether molecule because of electronic or steric
effects.

Crown or solvent complexation leading to loose ion pairs
can by itself cause a drastic change in the charge distribution
of the anion. A good example is found in reaction 4 studied by
Boche et al (19). At room temperature the sodium enolate tight
ion pair with

its olefinic nonafulvene structure is stable. At low temperature, stronger solvation forces cause formation of a loose ion
pair. This in turn promotes charge delocalization and stabil-
izes the CH_3CO-substituted sodium cyclononatetraenide structure
which is aromatic in character.

Loose ion pairs of such charge-localized oxyanion salts as
potassium t-butoxide may be difficult to form. This alkoxide is
a tetrameric aggregate in THF (20), and crown addition breaks it
down to the more reactive monomeric form. It is unlikely that
with benzo-15-crown-5 a 2:1 crown-K^+ loose ion pair can be
formed similar to that found with potassium picrate or potassium
fluorenyl. However, external complexation itself will slightly
stretch the O^-,K^+ bond, and this can have a profound effect on
the anion reactivity (21).

While crown complexation to a tight ion pair usually increases the interionic ion pair distance, it may shorten the

ionic bond when the ligand replaces part of the solvation shell of a loose ion pair. This in turn can decrease the fraction of free ions formed on ion pair dissociation. Hence, in reactions such as anionic vinyl polymerization or proton transfer reactions where free anions often are the reactive species, crown ether addition can lead to a decrease rather than an increase in reactivity. An example is found in the interaction of dibenzo-18-crown-6 and fluorenyl sodium in THF (22). The latter salt is a loose ion pair in THF with a free ion dissociation constant, K_d equal to $4.0 \times 10^{-5}M$. The crown ether converts the solvated loose ion pair into a crown-complexed loose ion pair (the U.V. absorption band remains at 373 nm, characteristic for a loose ion pair of Fl^-,M^+ (23)), but K_d drops to $3.3 \times 10^{-6}M$. Replacement of the bulky THF molecules around the Na^+ by the more planar crown molecule allows for a slightly shorter interionic distance, thereby decreasing the dissociation constant.

Polyamine interactions with carbanion pairs

Polyamine complexes of organolithium compounds have been extensively studied (8). Recent work by Fontanille et al (24) has shown that also in these systems two isomeric ion pair complexes can be formed. The spectrophotometric "ion pair probe" used was 9-propylfluorenyllithium (PFl^-,Li^+), which was complexed with tetramethyl ethylene diamine (TMEDA), hexamethyl triethylene tetramine (HMTT) and tetramethyl tetraaza cyclotetradecane

[TMEDA]   [HMTT]   [TMTCT]

(TMTCT) in cyclohexane or toluene as solvent. The amines are powerful Li^+ chelating agents and strongly catalyze the polymerization of polystyryllithium in cyclohexane by anion activation (8,25).

The U.V.-absorption spectrum of 9-propylfluorenyllithium in toluene is concentration dependent, showing at $10^{-3}M$ in the 320-400 nm region only one absorption band at 368 nm. It belongs to the $(PFl^-,Li^+)_2$ dimer which is stable under these conditions but dissociates on dilution into the monomeric ion pair PFl^-,Li^+. The absorption maximum of the latter species is 353 nm and the dissociation constant of the dimer at 25° equals $2.9 \times 10^{-6}M$ (26). On addition of minute amounts of an ether solvent, E, the 368 nm dimer breaks down or rearranges into a dietherate, viz., PFl^-,Li^+,THP_2, which is an externally solvated tight ion pair absorbing at 359 nm. The same absorption band appears on adding THP to the 353 nm monomeric ion pair. At higher THP or THF concentration two more solvent molecules are bound forming the loose $PFl^-||Li^+$ ion pair absorbing at 386 nm. The changes are shown in reaction 5.

$(PFl^-,Li^+)_2$ (368 nm)

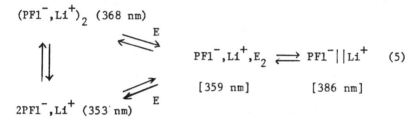

$$PFl^-,Li^+,E_2 \rightleftarrows PFl^-||Li^+ \quad (5)$$

[359 nm]   [386 nm]

$2PFl^-,Li^+$ (353 nm)

The dimer $(PFl^-,Li^+)_2$ does not dissociate on dilution when cyclohexane is the solvent (24). However, addition of small quantities of TMEDA forms the externally-complexed 1:1 PFl^-,Li^+, TMEDA tight ion pair, $\lambda_m 355$ nm. Although TMEDA is an excellent Li^+ chelating agent, even in pure TMEDA only a small fraction of loose ion pairs (~ 0.05) can be detected in the optical spectrum. Similar observations were reported earlier for the unsubstituted fluorenyllithium (2). Models suggest that formation of a complex of Li^+ with two TMEDA ligands in which all four nitrogen atoms can simultaneously interact effectively with the cation is hindered by the presence of the eight bulky methyl substituents. Nevertheless, a precipitate which slowly forms in pure TMEDA was shown to be a PFl^-,Li^+ $(TMEDA)_2$ complex (24). It dissolves in toluene with release of one TMEDA ligand and formation of the 359 nm externally complexed PFl^-,Li^+, TMEDA tight ion pair.

Addition of the polyamine HMTT to PFl^-,Li^+ in cyclohexane converts the 368 nm $(PFl^-,Li^+)_2$ dimer into a species absorbing at 357 nm. This is the only complex in solution when the ratio $r = HMTT/PFl^-,Li^+$ has reached the value 0.5. This suggests that two PFl^-,Li^+ ion pairs are complexed to one HMTT-ligand as shown below. Complex V derives its stability in part from the ion pair–ion pair interaction. By adding excess HMTT the 2:1 complex is converted into a 1:1 loose ion pair complex absorbing at 383 nm (386 nm in toluene) as shown in reaction 6.

$$(PFl^-,Li^+)_2 + HMTT \rightleftarrows (PFl^-,Li^+)_2 HMTT$$

368 nm   357 nm   (6)

$$(PFl^-,Li^+)_2 HMTT + HMTT \rightleftarrows 2\ PFl^-,HMTT,Li^+$$

357 nm   383 nm

It could not be ascertained whether the loose ion pair complex is monomeric, but on standing in cyclohexane it precipitates out of solution.

The cyclic polyamine TMTCT immediately converts the 368 nm dimer into a TMTCT-separated ion pair complex, λ_m 387 nm. The 1:1 complex precipitates from cyclohexane at 10^{-3} M, but is soluble in toluene. When TMTCT is added to a toluene solution of the 1:1 tight ion pair complex $PFl^-,Li^+,TMEDA$, the 357 nm band within a few minutes is replaced by the 387 nm band of the TMTCT separated ion pair complex:

$$PFl^-,Li^+,TMEDA + TMTCT \rightleftarrows PFl^-,TMTCT,Li^+ + TMEDA \qquad (7)$$

The second order rate constant for the reaction was found to be $k(25^o) = 250\ M^{-1}sec^{-1}$. The reaction is slow in comparison to exchange reactions involving crown ether complexes in ethereal solvents ($\underline{2},\underline{11}$). The rate constant for reaction 7 is most likely determined by the energy needed to remove the externally bound TMEDA ligand before TMTCT can interact to form the loose ion pair complex.

Intramolecular ionic interactions

Anionic polymerization initiated by electron transfer (e.g., sodium–naphthalene and styrene in THF) usually produces two-ended living polymers. Such species belong to a class of compounds called bolaform electrolytes ($\underline{27}$) in which two ions or ion pairs are linked together by a chain of atoms. Depending on chain length, counterion and solvent, intramolecular ionic interactions can occur which in turn may affect the dissociation of the ion pairs into free ions or the ligand–ion pair complex formation constants.

One of the first examples of intramolecular interactions in anionic polymerization was encountered in the propagation of two-ended polystyrylcesium in THF ($\underline{28}$). Intramolecular triple ions are formed which increase the ionic conductance but lower the propagation rate. Reactive free anions associate with ion pairs on the same chain to form less reactive triple ions.

Intramolecular triple ion formation was studied by Collins
and Smid using cesium bolaform salts of fluorenyl carbanions
(29). The cesium salts of α, ω-bis (9-fluorenyl) polymethylenes
(VI, n = 2,3,4 and 6), when dissolved in THF,

give a considerably higher conductance than the one-ended 9-
propylfluorenylcesium , due to reaction 8. The conductomet-
rically

$$Cs^+,Fl^-(CH_2)_nFl^-,Cs^+ \rightleftharpoons Fl^-,Cs^+,Fl^- + Cs^+ \qquad (8)$$
$$\qquad\qquad\qquad\qquad\qquad \lfloor(CH_2)\rfloor_n$$

$$Cs^+,Fl^-(CH_2)_nFl^- \rightleftharpoons Fl^-,Cs^+,Fl^- \qquad (9)$$
$$\qquad\qquad\qquad\qquad \lfloor(CH_2)\rfloor_n$$

determined cyclization constants, K_c, of reaction 9 were found
to be 3.3 for n = 2; 1.1 for n = 3; 10.8 for n = 4 and 8.0 for
n = 6. With Na^+ as counterion, cyclization is hindered by
sodium bound THF molecules. Two-ended living polymers with Na^+
as counterion generally behave like the one-ended species, at
least in ethereal solvents (4).
    Lithium salts of VI dissolved in toluene exhibit significant
intramolecular ion pair aggregation (26). While the dimer of
9-propylfluorenyllithium in toluene dissociates into the monomeric
ion pairs below 10^{-3} M salt concentration, the lithium salts of
VI remain intramolecularly aggregated. The 368 nm absorption
band characteristic for the aggregated species persists even
down to 4×10^{-6} M, and not even a shoulder is visible at 353 nm,
the absorption maximum of the monomeric ion pair. However, the
intramolecular aggregate breaks down or changes on addition of
small amounts of THP or THF. Externally solvated tight ion pairs
(λ_m 361 nm) are formed which convert into loose ion pairs
(λ_m 385 nm) on addition of excess THP or THF. Experiments with
THP reveal that in the second process two additional THP

molecules become bound to Li^+. The ratio K_1/K_2 of the ion pair
separation constants of reactions 10 are close to the statistical
factor 4,

$$Li^+,Fl^-(CH_2)_nFl^-,Li^+ + 2\ THP \underset{K_2}{\overset{K_1}{\rightleftarrows}} Li^+,Fl^-(CH_2)_nFl^-||Li^+ \tag{10}$$

$$Li^+,Fl^-(CH_2)_nFl^-||Li^+ + 2\ THP \rightleftarrows Li^+||Fl^-(CH_2)_nFl^-||Li^+$$

implying that the two ion pair separation processes are not
affected by the ion pair present at the opposite side of the
chain. The bolaform salt with n = 2 is the only exception.
This salt remains intramolecularly aggregated even in pure THP,
but the stronger solvating THF molecules convert the intramolecular aggregate (λ_m 368 nm) directly into a loose ion pair
(λ_m386 nm) without formation of the THF-solvated tight ion pair
(361 nm). The ratio K_1/K_2 (see equations 10) is only 0.32 for
the n = 2 salt when THF is added. Apparently, a considerable
amount of energy is required to break up the intramolecular
aggregate, but when this is accomplished by solvating one of
the Li^+ ions, the second Li^+ ion can be more easily solvated.
    A similar behavior is found when tetraglyme is added to a
THP or THF solution of the sodium salts of VI. The glyme
ligand converts the tight ion pair into a glyme-separated ion
pair (30). The ratio K_1/K_2 for the two reactions shown in 10
(but with Li^+ replaced by Na^+ and the two THP molecules by one
tetraglyme ligand) is again close to the statistical factor 4
except for the salt n = 2 in THP. For the latter system K_1/K_2 =
15.3, implying that the second Na^+ ion is more difficult to
separate once the first ion pair becomes a glyme-separated loose
ion pair (30). The same is found when this bolaform salt is
dissolved in 1,2 dimethoxyethane (DME). In this solvent the
monomeric analogue 9-alkylfluorenylsodium is at 25°C a loose
ion pair (2), but for the n = 2 bolaform salt an optical
spectrum is recorded which shows equal fractions of tight and
loose ion pairs. Apparently, when a DME or glyme ligand sepa-
rates one Na^+ from a Fl^- ion, the second cation is trapped between the two fluorenyl anions and is only externally complexed
by a DME or glyme ligand (30).
    With divalent counterions the two terminal carbanions of the
bolaform salts are associated with the same cation. The behavior
of the n = 2 salt again deviates from that of the other salts
(31). In THF at 25°C the compound di(9-n-butyl fluorenyl)barium
(the two carbanions are not connected by a chain) is a tight ion
pair while at −100°C the only stable species appears to be the
mixed tight-loose ion pair (reaction 11).

$$\text{BuFl}^-,\text{Ba}^{++},\text{Fl}^-\text{Bu} \xrightleftharpoons[]{\text{THF}} \text{BuFl}^-,\text{Ba}^{++}||\text{Fl}^-\text{Bu} \qquad (11)$$

$$25^\circ\text{C} \qquad\qquad\qquad -100^\circ\text{C}$$

$$\Delta H = -8 \text{ kcal/mole}$$

The second separation step leading to the fully separated ion pair requires more energy. The behavior for the barium bolaform salts (reaction 12) is similar except for n = 2 (<u>31</u>). For

$$\underset{25^\circ\text{C}}{\underset{\underset{}{\overset{}{\text{Fl}^-,\text{Ba}^{++},\text{Fl}^-}}}} \xrightleftharpoons[]{\text{THF}} \underset{-100^\circ\text{C}}{\text{Fl}^-,\text{Ba}^{++}||\text{Fl}^-} \qquad (12)$$

$$\Delta H = -8 \text{ kcal/mole}$$

this salt the tight ion pair (λ_m 361 nm) is stable down to -100°C and no 386 nm band can be detected in the spectrum.

Interaction of THF with Sr^{++} is considerably stronger than with Ba^{++} (<u>32</u>), and for the two salts with n = 4 and 6 (reaction 13) the ion pairs are already solvent separated at -30°C (<u>31</u>) (the 357 nm maximum is completely replaced by the 386 nm band).

$$\text{Fl}^-,\text{Sr}^{++},\text{Fl}^- \xrightleftharpoons[]{\text{THF}} \text{Fl}^-||\text{Sr}^{++}||\text{Fl}^- \qquad (13)$$

$$\Delta H = -12 \text{ to } -14 \text{ kcal/mole}$$

On the other hand, for strontium di(9-fluorenyl)ethane (n = 2) the tight ion pair spectrum (λ_m 361 nm for this salt) remains practically unchanged down to -90°. When this temperature is reached, a small fraction (~ 0.05) of loose ion pairs (λ_m 391 nm) can be detected in the optical spectrum. On further cooling, the 391 nm peak gradually increases with time, and when kept at about -100°C the only absorption maximum in the spectrum after one hour is that of the 391 nm loose ion pair. Following this conversion (an isosbestic point is observed), precipitation

occurs. The precipitate dissolves, and the spectrum again shows
only the 361 nm peak when the temperature is raised above -90°C.

In difluorenylstrontium itself the solvation is a stepwise
process, i.e., $Fl^-,Sr^{++},Fl^- \rightleftharpoons Fl^-,Sr^{++}||Fl^- \rightleftharpoons Fl^-||Sr^{++}||Fl^-$
(31). In the n = 2 bolaform salt the first separation step is
difficult, but once bound THF molecules force Sr^{++} to separate
from the first Fl^- ion, the cyclic structure probably opens up
due to the shortness of the $(CH_2)_2$ chain. This would leave a
free Fl^- ion on one end of the chain. Since the conductance of
the salt is known to be very low, this latter species most
likely will rapidly dimerize to form a non-conducting cyclic
aggregate consisting of loose ion pairs only, as shown in reaction 14. This aggregation shifts the equilibrium in favor of
the loose ion pairs.

$$Fl^-,Sr^{++},Fl^- \rightleftharpoons (Fl^-,Sr^{++}||Fl^-)$$

$$\underset{(CH_2)_2}{\rule{}{}} \quad slow \quad \underset{(CH_2)_2}{\rule{}{}}$$

$$(14)$$

$$\underline{Fl^-CH_2CH_2Fl^-}$$

$$\underline{Sr^{++}} \qquad \underline{Sr^{++}} \quad \overset{\leftarrow}{\rightarrow} \quad Fl^-CH_2CH_2Fl^-||Sr^{++}$$

$$\underline{Fl^-CH_2CH_2Fl^-}$$

The structure of the aggregate is of course speculative, but its
formation or that of higher aggregates may eventually cause the
observed precipitation. It is not impossible that for the n = 4
and 6 bolaform salts these types of aggregate are also formed,
rather than the cyclic structures of reaction 13, but this was
not further investigated. Concentration dependent studies may
shed more light on the mechanism of these interesting solvation
processes.

It is clear that the reactions with divalent cations are
considerably more complex than those involving monovalent
cations, especially when ligands are added to promote ion pair
separation (32). Complex rate phenomena can be expected in
anionic vinyl polymerization in the presence of divalent counterions. Some of these interesting systems will be described elsewhere by other investigators in this symposium.

Acknowledgement: Those investigations carried out in the
author's laboratory were financially supported by the National
Science Foundation and the Petroleum Research Fund administered
by the American Chemical Society. The work on the bolaform salts

with divalent counterions was carried out by Dr. C. Mathis at
the Centre de Recherches sur les Macromolécules (CRM) in
Strasbourg. The author gratefully acknowledges the invaluable
contributions of his coworkers and the hospitality of the CRM
staff during his stay at this Macromolecular Center.

Literature Cited

1. Szwarc, M., Ed.; "Ions and Ion Pairs in Organic Reactions",
    Wiley-Interscience, New York, Vol. I, 1972, Vol. II, 1974.
2. Smid, J., Angew. Chem. Intern. Ed. Engl. 1972, 11, 112.
3. Hogen Esch, T. E., Adv. Phys. Org. Chem. 1977, 15, 153.
4. Szwarc, M., "Carbanions, Living Polymers and Electron Trans-
    fer Processes", Wiley-Interscience, New York, 1968.
5. Smid, J., Ed.; "Ions and Ion Pairs and their Role in Chemical
    Reactions", IUPAC Symposium, Pergamon Press, Oxford, 1979.
6. Fox, M. A., Chem. Rev. 1979, 79, 253.
7. Gordon, J. E., "The Organic Chemistry of Electrolyte
    Solutions", Wiley-Interscience, New York, 1975.
8. Langer, A. W., Ed.; "Polyamine-Chelated Alkali Metal Com-
    pounds", Adv. in Chem. Ser. 130, American Chemical Society,
    Washington, D.C., 1974.
9. Takaki, U.; Hogen Esch, T.E.; Smid, J.,   J. Am. Chem. Soc.,
    1971, 93, 6760.
10. Chan, L.L.; Wong, K. H.; Smid, J., J. Am. Chem. Soc. 1970,
    92, 1955.
11. Wong, K. H.; Konizer, G.; Smid, J., J. Am. Chem. Soc. 1970, 92
    666.
12. Van Beylen, M. private communication.
13. Kobuke, Y.; Hanje, K.; Horiguchi, K.; Asada, M.; Nakayama,
    Y.; Furakawa, J., J. Am. Chem. Soc. 1976, 98, 7414.
14. Takaki, U.; Smid, J., J. Phys. Chem. 1972, 76, 2152.
15. Takaki, U.; Smid, J., J. Am. Chem. Soc. 1974, 96, 2588.
16. Pedersen, C. J.; J. Am. Chem. Soc. 1967, 89, 7017.
17. Bourgoin, M.; Wong, K. H.; Hui, J. Y.; Smid, J., J. Am.
    Chem. Soc. 1975, 97, 3462.
18. Ellingsen, T.; Smid, J., J. Phys. Chem. 1969, 73, 2712.
19. Boche, G.; Heidenhain, F., J. Am. Chem. Soc. 1979, 101, 738.
20. DiBiase, S. A.; Gokel, G. W., J. Org. Chem. 1978, 43, 447.
21. Smid, J.; Varma, A. J., Shah, S. C., J. Am. Chem. Soc. 1979,
    101, 5764.
22. Hogen Esch, T. E.; Smid, J., J. Phys. Chem. 1975, 79, 233.
23. Hogen Esch, T. E.; Smid, J., J. Am. Chem. Soc. 1966, 88, 307.
24. Helary, G.; Lefevre, L.; Fontanille, M.; Smid, J., J. Org.
    Metallic Chem., in press.
25. Helary, G.; Fontanille, M., Eur. Polym. J. 1978, 14, 345.
26. Takaki, U.; Collins, G. L.; Smid, J., J. Organomet. Chem.
    1978, 145, 139.
27. Fuoss, R. M.; Edelson, D., J. Am. Chem. Soc. 1951, 73, 269.
28. Bhattacharyya, D. N.; Lee, C. L.; Smid, J.; Szwarc, M.,
    J. Phys. Chem. 1965, 69, 612.

29. Collins, G. L.; Smid, J., J. Am. Chem. Soc. 1973, 95, 1503.
30. Collins, G. L.; Hogen Esch, T. E.; Smid, J., J. Sol. Chem.
    1978, 7, 9.
31 Mathis, C.; Francois, B; Smid, J., to be published.
32. Hogen Esch, T. E.; Smid, J., J. Am. Chem. Soc. 1972, 94, 9240.

RECEIVED February 12, 1981.

The Influence of Aromatic Ethers on the Association of the Polystyryllithium and 1,1-Diphenylmethyllithium Active Centers in Benzene

LEWIS J. FETTERS

Institute of Polymer Science, The University of Akron, Akron, OH 44325

RONALD N. YOUNG

Department of Chemistry, The University of Sheffield, Sheffield S3 7HF, U.K.

The influence of diphenyl ether and anisole on the association of the polystyryllithium and 1,1-diphenylmethyllithium active centers has been measured. Severe disaggregation of the polystyryllithium dimers, present in pure benzene, was found to occur at levels of ether addition at which several reliable kinetic studies reported in the literature unequivocally demonstrate a 1/2 order dependence upon polystyryllithium. These results indicate that a necessary connection between the degree of aggregation of organolithium polymers and the observed kinetic order of the propagation reaction need not exist.

The active centers of the polydienyl- and polystyryllithium species, which can be characterized as having polarized covalent carbon lithium bonds, appear (1-8) to form dimeric aggregates in hydrocarbon solvents at concentrations ($<10^{-2}$M) appropriate for polymerization. It is becoming more widely recognized that the kinetic consequences of the aggregation involving carbon-lithium species are but imperfectly understood regarding both the initiation and propagation processes.

It has been known for a considerable time (8,9) that the rates of diene propagation involving the lithium counter-ion in hydrocarbon solvents frequently depend upon the chain end concentration raised to a small power, such as 1/4 or 1/6. It has often been concluded that a small equilibrium concentration of unassociated chain ends is solely responsible for propagation and that the kinetic order of the reaction is a reflection of the extent of active chain end aggregation. The majority of results which seemingly show that the polydienyllithium species are dimeric, would seem to place this attractive kinetic rationalization in jeopardy insofar as the diene systems are concerned.

0097–6156/81/0166–0095$05.00/0

However, the polymerization of styrene in hydrocarbon solvents has been shown (8,10) to exhibit a 1/2 order dependency on
polystyryllithium concentration. Hence, the concept that only
the unassociated chain ends are reactive may be valid for this
system since association studies have shown (1,2,7,8) that the
polystyryllithium chain ends are associated as dimers. (Szwarc
has stated (11) that polystyryllithium is "probably tetrameric in
cyclohexane." This assessment, though, failed to consider the
light scattering results (7,8) of Johnson and Worsfold).

Several studies have appeared (12,13,14) in which the propagation reactions involving styryllithium were examined in mixed
solvent systems comprising benzene or toluene and ethers. The
kinetics were examined under conditions where the ether concentration was held constant and the active center concentration
varied. In most cases, the kinetic orders of the reactions were
identical to those observed in the absence of the ether. Thus,
in part, the conclusion was reached (13,14) that the ethers did
not alter the dimeric association state of polystyryllithium. The
ethers used were tetrahydrofuran, diphenyl ether, anisole, and
the ortho and para isomers of ethylanisole.

We have examined the influence of diphenyl ether and anisole
on the association of the polystyryllithium and the 1,1-diphenyl-
methyllithium active centers in benzene solution. The analytical
tool used was the vacuum viscometry method (1,2,3,5) which utilizes concentrated solutions of polystyryllithium and the terminated polymer in the entanglement regime. Our results show that
the presence of these ethers can alter the association states of
the foregoing active centers. These findings parallel previous
work (2) involving tetrahydrofuran.

Experimental Section

The procedures used for styrene and benzene purification are
given elsewhere (15) as is the technique used to study the association states (1,2,5). The viscometric method relies on the fact
that in the entanglement region, non-polar polymers in solution
and the melt exhibit Newtonian viscosities which are dependent on
$\overline{M}_w^{3.4}$. A compilation of $\eta-\overline{M}_w$ data presented elsewhere (16) reveals that for near-monodisperse polymers, the predominant value
of the exponent is 3.4. Furthermore, Doi (17) has recently ex-
tended de Gennes (18) tube model and has shown (17) that $\eta \propto \overline{M}_w^{3.4}$.
The value of the exponent means that even small changes in associ-
ation states can be, in principle, measured with a high degree of
accuracy. The concentrated solution viscosity measurements were
done at 20°C as were the polymerizations. The molecular weights
($\overline{M}_w$) of the base polystyrenes used in this work ranged from 1.2x
10^5 to 1.7×10^5 (based on GPC measurements) while the corresponding polymer concentrations ranged from ca. 42 to 30 percent.

The initiator was purified, by distillation (19), sec-butyllithium (Lithco). The ethers and 1,1-diphenylethylene were purified by sequential exposure to CaH_2, dibutyl magnesium (Lithco)

and polystyryllithium. This purification routine took about one
week to complete. GC analysis, after purification, failed to reveal impurities in these compounds. The polymerizations were
conducted in the presence of the ethers with benzene as the solvent. The polystyryllithium active center was converted to the
1,1-diphenylmethyllithium chain end by the addition of ca. 0.1
ml of 1,1-diphenylethylene.

The GPC analysis was conducted using a seven column Styragel
arrangement with tetrahydrofuran as the carrier solvent. Solution concentrations of 1/8% (w/v) were used. The flow rate was
1 ml min^{-1}. The characteristics of this column arrangement are
presented elsewhere (20,21). The Styragel columns were calibrated using eleven polystyrene standards. All of the polystyrenes prepared in this work were analyzed by gel permeation
chromatography. Without exception, all samples were found to
have values of M_z/M_w <1.07 and values of M_w/M_n <1.05. None of
these samples showed the presence of a low molecular weight tail,
i.e., chromatograms indicating Gaussian distributions were obtained.

Results and Discussion

A potential problem for this work is active center termination due to reactions with the ethers, i.e., metallation. It
has been reported (13) that polystyryllithium undergoes a moderately rapid termination reaction at 20°C in the presence of anisole or the ethylanisole isomers. These authors, however, did
not recognize that toluene, their solvent of choice, is known
(22,23,24,25,26) to act as a transfer agent in polymerizations
involving alkali metals. (More recent results (27,28,29) have
verified these observations regarding the lithium counter-ion.
n-Butyllithium has also been shown (30) to react with toluene.
In regard to this latter point, Cubbon and Margerison noted (31)
that adding n-butyllithium to toluene led to the formation of
solutions which "developed a yellow-orange color.") If the
spectrum of benzyllithium in toluene in the presence of anisole
resembles that in benzene where the λ max is reported (32,33) to
be 292 nm, the decay in absorbance with time noted (13) at 330
nm may be attributable to transmetallation involving toluene
rather than the foregoing aromatic ethers.

Chain termination proved not to be a problem during the
period over which our measurements were made. This was established by the constancy of the absorbance of polystyryllithium
at 334 nm and that of the 1,1-diphenylmethyllithium active center
at 440 nm. Furthermore, the flow times of the solutions were
found to remain constant over the measurement period (ca. 24 hrs.).
An additional indication that chain termination was not a significant problem was provided by gel permeation chromatography. As
mentioned previously all of the polystyrene samples exhibited
M_z/M_w and M_w/M_n ratios of 1.07 or less and none of the samples
showed the presence of a low molecular weight tail. This demonstrates that no detectable chain termination took place during
polymerization.

Figure 1 contains chromatograms of polystyrenes prepared anionically in the presence of anisole and diphenyl ether. The narrow molecular weight distributions of these samples demonstrate that no detectable termination took place during the polymerizations. This lack of a termination step, regarding anisole, is in agreement with the polymerization results (34,35,36,37) where this ether was used as a co-solvent.

The above comments should not be taken as claims that anisole and diphenyl ether cannot be metallated by organolithium species. For example, alkyllithiums are known (38,39,40) to react with anisole, usually in the ortho position. However, these reactions are generally slow, particularly at ambient temperature and when the ether is diluted with a hydrocarbon solvent. Our results merely indicate that active center deactivation via metallation of these aromatic ethers is not a serious problem during the time span of our measurements with species that are, at least, partially delocalized (33).

Table I contains the association values, N_W, for the polystyryllithium and 1,1-diphenylmethyllithium active centers in the presence of diphenyl ether or anisole. The influence on polystyryllithium association by the aromatic ethers is, as expected, less dramatic than observed (2) for tetrahydrofuran (THF) where the value for the equilibrium constant K, of the following, was evaluated to be about 2×10^2;

$$(SLi)_2 + 2nE \underset{\longleftarrow}{\overset{K}{\longrightarrow}} 2 (SLi \cdot nE) \qquad (1)$$

where SLi denotes polystyryllithium and E represents a complexing agent such as an ether. The equilibrium constant can be calculated by the following relation:

$$K = \frac{2(2-N_W)^2 C_0}{(N_W-1) E^2 n} \qquad (2)$$

where C_0 represents the total concentration of active centers. As was the case for tetrahydrofuran (2), reasonably uniform values of K for the aromatic ethers were obtained only for the case where n equals one.

The data of Table I shows that anisole has proportionally a greater effect on the association of the 1,1-diphenylmethyllithium active center than does diphenyl ether. This difference may, in part, be due to steric effects involving the latter ether and the 1,1-diphenylmethyllithium active center. Also, data of Table I indicate that the association state of the 1,1-diphenylmethyllithium chain end is nearly two.

Studies have been made (12,13,14) of the kinetics of propagation of styrene under conditions entirely analogous to those described for the measurements of association state. The archetype of this sort of investigation is the classic work of Worsfold and Bywater (10) on the polymerization of styrene by n-butyllithium in benzene. These authors extended (12) their

Table I

The Influence of Diphenyl Ether and Anisole
on the Association of the Poly(styryl) - and
1,1-Diphenylmethyllithium Active Centers

$\dfrac{[SLi]^a}{10^3}$	$\dfrac{[DPE]^b}{10^2}$	$\dfrac{[DPE]}{[SLi]}$	$N_w{}^c$	$\dfrac{K^d}{10^2}$	$N_w{}^e$	$\dfrac{K^f}{10^2}$
1.7			1.98		1.97	
2.1			1.99		1.98	
2.9	6.3	22	1.91	1.3	1.86	3.3
2.1	6.3	30	1.89	1.4	1.82	4.2
2.6	17	65	1.77	1.2	1.69	2.5
2.6	25	96	1.66	1.5	1.59	2.4
1.6	16	100	1.68	1.9	-g-	
1.8	20	111	1.66	1.6	1.58	2.7
2.3	28	122	1.64	1.2	-g-	
2.0	33	165	1.58	1.1	-g-	
1.5	33	220	1.50	1.4	1.41	2.3
1.9	44	232	1.48	1.1	-g-	
	$\dfrac{[\phi OCH_3]^h}{10^2}$	$\dfrac{[\phi OCH_3]}{[SLi]}$				
3.3	7.1	22	1.82	5.2	1.65	25
2.7	12	44	1.68	5.6	1.44	27
3.1	18	58	1.61	4.8	1.35	23
2.5	20	80	1.57	4.1	1.29	22
3.0	28	93	1.55	2.8	1.27	15
2.5	25	100	1.56	2.8	1.24	19

[a] Active center concentration.

[b] Diphenyl ether concentration.

[c] Association number for polystyryllithium.

[d] Based on Equation (2) for polystyryllithium.

[e] Association number for the 1,1-diphenylmethyllithium active center.

[f] Based on Equation (2) for the 1,1-diphenylmethyllithium active center.

[g] Diphenylethylene was not added to these runs.

[h] Anisole concentration.

work to the situation where tetrahydrofuran was present in benzene solution in constant concentration (10^{-3} molar); a one-half
order dependence of rate upon polystyryllithium concentration was
observed for an ether : lithium ratio over the range 1.2 to 17.5.
In an analogous study, Geerts, Van Beylen and Smets (13) found
that the propagation of styrene also exhibits a one-half order
dependence of rate upon active center concentration when anisole
or 4-ethylanisole was present over the range of [ether]: [lithium]
from 33 to 1030, and also in the presence of 2-ethylanisole for
the range 40 to 3980.

Comparison of the association numbers found from the viscosity measurements with the invariancy of the kinetic orders
from the value one-half clearly demonstrates the mistake (13,14)
of expecting the existence of a simple and necessary connection
between these two parameters. The assumption (14) that diphenyl
ether does not affect the degree of association of polystyryllithium even at the ratio of 150:1 can be seen to be incorrect.

A potentially valuable contribution to our understanding of
the generality of a connection between the order of the propagation rate with respect to the organolithium centers might come
from the study of o-methoxystyrene in a hydrocarbon solvent in
the absence of any added ethers or other complexing agents. Smets
and co-workers (41,42) using toluene as solvent reported that
when their results were plotted in the form log rate vs. log
[chain end], "a distinctly curved line was obtained, the order
with respect to the concentration of poly-o-methoxystyrene varying from 0.67 to 0.51 over a concentration range from 4.5×10^{-4}
(sic) to 1.8×10^{-2} mol/lit" (Table I of ref. 41 shows that the
lowest active center concentration was $5.3 \times 10^{-4} ML^{-1}$). This result was interpreted as indicating that "at the higher concentration the ion-pairs occur predominantly in the associated form
while increasing amounts of free-ion-pairs are present at lower
concentrations." Scrutiny of their data however, does not provide convincing evidence that their plot is non-linear. We have
replotted their data (Figure 2); and a least mean squares analysis shows that the gradient is 0.62 and that the relevant correlation coefficient is 0.9994. If there is indeed a connection
between the degree of aggregation and the observed kinetic order,
then it must be concluded that (a) the degree of aggregation is
constant over the concentration range 5.3×10^{-4} to 1.8×10^{-2} molar
and that (b) the number-average mean degree of aggregation, N_n,
is less than two, possibly 1/0.62 i.e., 1.6 (it can readily be
shown that $N_n = 2/(3-N_w)$ for systems containing near-monodisperse
molecular weight distributions).

Having assumed that the aggregated chain ends are incapable
of growth, Smets and co-workers (41,42) used a curve fitting procedure to deduce that the dissociation of the dimeric poly-o-
methoxystyrenyllithium is governed by the constant 10^{-3} molar.
Calculation shows that if this value is correct, the degree of
dissociation varies from about 0.15 to 0.61 over the range of

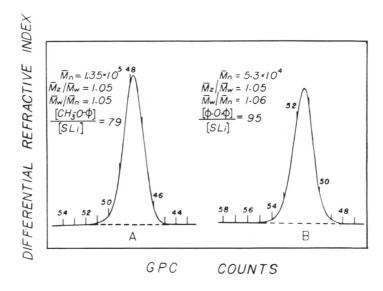

Figure 1.   *GPCs of polystyrenes prepared in benzene in the presence of anisole (A) and diphenyl ether (B).*

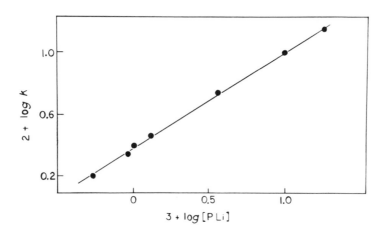

Figure 2.   *Polymerization rate vs. active center concentration for 2-methoxystyrene (41, 42). Rate units are in minutes.*

concentration studied. These results are clearly incompatible
with those outlined in the preceding paragraph.

In summary, we must conclude that in the case of ether-modi-
fied systems based on styrene studied to date, there is no direct
relationship between the degree of aggregation and kinetic order
in polystyryllithium. There is insufficient reliable data available to draw definite conclusions for reactions carried out in
pure hydrocarbons; a fresh study of the polymerization of o-methoxystyrene might shed new light on this situation.

In connection with the subject of the relation between
association state and kinetic order, it is germane to mention observations of Roovers and Bywater (45). They measured the dissociation constant for the tetramer $\rightleftharpoons$ dimer case for polyisoprenyllithium in benzene. The technique involved a study of the
electronic spectra at 272 and 320 nm. If the process they measured can be directly related to the association-dissociation
equilibrium, their results can be used to calculate the dissociation constant for the correct dimer $\rightleftharpoons$ monomer system. This
value is ca. 2×10^{-5} at 30.5°C. If this value is accepted, then
the situation is encountered where the degree of dissociation of
the polyisoprenyllithium chain ends varies from about 0.10 to
0.62 over a chain end concentration range of 10^{-3} to 10^{-5} molar.
However, over this active center concentration range, the kinetic
results (44,45, Figure 18 of 15) show a 1/4 order dependency of
rate on polyisoprenyllithium. Thus, the polyisoprenyllithium-
benzene system is also apparently one where a constant fractional
kinetic order is observed even though a variable fraction of the
active centers can exist in the unassociated state.

A report has appeared (46) which disputes a preliminary
statement of ours (47) regarding the effect of diphenyl ether on
the association state of polystyryllithium. This claim (46) was
based on only two measurements. We will comment briefly on those
measurements. (The authors of that report (46) refer to our work
(47) as having been done by "the Akron team" and "the Akron
group". No rational justification for the use of that peculiar
terminology exists since the first two authors of the preprint in
question (47) listed as their affiliation the University of
Sheffield. A portion of the work reported in the foregoing preprint (47) has been published, ref. 5 of this paper).

The authors of ref. 46 reported flow times for polystyryl-
lithium-benzene solutions before and after the addition of diphenyl ether; whereupon the active centers were terminated and
the flow times again measured. Table III of the note in question
(46) says that in pure benzene, N_w, is 1.96 and 2.0 — in apparent
agreement with the generally held belief that polystyryllithium
is exclusively dimeric in benzene. Following the addition of diphenyl ether to achieve the specified concentration (0.15M), the
authors in their Table III then reported values of N_w of 1.88 and
1.95 (based on their flow times T_2). From these values, it was
concluded that diphenyl ether does not influence the association
state of polystyryllithium.

However, the addition of diphenyl ether to achieve the afore-
mentioned concentration is accompanied by an increase in solution
volume of ca. 2.4%, i.e., a decrease in polymer concentration.
The data of Graessley and co-workers ($\underline{48},\underline{49}$) show that $\eta \propto c^{7.5} M^{3.4}$
(where c denotes polymer concentration) for polystyrene in butyl-
benzene. Thus, the apparently trivial dilution occasioned by the
addition of the diphenyl ether ought to have resulted in apparent
values for N_w of about 2.1 rather than the reported values of
1.96 and 2.0, unless the solution contained some polymer terminated by adventitious impurities. To place this anomaly in
sharper perspective, the results of Graessley, et al. ($\underline{48},\underline{49}$) predict that the two average flow times of column 1 (T_1) in Table
III of the note in question ($\underline{46}$) should be higher than those reported by about 400 seconds, i.e., 18 and 26% higher than the two
average values reported. This is based on the respective average
flow times of the terminated solutions and the differences in
polymer concentration existing between these active and terminated systems.

If the foregoing assessment is based on the initial flow
times, T_1, then the terminated flow times, T_3, of Table III of
ref. $\underline{46}$, would be about 20 and 15 percent lower than the values
listed. Thus, the association data given by these authors ($\underline{46}$)
do not accurately reflect the concentration change brought about
by the addition of diphenyl ether.

The data of Tables II and III of ref. ($\underline{46}$) also show ($\underline{48},\underline{49}$)
that the viscosities of the four polystyryllithium solutions
possessed viscosities ranging from about 3×10^3 to 1.7×10^5 poise
with three of these systems having viscosities greater than ca.
3.0×10^4 poise in the evacuated Ubbelohde viscometers needed for
these measurements. Solutions with viscosities in the range of
10^3 to 10^5 poise do not lend themselves to accurate flow time
measurements due to difficulties in filling the viscometer bulb,
the very slow rate of flow of these solutions between the fidu-
cial marks, and drainage problems, viz, the adherence of the
polymer solution to the viscometer interior. These problems are
not avoidable by the use of large diameter (>1 cm) tubing.
Hadjichristidis and Roovers ($\underline{50}$) used Ubbelohde viscometers to
study concentrated polyisoprene solutions and concluded that in
practice the technique is limited to systems having viscosities
less than 10^3 poise. (Unless the solutions are driven by an inert gas). Our experience is entirely in concordance with this
assessment, as is that of others ($\underline{51}$).

Finally, two comments made ($\underline{46}$) in reference to our preliminary statement ($\underline{47}$) should be assessed:

(a) "Purified, and not commercial, ether has to be used,
otherwise some carbanions are destroyed." Our purification procedure is presented in the experimental section of this paper.
The implication that we would use an unpurified ether is an un-
called for invention of the authors of ref. $\underline{46}$.

(b)   "Prolong (sic) action of the ether gradually destroys
the carbanions as revealed by the decrease of optical absorption
at 344 nm." We wish to note that this termination reaction seems
not to have influenced the viscosity results of Wang and Szwarc
(46) nor those of Yamagishi and co-workers (14) where species
analogous to diphenyl ether were used. These authors (46) seem,
though, to imply that termination has influenced our results regarding diphenyl ether. We have, as noted, considered this possibility in our work.

In summary, we wish to suggest that in the case of the tetrahydrofuran and aromatic ether modified systems based on styrene
studied to date, there does not appear to exist a direct relationship between the mean-degree of active center association and
kinetic order. We do not mean, with this assessment, to imply or
claim that the concept that only the unassociated active centers
of polystyryllithium in hydrocarbon solvents are the active entities is incorrect. However, our association results (this work
and ref. 2), combined with the available kinetic findings for
these ether modified polymerizations, seemingly indicate that reaction orders need not necessarily yield direct insight into the
states of active center association. In our opinion, greater
understanding of the apparent complex nature of these organolithium based reactions will be forthcoming by a more judicious
approach to the identification of the active centers in these
polymerizations; and approach, which in effect, was advanced by
Brown some years ago (52,53). Along this line, Kaspar and
Trekoval (54,55) have concluded from their kinetic analysis that
styrene can react directly with the polystyryllithium dimer in
benzene solution.

Acknowledgement: We wish to thank Prof. G. Berry, Prof. J. D.
Ferry, Prof. W. W. Graessley, Prof. N. Hadjichristidis, Dr. D.
Pearson and Prof. J. Schrag for their advice regarding the
practical limitations of Ubbelhode viscometers for the measure-
ment of concentrated polymer solutions. We also wish to thank
Dr. R. Quirk for his comments and information.

References:

1. Morton, M.; Fetters, L. J.   J. Polym. Sci., (1964), 2 Part A,
    3311.
2. Morton, M.; Fetters, L. J.; Pett, R.; Meier, J. F. Macromole-
    cules, (1970), 3, 327.
3. Fetters, L. J.; Morton, M. Macromolecules (1974), 7, 552.
4. Glaze, W. H.; Hanicak, J. E.; Moore, M. L.; Chandhuri, J.
    J. Organomet. Chem. (1972), 44, 39.
5. Al-Jarrah, M. M.; Young, R. N. Polymer (1980), 21, 119.
6. Hernandez, A.; Semel, J.; Broecker, H. C.; Zachmann, H. G.;
    Sinn, H. Makromol. Chem., Rapid Comm., (1980), 1, 75.
7. Worsfold, D. J. J. Polym. Sci., (1967), 5 Part A-1, 2783.

8. Johnson, A. F.; Worsfold, D. J. J. Polym. Sci., (1965), 3 Part A-1, 449.
9. Worsfold, D. J.; Bywater, S. Can. J. Chem., (1964), 42, 2884.
10. Worsfold, D. J.; Bywater, S. Can. J. Chem., (1960), 38, 1891.
11. Szwarc, M. Science, (1970), 170, 30.
12. Worsfold, D. J.; Bywater, S. Can. J. Chem., (1962), 40, 1564.
13. Geerts, J.; Van Beylen, M.; Smets, G. J. Polym. Sci., (1969), 7 Part A-1, 2805.
14. Yamagishi, A.; Szwarc, M.; Tung, L.; Lo, G.-Y.S. Macromolecules (1978), 11, 607.
15. Morton, M.; Fetters, L. J. Rubber Revs., (1975), 48, 359.
16. Fetters, L. J.,; Morton, M. J. Polym. Sci., Polym. Lett. Ed. (1981), 19, in press.
17. Doi, M. Polym. Prepr., Polym. Chem. Div., Am. Chem. Soc., (1981), 22 (1), 100.
18. de Gennes, P. G. J. Chem. Phys., (1971), 55, 572.
19. Worsfold, D. J.; Bywater, S. J. Organomet. Chem. (1967), 10, 1.
20. McCrackin, F. L. J. Appl. Polym. Sci., (1977), 21, 191.
21. Ambler, M.; Fetters, L. J.; Kesten, Y. J. Appl. Polym. Sci., (1977), 21, 2439.
22. Robertson, R. E.; Marion, L. Can. J. Res., (1948), 26B, 657.
23. Bower, F. M.; McCormick, H. W. J. Polym. Sci., (1963), 1 Part A, 1749.
24. Brooks, B. W. Chem. Commun., (1967), 68.
25. Krume, S. Makromol. Chem., (1966), 98, 120.
26. Krume, S.; Takahashi, A.; Nishikawa, G.; Hatana, M.; Kambara, S. Makromol. Chem., (1965), 84, 137, 147; ibid.,(1966) 98, 109.
27. Gatzke, A. J. Polym. Sci., (1969), 7 Part A-1, 2281.
28. Aukett, P.; Luxton, A. R. J. Oil Colour Chem. Assoc., (1977), 60, 173.
29. Cotton, J. P.; Decker, D.; Benoit, H.; Farnoux, B.; Higgins, J.; Jannik, G.; Ober, R.; Picot, C.; dos Claizeaux, J. Macromolecules (1974), 7, 863.
30. Gilman, H.; Gaj, B. J. J. Org. Chem., (1963), 28, 1725.
31. Cubbon, R. C. P.; Margerison, D. Proc. Roy. Soc., (1962), 268A, 260.
32. Waack, R.; McKeever, L. D.; Doran, M. A. Chem. Commun., (1969), 117.
33. Bywater, S.; Worsfold, D. J. J. Organomet. Chem., (1971), 33, 273.
34. Morton, M.; McGrath, J. E.; Juliano, P. C. J. Polym. Sci., (1969), No. 26 Part C, 99.
35. Fetters, L. J.; Morton, M. Macromolecules, (1969), 2, 453.
36. Morton, M.; Fetters, L. J.; Inomata, J.; Rubio, D.; Young, R. N. Rubber Chem. Tech., (1976), 49, 303.
37. Journe, J.; Widmaier, J. M. Eur. Polym. J. (1977), 13, 379.
38. Shirley, D. A.; Johnson, J. R.; Hendrix, J. P. J. Organomet. Chem. (1968), 11, 209.

39. Shirley, D. A.; Hendrix, J. P. J. Organomet. Chem., (1968),
     11, 217.
40. Fagley, T. F.; Klein, E. J. Am.Chem. Soc., (1955), 77, 786.
41. Geerts, J.; Van Beylen, M.; Smets, G.J. Polym. Sci., (1969),
     7 Part A-1, 2859.
42. Smets, G.; Van Beylen, M. 23rd Congress of Pure and Applied
     Chem., Vol. 4, Butterworths, London, 1971, p. 405.
43. Roovers, J. E. L.; Bywater, S. Polymer, (1973), 14, 594.
44. Margerison, D.; Bishop, D. M.; East, G. C.; McBride, P.
     Trans. Faraday Soc., (1968), 64, 1872.
45. Alvarino, J. M.; Bello, A.; Guzman, G. M. Eur. Polym. J.,,
     (1972), 8, 53.
46. Wang, H. C.; Szwarc, M. Macromolecules, (1980), 13, 452.
47. Al-Jarrah, M. M.; Young, R. N.; Fetters, L. J. Polymer Prepr.,
     Am. Chem. Soc., Div. Polym. Chem., (1979), 20, 793.
48. Graessley, W. W.; Hazleton, R. L.; Lindeman, L. R. Trans.
     Soc. Rheology (1967), 11, 267.
49. Graessley, W. W.; Segal, L. Macromolecules, (1969), 2, 49.
50. Hadjichristidis, N.; Roovers, J. E. L. J. Polym. Sci., Polym.
     Phys. Ed., (1974), 12, 2521.
51. Berry, G.; Ferry, J. D.; Graessley, W. W.; Hadjichristidis,
     N.; Pearson, D.; Schrag, J. private communications.
52. Brown, T. L. Adv. Organomet. Chem., (1965), 3, 365.
53. Brown, T. L. J. Organomet. Chem., (1966), 5, 191.
54. Kaspar, M.; Trekoval, J. Coll. Czech. Chem. Commun., (1976),
     45, 1976.
55. Kaspar, M.; Trekoval, J. Coll. Czech. Chem. Commun., (1980),
     45, 1047.

RECEIVED March 25, 1981.

The Photoisomerization of Allylic Carbanions

RONALD N. YOUNG

Department of Chemistry, The University of Sheffield, Sheffield S3 7HF, U.K.

The 1,3-diphenylallyl carbanion, which serves as a model for the propagating center derived from the monomer 1,3-diphenyl-butadiene, exists in solution as a mixture of tight (contact) and loose (solvent separated) ion pairs [1,2,3,4]. In solvents of low dielectric constant, such as ethers, the extent of dissociation into the free ions is negligible. The spectra of the loose ion pairs are indistinguishable from those of the free carbanions: the occurrence of dissociation is, of course, revealed by the consequent electrical conductivity. The perturbation of the molecular orbital energy levels arising from association into tight ion pairs is considerable [5]. The ground state of the tight ion pair is stabilised with respect to that of the loose ion pair to an extent that is much greater than that found for the similar comparison of the excited states, and in consequence, the absorption maxima of the tight ion pairs are blue-shifted by an amount that increases with decreasing cation radius. By way of illustration, Figure 1 shows the spectrum of 2-methyl-1,3-diphenylallyllithium in 2-methyltetrahydrofuran (MTHF): the absorption band at 468 nm is due to the loose ion pair and that at 532 nm to the tight pair. The conversion of the tight ion pair to the loose involves an exothermic solvation process and accordingly, the equilibrium between these species exhibits a corresponding temperature dependence.

The existence of two kinds of ion pair has been confirmed by NMR spectroscopy [4]. Early studies showed that the preferred conformation of the 1,3-diphenylallyl carbanion is trans,trans [4,6,7,8] but more recent work has shown [9] that there is, in fact, some 7% of the cis,trans isomer present at 20°C. On lowering the temperature, the proportion of cis,trans isomer present is decreased. The introduction of a substituent, such as a methyl group, into the allylic 2-position makes the cis, trans conformer much the most stable since the steric interaction is least [10]. Whereas the trans,trans conformer is planar the cis-trans is somewhat twisted and the ensuing reduction in charge delocalisation causes a blue-shift of the absorption maximum.

0097–6156/81/0166–0107$05.00/0

The values for the loose ion pairs are respectively 565 and
535 nm: the tight ion pairs manifest a similar dependence of
absorption maximum upon conformation.
    The diphenylallyl carbanion is conveniently formed by the
abstraction of an allylic proton from the corresponding diphenyl
propene. Under certain conditions it was found that the initial
product formed from trans 1,3 diphenyl-2-methyl propene was the
trans,trans anion which isomerised completely into the cis,trans
conformation within a few minutes [3]. Clearly, the abstraction
reaction proceeds under kinetic and not thermodynamic control.
    Recently we discovered [11] that photolysis of the diphenyl-
allyl carbanion with white light causes isomerisation of the
trans,trans conformer to the cis,trans (scheme 1). When the
source of the illumination is removed, conformational relaxation
proceeds at a rate which is markedly sensitive to the nature of
the ion pairing, being much slower for loose than for tight ion
pairs. The results of an extension of this work will be
presented in this paper together with an outline of the possible
relevance of this photochemical phenomenon to the stereochemistry
of the polymerisation of dienes.

Experimental

    Solvents were dried under high vacuum by stirring over
sodium potassium alloy along with a little benzophenone to serve
as indicator. The lithium salts of 1,3-diphenylallyl and
1,3-diphenyl-2-azaallyl were prepared by treating 1,3-diphenyl
propene or benzylidene benzylamine respectively with n-butyllithium; the Na and K salts were obtained by reduction of these
same precursors with the appropriate metal. All-glass reactors
were employed and were fitted with side-arms bearing 1mm
pathlength glass cells. An unsilvered Dewar vessel fitted with
optical windows which was partly filled with alcohol and mounted
in the cell compartment of the Perkin Elmer 554 spectrometer,
served as thermostat. The temperature was controlled to +0.2°C
by pumping coolant from a larger thermostat through a thin
coil of copper tubing immersed in the alcohol. Photolysis was
effected by means of a 50W quartz-halogen lamp situated some
5 cm from the sample: a typical exposure time was 30 to 60
seconds. The conformational relaxation in the dark following
photolysis was monitored from the change of absorbance with
time at the absorption maximum of either conformer.

Results

    Exposure of a solution of loose ion pairs of the diphenyl-
allyl anion to white light results in partial isomerisation of
the solute into the cis,trans conformation (scheme 1). Whereas
the trans,trans ion is planar, the cis,trans ion is non-planar
as a consequence of steric interaction between the hydrogen

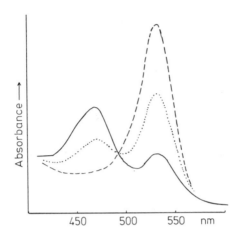

*Figure 1.   Absorption spectrum of 2-methyl-1,3-diphenylallyllithium in MTHF:
+21°C (—); −2°C (• • •); −40°C (- - -).*

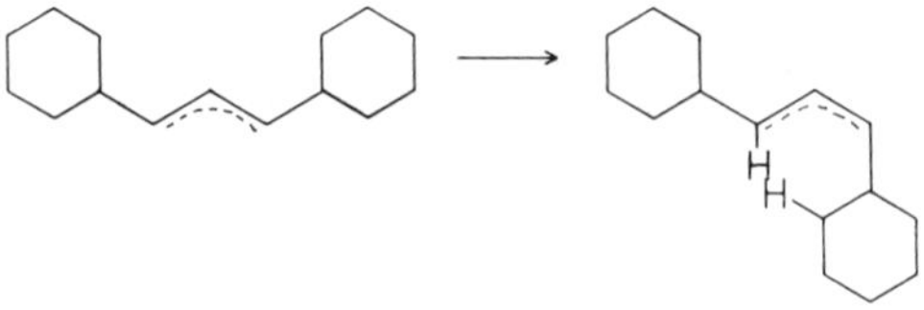

Scheme 1.

atoms shown. At temperatures above about -20° this process is in competition with thermal relaxation which converts the cis, trans product back to the trans,trans. At lower temperatures the rate of relaxation becomes negligible. On prolonged illumination (typically a minute with the sample located 5 to 10 cm distant from the 50W source) a photostationary state is established between these two conformers [12]. This equilibrium is strongly temperature dependent (Figure 2) favouring the cis,trans form at higher temperatures and the trans,trans at lower temperatures. As the temperature is lowered, the solutions begin to show an orange fluorescence whose intensity increases with decreasing temperature. In contrast, solutions of the tight ion pairs do not fluoresce. The excitation spectra of solutions containing both tight and loose ion pairs in both conformations match the absorption spectrum of the loose pair having the trans,trans conformation.

The temperature dependence of the photoisomerisation can be rationalised in terms of the scheme shown in Figure 3. Absorption of light causes the transition trans,trans → trans, trans*. At higher temperatures where sufficient thermal energy is available, the barrier to twisting (E_3) may be overcome; from the perpendicular state thereby reached, deactivation may lead to either the trans,trans or cis,trans ground states. At low temperatures the only path for the loss of excitation energy from trans,trans* is fluorescence, perhaps accompanied by non-radiative processes. Assuming that these have rate constants k_1 and k_2 respectively, essentially independent of temperature, and that that for crossing the barrier k_3 follows a normal Arrhenius law, then the intensity of the fluorescence should be proportional to

$$F = k_2/(k_1 + k_2 + A_3 \exp - E_3/RT) \qquad (1)$$

At very low temperatures, photoisomerisation becomes negligible, and the fluorescence intensity is proportional to

$$F_0 = k_2/(k_1 + k_2) \qquad (2)$$

Combining equations (1) and (2):

$$(F_0/F) - 1 = A_3(k_1 + k_2)^{-1} \exp - E_3/RT \qquad (3)$$

If this mechanism is indeed correct, then a plot of $\ln\{F_0/F - 1\}$ vs $1/T$ should be linear. Figure 4 shows that this is indeed the case and that E_3 is 19 kJ mol^{-1}. A more stringent test of the mechanism would, of course, be to determine the dependence of fluorescence lifetime upon temperature; such a study is in progress. It is interesting to note that even when a large mole fraction of the cis,trans conformer has been generated by photolysis at a suitable temperature, such a solution exhibits

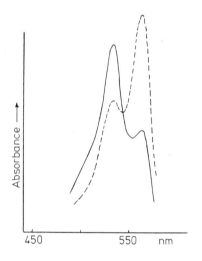

Figure 2. The photostationary state of 1,3-diphenylallylsodium in THF: $-50°C$ (—); $-90°C$ (---).

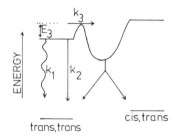

Figure 3. Energy level diagram showing routes for deactivation of excited trans, trans* conformer.

no fluorescence corresponding to the transition cis,trans* →
cis,trans. Apparently, there is no significant barrier to
twisting of the excited cis,trans conformer.

The related 1,3-diphenyl-2-azaallyl carbanion also exists
in solution as an ion pair. Although the absorption maximum in
the electronic spectrum is sensitive to the choice of solvent
and counterion, surprisingly, no case has been found to date of
a solution which simultaneously shows two absorption bands
attributable to the coexistence of tight and loose ion pairs.
The explanation of this observation is obscure. However,
photolysis does effect the transformation of the conformation
from trans,trans to cis,trans. No fluorescence has been
observed from solutions of this azaallyl ion. A further
difference from the diphenylallyl system is that the photostationary state is insensitive to temperature.

When a photoisomerised solution of either the diphenylallyl
carbanion, or of its 2-aza analog, is stored in the dark the
normal (thermal) conformational equilibrium is reestablished by
a process obeying first order kinetics. The reaction can be
conveniently monitored spectrophotometrically. In the case of
the diphenylallyl ion, the relaxation rate is much greater for
the tight ion pairs than for the loose. It seems probable that
the transition state may involve the association of the cation
specifically with one of the terminal allylic carbon atoms. Such
a species could be regarded as comprising a styrene moiety
together with a benzylic anion, the two being connected by a
single carbon-carbon bond about which the barrier to rotation
would be expected to be relatively low.

The Arrhenius parameters for the cis,trans → trans,trans
relaxation are collected in Table 1 for a range of combinations
of cation, anion and solvent. In the case of the relaxation of
the 1,3-diphenyl-2-aza-allyl carbanion, the absence of separate
absorption bands for the two kinds of ion pair prevents the
drawing of a general comparative conclusion. It would seem
probable that the lithium salt is loosely paired in THF but
tightly paired in diethyl ether from a simple consideration of
the relative solvating power of these solvents: this view is
apparently supported by the pronounced blue shift of the
absorption maximum in ether (λ_{max} 535 nm) with respect to that
in THF (λ_{max} 562 nm). It is, accordingly surprising, that the
Arrhenius parameters for these two systems are found to be so
similar. In the poorer solvating solvents diethyl ether and
MTHF, the E parameter is lower for the sodium salt than for the
lithium: that the lithium salt has a higher A factor might be
ascribed to the release of some bound solvent on forming the
transition state.

TABLE I

Arrhenius parameters[a] for thermal relaxation of conformation
from cis,trans to trans,trans 1,3 diphenyl-2-aza-allyl anion

Solvent	Li		Na		K	
	E	log A	E	log A	E	log A
Ether	79.7	14.8	66.4	11.7	62.9	11.5
MTHF[b]	89.2	15.2	73.8	12.4		
THF	82.5	14.8	82.9	13.8	56.8	9.9
DME[c]	78.0	13.7				
CH$_3$NH$_2$	80.4	13.1	85.3	13.9	73.6	12.5

[a]The activation energies E are expressed in kJ mol^{-1} and
A as sec^{-1}.
[b]MTHF is 2-methyltetrahydrofuran
[c]DME is 1,2-dimethoxyethane

Worsfold and Bywater [13] have demonstrated that the
stereochemistry of the propagation of polyisoprenyllithium in
hydrocarbon solvents is in part controlled by the rate at which
the cis active ends relax to the more stable trans forms
(scheme 2). The interesting possibility is accordingly raised
that by perturbing the cis ⇄ trans equilibrium of the propagating
centers of a diene photochemically during polymerisation, it
might be possible to alter the stereochemistry of propagation.
    Having made extensive studies of the behaviour of models of
the propagation center of 1,3-diphenylbutadiene this monomer
would be the most appropriate one to investigate. Unfortunately,
it seems that it cannot be isolated because of extremely facile
dimerisation via the Diels Alder reaction. A cursory
examination has, however, been made of the polymerisation of
1-phenylbutadiene by butyllithium in THF, yielding the following
results.

Temperature	% 3,4 addition in darkness	% 3,4 addition when illuminated
0°C	28	17
-37°	23	13

These results are encouraging and work is in hand to repeat and
extend these studies.

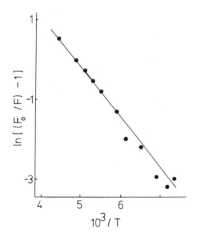

Figure 4. Analysis of temperature de-
pendence of fluorescence intensity using
Equation 3.

Scheme 2.

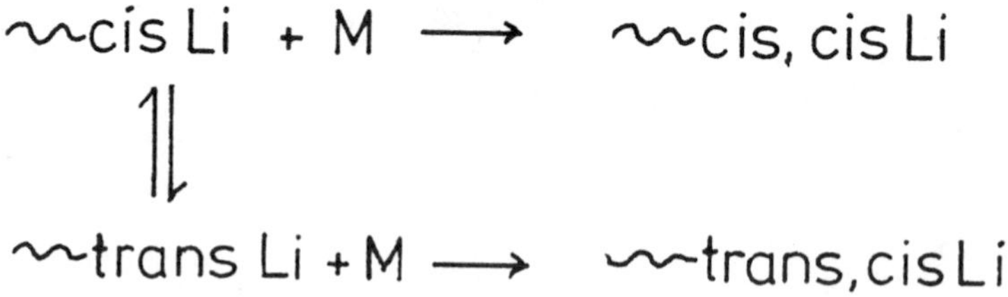

$$\sim\!\!\text{cis Li} + M \longrightarrow \sim\!\!\text{cis, cis Li}$$

$$\text{⇅}$$

$$\sim\!\!\text{trans Li} + M \longrightarrow \sim\!\!\text{trans, cis Li}$$

References

[1] J.W. Burley and R.N. Young, J. Chem. Soc. (B), 1019, (1971).

[2] J.W. Burley and R.N. Young, J. Chem. Soc. Perkin 2, 835, (1972).

[3] G.C. Greenacre and R.N. Young, J. Chem. Soc. Perkin 2, 1661, (1975).

[4] J.W. Burley and R.N. Young, J. Chem. Soc. Perkin 2, 1006 (1972).

[5] H.V. Carter, B.J. McClelland and E. Warhurst, Trans. Faraday Soc., 69, 2466, (1965).

[6] H.H. Freedman, V.R. Sandel and B.P. Thill, J. Amer. Chem Soc., 89, 1762, (1967).

[7] G.J. Heiszwolf and H. Kloosterziel, Recueil trav. Chim., 86, 1345 (1967).

[8] R.J. Bushby and G.J. Ferber, Chem. Commun., 407, (1973).

[9] R.J. Bushby and G.J. Ferber, J. Chem. Soc. Perkin 2, 1688, (1976).

[10] J.W. Burley and R.N. Young, J. Chem. Soc. Perkin 2, 1843, (1972).

[11] H.M. Parkes and R.N. Young, J. Chem. Soc. Perkin 2, 249, (1978).

[12] R.N. Young, H.M. Parkes and B. Brocklehurst, Makromol. Chem. Rapid Commun 1, 65 (1980).

[13] D.J. Worsfold and S. Bywater, Macromolecules, 11, 582, (1978).

RECEIVED March 5, 1981.

Solvation of Alkyllithium Compounds

Steric Effects on Heats of Interaction of Tetrahydrofurans with Polyisoprenyllithium and Polystyryllithium

RODERIC P. QUIRK

Michigan Molecular Institute, Midland, MI 48640

The enthalpies of interaction of tetrahydro-furan (THF) and 2,5-dimethyltetrahydrofuran (Me_2THF) with 0.03$\underline{M}$ benzene solutions of poly-(styryl)lithium (PSLi) and poly(isoprenyl)lithium (PILi) have been measured as a function of $\underline{R}$ ([THF]/[Li]) at 25°C using high dilution solution calorimetry. At low $\underline{R}$ values ($\underline{ca}$. 0.2) the enthal-py of interaction of THF with PILi (−5.8 kcal/mole) is more exothermic than with PSLi (−4.5 kcal/mole). However, the decrease in enthalpy for Me_2THF versus THF is larger for PILi (3.2 kcal/mole) than for PSLi (2.2 kcal/mole) at low $\underline{R}$ values. The enthalpies decrease rapidly with in-creasing $\underline{R}$ values for PILi, but are relatively constant for PSLi. It is suggested that THF inter-acts with intact dimer for PILi, but for PSLi this base coordinates to form the unassociated, THF-solvated species.

Lewis bases effect dramatic changes in microstructure, initiation rates, propagation rates, and monomer reactivity ratios for alkyllithium-initiated polymerizations of vinyl monomers (1-6). Some insight into the molecular basis for these observations has been provided by a variety of NMR, colligative property, and light-scattering measurements of simple and poly-meric organolithium compounds in hydrocarbon and basic solvents (7,8). In general, simple alkyllithiums exist predominantly as either hexamers (for sterically unhindered RLi) or tetramers (for sterically hindered RLi) in hydrocarbon solvents and as tetramers in basic solvents (9-12). Polymeric organolithium com-pounds such as poly(styryl)lithium exist as dimers in hydro-carbon solution and are unassociated in basic solvents such as tetrahydrofuran (13-15). The state of association of poly-(dienyl)lithiums in hydrocarbon solution is a subject of current

0097–6156/81/0166–0117$05.00/0

controversy (15,16), although the most recent results indicate
dimeric association for this species also (17,18). In spite of
this impressive wealth of data on association phenomena for
organolithium compounds, the fundamental nature of base–organolithium interactions responsible for changes in degree of association and for the dramatic effects of bases on reactions is not
understood.

We have measured the enthalpies of interaction of tetrahydrofurans with polymeric organolithiums to characterize the
specific nature of base–alkyllithium interactions and these
results are reported herein.

Experimental

Isoprene (99 + %, Aldrich) and styrene (99%, Aldrich) were
purified by initial stirring and degassing over freshly crushed
CaH_2 on a high vacuum line followed by distillation onto dibutylmagnesium (Alpha Inorganics). Final purification involved distillation from this solution directly into calibrated ampoules and
sealing with a flame. Benzene (Fisher, Certified ACS) was stored
over conc. H_2SO_4, washed successively with H_2O, dil. $NaHCO_3$
solution, and H_2O, followed by drying over anhydrous $MgSO_4$.
After filtration, further purification involved stirring under
reflux over freshly ground CaH_2, distillation onto sodium dispersion, and stirring and degassing on the vacuum line, followed by
distillation and storage over poly(styryl)lithium. Tetrahydrofuran (99%, Aldrich; and Burdick & Jackson, Distilled in Glass)
and 2,5-dimethyltetrahydrofuran (Aldrich) were stored over benzophenone ketyl and potassium–sodium amalgam and distilled in the
dry box immediately before use.

Polymerizations were carried out at $30°C$ in all glass,
sealed reactors using breakseals and standard high vacuum techniques (3). For the calorimetric measurements, a 1 liter sample
of a 0.03M solution of each polymeric lithium compound with $\bar{M}_n$
of ca. 4,000 was prepared in benzene solution using sec-butyllithium as initiator and transferred to the glove box.

Calorimetric measurements were performed in a recircu-
lated, argon atmosphere glove box using apparatus and procedures
which have been described in detail previously (19,20,21). Solutions of organolithium reagents were analyzed using the double
titration procedure of Gilman and Cartledge (22) with 1,2-
dibromoethane.

Results and Discussion

Poly(styryl)lithium. The calorimetric data obtained for
the interaction of tetrahydrofuran (THF) and 2,5-dimethyltetra-
hydrofuran (2,5-Me₂THF) with 0.03M solutions of poly(styryl)-
lithium in benzene as a function of R ([base]/[lithium atoms])
are shown in Figure 1. These data, as well as the corresponding
data for poly(isoprenyl)lithium, are referred to dilute

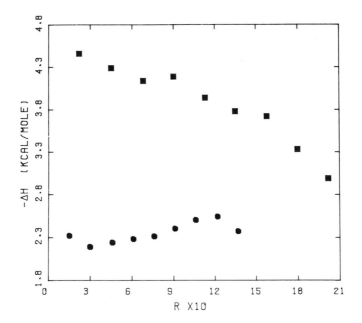

Figure 1.   Enthalpies of interaction of THF (■) and 2,5-dimethyltetrahydrofuran (●) with 0.03M solutions of polystyryllithium in benzene at 25°C: R = [base]/[Li] where [Li] is the concentration of carbon-bound lithium atoms in the solution.

solutions of the base in benzene at $25°C$ as the standard state
to correct for the heats of solution of the bases. The absence
of side reactions (which would consume poly(styryl)lithium and
give spurious enthalpies during the calorimetric measurements)
was determined by the double titration procedure (22) of the
base/poly(styryl)lithium solutions immediately after the calorimetric runs. For example, double titration analysis indicated
that the final concentration of active organolithium was 0.026_M
for a $0.027M$ solution of poly(styryl)lithium after a calorimetric run with tetrahydrofuran. The time required for a calorimetric run was generally less than 30 minutes. The accuracy and
precision of our calorimetric methods were determined at regular
intervals by comparison of our data with well-accepted values
for the heats of solution of standard substances (21). The
accuracy of our calorimetric data for heats of interaction of
bases with polymeric organolithium compounds was established by
demonstrating that our results are reproducible (± 0.1
kcal/mole) and independent of the source or method of purification of the tetrahydrofurans. The results are sensitive to the
amount of non-carbon bound lithium as shown by the data in
Figure 2. The effect of the presumably lithium alkoxide impurities is most apparent at R values greater than 1.0.

It is well established that poly(styryl)lithium is predominantly associated into dimers in hydrocarbon solutions (14,15),
while it is monomeric in tetrahydrofuran (14). Furthermore,
concentrated solution viscosity measurements have shown that the
equilibrium constant (K_{eq}) for the process shown in eq 1 [PSLi =
poly(styryl)lithium] has a value

$$\tfrac{1}{2}(PSLi)_2 + THF \underset{}{\overset{Keq}{\rightleftharpoons}} PSLi \bullet THF \tag{1}$$

of approximately 160 (14). Using this equilibrium constant, it
can be calculated that upon addition of tetrahydrofuran (1.2
mmoles) to 200 ml of a $0.03M$ solution of poly(styryl)lithium,
more than 98% of the tetrahydrofuran will be converted to PSLi•
THF. Therefore, it would be expected that the enthalpies of
interaction which we have measured refer to the process shown in
eq 1, i.e., the conversion by tetrahydrofuran of dimeric poly-
(styryl)lithium to solvated, monomeric poly(styryl)lithium.

The data in Figure 1 show no dramatic concentration dependence for the interaction of either tetrahydrofuran or 2,5-
dimethyltetrahydrofuran with poly(styryl)lithium. The absence of
distinct breaks within the range of R values from 0.2 to 2 can
be regarded as evidence that the initial base coordination
process (eq 1) is followed by successive coordination with other
tetrahydrofuran molecules as shown in eq 2.

$$PSLi \bullet THF + nTHF \rightleftharpoons PSLi \bullet (n + 1)THF \tag{2}$$

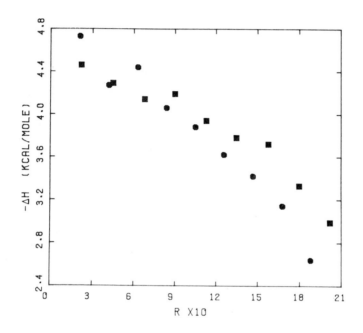

Figure 2. Enthalpies of interaction of THF with 0.03M solutions of polystyryl-lithium containing 1% (■) and 8% (●) noncarbon-bound lithium base impurities.

The relatively large difference in heats between tetrahydrofuran and 2,5-dimethyltetrahydrofuran (2.2 kcal/mole) indicates that the base coordination process for poly(styryl)lithium
is quite sensitive to the steric requirements of the base
(19,20).

Poly(isoprenyl)lithium. The calorimetric data obtained
for the interaction of tetrahydrofuran and 2,5-dimethyltetra-
hydrofuran with 0.03M solutions of poly(isoprenyl)lithium in
benzene as a function of R are shown in Figure 3. Double
titration analyses for 0.026M solutions of poly(isoprenyl)-
lithium after separate calorimetric runs with both tetyrahydro-
furan and 2,5-dimethyltetrahydrofuran indicated that the final
concentrations of active organolithium were 0.026M. Thus, no
decomposition is occuring during the calorimetric runs, at least
within the error limits of the double titration procedure (23).

The state of association of poly(dienyl)lithium compounds
in hydrocarbon solutions is a matter of current controversy
(15-18). Aggregation states of two (16,17,18) and four (15) have
been reported based on light-scattering and concentrated solution viscosity measurements. The most recent concentrated solution viscosity studies (16,17), which include results of various
endcapping and linking techniques, provide convincing evidence
for predominantly dimeric association of poly(isoprenyl)lithium
in hydrocarbon solution. The effect of tetrahydrofuran on the
degree of association of poly(isoprenyl)lithium has also been
examined by concentrated solution viscosity measurements (13).
These results indicate that the equilibrium constant for the
process shown in eq 3 [PILi = poly(isoprenyl)lithium] exhibits
an equilibrium

$$\tfrac{1}{2}(PILi)_2 + THF \underset{}{\overset{Keq}{\rightleftharpoons}} PILi \bullet THF \qquad (3)$$

constant of approximately 0.5. Thus, in contrast to the large,
favorable equilibrium constant for complexation of tetrahydrofuran with poly(styryl)lithium, the equilibrium constant for
poly(isoprenyl)lithium is less than one. This suggests that
considerable complexation might be occurring to the intact dimer
(aggregate) as shown in eq 4. Obviously, calorimetry can not by
itself define the complexation process. However, the effect of
structural

$$(PILi)_2 + THF \rightleftharpoons (PILi)_2 \bullet THF \qquad (4)$$

variations on complexation enthalpies can provide insight which
can be useful in deducing the nature of the complexation process.

The enthalpies of coordination of tetrahydrofuran with
poly(isoprenyl)lithium exhibit a dramatic concentration dependence as shown in Figure 3. The enthalpies decrease from -5.8

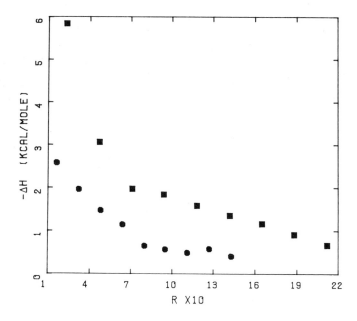

*Figure 3.   Enthalpies of interaction of THF (■) and 2,5-dimethyltetrahydrofuran
(●) with 0.03M solutions of polyisoprenyllithium in benzene at 25°C.*

kcal/mole at R = 0.2 to -3.1 kcal/mole at R = 0.5. A similar
concentration dependence is exhibited by the heats of interaction with 2,5-dimethyltetrahydrofuran. The surprisingly large
difference in enthalpies between tetrahydrofuran and 2,5-
dimethyltetrahydrofuran (3.2 kcal/mole) represents the largest
sensitivity to steric requirements of the base which we have
observed. In order to interpret these results, it is assumed
that whatever the coordination process is for a given organolithium compound with tetrahydrofuran, the same coordination process
is occurring for 2,5-dimethyltetrahydrofuran (i.e., observed
steric effects represent differences in degree and not in kind)
(19,20). On the basis of the steric effects for poly(styryl)lith-
ium versus poly(isoprenyl)lithium (see Table I), it can be
deduced that the coordination process for interaction of tetrahydrofurans with poly(isoprenyl)lithium involves primarily coordination to the intact dimer (aggregate). This conclusion is
based on the expectation that coordination to unassociated poly-
(isoprenyl)lithium would be less hindered than coordination to
unassociated poly(styryl)lithium.

Table I. Steric Effects on Enthalpies of Interaction

| Organolithium | $-\Delta H(kcal/mole)$[a] | | $\Delta\Delta H$[b] (kcal/mole) |
	THF	2,5-Me$_2$THF	
poly(isoprenyl)lithium	5.8	2.6	3.2
poly(styryl)lithium	4.5	2.3	2.2

[a]All enthalpies obtained at 25° by addition of 100μl (ca. 1
mmole) of base into 200 ml of 0.03M RLi ([base]/[Li] ca. 0.2).

[b]The difference in enthalpies for THF versus 2,5-Me$_2$THF for a
given RLi.

This supposition is supported by results for linking reactions
of polymeric organolithium compounds which indicate that the
steric requirements of a poly(styryl) chain end are larger than
those for a poly(dienyl) chain end (24,25). Since a larger
sensitivity to base steric requirements is exhibited by poly-
(isoprenyl)lithium and it is known that the coordination process
for poly(styryl)lithium involves coordination to give the unassociated species (eq 1), it is concluded that tetrahydrofuran
coordination with poly(isoprenyl)lithium must involve interaction with an associated species (presumably the dimer) to
explain the large sensitivity to the steric requirements of the
base.
        These results indicate the utility of high-dilution, solu-
tion calorimetry for examining and characterizing the nature of
base coordination with organolithium compounds. The large, exothermic enthalpies of coordination of bases with polymeric

organolithiums are consistent with the dramatic effects of bases on organolithium-initiated polymerizations of vinyl and diene monomers. Further work is in progress to determine the effect of the penultimate unit on the base coordination process and to characterize the coordination process for a variety of different bases.

Acknowledgement

The author wishes to express his gratitude to Mr. Dennis McFay for his help in obtaining the calorimetric results and to Mr. Wei-Chih Chen for his help in preparation of the polymeric organolithium solutions.

Literature Cited

1. Schoenberg, E.; Marsh, H.A.; Walters, S.J.; Saltman, W.M. Rubber Chem. Technol. 1979, 52, 526.
2. Bywater, S. "Comprehensive Chemical Kinetics," Elsevier, Amsterdam, 1976, Vol. 15; p 1.
3. Morton M.; Fetters, L.J. Rubber Chem. Technol. 1975, 48, 359.
4. Bywater, S. Prog. Polym. Sci. 1974, 4, 27.
5. Hsieh, H.L.; Glaze, W.H. Rubber Chem. Technol. 1970, 43, 22.
6. Bywater, S. Advances in Polymer Science 1965, 4, 66.
7. Wakefield, B.J. "The Chemistry of Organolithium Compounds," Pergamon Press, Elmsford, New York, 1974.
8. Szwarc, M. "Carbanions, Living Polymers, and Electron Transfer Processes," Interscience, New York, 1968.
9. Fraenkel, G; Beckenbaugh, W.E.; Yang, P.P. J. Am. Chem. Soc. 1976, 98, 6878.
10. Lewis, H.L.; Brown, T.L. ibid. 1970, 92, 4664.
11. Brown, T.L. Pure Appl. Chem. 1970, 23, 447.
12. West, P.; Waack, R. J. Am. Chem. Soc. 1967, 89, 4395.
13. Morton, M.; Fetters, L.J. J. Polym. Sci. Part A 1964, 2, 311.
14. Morton, M.; Fetters, L.J.; Pett, R.A.; and Meier, J.F. Macromolecules 1970, 3, 327.
15. Worsfold, D.J.; Bywater, S. ibid. 1972, 5, 393.
16. Fetters, L.J.; Morton, M. ibid. 1974, 7, 552.
17. Al-Jarrah, M.M.F.; Young, R.N. Polymer 1980, 21, 119.
18. Hernandez, A.; Semel, J.; Broecker, H.-C.; Zachmann, H.G.; Sinn, H. Makromol. Chem., Rapid Commun. 1980, 1, 75.
19. Quirk, R.P.; Kester, D.J. J. Organomet. Chem. 1977, 127, 111.
20. Quirk, R.P.; Kester, D.J. ibid. 1974, C23, 72.
21. Quirk, R.P.; Kester, D.J.; Delaney, R.D. ibid. 1973, 59, 45.
22. Gilman, H.; Cartledge, F.K. ibid. 1964, 2, 447.

23.   Crompton, T.R. "Chemical Analysis of Organometallic
      Compounds," Vol. 1, Academic Press, New York, 1973,
      Chapter 1.
24.   Bywater, S. Advances in Polymer Science 1979, 30, 89.
25.   Roovers, J.E.L.; Bywater, S. Macromolecules 1974, 7, 443.

RECEIVED February 12, 1981.

Anionic Polymerization of the Strontium Salt of One-Ended Living Polystyrene in Tetrahydrofuran and Tetrahydropyran

C. DE SMEDT and M. VAN BEYLEN

Laboratory of Macromolecular and Organic Chemistry, University of Leuven, Celestijnenlaan 200 F, B-3030 Heverlee, Belgium

ABSTRACT

Similarly to the previously reported barium salt, the strontium salt of one-ended living polystyrene (SrS_2) was prepared by converting on a strontium mirror dibenzylmercury into dibenzylstrontium. The latter was reacted with α-methylstyrene and the resulting oligomer was converted into the one-ended polystyryl salt by reaction with styrene. Conductance measurements in THF and THP in the temperature range of respectively -70°C and -40°C to +20°C indicated that in analogy with the barium salt, two equilibria are simultaneously established in both solvents:

$$SrS_2 \rightleftharpoons (SrS)^+ + S^- \qquad (K_1)$$

and $\qquad 2SrS_2 \rightleftharpoons (SrS)^+ + (SrS_3)^- \qquad (K_2)$

The ionic dissociation of strontium tetraphenylboride SrB_2 ($SrB_2 \rightleftharpoons SrB^+ + B^-$) was also investigated in THF in the temperature range of -70°C to +20°C and provided the respective Λ'_0 s and $K_{d_{SrB_2}}$. Viscometric measurements on mixtures of SrS_2 and strontium tetraphenylboride (SrB_2) in THF clearly indicate the formation of a mixed salt SrSB, which dissociates according to

$$SrSB \rightleftharpoons (SrS)^+ + B^- \qquad (K_b)$$

Conductance measurements on the mixed salt provided an accurate value for $\lambda^+_{SrS^+}$ and for K_b.

0097–6156/81/0166–0127$06.50/0

Kinetic measurements in THF in the absence and in the presence of variable amounts of added SrB_2 indicated that, similarly to BaS_2, propagation occurs mainly via the free S^- anions, the reactivity of the other species being negligible. Addition of SrB_2 slows down and eventually inhibits the propagation through formation of the mixed salt SrSB, the dissociation of which can no longer give free S^- anions.

Even in THP where the ionic dissociation of SrS_2 and therefore also the amount of free S^- anions was shown to be very small, propagation involving the free S^- anions still accounts for the observed kinetics and no propagation by species, other than S^- anions, could be detected experimentally.

The polymerization kinetics of alkali salts of living vinyl polymers in ethereal solvents, such as tetrahydrofuran (1), tetrahydropyran (2), dimethoxyethane (3), oxepane (4) and dioxane (5) have been studied and different comprehensive reviews or books dealing with anionic polymerization have been published (6). In all cases reported, except in the least polar solvents like dioxane, both ion pairs M^-, Cat^+ and free ions M^- were found to contribute to the propagation. In some cases even triple ions were shown to participate in the anionic propagation process (1a,7).

In ethereal solvents the alkaline-earth salts of living polymers (one-ended as well as two-ended) show a k_{app} ($=k_{obsd}/C$) inversely proportional to the carbanion concentration. B. Francois and coworkers proposed the formation of inactive aggregates of ion pairs $(S^-, M^{2+}, S^-)_n$ and a propagation proceeding through free ions (8,9,10). For the polymerization of the one-ended living Ba^{2+}, $(poly-S^-)_2$ in THF B. De Groof, M. Van Beylen and M. Szwarc presented another mechanism which fully accounts for the observed kinetics and conductances and which assumes the formation of M^{2+}, $(poly-S^-)_3$ triple ions next to M^{2+}, $(poly-S^-)$ and free poly S^- ions and a propagation mainly by the free polystyryl anions, the reactivity of other species being negligible or at least not observable (11). Other systems were however, reported in which the reactivity of species, other than free ions, cannot be neglected (12,13).

In the present communication the strontium salt of one-ended living polystyrene (SrS_2) was studied in tetrahydrofuran (THF) and tetrahydropyran (THP), in order to check the validity of the triple ion mechanism. The ionic dissociation of SrS_2 in THP was expected to be smaller than in THF and therefore it was thought that perhaps a contribution to the propagation from species, other than the free S^- anions, would be detectable.

The marked difference in behavior between two-ended and one-ended living polymers of alkaline-earth metals needs to be stressed. Whereas the rate of polymerization seems to be independent of the degree of polymerization in the case of one-ended polymers, such a dependence was disclosed in the case of the two-ended living polymers (8,9).

Experimental

1. Preparation of the strontium salt of one-ended living polystyrene (SrS$_2$)

The preparation of the strontium salt was carried out under high vacuum following the procedure described previously for the barium salt (11), with this modification that a greater excess (9.5 fold) of α-methylstyrene (α-MeS) was used at a temperature of 31°C. After 22 hr the remaining α-MeS was evaporated at the vacuum line together with the THF (10). After redissolving in THF, a 60-fold excess of styrene was added in 6 steps. An approximate molecular weight (M$_n$) of 6000 was determined for the single "S" arm by vapor-pressure osmometry after protonation of the salt, in good agreement with the calculated molecular weight (M$_{n,calc}$=6300). From GPC data a value of M$_w$/M$_n$ = 1.36 was obtained. To obtain the SrS$_2$ in THP, the THF was removed and the salt redissolved in THP. This procedure was repeated once to ensure complete removal of THF (14). The same ε-values as found in THF (see Table I) were used.

Table I

The main absorption maxima of the strontium salts in THF at 20°C

Sr(X)$_2$	λ_{max}, nm	$\varepsilon \times 10^{-4}$ (mol of X$^-$)$^{-1}$	Ref.
Sr^{2+}(PhCH$_2^-$)$_2$	332	1.46	this work
	322	1.2	a
Sr^{2+}(PhCH$_2$---αMeS$^-$)$_2$	334	1.27	this work
Sr^{2+}(poly-S$^-$)$_2$	348	1.23	this work
	348		b
	350		c
Sr^{2+}(PhCH$_2$CH$_2$CPh$_2^-$)$_2$	447		d

(a) K. Takahashi, Y. Kondo, R. Asami, J.Chem.Soc.(Perkin II), 577 (1978); (b) C. Mathis, L. Christmann-Lamande, B. François, J.Polymer Sci.(Pol.Chem.Ed.) 16, 1285 (1978); (c) C. Mathis, B. François, C.R. Acad.Sci.(Sér.C) 288, 113 (1979); (d) Preparation of SrS$_2$ by addition of styrene to the adduct of 1,1-diphenylethylene to Sr^{2+}(PhCH$_2^-$)$_2$ led to the same λ_{max} = 348 nm, though not all of the Sr^{2+}(PhCH$_2$CH$_2$CPh$_2^-$)$_2$ was converted into SrS$_2$.

2. Preparation of strontium tetraphenylboride

Two methods were used:

a) Strontium tetraphenylboride (SrB_2) was prepared by the method outlined in ref. 15. Since SrB_2 is less soluble than BaB_2, the precipitated salt on the filter was also recovered and purified.

b) A second method consisted of making first diphenyl strontium ($Sr(Ph)_2$) by reacting diphenyl mercury on a Sr-mirror in THF. After removal of the excess of Sr and Hg, the $Sr(Ph)_2$ solution was reacted with an equimolar amount of $B(Ph)_3$ in THF. The SrB_2 formed precipitated partially. It should be pointed out that conductance measurements with SrB_2 prepared according to method a or b led to the same results.

3. Purification of solvents and monomer

Tetrahydrofuran and tetrahydropyran were refluxed for 24 hr over Na-K alloy and distilled onto fresh alloy. Before use the solvents were distilled once more onto SrS_2 and subsequently distilled under vacuum into ampules. An all-glass apparatus equipped with breakseals was used for the latter operation. Styrene and α-methylstyrene were distilled under vacuum and dried twice over CaH_2 under high vacuum. Styrene was further purified by distilling it in the presence of SrS_2, while α-methylstyrene was further dried over Na-K alloy.

4. Conductance and kinetic measurements

All measurements were carried out under high vacuum. The details of the procedure and the apparatus used are given in ref. 11.

Results and Discussion

Conductance measurements

1. Strontium salt of one-ended living polystyrene (SrS_2) in THF

The conductance measurements cover a concentration range from 8×10^{-6} up to 7.5×10^{-4}M and a temperature range from $-70°C$ to $+20°C$. Adopting the procedure of Kraus and Bray (16) one finds a pronounced deviation from linearity, resulting in a constant value of Λ at high concentrations (fig. 1). In analogy with the Ba-salt this phenomenon can be explained by admitting that the ionization of SrS_2 is due not only to its dissociation into SrS^+ cations and free S^- anions (eq. 1) but also to the

formation of unilateral triple ions SrS_3^- (eq. 2), both ionic
equilibrium being simultaneously maintained:

$$SrS_2 \rightleftharpoons (SrS)^+ + S^- \qquad K_1 \qquad (1)$$

$$2\ SrS_2 \rightleftharpoons (SrS)^+ + (SrS_3)^- \qquad K_2 \qquad (2)$$

Thus, if the degree of dissociation (eq. 1) can be neglected (at
sufficiently high nominal concentrations C), the fraction of salt
converted into triple ions becomes independent of C.

Treatment of the data by the relation derived by Wooster
(17) for systems involving unilateral triple ions viz. $C\Lambda^2 = \Lambda^2{}_0 K_1 + (2\Lambda_0\lambda_0 - \Lambda^2{}_0)K_2 C$, where Λ_0 is the sum of the limiting
conductances of SrS^+ and S^- and λ_0 the sum of those of SrS^+ and
SrS_3^- respectively, yields a straight line (fig.2), confirming
the formation of unilateral triple ions. The constants K_1 and K_2
were calculated respectively from the intercept and the slope of
this straight line. Alternatively, as outlined in ref. 11, the
value of K_2 may also be calculated from the relation $\Lambda_1 = [K_2^{1/2}/(1 + K_2^{1/2})]\lambda_0$ which reduces to $\Lambda_1 = K_2^{1/2} \cdot \lambda_0$ if $K_2^{1/2} \ll 1$
and where Λ_1 is the equivalent conductance in the concentration
range where the fraction of triple ions and the equivalent
conductance remain constant and independent of C, whereas K_1 may
also be derived by determining, as in ref. 11, the concentration
C_i for which $[SrS_3^-] = [S^-]$, since then $K_1 = K_2 C_i$. The K_1-values
derived from the Wooster plot and the K_2-values obtained using
the equation $K_2^{1/2} = \Lambda_1/\lambda_0$ are the most reliable ones (11). The
respective K_1 and K_2 values are tabulated in Table II. K_1
obtained using the value of C_i and K_2 derived from the Wooster
plot are given in parentheses. The λ_0 and Λ_0 values needed for
these calculations are collected in Table III and were defined as
follows:

$$\Lambda_{0,SrS_2} = \lambda_{0,S^-} + \lambda_{0,SrS^+} = 2\lambda_{0,S^-} \ (=21.0 \text{ at } 20°C) \text{ see further}$$

part 4
conductance
of SrSB

$$\lambda_{0,triple} = \lambda_{0,SrS_3^-} + \lambda_{0,SrS^+} \ (=18.8 \text{ at } 20°C) \text{ where}$$

$$\lambda_{0,SrS_3^-} = \lambda_{0,S^-}/1.26 \ (18)$$

To obtain the values of Λ_0 and λ_0 at different temperatures we
used the following equation, rather than the Walden product which
is known to decrease with temperature (1b,19).

$$\frac{\lambda_0^-(20°C)}{\lambda_0^-(T°C)} = \frac{\lambda_0^+{}_{,Na^+}(20°C)}{\lambda_0^+{}_{,Na^+}(T°C)}$$

A Van't Hoff plot providing the corresponding thermodynamic
parameters for the ionic dissociation of SrS_2 (eq. 1) is shown in

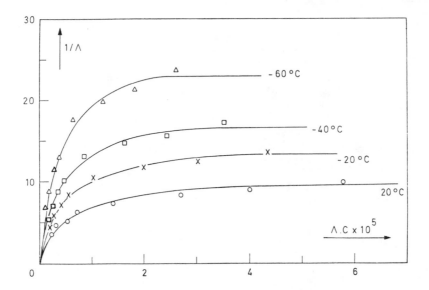

Figure 1. *Kraus and Bray conductance plot for* Sr^{2+}, $(poly\text{-}S^-)_2$ *at different temperatures in THF.*

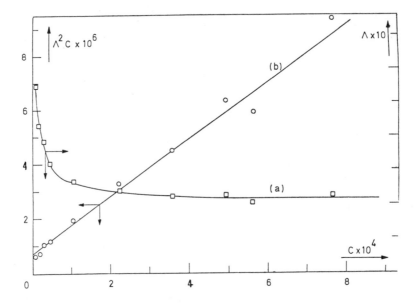

Figure 2. *Plot of equivalent conductance vs. concentration of* Sr^{2+}, $(poly\text{-}S^-)_2$ *in THF at* $20°C$ *(a); Wooster conductance plot in THF at* $20°C$ *(b).*

figure 3. (ΔH_1° = -1.87 kcal/mol and ΔS_1° = -46.4 e.u.). A curvature of the Van't Hoff plot such as observed for the triple ion formation constant K_2 has been reported in many other cases and may reflect the decrease in exothermicity of the dissociation at lower temperatures as expected on the basis of the increase of the dielectric constant with decreasing temperature. Why this is not the case for K_1 is difficult to say in cases where specific solvation occurs next to solvent polarization. If it were possible to go to lower temperatures a similar decrease of the exothermicity might eventually also show up for K_1.

Table II

Sr^{2+},(poly-S^-)$_2$ in THF and THP. Equilibrium constants K_1 and K_2 (THF) and K_2 (THP) at different temperatures.

T, °C	$K_1 \times 10^9$, M (THF)	$K_2 \times 10^5$ (THF)	$K_2 \times 10^6$ (THP)
20	1.6 (1.6)	3.3 (3.0)	1.2 (1.4)
15	1.9 (1.8)	3.4 (3.0)	1.2 (1.3)
10	2.0 (2.0)	3.5 (3.1)	1.2 (1.4)
5	2.2 (2.1)	3.7 (3.3)	1.2 (1.4)
0	2.3 (2.1)	4.1 (3.7)	1.2 (1.4)
-5	2.4 (2.3)	4.0 (3.6)	1.2 (1.4)
-10	2.6 (2.6)	4.3 (3.9)	1.2 (1.4)
-15	2.8 (2.7)	4.6 (4.1)	1.1 (1.3)
-20	3.1 (3.0)	4.9 (4.3)	1.1 (1.3)
-25	3.3 (3.2)	5.1 (4.5)	1.1 (1.3)
-30	3.6 (3.7)	5.6 (4.9)	1.1 (1.3)
-35	3.9 (3.9)	5.8 (5.0)	1.1 (1.3)
-40	4.1 (3.7)	5.9 (4.9)	0.92(1.1)
-45	4.4 (4.2)	6.0 (5.1)	
-50	4.8 (4.7)	6.1 (5.2)	
-55	5.4 (5.3)	6.9 (5.9)	
-60	6.2 (5.6)	7.0 (5.8)	
-65	6.5 (5.9)	7.1 (5.8)	
-70	7.3 (6.6)	7.5 (6.1)	

The SrS_2-salt is more dissociated in THF than the BaS_2-salt (11,15), probably due to greater solvation of the smaller Sr^{2+}

cation in the SrS_2-salt than of the Ba^{2+} cation in the
corresponding salt. For alkali ions the solvation energy
(specific and/or through solvent polarization) exceeds the
Coulombic interaction energy which binds the ions into ion pairs,
and therefore in THF the dissociation of ion pairs is exothermic
(1b). This is also the case with the SrS_2 salt, though the
solvation of the $(SrS)^+$ cation is much smaller than e.g. that of
the Na^+ cation. This is confirmed by comparing the corresponding
ΔH_1°-values (-1.87 and -8.2 kcal/mol respectively). In the case
of BaS_2 it was already shown that the gain in solvation energy of
the $(BaS)^+$ ion is smaller than the Coulombic interaction energy
(ΔH_1° = +0.9 kcal/mol) (11).

Furthermore, the entropy change ΔS_1° for the dissociation of
the alkaline earth polystyrene salts is substantially less
[$\Delta S_1^\circ, SrS_2$ = -46 e.u.; $\Delta S_1^\circ, BaS_2$ = -40 e.u. (11)] than for the

dissociation of the Na^+, S^- contact ion pair ($\Delta S^\circ \cong$ - 60 e.u.)
(14,1b). This smaller decrease is attributed to the fact that
upon dissociation of the $M^{2+}, (S-)_2^-$ salts, a S^- anion remains
bound to the M^{2+} cation preventing the M^{2+} to be reached from all
sides by the solvent molecules. The difference between $\Delta S_1^\circ, SrS_2$
and $\Delta S_1^\circ, BaS_2$ also accounts for the greater solvation   of the
smaller $(SrS)^+$ cation formed upon dissociation.

2. Strontium salt of one-ended polystyrene in THP

In THP, which has a lower dielectric constant than THF, and
in which conductance measurements were carried out at
temperatures ranging from -40°C to +20°C, the constant value of
the equivalent conductance was maintained throughout the whole
concentration range studied (3×10^{-5} to 8×10^{-4}M) (fig.4). From
these constant Λ_1 values, K_2 was obtained (see Table II). Since
the intercept of the Wooster plot (fig.4) is indistinguishable
from zero, it was impossible to determine K_1. The slope of this
plot gives K_2 (values given in parentheses). The relevant Λ_0 and
λ_0 values are listed in Table III and were obtained from those in
THF using the Walden rule. The viscosities of THF (19) and THP
(20) were taken from the literature. Our measurements in THP
thus confirm the mechanism proposed in the equations (1) and (2)
and clearly indicate the lower dissociation (eq.1) of $\bar{S}rS_2$ in THP
than in THF.

3. Strontium tetraphenylboride in THF

Conductance measurements on strontium tetraphenylboride
(SrB_2) were performed over a concentration range of 5×10^{-5}M to
2×10^{-6}M and at temperatures extended from -70°C to +20°C. In
Figure 5 the results are graphically presented as a $1/\Lambda$ vs. $C\Lambda$
plot giving an accurate intercept, as a result of the high
degree of dissociation of SrB_2 in THF. In this way the values of

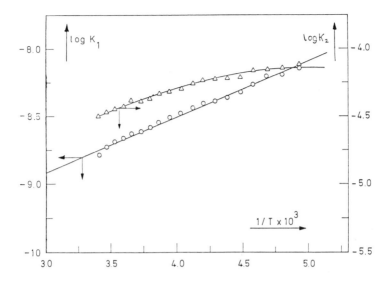

Figure 3.   Temperature dependence of K_1 (o) and K_2 ($\triangle$) for Sr^{2+}, (poly-S$^-$)$_2$ in THF.

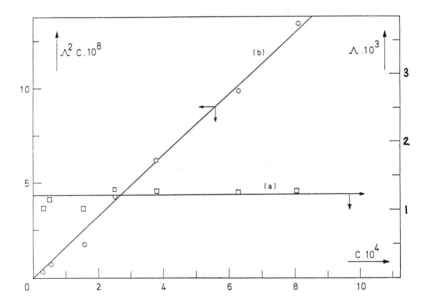

Figure 4.   Plot of equivalent conductance vs. concentration of Sr^{2+}, (poly-S$^-$)$_2$ in THP at 20°C (a); Wooster plot for Sr^{2+}, (poly-S$^-$)$_2$ in THP at 20°C (b).

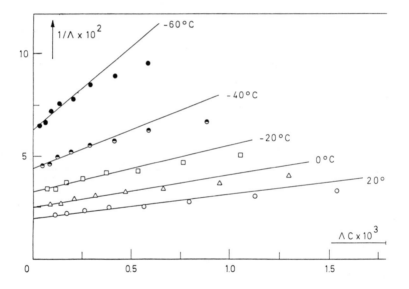

Figure 5.   Plot of 1/Λ vs. CΛ for Sr(BPh₃)₂ in THF at different temperatures.

Table III

Limiting equivalent conductances Λ_o and λ_o in THF and THP at different temperatures.

T °C	Λ_o, SrS_2 (THF)	λ_o, triple (THF)	Λ_o, SrS_2 (THP)	λ_o, triple (THP)
20	21.0	18.8	12.4	11.1
15	19.6	17.6	11.2	10.1
10	18.4	16.5	10.3	9.2
5	17.2	15.4	9.3	8.3
0	16.0	14.3	8.4	7.5
-5	15.0	13.5	7.6	6.9
-10	14.0	12.6	6.9	6.2
-15	13.0	11.7	6.2	5.6
-20	12.0	10.8	5.5	5.0
-25	11.2	10.0	5.0	4.4
-30	10.2	9.1	4.4	3.9
-35	9.4	8.4	3.7	3.3
-40	8.8	7.9	3.5	3.1
-45	8.0	7.2		
-50	7.3	6.6		
-55	6.5	5.7		
-60	5.8	5.2		
-65	5.2	4.7		
-70	4.6	4.1		

Λ_o and the corresponding K_d values for the dissociation were

$$SrB_2 \rightleftharpoons (SrB)^+ + B^- \qquad K_{d_{SrB_2}} \qquad (3)$$

determined. They are summarized in Table IV. In addition, Table IV shows the relevant thermodynamic parameters.

The enthalpy of dissociation of the contact ion pair of difluorenyl strontium ($SrFl_2$) was reported to be $\Delta H^\circ_{d,c} = -14.4$ kcal/mol at 20°C, that of the mixed tight-loose ion pair being much less negative ($\Delta H^\circ_{d,c-s.s} = -2.1$ kcal/mole) (21). In view of the fact that the SrB_2 and the BaB_2 salts are much more dissociated than the corresponding fluorenyl salts and that the enthalpy values of SrB_2 and BaB_2 are -0.7 kcal/mol and 0.0

Table IV

Dissociation constants and Λ_o-values of $Sr(B\emptyset_4)_2$ in THF at different temperatures

T °C	$K_{d_{SrB_2}} \cdot 10^5$ M	Λ_o cm^{2}mol$^{-1}\Omega^{-1}$	Λ_o^{+} * cm^{2}mol$^{-1}\Omega^{-1}$	$\eta.\Lambda_o.10$
20	3.05	51.0	12.1	2.49
15	3.22	48.0	11.3	2.46
10	3.31	45.4	10.9	2.46
5	3.63	42.7	10.2	2.45
0	3.85	40.1	9.5	2.44
-5	3.98	37.6	8.8	2.43
-10	4.00	35.4	8.5	2.44
-15	4.00	33.3	8.2	2.45
-20	4.15	31.1	7.8	2.46
-25	4.29	29.0	7.3	2.46
-30	4.52	26.8	6.8	2.46
-35	4.62	24.8	6.4	2.46
-40	4.69	22.9	6.1	2.46
-45	4.77	21.0	5.6	2.46
-50	4.99	19.2	5.2	2.46
-55	5.23	17.5	4.8	2.45
-60	5.11	15.9	4.5	2.46
-65	5.25	14.3	4.0	2.44
-70	5.25	12.8	3.7	2.44

*Λ_o^{+} : limiting equivalent conductance of $[Sr(B\emptyset_4)]^{+}$

Thermodynamic parameters of $Sr(B\emptyset_4)_2$ in THF

$\Delta H_d^{\circ} = - 0.69$ kcal/mol

$\Delta S_d^{\circ} = - 23$ e.u.

kcal/mol ([15]) respectively, we may conclude that these salts are at least of the mixed tight-loose type, for which there is but little change in the solvation state upon dissociaton. Moreover, since the entropy of dissociation for the contact ion pairs is of the order of -50 to -60 e.u., we may again say that the tetraphenylboride salts of Sr and Ba, showing an entropy change of respectively -23 and -22 e.u., are also for entropic reasons not of the contact type.

In fact, the Λ_o^{+}-value of $(SrB)^{+}$ being 12.1 and that of $(BaB)^{+}$ 26.1 may be taken to indicate a double solvent separated ion pair structure for the Sr-salt and a mixed contact-solvent separated pair structure for the BaB$_2$. The double solvated cation of the Sr salt represented as $B^-//Sr^{2+}//$ will upon dissocation indeed be less mobile than the smaller $B^-,Ba^{2+}//$ cation. The assumption outlined above is also reflected in the greater K_{d,SrB_2} ($=3.05\times10^{-5}$M) in comparison with K_{d,BaB_2} ($=1.65\times10^{-5}$M) at 20°C.

In conclusion, the following equilibria might be written for the dissociation of SrB_2 and BaB_2, respectively:

$$B^-//Sr^{2+}//B^- \rightleftharpoons B^- + //Sr^{2+}//B^- \qquad K_{d_{SrB_2}} \qquad (4)$$

$$B^-,Ba^{2+}//B^- \rightleftharpoons B^- + //Ba^{2+},B^- \qquad K_{d_{BaB_2}} \qquad (5)$$

4. The mixed salt SrSB in THF at 20°C

When mixed with SrB_2, the strontium salt of one-ended living polystyrene SrS_2, like the barium salt BaS_2 with added BaB_2 (15), forms a mixed salt SrSB according to the equilibrium:

$$SrS_2 + SrB_2 \rightleftharpoons 2 \text{ SrSB} \qquad K_a \qquad (6)$$

The formation constant K_a of the mixed salt SrSB was determined by measuring the viscosity of a solution of such mixture. Indeed the length of the SrS_2 molecule is twice as high as that of SrSB (and also of dead poly-S) so that we may consider SrSB equivalent to dead poly-S. To ascertain a sufficient viscosity, a SrS_2 salt of molecular weight of about 58,000 per "S"-arm was used ($M_w/M_n = 1.26$). Adopting the same method as outlined in ref. 15, a calibration curve was then constructed (fig.6, Table V), giving the specific viscosity of mixtures of living and rigorously dried dead polystyrene, obtained by termination of some of the investigated SrS_2, as a function of their composition r ($r=SrS_2/$ total poly-S) for a constant total weight concentration of polystyrene (83.5 gr/l). Subsequently the viscosity of a mixture of living SrS_2 and dead polystyrene of the same total concentration was measured without and with addition of SrB_2. Equivalent amounts of living SrS_2 and SrB_2, as were used in the case of barium, were not taken in this case, in view of the lower solubility of SrB_2. It should again be stressed that the molecular weight distribution of any mixture of living and dead polymer is the same as that of the corresponding mixture of SrS_2 and SrB_2 of the same specific viscosity, thus justifying the method used to determine the K_a value for any molecular weight distribution. The mixtures of SrS_2 and terminated S used in determining the calibration curve have the same molecular weight distribution as the mixture of SrS_2 and SrSB resulting from the addition of SrB_2. Indeed samples of the same stock solution were used to carry out these experiments. Thus, it was avoided that changes in molecular weight distribution would affect the

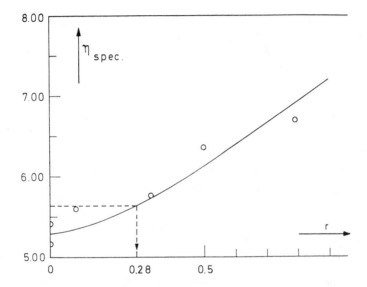

Figure 6. Viscosity of a solution containing SrS₂ and the terminated dead poly-S at various r values (r = [SrS₂]/total poly-S) but at constant weight concentration of total polystyrene (whether as SrS₂ or dead poly-S). The dashed line corresponds to a mixture of living strontium polystyrene ([SrS₂] = 4.3 × 10⁻⁴M), dead polystyrene ([2(terminated S)] = 2.9 × 10⁻⁴M), and a certain amount of SrB₂ (= 2.4 × 10⁻⁴M).

Table V

Viscosity data of the study of the mixed salt SrSB
in THF at 20 ° C

r	η_{spec}
0 (a)	5.42
0 (b)	5.16
0.09	5.60
0.33	5.76
0.50	6.36
0.79	6.69
r_m (c)	5.63

(a) obtained by termination of living SrS_2 with methanol

(b) obtained after complete spontaneous termination
    of the living SrS_2 solution

(c) ratio obtained from the calibration curve (fig.6)

    for the mixture of polystyrene (living + terminated)

    with SrB_2 (= 0.28)

results. The results are shown in fig.6 and from the r value
determined for the above described mixture, K_a was calculated to
be 100.

The mixed salt SrSB was assumed to dissociate according to
the equilibrium

$$SrSB \rightleftharpoons (SrS^+) + B^- \qquad K_b \qquad (7)$$

The conductance of a mixture $[SrS_2] = 2.19 \times 10^{-4}M$ and $[SrB_2] =
2.08 \times 10^{-4}M$ was measured and the corresponding $[SrSB]_o$
concentration and K_b could be calculated from the following set
of equations:

$$\frac{[SrSB]^2}{([SrB_2]_0 - 1/2[SrSB])([SrS_2]_0 - 1/2[SrSB])} = K_a = 10^2$$

$$[(SrS)^+][B^-]/[SrSB] = K_b$$

$$[(SrB)^+][B^-]/([SrB_2]_0-1/2[SrSB]-[(SrB)^+]) = K_d \quad = 3.05\times10^{-5}M$$
$$\qquad\qquad\qquad\qquad\qquad\qquad\qquad\qquad\qquad\qquad\qquad SrB_2$$

$$[(SrB)^+] + [(SrS)^+] = [B^-]$$

$$[B^-]\cdot\lambda^-_{o,B} + [(SrB)^+]\lambda^+_{o,SrB} + [(SrS)^+]\lambda^+_{o,SrS} = L.10^3.a$$

where L is the measured conductance and a is the cell constant. The following λ_0 values were used: $\lambda^-_{o,B} = 38.9$ ([19]), $\lambda^+_{o,SrB} = 12.1$ (see part 3), $\lambda^+_{o,SrS} = 10.5$ (see below). Seven concentrations were measured ($3.10^{-4}M \leftrightarrow 1.10^{-5}M$) and each time K_b was calculated resulting in a mean value of $K_b = 6.72\times10^{-5}M$ (see Table VI).

Table VI

Conductance of $Sr^{2+},(poly-S^-)(B\phi_4^-)$ in THF at 20°C.

*L x 10^6 Ω^{-1}	$[Sr^{2+}(poly-S^-)(B\phi_4^-)]$ 10^4, M	Λ* $cm^2 mol^{-1}\Omega^{-1}$	$(1/\Lambda).10^2$ $mol\,\Omega\,cm^{-2}$	$\Lambda c.10^3$ $cm^2 l^{-1}\Omega^{-1}$	Lx10^6 Ω^{-1}	K_b(calculated from L) x 10^5
353	3.54	17.9	5.59	6.33	369	7.55
220	1.80	21.9	4.56	3.95	232	6.68
140	0.905	27.8	3.60	2.51	148	6.86
99.9	0.562	31.9	3.13	1.79	106	6.96
68.9	0.348	35.5	2.81	1.24	73.4	6.76
40.7	0.180	40.6	2.46	0.73	43.6	7.24
21.9	0.091	43.1	2.32	0.39	23.7	5.83

Cell constant : 1.795×10^{-2} cm^{-1}

*L-values corrected for the conductance of the minute amount of remaining $Sr(B\phi_4)_2$

Alternatively Λ_o and K_b for the mixed salt may be calculated from a plot of $1/\Lambda$ vs. $[SrSB]_o\Lambda$ which yields a straight line (fig.7). The following values may be derived from this plot:

$1/\text{intercept} = \Lambda_o = \lambda^+_{o,SrS^+} + \lambda^-_{o,B^-} = 49.4$ and since $\lambda^-_{o,B^-} = 38.9$ (19) λ^+_{o,SrS^+} will be equal to 10.5, while the slope = $1/K_b \cdot \Lambda^2_o = 6.36$ yields $K_b = 6.44 \times 10^{-5} M$. The measured conductance (L) were corrected for a minute fraction of remaining $SrB_2(\sim 5\%)$.

It is noteworthy at this point that only dissociation of SrSB according to equation (7) was taken into account and not according to the equilibrium:

$$SrSB \rightleftharpoons (SrB)^+ + S^- \qquad K_c \qquad (8)$$

K_c should be much smaller than K_b because no propagation was detected in a quasi equivalent mixture of SrS_2 and SrB_2 (see kinetic measurements) thus the amount of S^- is undetectably small. Calculation of the equilibrium constant of the exchange reaction:

$$(SrB)^+ + S^- \qquad (SrS)^+ + B^- \qquad K_{i,ex} \qquad (9)$$

$$\text{yields} \quad K_{i,ex} = \frac{K_a \cdot K_b^2}{K_{d_{SrB_2}} \cdot K_1} = \frac{K_b}{K_c} = 8.50 \times 10^6$$

justifying our assumption that $K_b \gg K_c$

Comparing the two dissociations $SrB_2 \rightleftharpoons (SrB)^+ + B^-$ and $SrSB \rightleftharpoons (SrS)^+ + B^-$ characterized by the dissociation constants $K_{d_{SrB_2}} = 3.05 \times 10^{-5} M$ and $K_b = 6.44 \times 10^{-5} M$, respectively, and taking into account the fact that in K_d a statistical factor 2 appears but not in K_b, it seems that the S^- anion is tightly bound to the Sr^{2+} cation and thereby decreases the attraction of B^- in SrSB. This is not the case in the SrB_2 salt which is probably of the double solvent separated type (see 3) leading to a smaller value for $K_{d_{SrB_2}}$ than K_b.

Kinetic Measurements

Pseudo-first order kinetics were observed in each of the following cases.

1. Kinetics of the propagation of styrene polymerization initiated by SrS_2 in THF without added SrB_2

Though there is some scatter, the pseudo-first order rate constant k_{obsd} seems virtually independent of the concentration of the strontium salt (fig.8, Table VII) in the investigated concentration range ($>10^{-5}$M). Such a phenomena was already observed with one- and two-ended barium polystyrene in THF and THP ($\underline{8},\underline{9},\underline{11},\underline{13}$).

Table VII

$k_{observed}$ and $k_{apparent}$ of the polymerization of Sr^{2+}, $(poly-S^-)_2$ in THF at 20°C

$C \times 10^4$ M	$k_{obsd.} \times 10^2$ sec^{-1}	$\frac{1}{C} \times 10^{-4}$ M^{-1}	$k_{app.} \times 10^{-2}$ $M^{-1}.sec^{-1}$
0.11	3.36	8.79	29.54
0.13	3.80	7.99	31.02
0.21	4.11	4.66	19.16
0.22	3.97	4.47	17.76
0.75	5.07	1.33	6.77
1.33	2.25	0.75	1.70
2.44	5.91	0.41	2.42
2.50	4.08	0.40	1.63
5.20	3.45	0.19	0.66
5.98	2.79	0.17	0.47

If only the free ions contribute to the propagation to any measurable extent, the contribution of all other species to the polymerization being negligible, it can be shown($\underline{11}$) that $k_{obsd} = \{K_1/(K_1/C + K_2)^{1/2}\}k_-$ and since in most experiments $K_1/C \ll K_2$, a plot of the apparent propagation rate constant k_{app} ($=k_{obsd}/C$) vs. $1/C$ should give a straight line going through the origin indicating that no species other than free ions can be found to contribute to the propagation. The slope of this line is equal to $k_-.K_1/\sqrt{K_2}$ (fig.9). Thus it was calculated that slope = $\frac{k_-.K_1}{\sqrt{K_2}} = k_{obsd} = 4.1 \times 10^{-2}$ sec^{-1}, neglecting the two lowest concentrations of Table VII for which the condition $(K_1/C + K_2)^{1/2} \approx K_2^{1/2}$ is not fulfilled ($\underline{11}$). Calculations of k_{obsd} with the values of K_1 and K_2, determined from conductance measurements, and assuming $k_- = 122,000M^{-1}$ sec^{-1} (22,23) gives $k_{obsd} = k_-.K_1/\sqrt{K_2} = 3.\overline{4} \times 10^{-2}$ sec^{-1} in good agreement with the value obtained from the kinetic data.

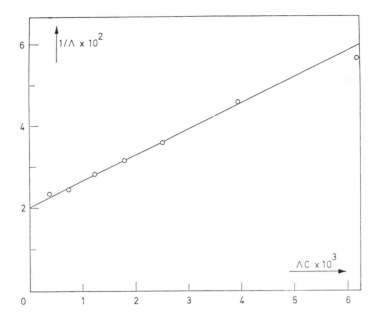

Figure 7. Kraus and Bray conductance plot for the mixed salt Sr^{2+}, (poly-S^-) (BPh_4^-) in THF at 20°C.

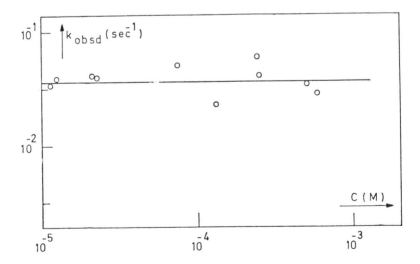

Figure 8. Plot of k_{obsd} of Sr salt of living polystyrene in THF vs. log of the concentration of Sr^{2+}, (poly-S^-)$_2$ at 20°C.

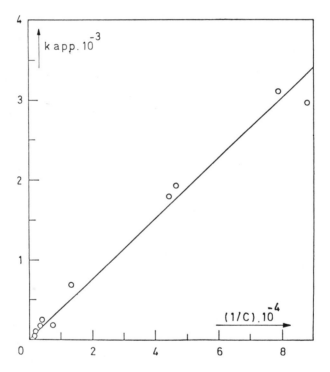

Figure 9.   The k_{app} $(= k_{obsd}/C$ vs. $1/C$ for Sr^{2+}, $(poly-S^-)_2)$ in THF at $20°C$.

The rate constant k_- is taken to be independent of the solvent and its temperature dependence can be represented by one common straight Arrhenius line for several solvents (22). Although there has been some controversy in the past as to its exact value, our results seem to indicate that the value given in ref.22 is to be preferred to that used previously in ref.11 (see also the remark given in ref.23). The value of k_- was calculated at 20°C from the following equation, given in ref.22

$$\log k_- = 8.0 - 3900/(4.57\ T) \tag{10}$$

2. Kinetics of the propagation of styrene polymerization initiated by SrS2 in THF in the presence of SrB2

A solution of $[SrS_2] = 1.2x10^{-5}M$ and $[SrB_2] = 1.7x10^{-5}M$ could not be found to propagate anymore. Considering the equilibria 1, 2, 6 and 7 it can be seen that upon addition of an approximately equimolar amount of SrB_2 or more the increase of $[SrS^+]$, resulting from equilibria 6 and 7, represses the formation of S^- and SrS_3^- anions in 1 and 2, making their concentration negligible. Therefore we may conclude that the propagation by SrS_2 and SrS^+ must be very small and not observable.

On the basis of our previous kinetic measurements in the absence of SrB_2 one may also conclude that there is no detectable propagation through triple ions SrS_3^- either. Moreover, a value of $k_{trp} < 100M^{-1}\ sec^{-1}$ is accepted (it will be seen later (24) why this upper limit of k_{trp} is chosen), and in our case the ratio $[SrS_3^-]/[S^-] = K_2[SrS_2]/K_1$ amounts to 10 at the highest concentrations of SrS_2 indicated in Table VIII, in which the results of kinetic measurements in the presence of varying relative amounts of SrB_2 are summarized. Therefore, taking k_- as 122,000$M^{-1}\ sec^{-1}$, the contribution of SrS_3^- to the observed propagation could not exceed 1% and it is justified to assume that propagation occurs entirely via the free S^- anions.

Since the ratio $[SrS_2]_0/[SrB_2]_0$ used in the experiments reported in Table VIII was large, virtually all SrB_2 was converted into SrSB. In this case one can write:

$$[(SrS)^+]^2/(2[SrB_2]_0 - [(SrS)^+]) = 6.44x10^{-5}M = K_b \tag{11}$$

Admitting, as explained above, that the observed propagation arises entirely from the growth of S^- anions, we can also write:

$$[(SrS)^+] = (k_- \cdot K_1)\ \frac{[SrS_2]}{k'_{obsd}} \tag{12}$$

Table VIII

Propagation of styrene polymerization initiated by $Sr^{2+},(poly-S^-)$
in the presence of $Sr(B\phi_4)_2$ (THF; 20°C)

$[SrS_2]_o \cdot 10^4$ M	$[SrB_2]_o \cdot 10^5$ M	$k'_{obsd} \cdot 10^3$ sec^{-1}	$[SrS_2] = ([SrS_2]_o - [SrB_2]_o)$ $\cdot 10^4$, M	$\dfrac{[SrS_2]}{k'_{obsd}} \cdot 10^2$ M.sec	$[(SrS^+)] \cdot 10^5$ M
0.769	0.196	8.08	0.749	0.928	0.371
3.70	0.348	19.5	3.67	1.88	0.634
3.89	0.412	19.7	3.85	1.95	0.739
2.42	0.712	5.15	2.35	4.57	1.20
5.28	1.27	10.4	5.15	4.63	1.95
0.812	1.24	1.17	0.688	5.89	1.91
3.80	3.58	1.18	3.44	29.2	4.30
1.93	3.17	0.447	1.61	36.1	3.94

A plot of $[(SrS^+)]$ (calculated from equation (11)) vs. $[SrS_2]/$
k'_{obsd} gives a straight line going through the origin with a
slope equal to $k_- \times K_1 = 1.8 \times 10^{-4}$ sec^{-1} (fig.10) resulting in a
K_1 value of 1.5×10^{-9}M in very good agreement with the K_1-value
determined from the conductance measurements.

3. Kinetics of the propagation of styrene polymerization
    initiated by SrS_2 in THP

The behavior of the SrS_2 salt in THP and THF looks very
similar. Again only the contribution of free S^- ions to the
propagation can be observed resulting in a straight line through
the origin in a plot of k_{app} vs. $1/C$ (fig.11,Table IX).
The slope is $k_- K_1 \sqrt{K_2} = k_{obsd} = 7.4 \times 10^{-5}$ sec^{-1}. Taking again
a value of $k_- = 122,000$ M^{-1} sec^{-1}, and thus assuming the
insensitivity of k_- of the solvent (22), and since $K_2 = 1.2 \times 10^{-6}$,
the value of K_1 (20°C) could be calculated to be 6.6×10^{-13}M.
(The condition $K_1/C \ll K_2$ is therefore fulfilled over the whole
concentration range used in these experiments).

The k_{obsd} and K_1 decrease by a factor of 5×10^2 and 2×10^3,
respectively, on going from THF to THP at 20°C. The decrease of
K_1 was within the expectations on lowering the dielectric
constant. The variation of k_{obsd} can also be entirely accounted
for by assuming a propagation occurring mainly through free ions
in THP (24), though the dissociation constant is significantly
lower in this solvent than in THF.

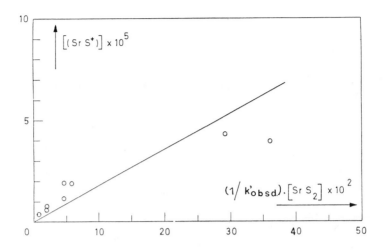

Figure 10. Plot of [(SrS⁺)], calculated from the relation $[(SrS^+)]^2/(2[SrB_2]_o - [(SrS^+)]) = 6.44 \times 10^{-5}$M, vs. $[SrS_2]/k'_{obsd}$. The first experiment (●) is clearly the least reliable one.

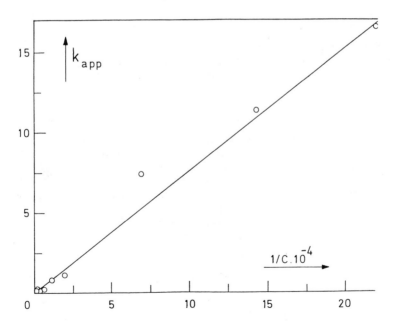

Figure 11. The k_{app} vs. $1/C$ for Sr^{2+}, $(poly\text{-}S^-)_2$ in THP at 20°C.

Table IX

$k_{observed}$ and $k_{apparent}$ of the polymerization of $Sr^{2+},(poly-S^-)_2$
in THP at 20°C

$C \times 10^4$ M	$k_{obsd} \times 10^5$ sec^{-1}	$\frac{1}{C} \times 10^{-4}$ M^{-1}	$k_{app} \times 10$ $M^{-1}sec^{-1}$
0.046	7.07	22.0	155.4
0.070	7.96	14.3	113.6
0.14	10.7	6.94	74.31
0.52	5.94	1.93	11.48
0.86	7.32	1.16	8.50
1.61	3.9	0.62	2.43
3.80	4.41	0.26	1.16
7.20	4.40	0.14	0.61

Acknowledgement

Financial support from the Belgian National Research
Foundation, the Belgian Ministry of Scientific Programming and
the Catholic University of Leuven, is gratefully acknowledged.

Literature Cited

1.   (a) D. N. Bhattacharyya, C. L. Lee, J. Smid, M. Szwarc, J.
      Phys. Chem., 69, 612 (1965); (b) T. Shimomura, K. J. Tölle,
      J. Smid, M. Szwarc, J. Am. Chem. Soc., 89, 796 (1967).

2.   (a) M. Van Beylen, M. Fisher, J. Smid, M. Szwarc, Macro-
      molecules, 2, 575 (1969); (b) A. Parry, J. E. L. Roovers,
      S. Bywater, Macromolecules, 3, 355 (1970); (c) G. Pizzirani,
      M. Di Maina, M. Palla, P. Giusti, Europ. Polym. J., 13, 605
      (1977).

3.   T. Shimomura, J. Smid, M. Szwarc, J. Am. Chem. Soc., 89,
      5743 (1967).

4.   G. Löhr, S. Bywater, Can. J. Chem., 48, 2031 (1970).

5.   D. N. Bhattacharyya, J. Smid, M. Szwarc, J. Phys. Chem., 69,
      624 (1965).

6. (a) M. Szwarc, "Carbanions, Living Polymers and Electron Transfer Processes" Interscience, New York (1968); (b) M. Szwarc, "Ions and Ion Pairs in Organic Reactions", Interscience, New York, p. 375 (1974); (c) S. Bywater, Prog. Polym. Science, 4, 27 (1974); (d) S. Bywater, "Anionic Polymerization of Olefins", in "Chemical Kinetics", Vol. 15, C. H. Bamford and C. F. H. Tipper, Ed. Elsevier, p.1 (1976).

7. (a) D. N. Bhattacharyya, J. Smid, M. Szwarc, J. Am. Chem. Soc., 86, 5024 (1964); (b) D. Honnoré, J. C. Favier, P. Sigwalt, Europ. Polym. J., 10, 425 (1974).

8. C. Mathis, Thèse (1976), Univ. Strasbourg.

9. C. Mathis, B. François, J. Polym. Sci. (Pol. Chem. Ed.), 16, 1297 (1978).

10. C. Mathis, L. Christmann-Lamande, B. François, Makromol. Chem., 176, 931 (1975).

11. B. De Groof, M. Van Beylen, M. Szwarc, Macromolecules, 8, 396 (1975).

12. A. Soum, M. Fontanille, J. Polym. Sci. (Pol. Chem. Ed.), 15, 659 (1977).

13. B. François, C. Mathis, IUPAC Macro Dublin (proceedings) I-4 (1977).

14. T. E. Hogen-Esch, J. Smid, J. Am. Chem. Soc., 88, 307 (1966) and 88, 318 (1966).

15. B. De Groof, W. Mortier, M. Van Beylen, M. Szwarc, Macromolecules, 10, 598 (1977).

16. C. A. Kraus and Bray, J. Am. Chem. Soc., 35, 1315 (1913).

17. Ch. B. Wooster, J. Am. Chem. Soc., 59, 377 (1937).

18. P. Huyskens, Ind. Chim. Belg., 37, 553 (1972).

19. C. Carvajal, K. J. Tölle, J. Smid, M. Szwarc, J. Am. Chem. Soc., 87, 5548 (1965).

20. D. Nicholls, C. Sutphen, M. Szwarc, J. Phys. Chem., 72, 1021 (1968).

21. T. E. Hogen-Esch, J. Smid, J. Phys. Chem., 79, 233 (1975).

22. L. L. Böhm, G. Löhr, G. V. Schulz, Ber. Bunsenges, 78, 1064 (1974).

23. The use of $k_- = 122,000$ $M^{-1}sec^{-1}$ gives also a better agreement between the kinetic and conductance results of the BaS_2 salt in THF. Taking, as in ref. 14, $K_1=1.1$ $10^{-10}M$ as the most reliable value and $K_2^{1/2}=3.4\times10^{-3}$ combined with the value $122,000M^{-1}sec^{-1}$ for k_- yields a value $k_{obsd}=3.9\times10^{-3}$ sec^{-1} in good agreement with the one mentioned in ref. 11 viz. $k_{obsd}=3.8\times10^{-3}$ sec^{-1}, which we feel is more realistic than the one accepted in ref. 15.

Using for the estimation of $\lambda^-_{o_{triple}}$ in the case of BaS_2 the relation $\lambda^-_{o_{triple}} = \dfrac{\lambda^-_o}{1.26}$ (18), as was also done in this paper, gives $\lambda_{o_{triple}} = 13.9 + 17.5 = 31.4$ and $K_2^{1/2}=3.8$ x 10^{-3} which in combination with the above accepted values of K_1 and k_- result in a value of $k_{obsd} = 3.5\times10^{-3}$ sec^{-1}, again in satisfactory agreement with the value reported in ref.11.

24. Using for SrS_2 in THP the full equation for $k_{obsd}/C(=k_{app})$ give in ref. 11 and taking $K_1/C \ll K_2$ one finds:

$$k_{app} = k_o + k_1K_2^{1/2} + k_{trp}\, K_2^{1/2} + k_- \cdot \frac{K_1}{K_2^{1/2}} \cdot \frac{1}{C}$$

The contribution of SrS_2 and SrS^+ being negligible,

$k_- \cdot \dfrac{K_1}{K_2^{1/2}} = 7.4.10^{-5}$ sec^{-1} and having in mind that at the highest measured concentration ($C=7.2.10^{-4}M$) no contribution to the propagation from species other than free ions, could be detected, an upper limit for the triple ion contribution could be determined

$$k_{trp} \cdot K^{1/2} < \frac{k_- \cdot K_1}{K_2^{1/2}} \cdot \frac{1}{7.2\times10^{-4}M}$$

or $k_{trp} < 100$ M^{-1} sec^{-1}

The value of k_{trp} in THF is probably not different.

RECEIVED July 22, 1981.

Dynamics of Ionic Processes in Low Polar Media

ANDRE PERSOONS and JACQUES EVERAERT

Department of Chemistry, University of Leuven, Leuven, Belgium

A thorough discussion is given of the field modulation
technique, a new stationary relaxation method based on electric
field perturbation of ionic equilibria. Concomitantly the
theory of electric field effect in ionic systems is reviewed
especially stressing their importance for conductance phenomena
in low polar solutions.
Some examples of conductance relaxation measurements in a
solution of tetraalkylammonium-salts are given. The results
confirm convincingly the applicability of the sphere-in-continuum
model as a basic model for ionic interactions: a complete
treatment of ionic processes can be given from the diffusion of
ions in a continuous medium.
Conductance relaxation is also shown to be critically
dependent upon aggregation equilibria affecting non-conducting
(ion-pairs) as well as ionic species. The relaxation behavior in
the presence of quadrupoles (ion-pair dimers) and triple ions is
thoroughly analyzed. The experimental results show the potential
of the field modulation techniques as a method for the
investigation of ionization processes, independent of conductance
measurements.

The thorough understanding of elementary ionic interactions
is a necessary prerequisite for a successful mechanistic
description of any chemical process involvling ionic species.
Conductance techniques are pre-eminently used to investigate
ionization phenomena. The data obtained yield, within the
framework of ionic solution theory, detailed information on the
distribution between conducting and non-conducting entities the
exact nature of which cannot always unequivocally assessed.
Although this distribution between "free" charges and ioniphores"

0097–6156/81/0166–0153$05.75/0

is a primary constraint the central problem in any analysis of
ionization mechanisms is the kinetic study of the interconversion
processes between the different species; for such a kinetic
investigation to be complete all the elementary processes should
be analyzed for their energetic and dynamic properties. Since
the elementary steps in ionic association-dissociation processes
are usually very fast - to the limit of diffusion- controlled
reactions-their kinetic investigation became only feasible with
the advent of fast reaction techniques, mainly chemical
relaxation spectrometric techniques.

In chemical relaxation spectrometry the information on the
kinetic parameters of fast processes is obtained from an analysis
of the response of a chemical system upon a fast perturbation in
an intensive thermodynamic variable defining the equilibrium of
the system. To attain an acceptable accuracy the response of the
system upon the perturbation should be of sufficient amplitude, a
requirement which is generally met only for chemical systems
where products and reactants are evenly distributed over the
equilibrium state. This requirement explains the relative
abundance of chemical relaxation studies on ionic processes in
aqueous, or aqueous-like media as compared to such studies in low
polar solvents. In the former media conducting species are
usually present in appreciable amounts while in the latter
ionization equilibria are almost completely shifted towards
non-conducting species, which is due to the long range of
coulombic interactions at low polarity. However a multitude of
organic reactions are carried out in solvents of low polarity and
it is therefore of the utmost importance to have a detailed
insight in ionic processes in these media in order to assess
unequivocally the intervention and role of ionic species in
organic reaction mechanisms.

To investigate the dynamic properties of ionic species
present at very low concentration in apolar solvents the field
modulation method is particularly suited (1-5). In this
relaxation technique, developed some years ago in our laboratory,
an ionic equilibrium is perturbed with an electric field and the
response of the system measured from the conductance properties
of the sample solution. Since both perturbation and detection
are carried out in the frequency domain this stationary
relaxation technique has a very good sensitivity and accuracy
which is due to the averaging inherent in repetitive
measurements.

In this paper we review the field modulation method
especially to indicate the applicability for the investigation of
organic reaction mechanisms where ionic species may intervene.
The potential of this relatively simple technique will be shown
from results obtained the the investigation of ionization
processes in low polar media. We will give first an introductory
summary of some aspects of electric field effects in ionic
equilibria since a knowledge of this topic is necessary

prerequisite for a proper understanding of the field modulation

Electric Field Effects in Ionic Equilibria

The basic assumption in conductance measurements is the independence of the sample resistance on electric field strength. However a deviation from the linear relation between current density and field strength will be observed if any field effect on the mobility and/or the number of free charge-carriers is present.

At high field strengths a conductance increase is observed both in solution of strong and weak electrolytes. The phenomena were discovered by M. Wien ($\underline{6}$-$\underline{8}$) and are known as the first and the second Wien effect, respectively. The first Wien effect is completely explained as an increase in ionic mobility which is a consequency of the inability of the fast moving ions to build up an ionic atmosphers ($\underline{8}$). This mobility increase may also be observed in solution of weak electrolytes but since the second Wien effect is a much more pronounced effect we must invoke another explanation, i.e. an increase in free charge-carriers. The second Wien effect is therefore a shift in ionic equilibrium towards free ions upon the application of an electric field and is therefore also known as the Field Dissociation Effect (FDE). Only the smallness of the field dissociation effect safeguards the use of conductance techniques for the study of ionization equilibria.

The energy dissipation of a system containing free charges subjected to electric fields is well known but this indicates a non-equilibrium situation and as a result a thermodyanmic description of the FDE is impossible. Within the framework of interionic attraction theory Onsager was able to derive the effect of an electric field on the ionic dissociation from the transport properties of the ions in the combined coulomb and external fields ($\underline{9}$). It is not improper to mention here the notorious mathematical difficulty of Onsager's paper on the second Wien effect.

A conceptual difficulty in the theoretical description of ionization equilibria is the distinction between "free" and "bound" ions (molecules or ionophores). Somewhat arbitrarily Onsager adopted Bjerrum's convention writing the distribution between free and bound ions as an association-dissociation equilibrium:

$$A^+b^- \; \underset{k_r}{\overset{k_d}{\rightleftharpoons}} \; A^+ \; + \; B^- \qquad K_d = k_d/k_r \qquad [1]$$

Where A^+B^- is the state of two bound ions, i.e. two oppositely charged ions at distances from each other where the electrostatic interaction of the two ions exceeds their thermal interaction with the surrounding medium. This picture introduces the Bjerrum

ion-pair concept together with the Bjerrum distance, i.e. the
critical distance at which the mutual electrostatic interaction
equals the thermal energy. The prevailing confusion about these
concepts is mainly due to the strict distinction between free and
not-free ions while conductance discriminates between species
contributing and not-contributing to conductance. An important
question is therefore if two oppositely charged ions at the
Bjerrum-distance, i.e. "bound" to each other, can still
contribute to the conductance. Since in low polar media the
Bjerrum-distance is rather large (125 A) in benzene at room
temperature) it is very plausible that such configurations
contribute to the conductance e.g. by an apparent increase in
mobility of a "free" ion passing by. Nevertheless it should be
noted that, especially in low polar solvents, the overwhelming
contribution ot he ion-pair state is from configurations with the
two ions very close to contact.

Applying the laws of Brownian motion to the distribution of
free ions and ion-pairs in the presence of a external electric
field Onsager calculated the rate constants of recombination and
dissociation for equilibrium [1] as:

$$k_r = 8\Pi qkT \ (\omega_+ + w_-) \qquad\qquad\qquad [2]$$

and

$$k_d = 8\Pi qkT \ (\omega_+ + \omega_-) \ K_d(E=0) \ F(2\beta q) \qquad\qquad [2b]$$

Here q is the Bjerrum-distance ($= -e_+e_-/2DkT$), k the Boltzmann
constance, ω the mechanical mobility of an ion $K_d(E=0)$ the
equilibrium constant in the absence of electric field and D and
dielectric constant of the medium. From these expressions we
see that the shift in dissociation constant upon the application
of an electric field is given by:

$$K_d(E)/K_d(E=0) = F(2\beta q) \qquad\qquad\qquad [3]$$

$F(2\beta q)$ is somewhat complex function which can however be written
in a clarifying series expansion:

$$E(2\beta q) = 1 + 2\beta q + 1/3(2\beta q)^2 + 1/18(2\beta q)^3 + \ldots \quad [4]$$

The argument $2\beta q$ is the ratio of two distances, the Bjerrum-
distance q and the distance $(2\beta)^{-1}$ given by:

$$2\beta = \frac{|e_+\omega_+ - e_-\omega_-|}{kT(\omega_+ + \omega_-)} \ |E| \ \overset{1:1}{=} \ \frac{e_0}{kT} \ |E| \qquad [5]$$

The dependence on the absolute value of the field strength is a
noteworthy feature of Onsager's theory; of course symmetry
considerations impose always a dependence of the FDE on an even

function of field strength. From the specification for a 1:1 electrolyte it is seen that $(2\beta)^{-1}$ is the distance at which two oppositely charged ions form a dipole, "a Bjerrum dipole" with an electrostatic interaction energy kT when oriented parallel to the field.

For a univalent electrolyte and retaining only the linear term in [4] the dependence of the dissociation constant upon electric field strength can formally be written as a Van 't Hoff type equation:

$$\frac{d\ln K_d}{d|E|} = \frac{e_o^3}{2Dk^2T^2} \qquad [6]$$

Eq. [6] clearly shows that the FDE is most pronounced in low polar media at low temperatures. A numerical calculation shows that the increase in dissociation constant is a few percent with relatively low field strengths in those media, e.g. 5% in benzene at room temperature, about 2% in tetrahydrofuran $-100C$ with a field of 1 kV.cm^{-1}.

As the equilibrium constant is mostly derived from conductance data the shift in ionization with field can also be determined from the measurement of the conductance increase upon application of a high electric field. Conductance increase and field dissociation are generally related as:

$$\frac{1 - \alpha}{2 - \alpha} \frac{\Delta K_d}{K_d} = \frac{\Delta G}{G} \qquad [7]$$

Eq. [7] shows that the relative change in conductance, $\Delta G/G$, is maximal at low degree of ionic dissociation α, a condition

The Field Modulation Technique

From the foregoing discussion of electric field effects in ionic equilibria it is clear that a solution of a weak electrolyte shows a non-linear behavior in conductance (or resistance) at high field strengths. With an interdisciplinary look at the field of electronics we note that such nonlinearities are at the heart of all modern electronic circuits and devices. We therefore can use a solution of a weak electrolyte subjects to high electric fields as an electronic device, which is the basic idea of the Field Modulation Tecnnique, the general principles we will discuss now.

If a sample solution of a weak electrolyte is subjected to an alternating electric filed of high amplitude the sample-conductance G becomes a periodic function of time which formally can be written as:

$$G = G + \Delta G \sin \omega_p t$$

Due to the dependence on the absolute value of the field strength (Eq. 6) a solution of a weak electrolyte has also rectifying properties for high alternating fields and ω_p is therefore <u>not</u> the frequency of the high alternating field:

$$\text{Since } |\sin \omega t| = 2/\Pi + 4/\Pi \cdot \Sigma \; \frac{\cos 2n \, \omega t}{1 - 4 \, n^2}$$

ω_p will be twice this frequency and the higher harmonics will also be present.

If this conductance is measured with an alternating, low amplitude signal V_m, of frequency W_M (from the same or another pair of electrodes in the sample cell) we see in the current frequency components which are <u>not</u> present in the signal sources:

$$I = G \cdot V_m = (\overline{G} + \Delta G \sin \omega_p t) \; V_M \sin \omega_M t \qquad [8a]$$

or

$$I = GV_M \sin \omega t + \frac{1}{2} V_M \Delta G \cos(\omega_p - \omega_M)t \; - \; \frac{1}{2} V_M \Delta G \cos(\omega_p + \omega_M)t \qquad [8b]$$

These new frequency components arise from a mixing between the high-field perturbation (ω_p) and the measuring signal (ω_M). We hence conclude that an electrolytic resistance can be used as a signal mixer if one of the signals has high amplitude. From eq. 8b it is also seen that the amplitude of the mixed signals is a unique function of the conductance increase due to the high electric field and therfore a direct measure of the amplitude of the FDE. This is the basic principle of the field modulation technique. So far we have considered the conductance increase as instantaneously upon the application of the high field. Since the conductance increase is due to an increase in number of ionic species generated in ionization processes it is evident that the conductance increase should have a finite rate determined by the rate of ionic equilibration. This introduces a phase lag between the high alternating field and the sample conductance increase at high frequencies and also a smaller amplitude of the conductance increase. The conductance increase shows therefore a dispersion with frequency characterized by a time constant. In the subsequent discussion we will see how this time constant is directly related to the relaxation time(s) of ionization equalibria present in solution.

Before proceeding to a complete calculation relating the conductance increase to the amplitude of the field dissociation effect and its temporal behavior we will give a short description of the actual circuit used for the field modulation technique (Figure 1) with reference to the previous discussion.

A sample cell containing three electrodes, preferably in a symmetric arrangement is used as a mixer for two signals derived from two signal generators. One signal is a square wave of relatively high amplitude which is further increased by the use of a step-up transformer. The other signal is directly applied through an isolation transformer to the electrodes of the sample cell. The application of the square wave field results in a modulation, hence the name field modulation, of the sample conductance at the same frequency as the square wave signal. Since the measuring signal is locked in phase and frequency on the square wave signal the frequencies ω_p and ω_M (eq. 8b) are equal resulting in a d.c. component in the mixed signal current. This d.c. current is used to charge a capacitor from which the amplitude can be measured as a d.c. voltage over the capacitor. This way of measuring the d.c. current has the advantage that, after completely charging the capacitor, no d.c. current flows in the sample avoiding thus polarization effects. Moreover, all signals in the circuit being a.c. signals, a d.c. signal can be measured with an extreme accuracy.

For obvious experimental reasons the square wave signal is not a pure square wave but is an envelope of a high frequency signal which is square wave modulated. Retaining only this square-wave envelope of the high-voltage, high frequency perturbing field, a complete mathematical analysis of the elctronic circuit gives for the d.c. voltage measured over the capicitor $c_M(2)$:

$$V_{d.c.} = \frac{V_M}{\pi} \frac{\Delta G}{G} \frac{1}{1 + \omega_{PRF}^2 \tau^2} = \frac{V_{d.c.\cdot max}}{1 + \omega_{PRF}^2 \tau^2} \qquad [9]$$

Where ω_{PRF} is the frequency of the square-wave-envelope or, equivalently, the frequency of the measuring signal (ω_M) $V_{d.c.}$max is the d.c. signal measured at low frequencies where the time-dependent behavior of the conductance increase is completely in phase with the high field signal. τ is the time constant for the conductance increase upon the application of the high field and it intimately related to the chemical relaxation of the ionization processes present in the sample solution. Eq. 9 is derived assuming a small perturbation, a necessary assumption to solve the nonlinear differential equations describing the circuit, and is recognized as a classical dispersion equation.

From the general discussion of the use of nonlinear properties to investigate dynamic phenomena it is clear that the field modulation techniques is but one example of a broader class of methods for the study of fast processes. A drawback, particular to electric field modulation, is the prohibitive heat dissipation in conducting systems. However, any forcing parameter imposing a conductance modulation could be used in principle as, for example, in the study of the dynamics of photoconductive phenomena.

Results and Discussion

The field modulation technique has been successfully applied
to a broad variety of systems showing a field dependent
conductance. Since a large amount of conductance data is
available for low polar solutions of tetraalkylammonium salts
these systems were studied most thoroughly. In the subsequent
discussion we will rely mainly on the results obtained for these
systems but the general conclusions apply directly to other
systems, e.g. fluorenyl-salts in ethereal solvents, metal
complexes in benzene, etc.

Conductance data clearly show that low polarity solutions of
tetraalkylammonium salts behave rather simply at low salt
concentration and the main properties can be completely described
by the simple equilibrium [1] between an ion-pari and the free
ions. For such an equilibrium the chemical relaxation time can
be writted as:

$$\tau^{-1}_{chem} = k_d + k_r(\overline{C_A} + \overline{C_B}) = (\text{electroneutrality}) = k_d + 2k_r\overline{C_A} \qquad [10a]$$

and for low degrees of dissociation, the usual situation in low
polarity solutions, this may be approximated by (with C_o the
total salt concentration):

$$\tau^{-1} = 2(k_d k_r)^{1/2} C_o^{1/2} = 2K^{1/2}_d k_r C_o^{1/2} \qquad [10b]$$

The relaxation time measured from the conductance modulation is
of course the same as this chemical relaxation time since the
ion-pair dissociation equilibrium is the only charge generating
process present in solution. These conclusions are completely
borne out by the experimental results: whenever the conductance
data indicate only the presence of the simple equilibrium [1] the
reciprocal relaxation time measured for the conductance
modulation is a linear function of the square root of the total
concentration as can be seen from Figure 2. For these cases it
becomes possible, from a knowledge of the dissociation constant
K_d and relaxation time, to determine through eq. 10b the rate
constants for ion-pair dissociation and for ionic recombination.
These data, some of which are presented in Table I, are extremely
valuable to investigate the applicability and validity of the
sphere-in-continuum model, which is basic to all electrolyte
solution theories. In this conceptual model ions are viewed as
(hard) spheres subjected to Brownian motion in a continuous
medium. An ion-pair is defined as two oppositely charged ions
within a given distance $\underline{a}$ which is the most critical parameter of

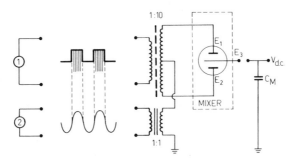

Figure 1.   Schematic circuit for the field modulation technique.

Generator 1 is a square wave (0.01–30 kHz) modulated high frequency (0.1–1 MHz) high voltage (0.1–ak V_{pp}) signal generator driving a step-up (1:10) ferrite core transformer. Generator 2 is a low-voltage (1–20 V_{pp}) sinusoidal oscillator synchronously locked on the square wave signal modulating the power oscillator 1. The sample cell contains, in a very symmetric arrangement, three electrodes, two of which (E_1 and E_2) are connected to the high-voltage transformer output terminals. Electrode E_3 is, over capacitor C_M (10 pF–22 nF), grounded for a.c. signals. The spacing between the Hv electrodes and the center electrode E_3 is of the order of 0.2–1 mm, depending on the particular cell design. The center tap of the high-voltage transformer is connected over a 1:1 isolation transformer to oscillator 2.

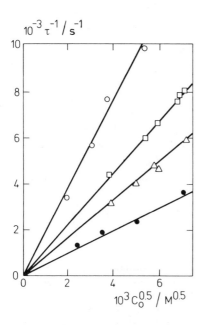

Figure 2.   Reciprocal conductance relaxation times as a function of the square root of TBAP concentration in media and concentration range where conductance indicates a preponderance of the simple ions. TBAP in diphenyl ether (D = 3,55) at 318 K (○). TBAP in benzene–chlorobenzene (60:40 v/v; D = 3, 55) at 298 K (△). TBAP in benzene–chlorobenzene (60:40 v/v; D = 3, 64) at 298 K and 350 atm (2). TBAP in benzene–chlorobenzene (60:40 v/v; D = 3, 36) at 323 K (0).

TABLE I

Rate Parameters for the Dissociation of
Tetrabutylammonium Salts (TBA-salts) in Low Polar Media

			Experimental[*]		Calculated	
D	$K_d/10^{-12}M$	$a/\overset{o}{A}$[**]	$k_r/10^{11}M^{-1}s^{-1}$	k_d/s^{-1}	$k_r/10^{11}M^{-1}s^{-1}$	k_d/s^{-1}

1. TBA-picrate in benzene-chlorobenzene mixtures at 298 K (15,16)

3.22	0.363	5.92	3.01	0.109	3.01	0.098
3.55	3.83	5.84	2.61	1.00	2.70	0.97
3.87	24.4	5.78	2.64	6.45	2.44	6.56

2. TBA-bromide in benzene-nitrobenzene mixtures at 298 K (17)

3.54	2.44	5.76	3.23	0.79	2.95	0.70
3.85	10.3	5.58	4.92	5.07	2.71	2.8

3. TBA-picrate in diphenylether at 318 K (3)

3.55	18.1	5.82	0.82	1.48	0.75	1.37

[*] for all experimental data the experimental error is
usually less than 4% and nowhere exceeds 7%

[**] calculated from corrected values for K_d obtained from
lnK_d vs. $1/D$ linear plots in case 1, otherwise directly
from K_d.

the model. This distance should be considered as the boundary were free ions cease to exist but is often taken as the distance of closest approach, i.e. the sum of the ionic radii. From this model the recombination and dissociation rate constants may be calculated as ($\underline{10},\underline{11}$):

$$k_r = \frac{4 \pi N_o e_o^2}{10^3 DkT} (D_A + D_B) / [1 - \exp(-e^2/DkT\underline{a}_o)] \qquad [11a]$$

and

$$k_d = \frac{3 e_o^2}{DkT\underline{a}^3} (D_A + D_B) / [\exp(e^2/DkT\underline{a}) - 1] \qquad [12a]$$

Where D_i is the diffusion coefficient of species i in the continuous medium with dielectric constant D and viscosity η. With reasonably good assumption of spherical ions of equal radii, and using the Stokes-Einstein relation for the diffusion coefficients these relations reduce to (since in low polar media $e^2 DkT\underline{a}_o < 1$):

$$K_r = \frac{8 \times 10^{-3}}{3} N_o E_o^2 \frac{1}{Dn\underline{a}} \qquad [11b]$$

and

$$k_d = \frac{2 e_o^2}{\pi} \frac{1}{Dn\underline{a}^4} \exp - \frac{e_o^2}{DkT\underline{a}} \qquad [12b]$$

These equations allow the calculation of the rate constants, within the assumptions made, from a knowledge of the distance $\underline{a}$. Some of these rate constants calculated with an $\underline{a}$-value derived from conductance data ($\underline{12}$) are given in Table I; the agreement between experimental and calculated values is a good indication that the association processes of ions in low polar media can be described as simple diffusion-controlled processes. The dissociation process can also be described as the diffusion-controlled separation of two spheres bound in an electrostatic interaction; we will call this a diffusion-controlled dissociation ($\underline{11}$). However, in view of the underlying assumptions made in the derivation of eqns. 11 and 12 and the uncertainty of the meaning of the distance $\underline{a}$, it is necessary to corroborate this important conclusion from other experimental data. Whatever the exact form of eqn. 11 and 12 and exact meaning of $\underline{a}$ it is clear that the temperature-dependence of diffusion-controlled recombination and dissociation (apart from the electrostatic work of charge-separation) shoud be mainly governed by the temperature-dependence of the viscosity. This is shown in Table II where it is seen that $\Delta^{\neq}_r = \Delta^{\neq}_d - \Delta H_d \approx \Delta H$ viscosity.

The mechanistic importance of the study of the temperature dependence of the rate constants is very clearly indicated in the study of the dynamics of fluorenyllithium in diethyl ether (5,13). Although the values of the rate constants fitted rather well with the values expected for a diffusional assocation-dissociation process at one temperature, the dissociation rate constant decreased with increasing temperature. This is a definite proof for the intervention of a solvent-separated ion-pair in the dissociation pathway.

Since the design of the measuring cell for field modulation studies is not very critical it was relatively easy to study also the pressure-dependence of ion-pair dissociation and ionic recombination (14). Here also the values (Table II) for the activation volumes show the essentially diffusion controlled aspects of the association-dissociation phenomena, since the calculated values, essentially the pressure dependence of the viscosity, and the experimentally determined values agree rather well.

At increasing salt concentrations the conductance of all solutions investigated shows the emergence of other equilibria involving the ions and ion-pairs: triple ions, quadrupoles . . . become stable species. Althought one expects from chemical relaxation theory the appearance of more relaxation processes, experimentally only one relaxation process, whatever the concentration of salt, is observed. This odd behavior is intimately related to the nature of equilibrium perturbation and observation of the response in the field modulation method. Indeed the application of a high electric field upon a weak electrolyte solution results not so much in a perturbation of the equlibrium between ion-pairs and free ions but rather of the equilibrium between a non-conducting state and the conducting species. This ionization mode is slow compared to other processes in each state since this mode involves charge-separation while the aggregation modes(s) only involves ion-dipole and/or dipole-dipole interactions. Such a pathway for ion-formation in low polar media is convincingly indicated from measurements of the FDE of tetraalkylammonium salts in benzene solution; the FDE is completely described by Onsager's theory although conductance measurements clearly show the presence of a variety of ionic and aggregated species (18).

Taking the presence of triple ions and quadrupoles into account the reaction scheme for ionization becomes very complex compared to the simple ion-pair dissocation [1]:

TABLE II

Activation Parameters for the Ionization of Tetrabutylammonium
Picrate (TBAP) in Benzene-Chlorobenzene Mixtures (15,16)

D_{298K}	ΔH_d	$\Delta_r^{\neq}$ ^{o}	$\Delta_d^{\neq}/kj \cdot mol^{-1}$	ΔV_d	$\Delta_r^{\neq}$	$\Delta_d^{\neq}/cm^3 mol^{-1}$
1. Experimental						
3.22	31.6	7.2±2.0	38.8	−51.8	19.6	−32.3
3.55	27.4	6.8±1.4	34.4	−48.5	18	−30.8
3.87	27.4	4.5±2.6	32.7	50.8	20.5	−30.6
2. Calculated $^{x,+}$						
	32.9	8.9	40.9	−53.0	18.91	−33.9
	29.8	8.7	37.7	−49.6	18.4	−32.0
	26.1	8.6	33.9	−48.6	18.0	−30.4

^{x} Physico-chemical parameters of the mixtures

D_{298K}	η/mP	$-(dD/dT)/K$	$(dD/dP)/atm^{-1}$	$(dln\eta/dT^{-1})/K$	$(dln\eta/dP/atm^{-1}$
3.22	5.45	5.07×10^{-3}	23.7×10^{-5}	1.19×10^3	7.37×10^{-4}
3.55	6.61	6.63×10^{-3}	27.1×10^{-5}	1.17×10^3	7.17×10^{-4}
3.87	6.78	8.24×10^{-3}	30.6×10^{-5}	1.15×10^3	6.98×10^{-4}

^{o} The errors on these small values are relatively large while
the errors for all other experimental values are typically
3-4%.

$^+$ for an exact agreement the temperature dependence of $\underline{a}$
should be accounted for; $dln\underline{a}/dT$ is [−0.34, −0.04 and 0.16]
x 10^{-3} for the three mixtures studied. $dln\underline{a}d/dP$ is always
found to be negligible.

$$
\begin{array}{ccccccccccc}
& & & \overset{+}{K_3} & & & & & \overset{+}{K_3} & & \\
B & + & A & + & 2AB \rightleftharpoons & B & + & ABA & + & AB \rightleftharpoons & ABA & + & BAB
\end{array}
$$

$$
\begin{array}{c}
k_d \quad k_r \qquad K_s \quad k_R \qquad \overset{+}{k_D} \quad k_T \qquad k_F \\
3AB \rightleftharpoons \qquad (AB)_2 + AB \\
k_r \quad k_d \qquad k_R^- \qquad k_D^- \quad k_T \qquad k_F
\end{array}
\qquad [13]
$$

$$
\begin{array}{ccccccccccc}
A & + & B & + & 2AB \rightleftharpoons & A & + & BAB & + & AB \rightleftharpoons & BAB & + & ABA \\
& & & & \overline{K_3} & & & & & \overset{+}{K_3} & &
\end{array}
$$

(A is the cation A^+ and B the anion B^- as in eq. 1; charges are
omitted here for clarity). Such a complex scheme is however
equivalent, from the viewpoint of ionization, with a simple
equilibrium between a non-conducting state (ion-pairs and
quadrupoles) and a conducting state (ions and triple ions). With
the simple assumption of fast equilibria within each state (the
"horizontal" equilibria in scheme 13) and noting that usually the
concentration of all species is negligible compared to the
ion-pair concentration C_0 the reciprocal relaxation time
associated with the slow ionization mode can be calculated as:

$$
\tau^{-1} = K_d + (k_D^- + k_D^+)4C_0/K_s + 9/16 \cdot k_F K_s (4C_0/K_s)^2 \; \frac{1}{1 + 4C_0/K_s} +
$$

$$
2 \; \frac{K_d^{1/2} C_0^{1/2} (1+C_0/\overset{+}{K_3})^{1/2} (1+C_0/\overline{K_3})^{1/2}}{1+2C_0/\overset{+}{K_3}+2C_0/\overline{K_3}+3C_0^2/\overline{K_3}\overset{+}{K_3}} \left[k_r + 2\left(\frac{\overset{+}{k_R}}{\overset{+}{K_3}} + \frac{\overline{k_R}}{\overline{K_3}} \right) C_0 + 3 \; \frac{k_T}{\overline{K_3}\overset{+}{K_3}} C_0^2 \right] [14]
$$

The derivation of this equation from general relaxation theory,
within the assumptions stated, is straightforward but tedious and
will be published elsewhere. At first sight equation 14 is so
involved that it may be everything but useful for any practical
purpose. Fortunately, depending on the (relative) stability of
the different aggregated species and on the total concentration,
different limiting formes of eq. 14 are very illuminating.
First, and gratifying, at low concentration and/or low stability
of the aggregated species eq. 14 reduces to:

$$\tau^{-1} = k_d + 2K_r K_d^{1/2} c_o^{1/2} \qquad [10]$$

which is indeed the expression obtained for the simple ion-pair dissociation. Moreover, as long as the concentration of quadrupoles is negligible compared to the ion-pairs the first term in eq. 14, which describes the dissociation, reduces always to k_d or depending on the relative magnitude of k_d, $K_D^{\pm}$ and k_F to a concentration-dependent erm (which is always small compared to the recombination term).

Focusing on the recombination term in eq. 14 we see that this term reduces to:

$$2k_T \left[\frac{K_d}{K_3^- K_3^+} \right]^{1/2} c_o^{3/2} \qquad [14a]$$

at high ion-pair concentration and/or high stability of the triple ions since then $C_o/K_3^{\pm} < 1$ and $k_r \approx k_R^{\pm} \approx k_T$ assumed to be diffusion-controlled recombination rate constants. This is intuitively clear since at these conditions the main ionic recombination process will be between the anionic and cationic triple ions, the concentration of which depends on the 3/2 power of the total ion-pair concentration. Reasoning along the same lines we note that for the case of unilateral triple ion formation (e.g. $K_3^+ \dashrightarrow \infty$) at high ion-pair concentration, or for the condition $K_3^+ > C_o > K_3^-$, the recombination term would be given as:

$$2k_R^- \left[\frac{K_d}{K_3^-} \right]^{1/2} c_o^1 \qquad [14b]$$

A linear concentration dependence of the reciprocal relaxation time upon total ion-pair concentration would therefore point to a main recombination process between a simple ion and a triple ion.

Considering these different limiting forms of the recombination term an important tentative conclusion emerges: the concentration dependence of the reciprocal relaxation time is a direct measure of the main ionic recombination process and yields therefore information on the ionic species present in solution. A linear dependence on total ion-pair concentration would therefore indicate unilateral triple ion formation or, if both kinds of triple ions are present as indicated by conductance, a sufficient difference in their stability. At this point it should be noted that the usual method of Fuoss and Draus (19) to determine stability constants for triple ions assumes an equal stability. In many investigations this assumption is often

implicitly accepted even with contra-indications such as a
difference in size of the ions.

Experimentally all the limiting cases of the recombination
term are encountered in all systems investigated up to now. In
solutions of tetraalkylammonium salts in benzene the reciprocal
relaxation time is even dependent on the square of total
concentration (but here the quadrupoles may be comparable in
concentration with ion-pairs) which would indicated a
preponderance of quintuple and triple ions in the ionic
recombination.

However a very puzzling reature of the experimental results,
as can be seen directly in Figure 3, is the relatively important
intercept which is always measured for systems where the
reciprocal relaxation time has a higher than square-root
dependence on concentration. This is in contradiction with the
dissociation term of the general eq. 14 which always gives a
vanishingly small monomolecular constant for systems where the
aggregation into quadrupoles is small. K_s is of the order
$1-10^{-4}$M for tetrabutylammonium salts in low polar solutions, the
smaller values applying to benzene-solution (20) and somewhat
depending on size and form of the ions; the extent of the
dimerization of ion-pairs with picrate is usually small as
indicated from conductance. The relaxation behavior indicates
a situation as if each aggregated form dissociates into the
corresponding ions independently of the other aggrgated species
present, indeed if this would be the case the intercept (which
has the dimensions of a reciprocal time) could be explained as
the sum of the dissociaiton rate constants of all monomolecular
dissociation processes present. This would be so if the
different aggregated species were not in equilibrium or, which is
the same, would equilibrate more slowly than the ionization
processes. However this assumption is untenable as confirmed by
some measurements on the relaxation of ion-pair-quadrupole
equilibria (21). It is therefore not a trivial problem to define
which relatively fast monomolecular process present in solution
gives rise to the intercept. A satisfactory explanation could
only be given from a detailed numerical calculation of the extent
of perturbation of the ionization mode upon application of the
electric field. The concentration of ionic species, which is
directly proportional to the conductance, is very small in these
low polar media and since the relative change in conductance is
of the order of 1% with field the absolute change in ionic
concentration is exceedingly small. As a result the change in
concentration of the aggregated species vanishes almost
completely. Otherwise stated, the aggregation equilibria are not
perturbed by the application of the electric field which
formally, if we consider the aggregation mode(s), uncouples these
equlibria! This means that the dissociation term in eq. 14
should be replaced by the sum of all monomolecular rate constants
pertaining to the dissociation of all aggregates present in

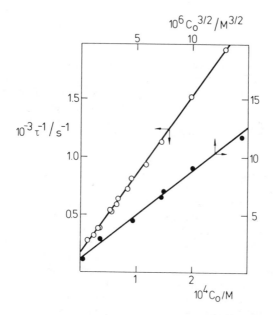

Figure 3.   Dependence of the reciprocal conductance relaxation time on TBA–salt concentration in media and concentration range where conductance indicates an important fraction of triple ions. TBA–bromide in benzene–nitrobenzene (3, 22 vol %; D = 2, 90) at 298 K (○). TBA–picrate in benzene–chlorobenzene (16 vol %; D = 2, 78) at 298 K (0).

solution, i.e. the interacept should be given by $k_d + k_D^+ + k_D^-$ if only ion-pairs and quadrupoles are present in solution.

The foregoing discussion allows an analysis of the relaxation behavior observed in solutions of tetrabutylammonium bromide in a benzene-nitrobenzene mixture (3.22 vol% nitrobenzene, D = 2.9) (17). As shown in Figure 3 the reciprocal relaxation time is proportional to the 3/2 power of the concentration of TBAB ion-pairs over the concentration range $(0.5 - 7) \times 10^{-4}$M/ In view of the foregoing discussion this would imply that the ionic recombination process is between the two triple ions which therefore would be the main ionic species present in solution. The intercept can be identified as $k_D^- + k_D^+ (=2k_{dau})$. With help of eq. 12 we can estimate a mean value of the interionic distance between triple and simple ion $(\underline{a}_{st})$ from the mean dissociation rate constant and, knowing the sum of the hydrodynamic radii of the simple ions (Table I), we estimate the sum of the hydrodynamic radii of the triple ions $(\underline{a}_{tt})$ as 15 A. With this value the recombination rage constant k_T can be calculated. These calculations are summarized in Table III. In this way we obtain, with the ion-pair dissociation constant determined from conductance data, a value of 5.3×10^{-5} for the average stability constant of the triple ions (K_3 from eqn. 13). This should be compared with 1.24×10^{-4}M as derived from conductance indicates an important fraction of simple ions over the largest part of the concentration range studied.

As a second example we turn to the analysis of the conductance relaxation observed in tetrabutylammonium picrate in benzene-chlorobenzene over the concentration range $0.1-2 \times 10^{-4}$M as shown in Figure 3. A qualitative, but important difference with the previous example is the linear relation between reciprocal relaxation time and total TBAP-concentration. Applying the previous picture this would imply a noteworthy difference in stability between the two triple ions resulting in an ionic recombination process between a triple ion and simple ion. Such a difference in stability would be very plausible judging from the important difference in size and shape of the two ions which would make the triple anion more stable as a structure where the two picrate ions penetrate along "threefold axes" in two cavities formed by the alkyl-limbs. Applying eq. 14b the intercept measured is therefore k_D^- and the slope $2k_R^- (K_d/K_3^-)^{1/2}$. Following the same strategy as in the previous example we calculate for k_R^- a value which is very reasonable for a diffusion-controlled recombination rate constant and with K_d', again from conductance, a value of 2.07×10^{-5}M is obtained for K_3^-. This should be compared with the value of 1.44×10^{-4}M obtained from a Fuoss-Kraus plot (19) of the conductance data. However, apart from the uncertainly in the limiting equivalent conductance of simple and triple ions, we have here a very important discrepancy between conductance and relaxation data: conductance indicates the existence of the two triple ions in

TABLE III

Relaxation Behavior of the Ionization Equilibria of
Tetrabutylammonium (TBA) Salts

1. TBA-bromide in benzene-nitrobenzene (3.22 vol% D=2.9)

$$\tau^{-1} = (1.06 \pm 0.2) \; 10^3 + (7.68 \pm 0.13) \; 10^8 c_o^{3/2} \quad 298K$$

eq 14a:

K_d/M	K_{Dav}/s^{-1}	$\xrightarrow{eq.12}$	$a_{st}/\overset{o}{A}$	$\rightarrow$	$a_{tt}/\overset{o}{A}$	$\xrightarrow{eq.11}$	$k_T/M^{-1}s^{-1}$	$\rightarrow (K_3\overline{K_3})^{1/2}/M$
$2.2 \; 10^{-4}$	530		10.3	15			$1.37 \; 10^{11}$	$5.2 \; 10^{-5}$

2. TBA-picrate in benzene-chlorobenzene (16 vol% D=2.78

$$\tau^{-1} = (1.69 \pm 0.1)10^2 + (6.88 \pm 0.14)10^6 c_o^1 \quad 298K$$

eq 14b:

K_d/M	$K_{\overline{D}}/s^{-1}$	$\xrightarrow{eq.12}$	$a_{st}/\overset{o}{A}$	$\xrightarrow{eq.11}$	$k_R/M^{-1}s^{-1}$	K_3/M
$5.55 \; 10^{-15}$	169		10.1		$2.1 \; 10^{11}$	$2.07 \; 10^{-5}$

	ΔH_{slope}	$\Delta H_{int}/kJ \cdot mol^{-1}$	ΔV_{slope}	$\Delta V_{int}/cm^3 mol^{-1}$
experimental:	24.9	34.3	2.3	-16.1
calculated: (see text)	26.4	<41	-1.4	-14

comparable stability and relaxation data points to a
preponderance of one triple ion.

Since the uncertainty of the numerical values in K_3 makes
the test of eq. 4b somewhat hazardous, the temperature, and
pressure, dependence of the conductance relaxation was
investigated. Temperature , and pressure, dependence of
intercept and slope of the experimental reciprocal relaxation
time vs. concentration is given in Table III. If the reciprocal
relaxation time indeed is functionally described by eq. 14 b then
we are able to calculate these temperature, and pressure,
dependences from previously obtained experimental data.

The negative volume-change of the intercept (ΔV_{int}) is fully
compatible with the identification of the intercept as k_D^-.
Moreover, from the a_{st} value derived, with eq. 12, we calculate
$\Delta_D^{\neq}$ as -14 cm^3. mol^{-1}, in excellent accordance with the
experimental value. The experimental energy of activation for
the dissociation of the dimer is again comparable with values
found for the dissociation of an ion-pair. The temperature
dependence of the slope should be, if eq. 14b is applicable,
($\Delta_R^{\neq} + 1/2\Delta H_d - 1/2 \Delta H_3$).

Since the activation energy for ionic recombination is
mainly due to viscosity we use the activation energy for viscous
flow (10kJ.mol^{-1}). ΔH_d and ΔH_3 were determined from conductance
as 44.2kJ.mol^{-1} and 11.4kJ.mol^{-1}. From the data presented in
Table III it is clear that the temperature dependence of the
slope is very satisfactorily described by $\Delta H_R^{\neq} + 1/2(\Delta H_d - \Delta H_3)$.
Another, and rather critical, test for the applicability of eq.
14b is the effect of pressure since the slope of eq. 14b is
largely pressure independent so that we ask here for a
compensation of rather large effects. From Table III we indeed
see an excellent accordance between the experimental value and
the pressure-dependence calculated from the activation volume of
viscous flow (+20.3 cm^{3}mol^{-1}), ΔV_d (-57.3 cm^{3}mol^{-1}) and
ΔV_3(-13.9 cm^{3}mol^{-1}); the difference between the small
experimental and calculated values is entirely with the
uncertainties of compressibility - corrections and experimental
errors.

The temperature- and pressure-dependence of the conductance
relaxation in TBAP solutions in benzene-chlorobenzene (16 vol%)
corroborates importantly the description of the phenomena by eq.
14b implying an ionic recombination process between the triple
ion and a simple ion. Apparently this picture is still in
conflict with conductance data which who the presence of two
kinds of triple ions. This discrepancy remains as yet
unresolved.

Conclusion

The foregoing discussion and results clearly indicate the potential of the field modulation technique as an independent tool for the investigation of ionic processes in low polarity, or better, in low conducting media. Due to the specificity of equilibrium-perturbation and response-detection in this method, field modulation measurements isolate the relaxation mode of ionization allowing the study of only this mode and, eventually, slower modes coupled to the ionization equilibria. In media where the ionic concentration is small - a necessary condition for the method- the electric field will not perturb the aggregation equilibria, or equilibria not involving a charge separation.

Relaxation measurements in systems where only a simple ion-pair equilibrium is present allowed for a critical test for the simple and apparently crude sphere-in-continuum model which is the basic model for all our theoretical considerations of ionization processes. The good accordance between all experimental results and their values calculated from the sphere-in-continuum model indicate the wide applicability of this simple picture. The association-dissociation processes of ions and ion-pairs in the low polar media studied are therefore simple diffusion-controlled phenomena, a conclusion corroborated by the fact that the activation parameters can also be calculated from the sphere-in-continuum model. However a complete agreement for the energy-parameters can only be obtained if a temperature dependence of the hydrodynamic radii of the ions - and therefore of the interionic distance in the ion-pair - is accepted. This should be seen as a minor correction to the hard-sphere-in-continuum model and not, as is often done, as an indication of specific interactions. Indeed in the solvents studied it is rather more acceptable to have a small change in hydrodynamic radii with temperature since the important physical parameters of the medium - dielectric constant and viscosity - change also with temperature. The more complex relations between conductance relaxation time and ion-pair concentration found at lower polarity, or high concentration, are thoroughly analysed as pertaining to an overall-ionization process between a non-conducting state, in which several fast equilibria may be present, and another state in which all ionic species are rapidly interconverting. Contrary to the usual behavior in chemical relaxation the fast aggregation equilibria in the non-conducting state are virtually uncoupled from each other due to the vanishingly small perturbation of this state. From an analysis of the temperature- and pressure-dependence this picture describes consistently all results obtained in the relaxation experiments. However, some discrepancies with the data obtained for the ionic equilibria still exist. These discrepancies may

point to the necessity of a critical assessment of the methods
used in the treatment of conductance data.

Literature Cited

1. L. De Maeyer and A. Persoons in "Techniques of Chemistry"
    Vol VI, Part 2, p. 211. A. Weissberger and G. Hammes, Ed.,
    Interscience, New York (1973).

2. A. Persoons, J. Phys. Chem., 78, 1210 (1974).

3. F. Nauwelaers, L. Hellemans and A. Persoons, J. Phys. Chem.,
    80, 767 (1976).

4. A. Persoons and L. Hellemans, Biophys. J., 24, 119 (1978).

5. A. Persoons and M. Van Beylen, Pure Appl. Chem., 51, 887
    (1979).

6. M. Wien, Ann. PHys., 83, 327 (1927); ibid., 85, 795 (1928);
    Physik. Z., 28, 834 (1927) ibid., 29, 751 (1928).

7. M. Wien and J. Schiele, Physik. Z., 32, 545 (1931).

8. H. C. Eckstrom and C. Schmeltzer, Chem. Rev., 24, 367
    (1939).

9. L. Onsager, J. Chem. Phys., 2 599 (1934).

10. P. Debye, Trans. Electrochem. Soc., 82 265 (1942).

11. M. Eigen, Z. Phys. Chem. (Frankfurt am Main), 1, 176 (1954).
12. R. M. Fuoss, J. Am. Chem. Soc., 80, 5059 (1958).

13. H. Uytterhoeven, A. Persoons and M. Van Beylen, manuscript
    in perparation.

14. J. Everaert and A. Persoons in "Techniques and Applications
    of Fast Reactions in Solution" p. 367. W. J. Gettins and E.
    Wyn-Hones Ed., D. Reidel, Dordrecht (1979).

15. K. Clinckspoor, J. Everaert and A. Persoons in "Nonlinear
    Behavior of Molecules, Atoms and Ions in Electric, Magnetic
    or Electromagmetic Fields" p. 381. L. Neel Ed., Elsevier,
    Amsterdam (1979).

16. J. Everaert, Ph.D. thesis, University of Leuven (1979).

17. K. Clinckspoor, M. Sc. thesis, University of Leuven (1978).

18. T. Honda, J. Everaert and A. Persoons in ref. 15, p. 369.

19. R. M. Fuoss and C. A. Krauss, J. Am. Schem. Soc., $\underline{55}$, 2387 (1933).

20. K. Bauge and J. W. Smith, J. Chem. Soc., 4244 (1964).

21. R. Nackaerts, M. De Maeyer and L. Hellemans, J. Electrostatics, 7, 169 (1979).

RECEIVED May 12, 1981.

NMR Characterization of Ion Pair Structure

D. J. WORSFOLD

National Research Council of Canada, Ottawa, Canada K1A OR9

An account is given of the NMR methods which
may be applied to study the structure of the ion
pair at the active chain end in anionic polymerizations, and the results of some of these studies
when applied to styrene and diene polymerizations.
Studies by means of spectral patterns and coupling
patterns, the displacement of the NMR lines from
those in the parent hydrocarbon, carbon-carbon
coupling constants, and coalescence temperatures
are described. Some special effects can be
illustrated by the effect of selective deuteration.
The picture that emerges is that although in most
cases there is little change in hybridization or
delocalization of charge with change of counter
ion, there is marked change in the rigidity of
carbanion with different counter ions.

Ionic polymerizations are remarkable in the variety of
polymer steric structures that are produced by variation of the
solvent or the counter ion. The long lived nature of the active
chain ends in the anionic polymerization of diene and styrene type
monomers lends itself to studies of their structure and properties
which might have relevance to the structure of the polymer
produced when these chain ends add further monomer. One of the
tools that may be used in the characterization of these ion pairs
is the NMR spectrometer. However, it should always be appreciated
that the conditions in the NMR tube are frequently far removed
from those in the actual polymerization. Furthermore NMR observes
the equilibrium form on a long time scale, and this is not
necessarily that form present at the moment of polymerization.
Several NMR techniques may be used in these studies including
(a) spectral patterns (b) chemical shifts (c) coalescence temperature (d) coupling constants (e) special effects such as those
caused by deuteration. Here these techniques are applied to the

0097–6156/81/0166–0177$05.00/0
Published 1981 American Chemical Society

allyl carbanion, the parent of the diene polymerizing carbanion, and the results compared with those from models of the diene and styrene polymerizing systems.

Allylalkalimetal Compounds

(a) The ^{13}C spectral pattern of all the compounds, from Li to Cs, in THF solution consists of two lines, Table I.[1,2,3] This indicates that the two end carbons are equivalent, and the compound is either a symmetrical ion pair or a rapidly equilibrating mixture of two identical asymmetric species.

Table I

^{13}C Shifts of Allylalkalimetal Compounds in THF at 0°

M	Li	Na	K	Rb	Cs	Na+222
δ_α	50.8	48.7	52.5	54.5	58.7	51.8
δ_β	146.0	145.1	143.8	143.9	143.8	147.9
$J_{C_\alpha-H}$	146	148	148	149	145	
$J_{C_\beta-H}$	133	132	132	131	132	

The ^1H spectral pattern, Table II[4,5,6] is more complex and each of these compounds has two forms. There is a higher temperature form with an A_4X pattern indicating equivalence of all 4 end protons, and rapid rotation about the C-C bonds. At lower temperature the spectral pattern becomes an AA' BB'X pattern, indicating that rotation about the C-C bonds is slow on the NMR time scale, but that the two end groups are still identical. This latter form is explicable in terms of a planar molecule with extensive charge delocalization.

Table II

^1H Shifts of Allylalkalimetal Compounds in THF at -30°

M	Li[5]	Na	K	Rb	Cs
δ_α'	1.78	1.75	1.94	2.01	2.18
δ_α	2.24	2.53	2.32	2.26	2.04
δ_β	6.38	6.46	6.28	6.14	5.84
$J_{\alpha-\alpha'}$	1.6	2.6	2.6	2.4	2.6
$J_{\alpha-\beta}$	15.2	14.5	14.7	14.7	14.7
$J_{\alpha'-\beta}$	8.6	8.6	8.2	9.6	8.8

α' is hydrogen cis to β hydrogen

(b)   Both ^{13}C and ^{1}H chemical shifts are sensitive to charge on
the carbon atoms. There has been established a linear
relationship between charge and change in shift from the parent
hydrocarbon for a series of cyclic aromatic ions[7], and this
is thought to have wider application.

In the allylalkalimetal compounds the ^{13}C absorptions are far
removed from the parent's absorptions, and an attempt may be made
to relate this to charge. The sum of the shift differences from
propene for allylpotassium, after making allowance for the change
in hydridization of the α carbon, is -115 ppm[1] and should
represent the effect of 1ε charge in this system. This is
considerably smaller than the 160 ppm found for aromatic systems.
Moreover, the summation of these shift changes varies considerably
over the series Li to Cs (Table III). Li is somewhat out of order,
but if the variation is caused by closeness of approach of the two
ions, peripheral solvation of this intimate ion pair could cause
an effective increase in charge separation. However, the addition
of the cryptand 222 to allylsodium, which would presumably give a
solvent separated ion pair, only has a minor effect on the
adsorptions. The decrease in $\Sigma\Delta\delta$ with the higher members (Rb, Cs)
is opposed to that expected from the simple electrostatic
consideration of greater charge separation along the series. The
decrease is possibly caused by the polarizability of these large
ions reducing the effective charge on the allyl moiety. A similar
shift is found in other series such as the cyclopentadienyl
alkalimetal compounds (unpublished observations).

The average shift of the α protons in the ^{1}H spectra remains
fairly constant except for allyllithium. No reason can be given
for the drift in the values for the individual protons.

Table III

Sum of ^{13}C Shift Differences between Propene and Allylakalimetal
Compounds.

M	Li	Na	K	Rb	Cs	Na+222
$\Sigma\Delta\delta_d$	-116	-123	-115	-111	-103	-112.4

(c)   As noted above, all the allylalkalimetal compounds exhibit
two forms of ^{1}H spectra at different temperatures as the rotation
about the C-C bond increases in rate and changes from slow to fast
on the NMR time scale. This is also reflected in the ^{13}C NMR
spectrum if it is observed in the proton coupled form because of
the corresponding changes in the C-H coupling patterns. The
coalescence temperature of the α protons absorptions in the ^{1}H
spectra changes greatly with counter ion, Table IV, and the
activation energy may be calculatd to increase from 10.5 Kcal for
Li, 11.5 for Na, and 14.3 for K. This variation in freedom of
rotation could be caused by increasing doublebond character in the

C-C bonds, or by greater steric hindrance with increased counter
ion size. The latter explanation is preferred because there would
appear from the other evidence that in fact little change in
hybridization occurs over this series.

Table IV

Coalescence Temperatures of α Proton of Allylalkalimetal Compounds

M	Li	Na	K	Rb	Cs
T_c	-50	+5	+65	> 70	85

(d) Enriching the terminal carbon with ^{13}C enables the C_α-C_β
coupling constant to be measured, Table V. C-C coupling constants
vary considerably[8,9,10] with the largest factor being the
bond order. The values found here are fairly typical of
delocalized systems such as aromatic systems. But also the values
are close to the mean of a pair of rapidly equilibrating single
and double bonds.

Table V

^{13}C Coupling Constants in Allylalkalimetal Compounds at 0°

M	Li	K	Cs
J_{c-c}	55.9	59.8	61.1

(e) Attempts have been made to differentiate between rapidly
exchanging symmetrical carbenium ions and delocalized ions by
measuring the effect deuteration of one end will have on the ^{13}C
shift of the other protonated end.[11,12,13] In a delocalized
system such effect would be minimal, but in an exchanging system
depending on the K_H/K_D ratio quite large effects could be
expected. The results found here, Table VI, are more consistent
with those expected from a delocalized system, unless the
K_H/K_D ratio is exceptionally low. Unfortunately K_H/K_D ratios are
not well known for carbanions.

Table VI

Effect of Deuteration in Allylalkalimetal Compounds at 0°C

M	Li	Na	K
$\Delta\delta_{c_\alpha}$ ppm	0.56	0.44	< 0.25

Model diene polymerization systems

NMR spectra have been taken on diene polymeric systems by
Morton et al.[14], and of models formed by the addition of
t-butyllithium to dienes by Glaze et al.[15] and Bywater et
al.[16] Compounds with other counter ions have been made from
these.
(a) The ^{1}H spectral patterns show that the addition of monomer to
the alkalilithium compound is made at the 1 position (4 for
isoprene). The positions of the other NMR absorptions, however,
show that these compounds differ greatly from the conventionally
written 1:4 addition product, although it is convenient to talk of
them in that way. Under some conditions two forms can be
identified in the NMR spectra that can be associated with cis and
trans isomerization about the central bond of the four carbon
skeleton, Table VII. The rotation about this bond appears always
slow on the NMR time scale (except for the magnesium compounds),
although it can be shown to be far more rapid than that associated
with a normal double bond. There is also evidence for restricted
rotation about the end C-C bond.

Table VII

% Trans Structure of 5,5-Dimethylhexen-2-yl M in Solution

M	Solvent	% Trans
Li	C_6D_6	77
Li	DEE	75
Li	THF	35
Na	THF	23
K	THF	c.a.10
Rb	THF	<5
Cs	THF	–

Residue in cis form.

(b) The ^{13}C NMR shifts may be treated in the same fashion as were
those of the allylalkalimetal compounds. The shifts may be
compared with those of the parent hydrocarbon (5,5-dimethylhexene-
2 in the case of butadiene) and after allowance is made for a
hybridization change on the α carbon, the charge may be calculated
on the allyl positions assuming the same 115 ppm per ε. Table
VIII lists the values for the butadiene compound, similar figures
are obtained for isoprene. It is seen that the total charge
calculated appears close to 1 ε in all cases even for the Li
compound in benzene solution. There also appears the same drift
downward of apparent charge for the Rb and Cs compounds as
occurred for the allylalkalimetal compounds themselves.

Table VIII

Distribution of charge in 5,5-Dimethylhexen-2-yl M.

M	Solvent	α	β	γ	Σ
Li	C_6H_6	-.79	.13	-.22	-0.88
	DEE	-.69	.13	-.35	-0.91
	THF	-.69	.15	-.40	-0.94
Na	THF	-.65	.12	-.49	-1.02
K	THF	-.59	.11	-.53	-1.01
Rb	THF	-.55	.11	-.53	-0.97
Cs	THF	-.51	.12	-.52	-0.91

The distribution of charge changes radically with conditions. For the largest counter ions there appears to be an almost even distribution of charge over the α and γ positions, but with the ion Li^+, particularly in the nonsolvating solvent benzene, the charge and presumably also the Li atom are positioned closer to the end carbon atom. The small effect of temperature on the shifts would appear to discourage any attempt to explain the intermediate positions of some of these shifts in terms of ionic-covalent equilibrium, or 1:2 - 1:4 equilibria.

(c) The coalescence temperature in THF for the α hydrogens in these compounds is observable below -50° for both isoprene and butadiene compounds with the Li counter ion, about -20° and above 20° for the Na and K derivatives of butadiene respectively. It would appear that the restriction to rotation is similar in this bond in this series and the allyl series, suggesting comparable bonding.

The central bond does exhibit slower rotation, at least in the lowest member where it can be followed, as no coalescent temperature has been observed. This is at least partly caused by steric hindrance of the t-butyl group. The rate of cis/trans isomerization was found to increase greatly when the t-butyl was replaced by n-butyl or another isoprene unit in the isoprene based derivative of lithium in hydrocarbon solvents; the half life decreased to seconds at room temperature. The rate of cis/trans isomerization in THF is also not very slow, because after the preparation of these compounds from mixed cis/trans starting materials at -20 with a reaction time of $\frac{1}{2}$ hour, only the lower energy cis form is observable in some cases. It seems likely that the difference in rates of rotation about these two bonds decreases along the series to Cs.

(d) Coupling constants have been measured for the end C-C bond in the isoprene derivative with a lithium counter ion. The values found, 46 and 36 in THF and benzene respectively, are difficult to interpret. They fall considerably below those of the allylalkali-metals, and the benzene solution figure is even below that expected of a normal covalent bond in this position.

(e)   A number of butadienes have been prepared with deuterium in selected positions, and formed into model compounds with Li or Cs and examined by NMR.   In the case of the Li compounds the solvents used where THF and benzene.   There was no evidence of extra peaks appearing in say the γ region when the deuterium was on the end position, even though 50% of the α carbons would be deuterated. This is in good agreement with the delocalized nature of these compounds, although again lack of information on K_D/K_H ratios is inhibiting.

Model styrene systems

These have been described before with both styrene and α-methylstyrene models.[17,18]   The spectral patterns fit very well the expected structures of planar benzyllic carbanions, and the marked upfield shifts of the aromatic carbons compared with the parent hydrocarbons are interpretable in terms of the charge residing on the rings.   The α carbon shows, however, a marked downfield shift which as in the allyl compounds is explained in terms of a change in hybridization of this carbon to sp_2 compensated by the upfield movement caused by the charge on this carbon.   Some $^{13}C-^{13}C$ coupling constants have been measured for ^{13}C enriched model compounds of α-methylstyrene, the compound with t-butyllithium and the dimer formed by reaction with potassium in THF.   The C-C coupling constant between C_α and C_1 on the ring was found to be very high, over 70 Hz, much higher than the allylic cases and similar to that found in true double bonds, consistent with the suggested sp_2 hybridization of the C_α. Nevertheless there is still rotation about this bond as is shown by the presence of two sets of o and m resonances in the ^1H and ^{13}C spectra of the Li compound which coalesce at higher temperatures, about 50° for the α-methylstyrene models, and 0° for the corresponding sytrene models.   These temperatures are higher than those for the allyl compounds, but still represent fast rotation.

Thus by NMR it is possible to illustrate the variation of the charge distribution under various conditions of the apparently delocalized systems.   It is also possible to illustrate the various rates of rotation and isomerization about the various bonds, and the structure of the ions.   It is these factors coupled with the rates of addition of monomer which determine the structure of the polymer formed.

Literature Cited

1. Bywater, S. and Worsfold, D.J.   ACS Symposium Series, 1979, 89, 103.
2. van Dongen, J.P.C.M., von Dijkman, H.W.D., and de Bie, M.J.A. Recl. trav. chim. Pays Bas, 1974, 29, 93.
3. O'Brien, D.H., Russell, C.R. and Hart, A.J.   Tetrahedron Lett., 1976, 37.

4. Brownstein, S., Bywater, S. and Worsfold, D.J.   Unpublished data.
5. West, P., Purmont, J.I. and McKinley, S.V.   J. Am. Chem. Soc., 1968, 90, 797.
6. Thompson, T.B. and Ford, W.T.   J. Am. Chem. Soc., 1979, 101, 5459.
7. Spiesecke, H. and Schneider, W.G.   Tetrahedron Lett., 1961, 468.
8. Stothers, J.B.   "Carbon-13 NMR Spectroscopy", P.370, Academic Press N.Y., 1972.
9. Becker, G., Lütthe, W. and Schrumpf, G.   Angew. Chem. (Int. Ed.), 1973, 12, 339.
10. Hansen, P.E., Poulsen, O.K. and Berg, E.   Org. Mag. Res., 1979, 34, 43.
11. Saunders, M., Telkowski, L. and Kates, M.R.   J. Am. Chem. Soc., 1977, 99, 8070.
12. Saunders, M., Kates, M.R., Wiberg, K.B. and Pratt, W.   J. Am. Chem. Soc., 1977, 99, 8072.
13. Saunders, M. and Kates, M.R.   J. Am. Chem. Soc., 1977, 99, 8071.
14. Morton, M., Falvo, L.A. and Fetters, L.J.   Polymer Letters, 1972, 10, 561.
15. Glaze, W.H., Hanicak, J.E., Chaudhuri, J., Moore, M.L. and Duncan, D.P.   J. Organomet. Chem., 1973, 51, 13.
16. Bywater, S. and Worsfold, D.J.   J. Organomet. Chem., 1978, 159, 229.
17. Bywater, S., Lachance, P. and Worsfold, D.J.   J. Phys. Chem., 1975, 79, 2148.
18. Bywater, S., Patmore, D.J. and Worsfold, D.J.   J. Organomet. Chem., 1977, 135, 145.

RECEIVED February 9, 1981.

13

Polymerization of Methyl Methacrylate Initiated by *t*-Butyl- and Phenylmagnesium Compounds

Factors Influencing the Nature of the Active Centers

P. E. M. ALLEN, M. C. FISHER, C. MAIR, and E. H. WILLIAMS

Department of Physical and Inorganic Chemistry, University of Adelaide, P.O. Box 498, Adelaide, South Australia 5001

In previous papers (1,2) we have argued that the polymerization of methyl methacrylate (MMA) initiated by butylmagnesium in tetrahydrofuran (THF) — toluene mixtures is pseudo-anionic. The eneidic mechanism in which persistent and independent growth centres of different reactivity and stereochemistry, demonstrated by the polymodal molar-mass distributions of the polymers (1), is not consistent with the growing species being genuine ions, whether paired or free. Bimodal distributions can arise in a truly anionic polymerization from single and double-ended chains (3) but it is difficult to explain trimodal distributions (1). Evidence that the peaks of the distribution differ in steric triad frequency indicated that the growth centres responsible for these peaks are different chemical species. If these different centres were ionic, the lability of ion:ion-pair and ion-pair:ion-pair equilibria would ensure that a given centre did not maintain its chemical identity throughout its lifetime; it would act as a Coleman–Fox multistate centre (4) and the polymer formed would be a stereoblock having a broad unimodal distribution. In the systems we have studied exchange between active centres of different types is normally frozen. A particular type of centre may itself operate as a labile Coleman-Fox multistate centre leading to stereoblock polymerization (5), but the general immutability of the different types of centre is affirmed by the polymodal molar-mass distributions (1,5).

Immutability of the active centres is most easily understood if it is assumed that the growth sites are covalent magnesium-carbon bonds. This hypothesis also provides a ready explanation (2) of the results of Yoshino which provided the earliest evidence for the stereochemical persistence of the active centres (6) and the earliest suggestion that the tetrahedral symmetry of a methine C atom in a poly(acrylate) chain is already present when that atom is the active chain end (7). Magnesium is believed to exist in a tetrahedral coordination state in solution so the different active centres may be assumed to have the general configuration (I)

0097–6156/81/0166–0185$05.00/0

(I)

The three vacancies in the coordination shell of Mg in I are
filled by a covalent bond to either an alkyl group (R) or halide
(X) and two coordinated ligands which may be bridging X or R, if
the species is associated, or THF, monomer, or the carbonyl
groups of the ultimate, penultimate, or antipenultimate monomer
residues of the growing chain. It is in the arrangement of these
groups and ligands around the magnesium atom of an active centre
that its stereospecificity has been and must be sought.   The
various hypotheses have been discussed and full references given
elsewhere (2).

     Kinetic analysis (1) indicates that the concentration of
active centres is very low, probably in the μM–mM range.   This
being so, spectrometric methods are unlikely to provide direct
evidence of the covalent nature of the active centres or their
detailed structures.   Such methods are valid for investigating
the structure of the initiator, but the nature of the active
centre must be deduced from indirect evidence; kinetic analysis,
for instance, has shown that in the syndiotactic–like or stereoblock polymerizations initiated by n–butylmagnesium compounds
monomer is coordinated to the active centre (8).

     In the present paper we report results obtained with t–butyl-
and phenyl–magnesium compounds which under suitable conditions
give highly isotactic polymers. We are in particular looking to
see the extent to which the character of the polymerization is
predetermined during the initiation stage when the stable and
persistent growth centres are formed, and whether monomer is coordinated to the active centre in these cases.

Nominal Solvent and Solvation–State of the Initiator

     The influence of solvating solvents, in our case THF, has
been discussed previously (1).   There is some ambiguity about the
state of solvation of the initiator in polymerization in non-
polar solvents, in our case toluene.   These systems, in which
highly isotactic polymers are formed, are sometimes described as
"ether–free" or "de–etherated".   The basic procedure is to pump

off the ether in which the reagent is prepared and redissolve the
reagent in toluene. The ether removal may be enhanced by repeated evaporation and redissolving in toluene or by heating. The
most drastic treatment reported (9) involves heating the reagent
for 8 hours at 343 K under high vacuum (10^{-3} Pa). Not all sources
specify the treatment. The ether component (THF in our experiments) is of paramount importance in determining the kinetics and
stereospecificity of the polymerization (1,2). It assumes a vital
role in extant mechanisms proposed to explain the stereospecificity of the polymerizations (e.g. 10). We prepared initiator
solutions in toluene according to the most drastic recipe for THF
removal (9) and used 90 MHz ^{1}H NMR to check whether any THF remained.

Solutions of PhMgBr prepared in THF were evaporated to dryness under high-vacuum. The residue was heated at 353 K for
eight hours under high-vacuum and treated with d^8-toluene. The
toluene solution was filtered from some insoluble residue. THF
resonances were observed; their integrated intensities corresponded to a mole ratio of THF:Ph-Mg groups of 1·2:1. Further experiments with successively less exhuastive procedures showed that
only 20 minutes heating at 353 K under high vacuum sufficed to
attain the monoetherate composition.

The α and β CH_2 resonances of THF in monoetherate initiator
solutions in toluene were shifted downfield by 0·12 and 0·62 p.p.m.
respectively (11). This confirms that the THF in these solutions
is, at least, predominantly coordinated. On addition of THF in
excess of the monoetherate stoichiometry the resonances shifted
towards their field positions in THF-toluene solutions. Separate
peaks corresponding to free and coordinated THF were not observed.
It was concluded that the equilibrium between these species is
labile.

In further experiments the residues were redissolved in THF.
The NMR spectra were essentially similar to those of solutions of
PhMgBr prepared in THF. The de-etheration procedure does not
therefore bring about any irreversible changes in composition of
the reagent.

Experiments with t-BuMgBr gave similar results. One THF per
t-Bu group is tightly bound and survives and remains when the
residue is dissolved in toluene. The de-etheration procedure can
be reversed by redissolving the residue in THF.

It is well established that THF is in general deleterious to
the formation of isotactic polymer, e.g. (1), and the same is
true with t-BuMgBr and PhMgBr. However, the most efficient initiation of isotactic growth centres in toluene solution was not
observed with the minimal monoetherate compositions, but when the
THF concentration was slightly in excess (11). Since the revers-
ibility of THF removal under heat-treatment has been demonstrated
this effect must be due to the presence of a trace of excess THF,
rather than decomposition of initiator during the heat treatment
necessary to produce the monoetherate. The monoetherate composition is consistent with the solute being $R_2Mg(THF)_2$ but if the

solute is RMgX, tetrahedral coordination can only be maintained
with the dimer $(RMgX)_2(THF)_2$. The proton NMR spectra of toluene
solutions of tBuMgBr and PhMgBr were very similar to those of
tBu_2Mg and Ph_2Mg (11) and it must be concluded that the latter
species predominate — a conclusion supported by halide assay of
the toluene solutions.

However from the point of view of initiation the solutes cannot be regarded as dialkyl magnesiums; tBu_2Mg in toluene does not
yield isotactic polymer, tBuMgBr and tBuMgCl in toluene yield
polymer in which syndio- and hetero-tactic triads cannot be
detected by 90 MHz NMR (Figure 1). It is suggested that the small
excess of THF is required to ensure that sufficient halide, an
essential component of a highly isotactic growth centre, passes
into solution. The optimum amount corresponded to approximately
twice the amount corresponding to the monoetherate composition.
In polymerizing solution this would amount to a mol fraction:
$X_{THF} \approx 0.0003$.

The Importance of Exact Specification of Initiation Conditions

One problem encountered in the field is the apparent irre-
producibility of the results of different workers, even those in
the same laboratory. This is particularly the case with molar
mass distribution and steric triad composition. The explanation
of these apparent inconsistencies comes with the realization that
the mechanisms are eneidic and the polymer properties are primarily determined by independent active centres of different reactivities and stereospecificities whose relative proportions are
set at the initiation step, which is completed in the first
seconds of the polymerization. The irreproducibilities arise
from irreproducibilities in the initiation step which had not
been thought relevant. Ando, Chûjô and Nishioka (12) noted that
these rapid exothermic reactions tend to rise very significantly
above bath temperature (we have confirmed this effect) and allowed for this in considering the stereochemistry of the propagation
reaction. However our results show that the influence on the
initiation reactions can have a more far-reaching effect.

Our standard procedure in preparative experiments of isotactic polymer is that used in earlier work, where the temperature during initiation was carefully controlled. We initiated
the polymerization by mixing initiator solution and monomer
solution already thermostatted to a predetermined temperature,
usually that chosen for the polymerization. At 230 K with
t-BuMgBr in THF-toluene mixtures this led to a trimodal distribution with the high molar-mass peak in the 10^6 range (1,2). More
recent work, shown in Figure 2, has confirmed that trimodal distributions arise when the mol fraction of THF is below ca 0.3.
At one stage experimental convenience led us to distil the monomer onto frozen initiator solution so that initiation took place
at the melting temperature of the monomer-toluene-THF mixture.
Even though the temperature rose within a few minutes to 230 K

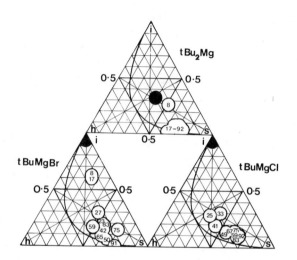

Figure 1. Isotactic, heterotactic, and syndiotactic triad frequencies (i, h, and s) in poly(methyl methacrylate) polymerized and initiated at 225 K by t-butylmagnesium bromide (left), t-butylmagnesium chloride (right), and di-t-butylmagnesium (top) with initial mole fraction of monomer: $X_{MMA} = 0.10$.

Toluene-soluble initiator, $X_{THF} \approx 0.003$ (●); toluene-THF solutions with percent THF ($100X_{THF}$) indicated for each point (○). The apex i corresponds to i = 1.00, the base opposite i to i = 0. The i content of any sample is given by the perpendicular distance from the base opposite i. The curves indicate the permitted values of the Bernoullian distribution of i, h, and s.

and the bulk of the propagation carried out at this temperature, the polymer was entirely different, being monomodal and of moderate molar mass at a solvent composition which would have led to a trimodal distribution if initiation had occurred at 230 K.

The effect on the chain configuration of the polymers was just as drastic. t-Butylmagnesium bromide in toluene solution at 230 K produced a polymer of steric triad composition i:h:s ≈ 1:1:1 if initiated at the melting point of toluene, and i:h:s ≈ 99:0:1 if initator solution at 230 K was mixed with monomer solution at 230 K.

We have ourselves criticised (1) Petiaud and Quang-Tho Pham (13) for assuming in their analysis of the steric triad compositions of their polymers that they had a coherent unimodal molar-mass distribution. Our results on a very similar system produced polymer of trimodal distribution where each peak seemed to be independent and had different stereochemistry. However our recent work suggests that it is most likely that Petiaud and Pham did in fact produce coherent monomodal samples. Nevertheless incomplete specification of their initiation conditions makes it impossible to be sure. As indicated below, under some circumstances even the order of addition of reagents is an experimental variable.

Medium Effect on Molar-Mass Distribution of Polymers Produced by Initiation and Propagation at 250 K Using t-Butylmagnesium Compounds

Figure 2 shows gel permeation chromatograms of PMMA initiated and polymerized at 250 K by t-butylmagnesium bromide in various toluene-THF mixtures. The top curve (X_{THF} ≈ 0.003) refers to a toluene solution of initiator prepared by dissolving the residues remaining after THF had been removed by pumping on a high-vacuum line with brief warming. The residual THF corresponded to the optimum amount for the production of isotactic growth centres. The polymers produced, provided the mol fraction of monomer did not exceed X_{MMA} < 0.20, were isotactic ($i = 1·00$) within the sensitivity of our assay. As may be seen even though the propagation reaction takes a unique stereochemical mode, two independent growth centres are operating since the distribution is bimodal.

The polymers produced, after precipitation by methanol, were not fully soluble in THF, chloroform, toluene or any solvent compatible with "Styragel" columns. However after quick heating to a temperature just below T_m and quick cooling, samples in the form of a thin film dissolved readily in THF and other solvents. This, together with the observation that G.P.C. traces of the whole samples were indistinguishable from those of the THF-soluble fraction of the sample prior to heat-quench treatment, led to the conclusion that the insoluble material in the untreated polymers was crystalline. NMR spectra of the treated samples showed no detectable syndio- or hetero-tactic triad peaks, confirming that no significant degradation had occurred.

When $X_{THF} \approx 0.10-0.25$ (some traces have been omitted from Figure 2 for clarity) trimodal distributions arise. At high THF ($X_{THF} > 0.50$) the product is monomodal of low molar mass.

Since the chain structure of the polymer changes throughout Figure 2, we have not applied calibration procedures, having neither suitable standard PMMAs of appropriate chain structure nor a sound basis for applying a universal method. The chromatograms are aligned according to elution volume and a mixture of standard polystyrenes is included to provide a qualitative reference.

By way of contrast di-t-butylmagnesium gave polymer of broad monomodal distribution of molar mass $\sim 10^4$ over the whole range of THF concentration.

Medium Effect on the Steric Dyad and Triad Composition of Polymers Produced by Initiation and Propagation at 250 K and 225 K Using t-Butylmagnesium Compounds

The influence of THF is shown in Figure 1, where the frequencies of the iso-, syndio- and hetero-tactic triads, i, s and h, of each sample can be read off along the appropriate median (e.g. the apex i corresponding to i = 1·0, the base opposite i corresponding to i = 0). The curve represents the Bernoullian triad distribution ($h = 2i^{1/2}s^{1/2}$). The black circles refer to samples prepared in toluene solution containing the optimum trace of THF ($X_{THF} \approx 0.003$) for initiation of isotactic growth centres. It will be seen that an excess of this, as indicated by mol percent ($100X_{THF}$) on the open circles, led to syndiotactic-like polymers. Except for a few samples having very low isotactic content, all polymers lie to the right of the Bernoullian line and have a stereoblock character. While toluene solutions of t-butylmagnesium bromide and chloride give polymers which were isotactic within the sensitivity of our assay (i = 1·00 ± 0.01), toluene solutions of di-t-butylmagnesium gave stereoblock polymer with i = 0·33.

The analysis shown in Figure 1 is however incomplete. The medium is a three-component system consisting of two polar components (monomer and THF) and one non-polar (toluene). Correlations in terms of two components only lead to incomplete characterization and, if extrapolated, invalid conclusions. In Figure 3 the meso (isotactic) dyad frequencies of polymers produced by t-butylmagnesium bromide at 225 K are shown as a function of the mol fractions of the three components of the medium. It is apparent that monomer itself acts as a polar solvent component decreasing the isotactic content in comparison to toluene. It should be noted that the polymerizations carried out at X_{MMA} = 1.00 were heterogeneous; monomer was run on to solid monoetherate initiator. In all other cases filtered solutions of initiator were used.

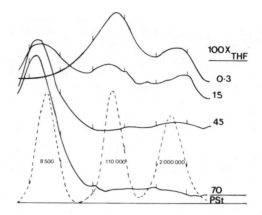

Figure 2.   GPCs of THF solutions through 10^3, 10^4, 10^5, and 10^6 Å μ-Styragel columns in series.

Poly(methyl methacrylate), initiated and polymerized at 250 K by t-butylmagnesium bromide in toluene–THF solution (——). Mole fraction of monomer, $X_{MMA} = 0.10$M. X_{THF} is indicated in each case. A mixture of standard polystyrene samples of indicated molar mass (— — —). All traces are aligned so that the elution volumes correspond.

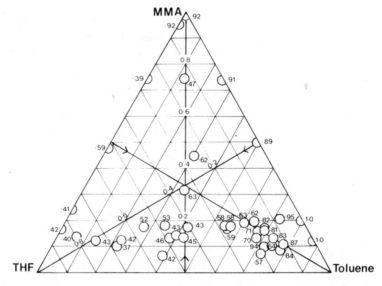

Figure 3.   Meso (isotactic) dyad frequency (m) of poly(methyl methacrylate), prepared at 250 K initiated by t-butylmagnesium bromide, as a function of solvent composition. The meso-frequency m is indicated for each point. The solvent composition, given in terms of mole fractions of THF, X_{THF}, of toluene, $X_{toluene}$, and of monomer, X_{MMA}, are indicated by the scale on each median, with each apex corresponding to the pure component indicated.

Kinetics of the Polymerization Initiated by t-Butylmagnesium Bromide in Toluene Solution

A number of mechanisms proposed to explain the stereospecificity in polymerizations involve complexing of the monomer to a metal prior to addition to the chain (2). Kinetic evidence has shown that such a mechanism does occur with n-BuMgBr, n-Bu$_2$Mg and s-BuMgBr in THF-toluene solution. These polymerizations follow an internal zero-order rate equation. Bateup (1,8) proposed that the mechanism is

$$\text{\sim\sim Mg} \overset{/}{\underset{\backslash}{\lhd}} \; + \; M \; \underset{k_{-c}}{\overset{k_c}{\rightleftarrows}} \; \text{\sim\sim Mg} \overset{\nearrow^M}{\underset{\backslash}{\lhd}} \; \overset{k_p}{\longrightarrow} \; \text{\sim\sim MMg} \overset{/}{\underset{\backslash}{\lhd}} \qquad (1)$$

$$[P] \qquad\qquad\qquad\qquad [C]$$

Initiation is rapid but of low efficiency (f) so once initiation is complete

$$[P] \; + \; [C] \; = \; f[RMgX]_0 \qquad\qquad (2)$$

where $[RMgX]_0$ is the initial concentration of initiator. The zero-order rate equation occurs when the steady-state of the complex C is high and the second step of Reaction 1 becomes rate controlling, i.e.

$$\frac{k_p + k_{-c}}{k_c[M]} \longrightarrow 0 \qquad -\frac{d[M]}{dt} \longrightarrow fk_p[RMgX]_0 \qquad (3)$$

Yerusalimskii has demonstrated a similar mechanism for acrylonitrile polymerization initiated by Grignard reagents (13). We were unable at the time to investigate the isotactic polymerization initiated by t-BuMgBr in toluene as it was too fast for classical techniques. We have now developed techniques which can follow this reaction.

In Figure 4 we show the results of a run followed by a novel NMR technique. The proton resonances of relevant species have been scanned at predetermined intervals after mixing the reagents. A T_1-programme was modified to store the data and display it on recall in the form shown. Points to note are the progressive decrease of the vinyl resonances and the growth of the polymer (p) methoxy and α-methyl peaks at the expense of the corresponding monomer (m) peaks. There is also a small sideproduct methoxy peak just down-field of the polymer peak. It is present at 2 min. and stays constant thereafter, eventually being obscured by the spreading base of the polymer peak. Most noteworthy of all however is the persistence of unreacted t-BuMg peaks, undiminished from 2 min. to complete conversion. We have previously reported indirect evidence (5) that in polymerizations initiated by n-BuMgBr 20% of the n-BuMg survive the polymerization. Details of the technique have been reported elsewhere (10), as have the raw kinetic data. However the overall

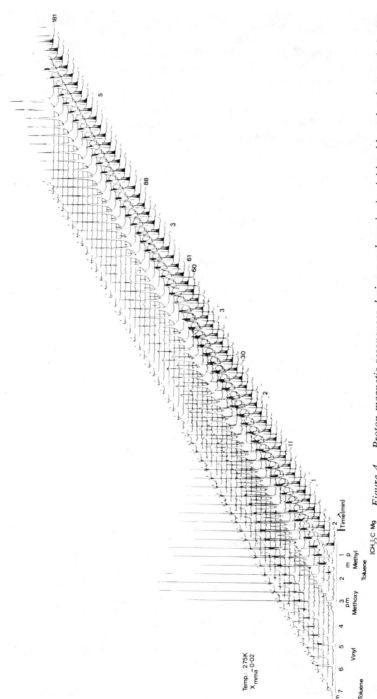

Figure 4. Proton magnetic resonances during a polymerization initiated by a d_8-toluene solution of de-etherated t-butylmagnesium bromide at 225 K with $X_{MMA} = 0.02$, $X_{C_7D_8} = 0.98$, $X_{THF} \approx 0.003$. Polymer and monomer peaks in αCH_3 and OCH_3 regions are indicated by p and m. The toluene peaks are protonated impurity in the d_8 solvent. The chemical shift is given in parts per million from TMS and t is time from mixing.

result is relevant to the present discussion.

At 273 K and X_{MMA} = 0.02 the polymerization followed first order kinetics. At 225 K (Figure 4) the conversion curve was indistinguishable from a zero-order plot up to 40 percent conversion, but if the whole curve was examined the internal order was seen to lie between zero and one. At 250 K the internal order also lay between zero and one but the fit was nearer to the first-order plot in monomer. The kinetics are consistent with a Bateup-Yerusalimskii mechanism. At low temperatures the limiting condition of Equation 3 is approached. As the temperature rises the stationary state concentration of the complex decreases and the mechanism shifts to its other limit

$$\frac{k_c[M]}{k_p + k_{-c}} \longrightarrow 0 \qquad -\frac{d[M]}{dt} \longrightarrow f\frac{k_c k_p}{k_{-c} + k_p}[RMgX]_0[M] \qquad (4)$$

which corresponds to first-order monomer decay.

No mechanistic significance can be attached to the change of order being observed in the present experiments and not in the earlier ones. The kinetic experiments with n-butylmagnesium compounds were carried out at monomer concentrations 10 times higher than permissible in an NMR experiment. This not only favoured the limiting condition of Equation 3, but increasing viscosity restricted measurements to ca. 40 percent conversion. The deviation from zero order kinetics at 225 K in the present work was not apparent until this conversion was exceeded.

Effect of Order on Addition of Reagents

In the kinetic experiments a special method of mixing was required. This revealed a further experimental variable: order of addition. The initiator solution was frozen into the bottom of the NMR tube. Monomer was frozen on the upper walls of the tube. On initiation the monomer melted first and ran down onto the frozen initiator, the initiator solution thus melted into a high concentration of monomer. If however molten initiator solution was run on to frozen monomer the rate of polymerization was drastically reduced. The concentration of unreacted t-BuMg groups was also significantly reduced. The efficiency of initiation of active polymerization sites was much lower when this event occurs at high initiator and low monomer concentration. This is again consistent with the Bateup mechanism (8) where the order with respect to monomer is higher for the initiation of polymerization chains than for the formation of non-polymeric side-products.

In experiments on a larger scale, where the temperature of both solutions was controlled and adequate turbulence during mixing ensured, the order of mixing did not affect the polymerization.

Conclusions

1. As previously proposed (1,2) stable, persistent active centres of different reactivity and stereospecificity may coexist independently during the course of the polymerization.

2. Stringent control of the initiation conditions (temperature, solvent-composition, order and, possibly, rate of addition of reagents) must be maintained if reproducible stereochemistry and kinetics are to be achieved, because the proportions of the different active centres are established during the initiation stage.

3. Preliminary results on the kinetics of the polymerization and the efficiency of initiation of the isotactic polymerizations initiated by t-BuMgBr in toluene solution are consistent with the Bateup mechanism proposed for the stereoblock and syndiotactic-like polymerizations initiated by n-BuMgBr in THF-rich solution — a mechanism which involves initiation and propagation through monomer — active centre complexes (5,8).

4. The most drastic treatments applied to remove THF do not yield a completely de-etherated toluene solution of initiator: one complexed THF remains for every RMg group present.

5. The optimum conditions for isotactic polymerization require a trace of THF in excess of the mono-etherate stoichiometry.

6. In interpreting the influence of the medium on chain configuration of the polymers, the medium must be specified as a three-component system: polar monomer, polar solvent, nonpolar solvent.

7. Isotactic polymerization requires the presence of halide, and is critically dependent on the nature of the alkyl or aryl group. At least one of each of these groups must be present in the vicinity of the active centre. The requirement of a minimum amount of THF for the formation of an isotactic growth site implies that this must also be coordinated at this site. We have evidence that monomer is complexed at the active centre, which has been postulated for many years (10,15). There is also the widely supported hypothesis (10,16,17) that chelation of polymer carbonyls to the metal atom plays a role in the stereo control of the propagation reactions. Yoshino (18) has shown that the effect does occur with PhMgBr though it is not a prerequisite for isotactic polymerization as is sometimes assumed. Even if tetrahedral Mg in the active centre I is permitted to increase its coordination number in the transition state, there is no room for all these ligands in a monomeric species. The active centre must therefore be octahedral, which is unlikely, or associated. The very low concentration of active centres is consistent with them being associated species.

8. The conclusion that the mechanism is pseudoanionic is restricted to organomagnesium initiated systems. The gist of our argument is that these polymerizations behave differently from the true anionic mechanisms initiated by (typically)

carbanionic salts of Group IA and lower Group IIA metals. Nevertheless it is interesting to note a recent report (19) of a butyllithium initiated system which produces trimodal poly(ethyl methacrylate) with peaks showing significant differences in steric triad frequency as well as molar mass. If pseudoanionic polymerization occurs anywhere else in the periodic table, beryllium and lithium are the most likely candidates.

Acknowledgement

This work was supported by the Australian Research Grants Committee.

Literature Cited

1 Allen, P.E.M.; Bateup, B.O. Eur.Polym.J., 1978, 14, 1001
2 Allen, P.E.M. J.Macromol.Sci.,Chem., 1980, A14, 11.
3 Chaplin, R.P.; Yaddehige, S.; Ching, W. Eur.Polym.J., 1979, 15, 5.
4 Coleman, B.O.; Fox, T.G. J.Chem.Phys., 1963, 38, 1065.
5 Bateup, B.O.; Allen, P.E.M. Eur.Polym.J., 1977, 13, 761.
6 Yoshino, T.; Komiyama, J. J.Am.Chem.Soc., 1966, 88, 176.
7 Yoshino, T.; Iwanaga, H.; Kuno, K. J.Am.Chem.Soc., 1968, 90, 2434.
8 Allen, P.E.M.; Bateup, B.O. J.Chem.Soc.,Faraday Trans.1, 1975, 71, 2203.
9 Tsvetanov, Ch.B. Eur.Polym.J., 1979, 15, 503.
10 Fowells, W.; Schuerch, C.; Bovey, F.A.; Hood, F.P., J.Am.Chem. Soc., 1969, 89, 1399.
11 Allen, P.E.M.; Fisher, M.C.; Mair, C.; Williams, E.H. presented at 11th Australian Polymer Symposium, Lorne, Vic., February 1980, submitted to J.Macromol.Sci.Chem.
12 Ando, I.; Chûjô, R.; Nishioka, A. Polym.J., 1970, 1, 609.
13 Petiaud, R.; Pham, Q.-T. Eur.Polym.J., 1976, 12, 455.
14 Yerusalimskii, B.L. Polym.Sci.USSR, 1971, 13, 1452.
15 Cram, D.J.; Kopecky, K.R. J.Am.Chem.Soc., 1959, 81, 2748.
16 Glusker, D.L.; Lysloff, I.; Stiles, E. J.Polym.Sci., 1961, 49, 315.
17 Bawn, C.E.H.; Ledwith, A. Quart.Rev., 1962, 16, 361.
18 Yoshino, T.; Komiyama, J.; Iwanaga, H. J.Am.Chem.Soc., 1967. 89, 6925.
19 Hatada, K.; Kitayama, T.; Sugino, H.; Umemura, Y.; Furomoto, M.; Yaki, H. Polym.J., 1979, 11, 989.

RECEIVED March 5, 1981.

Anionic Polymerization of ε-Caprolactone for Block Copolymer Synthesis

YUYA YAMASHITA

Department of Synthetic Chemistry, Faculty of Engineering,
Nagoya University, Chikusa-ku, Nagoya 464 Japan

The importance of tailored blockcopolymers has been demonstrated by extensive studies on morphology and colloidal properties successfully developed in recent years. The living polymer
technique pioneered by Szwarc has been widely used for their
syntheses. The living tendency frequently observed in ring-
opening polymerization also provided novel methods for the preparation of tailored blockcopolymers, which were not obtained by
other methods. Anionic polymerization of ε-caprolactone was
studied to clarify the mechanism of the polymerization and to
find optimum condition for the preparation of blockcopolymer. It
was found that considerable amounts of cyclic oligomers were
formed in equilibrium system using dissociated catalyst such as
potassium. It was necessary to select associated catalyst such
as lithium to obtain blockcopolymer without oligomer formation
under kinetically controlled living system.

Equilibrium Cyclic Oligomer Formation

Anionic polymerization of ε-caprolactone has been studied in
several laboratories and it was found that considerable amounts of
oligomers were found as by-products.[1,2] We have studied   the
formation of oligomers in the anionic polymerization of ε-caprolactone by gpc technique.[3] In view of the very facile intra- and
intermolecular transesterification reactions in this system[2,4],
the product distribution seems very interesting to check the
validity of the thermodynamic equilibrium.

By using potassium tert-butoxide in THF, the system is essentially represented by the living ring-chain equilibrium and the
oligomers, isolated and identified to be cyclic, reach   equilibrium distribution shortly after the introduction of the initiator
solution.

$$\left[\begin{array}{c} C(CH_2)_5O \\ \| \\ O \end{array} \right] \xrightarrow{KO^tBu} {}^tBuO \left[\begin{array}{c} C(CH_2)_5O \\ \| \\ O \end{array} \right]_n -\underset{\underset{O}{\|}}{C}(CH_2)_5O^-K^+$$

0097–6156/81/0166–0199$05.00/0

$$^{t}BuO\left[\begin{array}{c}C(CH_2)_5O\\\|\\O\end{array}\right]_n\begin{array}{c}C(CH_2)_5O^-K^+\\\|\\O\end{array} \quad \rightleftharpoons$$

$$^{t}BuO\left[\begin{array}{c}C(CH_2)_5O\\\|\\O\end{array}\right]_{n-x}\begin{array}{c}C(CH_2)_5O^-K^+\\\|\\O\end{array} + \left[\begin{array}{c}C(CH_2)_5O\\\|\\O\end{array}\right]_x$$

Typical gpc chromatograms are shown in Figure 1. The number on each peak indicates the degree of polymerization of the corresponding cyclic oligomers. Surprisingly, the reaction was very fast and the monomer was consumed in a few minutes and the product distribution changed very little after standing for a long time. The product distribution depends on monomer concentration, and little on reaction temperature. In dilute solution, cyclic dimer is the main product and its yield reaches as high as 70%. No polymer was formed in dilute solution. However, in rather concentrated solution, cyclic oligomers of various ring size coexist and also considerable amount of polymer was observed. The reaction mixture remained living after reaching equilibrium, because addition of monomer, dilution with solvent shifted the product distribution to a new equilibrium. The equilibrium concentration of each oligomers as a function of total monomer unit concentration is shown in Figure 2. It is evident that equilibrium formation of oligomers cannot be neglected even in bulk polymerization and oligomer can consist more than one third of the products in solution polymerization.[2]

According to the Jacobson-Stockmayer theory[5,6], the molar cyclization equilibrium constant is defined by

$$-M_n- \quad \rightleftharpoons \quad -M_{n-x}- \quad + \quad c-M_x$$

$$K_x = \frac{[-M_{n-x}-][c-Mx]}{[-M_n-]} = \frac{[c-M_x]}{p^x}$$

where $-M_n-$ and $-M_{n-x}-$ are the linear n and n-x mer, and c-M_x the cyclic x mer, and p is the ratio of the concentration of the living linear chains of n+1 mer to that of n mer. Since p should approach unity at the condition n≫1, K_x is nearly equal to the limiting equilibrium concentration of the corresponding cyclic x mer which formed in more concentrated solution. Because the product distribution of cyclic oligomer changed very little from equilibrium at -78,0 and 50°C, the equilibrium seems to be determined by the differences in the entropy of each oligomers. Log-log plots in Figure 3 show that K_x decreases in proportion to -2.5 power of x, in accord with the Jacobson-Stockmayer theory which assumes Gaussian chain statistics for oligomers without ring strain. The equilibrium concentration of trimer is lower than expected, suggesting some unfavored ring conformation. The increase of the trimer fraction was observed, though very slightly,

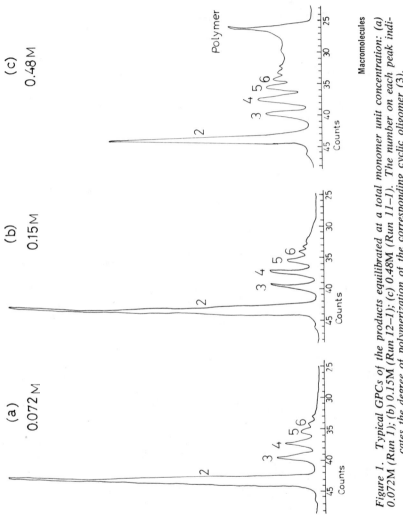

Figure 1. Typical GPCs of the products equilibrated at a total monomer unit concentration: (a) 0.072M (Run 1); (b) 0.15M (Run 12–1); (c) 0.48M (Run 11–1). The number on each peak indicates the degree of polymerization of the corresponding cyclic oligomer (3).

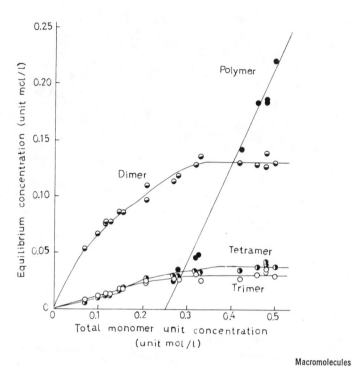

Macromolecules

Figure 2. Equilibrium concentrations of cyclic dimer ($\ominus$), trimer ($\bigcirc$), tetramer ($\circledcirc$), and linear polymers ($\bullet$) as a function of total monomer unit concentration at 0°C (3).

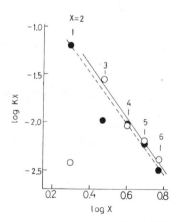

Figure 3. Log–log plots of molar cyclization constant (K_x) against ring size (x) for δ-valerolactone ($\bigcirc$) and ε-caprolactone ($\bullet$). The slopes of the straight lines are −2.5 (8).

Polymer Bulletin

in an experiment at 50°C, and further confirmed by an experiment at 100°C in a cationic system where trimer exceeded tetramer concentration. Equilibrium data for six-membered δ-valerolactone[8] was included in Figure 3. All the oligomer except dimer followed the Jacobson–Stockmayer theory. Thus nearly all oligomers are formed in equilibrium following the decreased entropy of ring-closure compared with linear polymers.

We have further examined the effect of initiators and solvents on the equilibrium product distribution. The use of lithium tert-butoxide instead of potassium tert-butoxide and the change of solvent from THF to toluene showed no effect on the product distribution of ε-caprolactone oligomers.[2] Also we are examining the equilibrium oligomer distribution in cationic system such as trifluoromethane sulfonic acid catalyst.[7]

Kinetics of Anionic Polymerization and Oligomerization

In the anionic polymerization of ε-caprolactone in THF with potassium tert-butoxide, monomer consumption is very fast and monomer could not be detected in any appreciable amount in the equilibrated mixture obtained after several minutes from the initiation. Upon terminating the reaction at 6 seconds after the introduction of the initiator solution, 6% monomer remained un-reacted accompanied with polymers (68%) and oligomers (26%). From the integrated rate equation assuming a living system with rapid initiation, kp at 0°C is calculated to be 120 1/mol•sec.[3]

$$\ln [M]_0 \ / \ [M] = kp [I]_0 t$$

This value seems reasonable considering the attack of the alkoxide anion on electrodeficient carbonyl group. Considering the great tendency of association of alkoxides, the kp value should not be ascribed to an dissociated species.

Attempts to obtain a living polymer without forming cyclic oligomers were not successful in potassium catalyst system, because the backbiting reaction was so fast that thermodynamic equilibrium was established after several seconds from the initiation. On the other hand, polymerization with lithium tert-butoxide in THF proceeds rather slowly without forming cyclic oligomers at least in the initial stage. The change of product distribution is shown in Figure 4.[2] It is clear that monomer is first converted to linear polymers and backbiting reaction from the polymer end occurs slowly to form cyclic oligomers. We have depolymerized isolated linear polymers with lithium tert-butoxide in THF for several days, and obtained equilibrium distribution of cyclic oligomers. (Figure 5) Thus the mechanism of oligomer formation was proved to consist of the backbiting reaction from the active polymer end.

The rate of the propagation reaction was calculated from monomer consumption in Figure 4. kp at 0°C is 0.45 1/mol•sec.

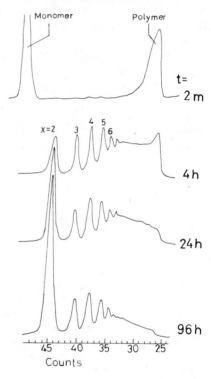

Figure 4. Change of product distribution (GPC) with polymerization time $[M]_o = 0.24$M, $[I]_o = 4.4 \times 10^{-3}$M *(9)*.

Macromolecules

Thus kp for lithium counterion is 1/300 of kp for potassium counterion. The low reactivity and association of lithium alkoxide
was reported in the anionic polymerization of epoxides.[10] We
have found that two fold increase of the lithium initiator concentration has led to a decrease of the kp nearly to one half.
This indicates that the kinetic order with respect to the initiator would be near to zero, suggesting a very high degree of
association of the active species. Thus the propagation reaction
appears to proceed in practice through a very minor fraction of
monomeric active species in case of lithium catalyst.

Polymerization and depolymerization in the present system
should be represented as follows.

$$P_n^* + M \xrightarrow{k_{p,1}} P_{n+1}^*$$

$$P_n^* \underset{k_{p,x}}{\overset{k_{d,x}}{\rightleftharpoons}} P_{n-x}^* + Mx \qquad \text{(for } x \geq 2)$$

$$K_x = k_{d,x} / k_{p,x} \qquad \text{(for } x \geq 2)$$

here $k_{p,x}$ and $k_{d,x}$ are the rate constants for propagation and
depropagation of cyclic x-mer. From this scheme, the rate of
formation of cyclic x-mer is expressed as follows.

$$d[Mx] / dt = (k_{d,x} - k_{p,x}[Mx])[I]_0$$

On integration with the initial condition [Mx]=0 at t=0,

$$\ln \frac{K_x}{K_x - [M]_x} = k_{p,x}[I]_0 t \qquad \text{(for } x \geq 2)$$

Applying this equation to the depolymerization experiment
(Figure 5) of the polymer with lithium tert-butoxide, $k_{p,x}$ can
be obtained. $k_{d,x}$ was calculated by using the value of k_x.
These values are shown in Table I.[2] It is to be noted that $k_{d,x}$
is almost constant for cyclic oligomers, while $k_{p,x}$ increases
with the increase of x, and $k_{p,2}$ being very low. Thus the high
yield of the dimer at equilibrium is due to concentration by
its low reactivity in polymerization. The value of $k_{p,x}$ and $k_{d,x}$
in case of potassium counterion is roughly calculated to be 1000
times larger than the value in Table I.

In summary, we have compared the reactivity of macrocyclic
compounds to conclude that the larger the ring the higher is its
reactivity and the lower its equilibrium concentration and that
the cyclic dimer is the least reactive among the cyclic oligomers
to reach the highest equilibrium concentration.

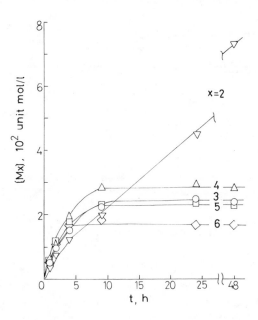

Figure 5.   Formation of cyclic dimer (∇), trimer ($\bigcirc$), tetramer ($\triangle$), pentamer ($\square$), and hexamer ($\diamond$) as a function of depolymerization time. $[M]_T = 0.23M$, $[I]_0 = 4.5 \times 10^{-3}M$ (9).

Blockcopolymer Synthesis

Successive addition of monomers to the end of macromolecular initiator is the usual technique for the synthesis of tailored blockcopolymers. Anionic polymerization of pivalolactone[11], α-pyrrolidone[12] and the NCA of γ-methyl-D-glutamate[13] was started from the end group of a prepolymer consisting carboxylate group or acyl lactam group or amino group. Living polymer of ε-caprolactone was expected to be formed by the initiated polymerization from polymer carbanion under kinetic controlled condition.

All the monomers and solvents were purified with special care for the manipulation under high vacuum system. The synthesis of the blockcopolymer was carried out by the following scheme.

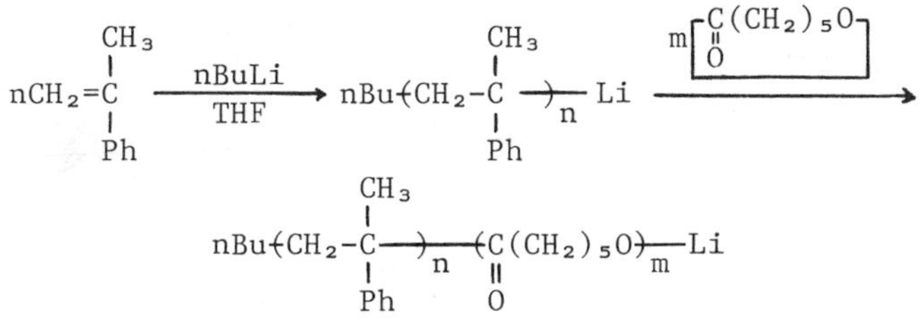

$$
nCH_2{=}\underset{\underset{Ph}{|}}{\overset{\overset{CH_3}{|}}{C}} \xrightarrow[\text{THF}]{\text{nBuLi}} nBu{\left(CH_2{-}\underset{\underset{Ph}{|}}{\overset{\overset{CH_3}{|}}{C}}\right)_n}Li \xrightarrow{m\left[\overset{\overset{C}{\overset{\|}{O}}(CH_2)_5O}{}\right]}
$$

$$
nBu{\left(CH_2{-}\underset{\underset{Ph}{|}}{\overset{\overset{CH_3}{|}}{C}}\right)_n}{\left(C(CH_2)_5O\right)_m}Li
$$

n-Butyllithium in n-hexane was added to the THF solution of α-methylstyrene at room temperature. After the development of the red color of the carbanion, the reaction mixture was cooled to dry-ice temperature by stirring. After several hours, an aliquot of the polymer solution was separated to confirm the complete conversion of α-methylstyrene to reach the expected molecular weight with narrow distribution from the analysis of the precipitated polymer. THF solution of ε-caprolactone was added to the red living polymer solution at -78°C. The color of the solution changed from orange to colorless and the polymerization was continued for several hours at 0°C. The polymerization was completed in several minutes, and then depolymerization reaction began to form cyclic oligomers by the backbiting reaction from the polymer end. The reaction mixture was precipitated by methanol, and trace amount of α-methylstyrene homopolymer was extracted with cyclohexane. The variation of polymer structure with reaction time is shown in Table II and Figure 5.

In number 1, increase of molecular weight by successive addition of ε-caprolactone occurred in a few minutes. The molecular weight distribution is not sharp enough owing to the backbiting reaction at 0°C. In number 2, complete conversion of ε-caprolactone increased the molecular weight, but trans-esterification occurred to increase the molecular weight distribution. The formation of cyclic oligomers was observed in methanol solution. After several hours, the decrease of molecular weight accompanied with the accumulation of oligomers by backbiting reaction became significant and finally equilibrium was attained.

In conclusion, tailored blockcopolymers can be synthesized under kinetically controlled condition.

Table I

Molar Cyclization Equilibrium Constant (Kx) and the Rate Constant
of Propagation ($k_{p,x}$) and Depropagation ($k_{d,x}$) for Cyclic x-mer
at 0°C with LiO-t-Bu (4.5x10^{-3}M)

x	Kx mol/l	$k_{p,x}$ 1/mol·sec	$k_{d,x}$ sec^{-1}
1	0	0.45	0
2	0.065	0.0016	0.00010
3	0.0107	0.010	0.00011
4	0.0098	0.010	0.00010
5	0.0060	0.014	0.0008
6	0.0034	0.020	0.0007

Table II

Blockcopolymerization of ε-Caprolactone by Poly-α-Methylstyrene
Lithium in THF at 0°C

No	Time(hr)	$\overline{Mn}$ a)	$\overline{Mw}$ a)	$\overline{Mw}/\overline{Mn}$
0b)	0	3920	4420	1.13
1c)	0.13	13100	18200	1.39
2	0.83	13700	21600	1.58
3	5	11500	17600	1.53
4	24	10700	11500	1.45
5	120	10600	14100	1.33
6	168	9570	13200	1.38

$[\alpha MeSt]_0$=0.41 mol/l $[nBuLi]_0$=1.12x10^{-2} mol/l
$[ε-CL]_0$=0.58 mol/l
a) by gpc
b) prepolymer as an initiator
c) residual ε-caprolactone was observed

Literature Cited

1. Tabuchi, T., Nobutoki, K., and Sumitomo, H., Kogyo Kagaku Zasshi, 71, 1926 (1968), Nobutoki, K., Sumitomo, H., Bull. Chem. Soc. Japan, 40, 1741 (1967)
2. Perret, R., and Skoulious, A., Makromol. Chem., 152, 291 (1972)
3. Ito, K., Hashizuka, K., and Yamashita, Y., Macromolecules, 10, 821 (1977)
4. Deffieux, A., and Boileaux, S., Macromolecules, 9, 369 (1976)
5. Jacobson, H., and Stockmayer, W. H., J. Chem. Phys., 18, 1600 (1950)
6. Flory, P. J., and Semlyen, J. A., J. Am. Chem. Soc., 88, 3209 (1966)
7. Yamashita, Y., Polymer Preprints (ACS), 20, 126 (1979)
8. Ito, K., Tomida, M., and Yamashita, Y., Polym. Bull., 1, 569 (1979)
9. Ito, K., and Yamashita, Y., Macromolecules, 11, 68 (1978)
10. Nemma, S., and Figueruelo, J. E., Eur. Polym. J., 11, 511 (1975)
11. Yamashita, Y., and Hane, T., J. Polym. Sci., Polym. Chem. Ed., 11, 425 (1973)
12. Yamashita, Y., Murase, Y., and Ito, K., J. Polym. Sci., Polym. Chem. Ed., 11, 435 (1973)
13. Yamashita, Y., Iwaya, Y., and Ito, K., Makromol. Chem., 176, 1209 (1975)

RECEIVED July 22, 1981.

Original Anionic Pathway to New PA(PO)₂ Star-Shaped Block Polymers Based on Polyvinyl or Polydiene Hydrocarbons and Polyoxirane

R. JERÔME and Ph. TEYSSIE

Laboratory of Macromolecular Chemistry and Organic Catalysis,
University of Liege, Sart Tilman, 4000 Liege, Belgium

G. HUYNH–BA

E. I. du Pont de Nemours and Company, Inc., Polymer Products Department,
Experimental Station, Wilmington, DE 19898

Naphthalene-terminated polyvinyl aromatics and
polyisoprene were obtained successfully. These
functional polymers were metalated by potassium
in THF at 25°C. The formation of a stable dinegative ion is observed unless the naphthalene
is directly attached to the end of the polyvinyl
aromatics, in which case a few isoprene units
can be advantageously inserted between the
naphthalene end group and the polyvinyl aromatics.
The polymeric and stable dinegative ion polymerizes
oxirane by both anionic sites and forms three-
branched starshaped block copolymers.

Thanks to their multiphase constitution, block copolymers
have the originality to add advantageously the properties of
their constitutive sequences. These very attractive materials
can display novel properties for new technological applications.
In this respect, thermoplastic elastomers are demonstrated
examples (1, 2, 3); they are currently used without any modification as elastic bands, stair treads, solings in the footwear
industry, impact resistance or flexibility improvers for polystyrene, polypropylene and polyethylene whereas significant
developments as adhesives and adherends are to be noted (4, 5).
   In solution, block copolymers display interesting colloidal
and interfacial properties. They can be used as emulsifying
agents in water-oil and oil-oil systems (6). In the later case,
the oil phases are solid and they give rise to polymeric alloys
(7) or they are liquid and they allow the preparation of latexes
in organic medium (8). However, the molecular structure of block
copolymers based on polybutadiene PB (70%) and polystyrene PS
behave as thermoplastic elastomers when engaged in multiblock
(PB-PS)ₙ or triblock (PS-PB-PS) structures but never when implied
in inverse triblock or diblock arrangements. Similarly the

0097–6156/81/0166–0211$05.00/0
© 1981 American Chemical Society

surface activity of block copolymers based on polyoxirane (PO) and polystyrene seems to depend on their molecular parameters (8).

The actual knowledge of the basic molecular structure property relationships relies mainly on the availability of well defined linear architectures: the di, tri- and multiblock copolymers. New and well controlled molecular structures could undoubtedly provide a deep understanding of the behavior of block copolymers and a more efficient mastering of their applications.

The purpose of this contribution is to show how the design of a new dianionic species can lead to interesting advances in block copolymerization and especially to original PA(PO)$_2$ star-shaped block copolymers. PA is a hydrophobic block: polystyrene, polytertiarybutylstyrene (PTBS) or polyisoprene (PI). The surface activity of this novel and well mastered molecular architecture is considered and compared as far as possible with the behavior of the corresponding PA-PO diblock copolymers.

$$PA(PO)_2 = \quad \text{PA} \quad O \quad \text{PO} \quad \text{PO}$$

Metalation of Naphthalene by Potassium in Tetrahydrofuran at Room Temperature

A polymer PA terminated with a stable dinegative ion able to polymerize oxirane is an ideal pathway towards PA(PO)$_2$ star-shaped block copolymers.

Naphthalene is known to form a stable lithium dianion at -80°C in tetrahydrofuran (THF) at concentrations lower than 0.5 mol.l^{-1} (9-12). Unfortunately organolithium compounds are unable to polymerize oxirane (13). Naphthalene can also be metalation by sodium and potassium in THF but no experimental evidence for a dianion of naphthalene sodium or potassium is to be found. Although naphthalene metalation by sodium is thoroughly described (9, 14, 15), very few results about potassium are published (16, 17). As the reducing power of the alkali metals decreases from lithium (Li/Li$^+$ = 3.02v) to potassium (K/K$^+$ = -2.92v) and finally to sodium (Na/Na$^+$ = 2.71v) (18), it is attractive to study in more detail the naphthalene metalation by K in the THF.

There is a great similarity in the course of the naphthalene metalation by Li at -80°C on one hand and by K at RT on the other hand. The titration of the carbanions formed (19) proves the presence of two anions per naphthalene only at naphthalene concentrations lower than 0.03 mol.l^{-1} for K at RT (Table I) and 0.5 mol.l^{-1} for Li at -80°C (9). In both cases, the dianionic species are able to transfer one of their electrons to neutral naphthalene with formation of the well-known naphthalene radical anion (NRA) (10, 19). The naphthalene metalation by K at RT proceeds in two successive steps: initial reduction to the NRA (Figure 1A) and final formation of the dianionic species at naphthalene concentration lower than 0.03 mol.l^{-1} (Figure 1B). The assumed potassium dianion is still stable at RT after 100 hrs; however, in the presence of an excess of neutral naphthalene, the absorption of the NRA is again observed (Figure 1C).

The hydrolysis products of the potassium naphthalene dianion have been analyzed by proton NMR spectroscopy. As reported elsewhere (19), they agree with eq [1], whereas the ratio between the forms I and II is close to 3.

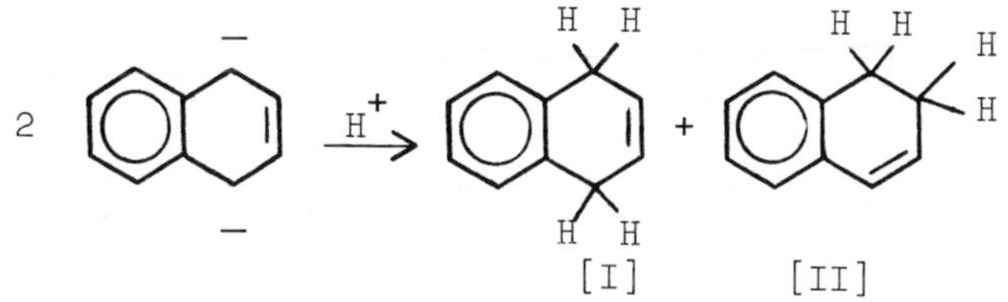

$$[1]$$

On the basis of all these experimental observations the following reaction scheme can be proposed for the metalation of naphthalene by K in THF at RT (eq [2]).

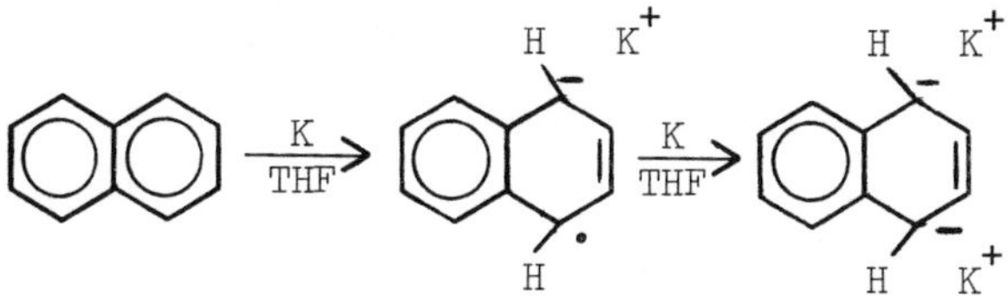

$$[2]$$

Eq [3] takes into account the behavior of the potassium naphthalene dianion in the presence of an excess of naphthalene.

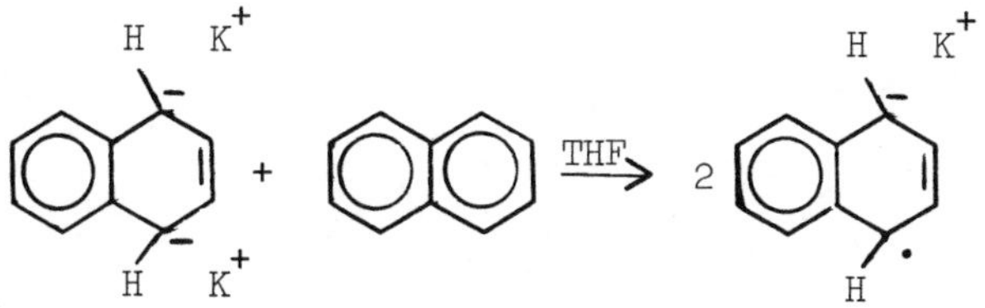

$$[3]$$

Metalation of β-Ethylnaphthalene by Potassium in THF at Room
Temperature

Once fixed at the end of a polymer PA, the naphthalene is β
substituted; to assess the influence of the substitution on the
metalation of naphthalene, β-ethylnaphthalene (EN) was studied
as a model compound.

In contrast to the potassium naphthalene dianion, EN gives
rise to a dianionic species at concentrations higher than 0.03
mol.l^{-1} (Table II). Furthermore the EN dianion is unable to
transfer one electron to another naphthalene molecule. These
sharp differences are to be attributed to the participation
of the β-ethyl group in the metalation process.

The EN metalation by K in THF at RT has been analyzed by UV
spectrophotometry (Figure 2). ENRA (Figure 2A) and potassium EN
dianion (Figure 2, B to be compared to Figure 1B) are successively formed, but a dianionic species characterized by spectrum C
(Figure 2) is finally formed which is stable in the presence of
pure naphthalene. The absorption at 435nm (Figure 2C) corresponds to the value reported for the dihydronaphthalene monoanion
(DHNA) 433-435nm (20). DHNA is easily hydrolyzed into dihydronaphthalene (DHN) which can be again metalated to give DHNA (eq
4).

DHNA   DHN   DHNA   [4]

Similarly the hydrolysis product of the stable potassium EN
dianion (Figure 2C) is metalated by K in THF at RT and the
characteristic of DHNA is again observed at 435nm (Figure 2D).
However, the spectra are slightly different. Hence, the EN
dianion is accordingly assumed to be constituted by a DHNA anion
and also by an extra cyclic carbanion resulting from an intramolecular hydrogen transfer (eq 5, 6). For clarification, the
1,2 dihydro structure is omitted.

III   [5]

TABLE I

Naphthalene Metalation by Potassium

in THF at Room Temperature

Naphthalene Concentration $[N] m.l^{-1}$	Carbanions Concentration $[C^-] m.l^{-1}$ (a)	$\dfrac{[C^-]}{[N]}$
4.5×10^{-3}	0.9×10^{-2}	2.0
9.5×10^{-3}	1.9×10^{-2}	2.0
2.0×10^{-2}	4.0×10^{-2}	2.0
3.0×10^{-2}	5.9×10^{-2}	1.97
7.0×10^{-2}	1.1×10^{-1}	1.57
1.1×10^{-1}	1.1×10^{-1}	1.00

(a) Final constant values

TABLE II

Metalation of β-ethylnaphthalene (EN)

by K in THF at Room Temperature

EN $(m.l^{-1})$	Time of metalation (hours)	[Carbanions] $(m.l^{-1})$	Nr of carbanions per EN
0.160	120	0.32	2.0
0.125	96	0.24	1.9
0.015	33	0.03	2.0

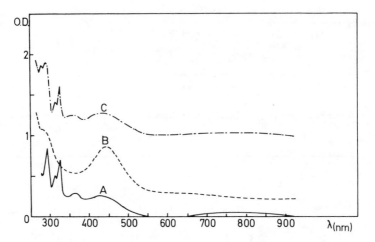

Figure 1. Visible and UV spectra of (A) naphthalene radical anion (NRA) (THF, 25°C); (B) naphthalene dianion (OD values shifted upward by 0.2 unit); (C) naphthalene dianion after neutral naphthalene addition (OD values shifted upward by 1 unit; the NRA concentration is the same in A and B).

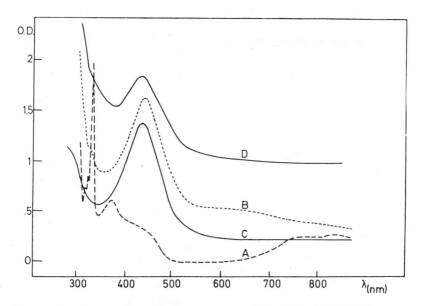

Figure 2. The β-ethylnaphthalene metalation by K in THF at RT. UV spectra after (A) 10 min, NRA spectrum; (B) 10 h, naphthalene dianion spectrum (OD shifted upward by 0.4 unit); (C) 50 h, isomerized β-substituted naphthalene dianion (OD shifted upward by 0.25 unit); (D) compound corresponding to Spectrum C, water deactivated, purified, and again metalated by K (OD shifted upward by 1 unit).

$$[6]$$

Forms III and IV must be considered to describe the naphthalene type dianion which is first formed during the EN metalation (Figure 2, B). The proton NMR analysis of the hydrolysis products (V and VI) of the isomerized EN dianion (Figure 2, C) agrees with the presence of the ethyl group on the dihydronaphthalene ring (19) and supports the existence of the forms IV and VII. Again metalated by K, the DHN entity (VI) gives rise to a DHNA (VIII);(eq 7) the UV absorption of which (Figure 2, D) is slightly modified in the isomerized EN dianion (VII Figure 2, C) by the extra cyclic carbanion.

$$[7]$$

From the experimental results, the following mechanism is proposed for EN metalation (eq 8)

$$[8]$$

Synthesis of Naphthalene Terminated Polymers

The use of lithionaphthalene and lithio β-methylnaphthalene as functionally substituted anionic initiators is very disappointing (21, 22). The deactivation of living polyanions onto α-bromonaphthalene or bromo-β-methylnaphthalene is another unsatisfactory approach due to metal halogen interconversion (eqs 9, 10) (23).

$$[9]$$

$$IX + PS^-Li^+ \longrightarrow LiBr + PS-PS \qquad [10]$$

To limit this process, Richards has proposed to transform the living polyanion into its Grignard analogue which is much less reactive but still able to interact with the usual bromo-reagents (24). Equations 11 and 12 summarize the Grignard polymer formation and its reaction with β-bromomethylnaphthalene

$$PS^-Li^+ \ + \ MgBr_2 \ \longrightarrow \ PS\text{-}MgBr \ + \ LiBr \qquad [11]$$

$$PS\text{-}MgBr \ + \ Br\text{-}CH_2 \text{⬡⬡} \ \longrightarrow \ PS\text{-}CH_2 \text{⬡⬡} \ + \ MgBr_2 \quad [12]$$

By this pathway the degree of polymer coupling is reduced to about 5 to 10% (Figure 3) and the titration of the naphthalene end group indicates a functionality higher than 90% (21, 22, 25).

Metalation of Naphthalene Ended Polydiene by Potassium in the At RT

The metalation of naphthalene β-substituted by both an ethyl group or polyisoprene chain (PIP) is completely similar as established by the titration of the carbanions formed and the UV analysis of the reaction medium (21, 22, 25). Accordingly the naphthalene radical anion, naphthalene dianion and its further isomerization by hydrogen transfer are successively observed and the final stage of the metalation can be represented by the following structure:

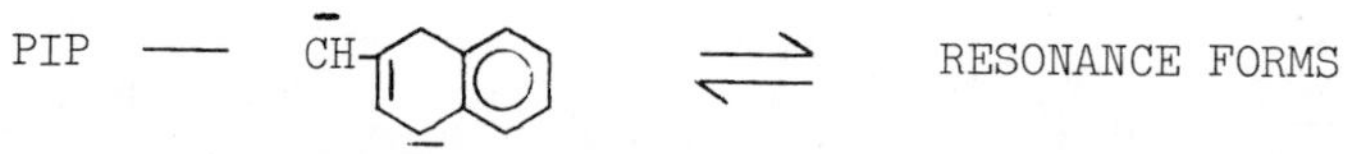

$$PIP \text{———} CH \text{⬡⬡} \ \rightleftharpoons \ RESONANCE \ FORMS$$

Metalation of Naphthalene Ended Polyvinyl Aromatic Chains by Potassium in THF At RT

When naphthalene is substituted by a polyvinyl aromatic chain polystyrene PS, poly(p-tert-butylstyrene) PTBS or poly(α-methylstyrene) PMS the course of the metalation is deeply modified (21, 22, 25).

Four carbonions can be formed per naphthalene end group (Figure 4) whereas the naphthalene end group is completely released at the end of the metalation. This phenomenon has been thoroughly studied and will be published elsewhere (22). Briefly when naphthalene is conjugated to an aromatic nucleus the potassium dianion is unstable and the naphthalene end group is released whereas the polyvinyl anion is freed and finally isomerized; eqs 13 to 17 summarize these observations.

$$\text{wwwCH-CH}_2\text{(naphthyl)} \xrightarrow{K} \text{wwwCH-CH}_2\text{(naphthyl radical)} \qquad [13]$$

with (X), R substituents

$$(X) \xrightarrow{K} \text{wwwCH-CH}_2\text{(naphthyl anion) (XI)} \qquad [14]$$

$$(XI) \longrightarrow \text{wwww}\bar{C}H\text{ (XII)} \quad + \quad {}^-CH_2\text{(naphthyl) (XIII)} \qquad [15]$$

$$2\,[XII] \xrightarrow{\text{Rearrangement}} \text{wwCH}_2 + H^- + \text{wwCH}_2\text{-}\bar{C}\text{-CH=CH} \qquad [16]$$

$$XIII \xrightarrow{K} \bar{C}H_2\text{(naphthyl)} \xrightarrow{K} \bar{C}H_2\text{(naphthyl)} \qquad [17]$$

Similar instability of conjugated anionic systems has already been observed; Lagendijk and Szwarc have reported that di(α-naphthyl)ethane dianion (NN^{-2}) in ether solvent undergoes scission of the central C-C bond with formation of two monoanion ($NN^{-2} \rightarrow 2N^-$) (26). To suppress the deleterious effect of conjugation and to obtain a stable dianion terminated PS or PTBS, at least one isoprene unit is to be inserted between the polyvinyl aromatic chain and the naphthalene end group; this opportunity has been successfully applied.

Star-Shaped Block Copolymerization

In the first step the oxirane polymerization initiated by the isomerized EN dianion has been studied. The results reported in Table III and the low polydispersity observed (about 1.1; Figure 5) are conclusive for a living anionic system. Moreover the initiation rate of which is higher than the propagation one.

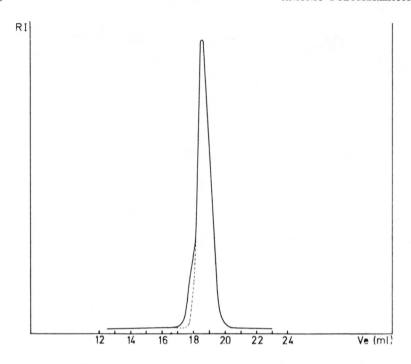

Figure 3. *GPC of a naphthalene-terminated polystyrene ($\overline{M}_n = 3400$) obtained by reaction of polystyrylmagnesium bromide with β-bromomethylnaphthalene.*

TABLE III. Oxirane Polymerization Initiated by the
Stable Potassium β-ethylnaphthalene Dianion[a]

Sample	[Initiator] (m.l^{-1})	Conversion (%)	$\overline{M}_n$ theor. (b)	$\overline{M}_n$ exper.
PO 1	16.3×10^{-4}	100	2.6×10^3	2,700
PO 2	8.2×10^{-4}	93	5.0×10^3	5,200
PO 3	5.4×10^{-4}	95	7.7×10^3	7,500

(a) [Oxirane] = 0.1 m.l^{-1}, time: 72 hrs, temperature: 25°C solvent: THF (80%) + Bz (20%).

(b) $\overline{M}_n$ calculated on the basis of the monomer over initiator molar ratio.

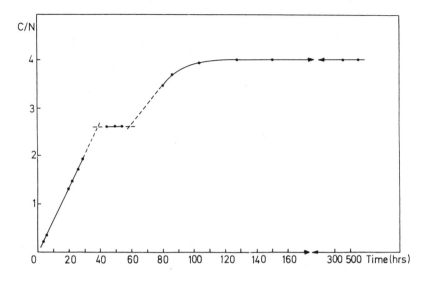

Figure 4. *Kinetics of the metalation of naphthalene-terminated PTBS ($\overline{M}_n$ = 3400) by K (THF, 25°C, [naphthalene] = 3 × 10⁻² mol/L); C/N = number of carbanions generated per naphthalene end group.*

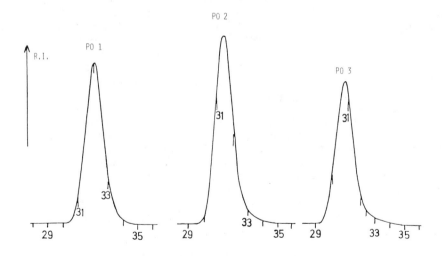

Figure 5. *GPC of polyoxirane using potassium β-ethylnaphthalene dianion as initiator.*

The oxirane polymerization can be described by eqs 18 and 19:

$$\overline{C}H\text{-}CH_3 \quad \xrightarrow[k_{a1}]{C_2H_4O} \quad \overset{CH_3}{\underset{}{CH}}\text{-}CH_2\text{-}CH_2\text{-}O^- \qquad\qquad [18]$$

XIV

$$XIV \xrightarrow[k_{a2}]{C_2H_4O} \quad \overset{CH_3}{CH}\text{-}CH_2\text{-}CH_2\text{-}O^-$$

$$\xrightarrow{k_{prop}} \quad CH_2\text{-}CH_2\text{-}O^- \qquad\qquad [19]$$

$$\overset{CH_3}{CH}\text{-}CH_2\text{-}CH_2\text{-}O\text{-}CH_2\text{-}CH_2\text{-}O^-$$

From the kinetics of the oxirane polymerization initiated by
alcoholate (16) and by fluorenyl potassium (27) and as fluorenyl
and dihydronaphthalene mono anion (28) have approximately the
same basicity, the k_{a2} over k_{prop} ratio may be estimated to 20.
Therefore, the length of the two growing polyether chains must
be largely independent on the nature of the initiating site.

In the second step, oxirane has been added at -30°C to an
isomerized and stable naphthalene dianion fixed at the end of the
PA chain. Oxirane is completely polymerized at RT, and the crude
product obtained is separated into block copolymer, starting
homopolymer PA and homopolyether. The separation scheme is
described in Table IV and the results obtained for a great lot
of PA (PIP, PS, and PTBS) based copolymers agree with a ratio
of about 90% block copolymer, 10% homo PA and only traces of
homopolyether.

The composition of the star-shaped block copolymer is easily
determined by proton NMR analysis; from this and the mean number
average molecular weight ($\overline{M}n$) of the sequence PA, $\overline{M}n$ of the polyether component can be calculated. The later is very similar
to the value from membrane osometry. Hydroxyl end group of
$PA(PO)_2$ star-shaped block copolymers have been titrated and their
mean number per copolymer (1.85) agrees with the presence of two
polyoxirane branches. On the average, the polydispersity of the
star-shaped block copolymers varies between 1.2 and 1.3 (Figure
6).

The molecular parameters of a series of well characterized
$PA(PO)_2$ block copolymers can be found in Table V.

TABLE IV

Separation of PTBS (PO)$_2$ Copolymers

from Corresponding PTBS and PO Homopolymers

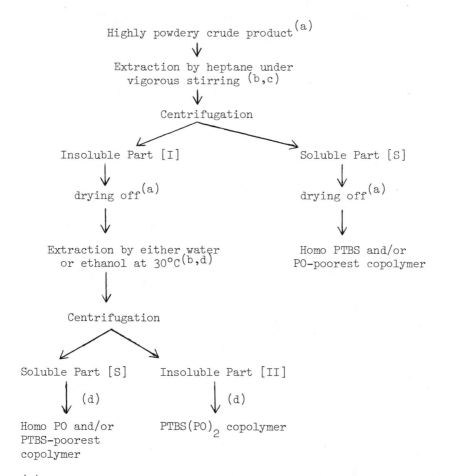

Highly powdery crude product[a]

Extraction by heptane under
vigorous stirring [b,c]

Centrifugation

Insoluble Part [I]   Soluble Part [S]

drying off[a]   drying off[a]

Extraction by either water   Homo PTBS and/or
or ethanol at 30°C[b,d]   PO-poorest copolymer

Centrifugation

Soluble Part [S]   Insoluble Part [II]

(d)   (d)

Homo PO and/or   PTBS(PO)$_2$ copolymer
PTBS-poorest
copolymer

(a) The polymer is dissolved in benzene and the solution frozen.
    Benzene is removed under vacuum and a powdery polymer is
    obtained.
(b) Extraction is repeated three times with fresh solvent.
(c) The separation scheme is also applicable to PS(PO)$_2$
    copolymers but diethyl ether is used instead of heptane.
(d) Higher molecular weight copolymers are extracted by ethanol.

TABLE V
Molecular Parameters of the PA(PO)$_2$ [a]
Star-shaped Block Polymers

Copol. [b]	$\overline{Mn}$ x 10^3 Copol. [c]	$\overline{Mn}$ A x 10^3 [c]	$\overline{Mn}$(2xPO) x 10^{3} [d]	A % [e]	PO % [e]
I2	4.8	2.1	2.7	44	56
S1	6.3	3.0	3.3	48	52
S2	15.0	4.8	10.2	32	68
S3	18.0	4.8	13.2	27	73
S4	25.0	4.8	20.0	19	81
B11	5.7	1.9	3.8	33	67
B12	7.6	1.9	5.7	25	75
B21	4.5	3.0	1.5	67	33
B22	4.9	2.7	2.2	55	45
B23	7.5	3.0	4.5	40	60
B24	8.8	3.0	5.8	34	66
B25	12.0	3.0	9.0	25	75
B26	20.0	3.0	17.0	15	85
B27	60.0	3.0	57.0	5	95
B32	16.7	6.5	10.2	39	61
B33	20.0	6.5	13.5	33	67
B41	19.0	9.6	9.4	50	50
B42	42.0	9.6	32.4	23	77
B51	53.0	15.0	38.5	28	72

(a) A is a polyisoprene (PIP), polystyrene (PS) or poly p-
    tertbutylstyrene (PTBS) sequence.
(b) I or PIP(PO)$_2$ copol. - S or PS(PO)$_2$ copol., and B or
    PTBS(PO)$_2$ copol.
(c) $\overline{Mn}$ determined by vapor or membrane osmotic pressure.
(d) $\overline{Mn}$ calculated from $\overline{Mn}$ copol. and $\overline{Mn}$ A.
(e) Calculated from ^{1}H NMR analysis.

Surface Activity of the PA(PO)$_2$ Star-Shaped Block Copolymers

Experimental evidences of the influences of the new PA(PO)$_2$ molecular architecture on the surface activity are gathered hereafter for the copolymers based on both PA and PO blocks.

Riess demonstrated recently that poly(styrene-b-oxirane) copolymers could act as non-ionic surfactants and lead to water/ toluene microemulsions (29, 30). Using isopropanol as cosurfactant, both O/W and W/O microemulsions are obtained (31). This is a very important conclusion, since PO based diblock copolymers give rise only to O/W microemulsions under the same experimental conditions (8, 31). In this respect the "branched structure" of the PO hydrophilic component could favor a decrease in the packing density of the inverse micelle forming molecular and explain the different behavior of the linear and star-shaped PS/PO block copolymers in the W/O microemulsification process.

Similarly low molecular weight (+5000) copolymers containing 50% polyoxirane (I$_2$, S$_1$ and B$_{22}$, Table V) have the originality to generate both O/W and W/O stable water toluene emulsions (32). The effect of the molecular architecture of block copolymers on the stability of water toluene emulsions is shown in Table VI. The criterion of stability (R) is the constant limit value of the emulsified volume percentage at 20°C (32, 33). It appears that 7/3 water toluene emulsions are more efficiently stabilized by star-shaped block copolymers whereas linear block copolymers give better results for 3/7 water toluene emulsions; the stability of the 1/1 water toluene emulsions seems to be insensitive to the molecular architecture of PTBS/PO block copolymers.

As far as water/toluene interfacial tension is measured, it appears that the saturation of the interface is reached more quickly with PTBS (PO)$_2$ star-shaped block copolymers (Table VII); this molecular architecture seems to be more efficient to fill in the interface (34).

Figure 7 compares the water/toluene interfacial tensions measured in the presence of various commercial surfactants and PO/PS based diblock (8) and star-shaped copolymers: the higher activity of the star-shaped block copolymers over a broad range of concentrations is clearly put in evidence.

Conclusion

The knowledge and good control of new dianionic species like β substituted naphthalene dianions is a very attractive tool to tailor block copolymers with the new molecular structure. This approach can be applied to develop new materials enjoying original and useful sets of properties.

TABLE VI

Dependence of the Stability (R) of Water-Toluene

Emulsions on the Molecular Composition of PTBS $(PO)_2$

Copolymers of Constant Molecular Weight

Copol. (1 %)	$\overline{Mn}$ x 10^3	PO %	R(%) at different water-toluene volume-ratios (a)		
			3/7	1/1	7/3
B_{26}	20	85	41	42	42
B_{33}	20	67	39	63	94
B_{41}	19	50	42	$65^{(a)}$	67

(a) W/O emulsions, all the other emulsions being O/W.

TABLE VII

Toluene/Water Interfacial Tension of Diblock

and Star-shaped Copolymers at Different Concentrations (20°C)

P O %	Diblock copol.(a)			Star-shaped copol.(b)		
	0.04%	0.4 %	1 %	0.04%	0.4 %	1 %
33(B21)				6.0	4.1	3.0
41	1.4	0.8	1.0			
45(B22)				0.6	(+)	(+)
57	7.2	6.0	1.2			
66(B24)				4.7	4.0	3.9
68	8.3	4.4	3.2			

(+)   too low to be measured
(a)   based on a PTBS sequence the $\overline{Mn}$ of which is 2.700
(b)   see Table V for their molecular parameters

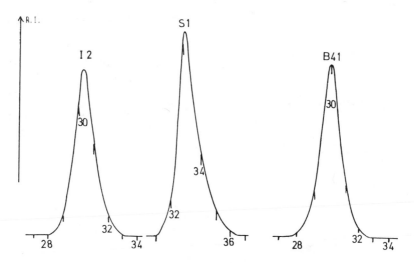

Figure 6. *GPC of PA(PO)₂ star-shaped block polymers I2, S1, and B41 (see Table V for explanation).*

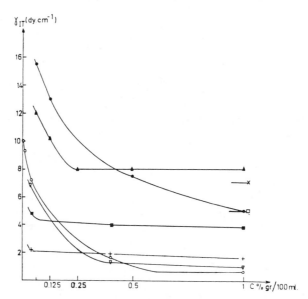

Figure 7. *Comparison of the toluene–water interfacial tension obtained from commercial surfactants and polyoxirane-based block copolymers with different molecular architecture.*

(▲) EPOL (PEO laurate; $\overline{DP} = 6$); 25°C (A) (●) PS/PEO ($\overline{M}_n = 17,700$, 63% POE), 25°C (A) (X) Tween 80 (sorbitol monooleate –PEO; $\overline{DP} = 20$), 20°C (B) (□) Span 80 (sorbitol monooleate–PEO; $\overline{DP} = 4.3$); 20°C (C) (■) PTBS/(PEO)₂ ($\overline{M}_n = 8800$, 66% PEO); 20°C (C); (▽) PTBS/(PEO)₂ ($\overline{M}_n = 7600$, 75% PEO); 20°C (C); (+) PIP(cis)/ (PEO)₂ ($\overline{M}_n = 4800$, 56% PEO); 20°C (C); (○) PS/(PEO)₂ ($\overline{M}_n = 18,000$, 73% PEO; 20°C (C); (A) measured with a Dognon–Abriba tensiometer; (B) pendent drop method— aqueous solution/toluene; (C) sessile drop method—toluene solution/water.

Acknowledgements

The authors are very much indebted to Imperial Chemical Industry for efficient support through the first European Joint Project, including a Fellowship to H. B. G. They also want to express their gratitude to Dr. S. Boileau (Paris VI University) for very fruitful discussions and Professor J. E. McGrath (Virginia Polytechnic Institute and State University - Chemistry Department - Blacksburg, Virginia - 24061 - USA) for attention paid to this contribution.

Literature Cited

1. a) Holden, G., Bischop, E. T. and Legge, N. R., J. Polym. Sci. Part C, (1969), 26, 37.
    b) McGrath, J. E. and Noshay, A., "Block Copolymers", Acad. Press, (1977).
2. Aggarwal, S. L., Polymer, (1976), 17, 938.
3. Bonart, R., Polymer, (1979), 20, 1389.
4. Gent, A. N. and Hamed, G. R., Rubber Chem. Technol., (1978), 51, 354.
5. Carter, A. R. and Petit, D., "Developments in Adhesives", Wake, W. C. Edit., Applied Science Publ., (1977), p. 223.
6. Reiss, G., Periard, J. and Banderet, A., "Colloidal and Morphological Behaviour of Block and Graft Copolymers", Molau, G. E. edit., Plemum Press, N.Y. (1977), p. 173.
7. Schmitt, B. J., Angew. Chem. Trt. Ed. Engl., (1979), 18, 273.
8. Nervo, J., Ph.D. Thesis, Mulhouse, France (1976).
9. Burgess, F. J., Cunliffe, A. V. and Richards, D. H., Europ. Polym., J., (1974), 10, 645.
10. Smid, J., J. Amer. Chem. Soc., (1965), 87, 655.
11. Fujita, T., Suga, K. and Wantanabe, S., Synthesis, (1972), 630.
12. Carnahan, J. C., Jr., and Closson, W. D., J. Org. Chem., (1972), 37, 4469.
13. Steiner, E. C., Pelletier, R. R., and Trucks, R. O., J. Am. Chem. Soc. (1964), 86, 4678.
14. Szwarc, M., "Progress in Physical Organic Chemistry", Streitwieser, A. and Taff, R. W. Edit., John Wiley, Interscience, (1968), 6, 313.
15. Hsieh, H. L., J. Organomet. Chem., (1967), 7, 1.
16. Kazanski, S. K., Solovyanov, A. A. and Entelis, S. G., Europ. Polym. J., (1971), 7, 1421 and J. Polym. Sci. URSS, (1972), 14, 1186.
17. Buschow, K. H. J. and Hoijtink, G. J., J. Chem. Phys., (1964), 40, 1501.
18. Hoijtink, G.J. deBoer, E. Van Der Maij, P. H. and Weyland, W. P., Rec. Trav. Chem., Pay-Bas., (1956), 75, 487.
19. Huynh-ba, G., Jerome, R. and Teyssie, Ph., J. Organomet. Chem., (1980), 190, 107.

20. Bank, S. and Bockrath, B., J. Amer. Chem. Soc., (1979), 93, 430.
21. Huynh-ba, G., Ph.D. Thesis, University of Liege, (1978).
22. Huynh-ba, G., Jerome, R. and Teyssie, Ph., J. Polym. Sci., Polym. Chem. Edit., accepted for publication.
23. Gilman, G., Cecil, G. B. and Ingham, R. K., J. Org. Chem., (1957), 22, 685.
24. Richards, D. H. and Burgess, F. J., Polymer, (1976), 17, 1020.
25. Teyssie, Ph., Jerome, R. and Huynh-ba, G., Polymer Prepr. Div. Polym. Chem., A.C.S., (1979) 20(2), 344 and (1980), 21(1), 77.
26. Lagendijk, A. and Szwarc, M., J. Amer. Chem. Soc., (1971), 93, 5359.
27. Lassale, D., Ph.D. Thesis, Paris, (1972).
28. Streitwieser, A. and Hammons, J. H., "Progress in Physical Organic Chemistry", John Wiley (1965), 3, 41.
29. Riess, G. and Nervo. J., Information Chimie, Fr., (1977), 170, 185.
30. Riess, G., Nervo, J. and Rogez, D. Polym. Prepr. Div. Polym. Chem. A.C.S., (1977), 18, 329.
31. Huynh-ba, G., Jerome, R. and Teyssie, Ph., Colloid Polym. Sci., (Kolloid-Z.u.Z. Polymere), (1979), 257, 1979.
32. Huynh-ba, G., Jerome, R. and Teyssie, Ph., J. Appl. Polym. Sci., (1981), 26, 343.
33. Hallworth, G. W. and Larless, J. E., J. Pharm. Parmacol., (1972), 24, suppl. 71 and 87.
34. Huynh-ba, G., Jerome, R. and Teyssie, Ph., J. Polym. Sci., Polym. Phys. Edit., (1980), 18, 2391.
35. Current Address.

RECEIVED March 25, 1981.

16

Ion Pair Structure and Stereochemistry in Anionic Oligomerization and Polymerization of Some Vinyl Monomers

T. E. HOGEN-ESCH, W. L. JENKINS, R. A. SMITH, and C. F. TIEN

Chemistry Department, University of Florida, Gainesville, FL 32611

We have recently (1,2) reported the stereoregular anionic oligomerization of 2-vinylpyridine and various alkylacrylates in THF and similar low dielectric constant media (Eq. 1):

$$CH_3CH_2R \xrightarrow{R'M} CH_3\bar{C}HR,M^+ \xrightarrow[-78°/THF]{R} CH_3[CHR-CH_2]_n^-CHR^-,M^+$$

$$[1] \qquad\qquad [2] - [5]$$

$$\xrightarrow[-78°/THF]{CH_3I} CH_3[CHR-CH_2]_n-CHRCH_3 \qquad\qquad (Eq. 1)$$

$$[6] - [9]$$

a) R = 2-pyridyl, b) R = 4-pyridyl, c) R = carboalkoxy.
[2],[6] n = 1; [3],[7], n = 2; [4],[8] n = 3; [5],[9] n ⩾ 4.
M = Li, Na, K, Rb

One of the interesting features of these reactions is the potential for comparing the stereochemistry of oligomerization with that of alkylation protonation or other organic reactions of general interest. For instance, Table 1 shows that the methylation of the Li and Na salts of anion [2a] is highly stereoselective,

Table 1

Methylation Stereochemistry of Anion [3]
As a Function of Cation, and Solvent or
Coordinating Agent[f]

Cation	Solvent/Coordinating Agent	Electrophyle	T°C	% Meso
Li	THF	CH_3I	-78°	>99
Li	THF	CH_3I	-30°	98
Li	THF	CH_3I	0°	95
Na	THF	CH_3I	-78°	96

0097–6156/81/0166–0231$05.00/0

Table 1 (Cont'd.)

Cation	Solvent/Coordinating Agent	Electrophyle	T°C	% Meso
K	THF	CH_3I	-78°	70
Rb	THF	CH_3I	-78°	57
Li	THF/Pyridine[a]	CH_3I	-78°	83
Li	THF/TG[b]	CH_3I	-78°	93
Li	THF/12-crown-4[c]	CH_3I	-78°	92
Li	Toluene[d]	CH_3I	-78°	>99
Li	Toluene[g]	CH_3I	-78°	60
Li	THF	CH_3SO_2F	-78°	--e
Li	THF	$(CH_3)_2SO_4$	-78°	>99
Na	THF/18-crown-6[c]	CH_3I	-78°	58

a. 50% by volume, b. Triglyme 20% excess over carbanion,
c. 20% excess of crown ether, d. Trace of THF present,
e. Mixture of pentane and butane products, f. About .2M
g. Ethyl carbanion salt prepared in toluene without the
   presence of THF.

but the stereoselectivity decreases sharply with increasing
cation size and -coordination. The results also suggest that
the methylation stereochemistry is kinetically determined, and
this is confirmed by epimerization (k-tBuO/DMSO) of the meso-
pentane which yields approximately equal proportions of the
meso- and racemic 2,4-di-(2-pyridyl)pentane. The place of substitution in both the pyridine rings of [2] is a key factor in
the stereoselectivity. The methylation of the Li salt of [2b]
does yield approximately equal proportions of meso and racemic
[6b] under these conditions, and the methylation of the 1-
lithio-1-(4-pyridyl)-3(2-pyridyl)butane likewise yields a 50/50
mixture of diastereomers.

A comparison of the stereochemistry of the 2-vinylpyridine
addition to [2a] with that of methylation addition is particularly instructive. Table 2 shows the stereochemistry of formation of [7a] as a function of counterion size and coordination.

Table 2

Stereochemistry of Formation of 2,4,6-Tri(2-pyridyl)-heptane
As a Function of Cation and Cation Coordination in THF at -78°C

	I(m,m)	H(mr)	S(rr)
Li, THF[d]	⩾95(98.0)[a]	⩽5(2.0)	⩽5(~0)
Na, THF	⩾95(92.2)	⩽5(7.7)	⩽5(0.1)
Na, THF; Na, Ce, THF[b]	~50(56.0)	~50(44.0)	

Table 2 (Cont'd.)

	I(m,m)	H(mr)	S(rr)
K, THF	50.0(49.0)	42.6(42.0)	7.4(9.0)
t-BuOK, DMSO[c]	25.0(25.0)	55.0(50.0)	20(25.0)

a. Values in brackets calculated using Table 1.
b. 18-Crown-6 present during methylation only.
c. Epimerization of isotactic trimer for 300 hrs. at 25°C.

These results indicate that for the 4 systems examined, the methylation and 2-vinylpyridine addition stereochemistry are identical. Thus predominantly isotactic [7a] is formed with Li and Na as counterions. However, when the methylation of [3a] is carried out in the presence of "crowned" Na ion, approximately equal proportions of isotactic and heterotactic [7a] are obtained in accordance with the 50% stereoselectivity of methylation of the corresponding Na/crown complex of [2a] (Table 1).

The results for the potassium system are likewise closely similar to the distribution of stereoisomers calculated by equating the stereoselectivity of 2-vinylpyridine addition with that of methylation. With Li as counterion, the oligomerization of 2-vinylpyridine continues to be highly stereoselective, at least up to ten monomer units and most likely considerably higher (3).

The dimerization of methyl- isopropyl- and tert butyl-acrylates under the conditions of Equation 1 is less stereoselective (Table 3) (2). In less polar media such as toluene, however, the methylation stereoselectivity is enhanced.

Table 3

Stereochemistry of Formation of 2,4-Dicarboalkoxy-pentanes Prepared According to Equation (1) at -78°C

Initiator[a]	Solvent	Ester	% Meso
Li ICHA[b]	THF	Methyl	57
Li DIPA[c]	THF	Isopropyl	61
Li DIPA	Toluene	Isopropyl	68
Li DIPA	Toluene	Isopropyl	82
Li DIPA	THF	t-Butyl	62

a. Initiator is reacted with the propanoate ester to produce [1c].
b. Lithio isopropylcyclohexylamide
c. Lithio diisopropylamide.

Our experimental findings are summarized as follows:
1) The formation of symmetrical oligomers [6]-[9] of 2-vinyl pyridine in the presence of Li (or Na) ion is highly

isotactic (>95%). Thus no minimum chain length is required for stereoselective addition of 2-vinylpyridine.

2) The stereochemistry of methylation of and addition of 2-vinylpyridine to the "living" 2-vinylpyridine oligomers is identical.

3) The size and coordination of the cation as well as the position of the nitrogen atom in both ultimate and penultimate pyridine groups are key parameters in the stereoselectivity of methylation and 2-vinylpyridine addition.

4) The observed stereoselectivity is kinetically controlled.

The mechanism proposed to account for these observations involves two diastereomeric ion pairs [10] and [11] (1).

[10]   [11]

[10a]   [11'a]

It was shown (4, 5) that in these systems small cations are coordinated with the nitrogen lone pair of the penultimate 2-pyridine ring. Models [10a] and [11'a] incorporating this feature (6) indicate that [10a] is preferred over [11'a] on account of CH_3 butane gauche and CH_3-nitrogen lone pair interactions in [11'a]. A space filling model of [10a] (Fig. 1) shows, moreover, that an exchange of CH_3 and H of the penultimate asymmetric center (generating [11'a]) results in unfavorable interactions between CH_3 group and cation.

Thus in the presence of small cations where coordination with the penultimate 2-pyridyl group is most likely, the stereoselectivity of these systems is readily understood by a cation side (syn) attack of electrophyle on the preferred diastereomer [10]. With large or more extensively solvated cations such as Rb+ or (Na, CE)+, coordination with the

Figure 1. Space-filling model of ion pair diastereomer [10a]

penultimate 2-pyridyl group is weakened, and the unchelated anions are expected to predominate. In such a case [10] and [11] are expected to be about equally stable so that the stereoselectivity is dramatically decreased as is observed.

The proposed mechanism is consistent with the experimental results obtained so far. For instance the effects of cation size and -coordination are readily understood in terms of the chelated complex [10a] as is the absence of stereoselectivity in the case of the corresponding 4-pyridyl derivatives where coordination with cation in the manner shown above is impossible. The similarity between vinyl addition and methylation stereochemistry is likewise consistent with the proposed mechanism. Thus it is the equilibrium between [10] and [11] that is primarily responsible for the observed stereochemistry. Work on the stereochemistry of other electrophylic reactions of [2a] is in progress.

The lesser stereoselectivity observed for the acrylate dimers [2c] in THF is readily understood in terms of the lesser coordinative ability of the carboalkoxy group compared to that of pyridine (7) (Table 3).

Table 4

Stereochemistry[a,b] Of Polymerization At -78°C Of 2-(2-Pyridyl)-Propene As A Function of Cation and Cation Coordination

	I	H	S
Li THF	5.8	24.7	69.5
Li Tol	1.6	17.8	80.6
Li [2.1.1], THF	3.3	16.7	80.0
Li [2.2.1], THF	2.6	16.8	80.6
Cs THF	6.2	31.5	62.3

a. Tentative assignment (see text) based on alphamethyl ^{1}H absorption.
b. Spectra taken at 100 MHz at 120°C in o-dichloro benzene with C_6D_6 as internal lock.

Replacing THF with toluene allows increased coordination of cation and ester group, and thus enhances stereoselectivity.

The applicability of the proposed mechanism to the alpha methyl derivative of 2-vinylpyridine, 2-(2-pyridyl)propene is of considerable interest because of its similarity to the analogous methylmethacrylate. Thus the stereochemistry of oligomerization of this monomer was investigated (Eqn. 2).

$$(CH_3)_2\underset{\substack{|\\2\text{-py}}}{C^-}, Li^+ \quad \xrightarrow[-78°/]{\underset{2\text{-py}}{\overset{CH_3}{\Longrightarrow}}} \quad \xrightarrow[THF]{CH_3I} \quad CH_3-\underset{\substack{|\\2\text{-py}}}{\overset{\overset{CH_3}{|}}{C}}-[CH_2-\underset{\substack{|\\2\text{-py}}}{\overset{\overset{CH_3}{|}}{C}}]_n-CH_3 \quad (Eqn. 2)$$

[12] - [16]

[12], n = 1; [13], n = 2; [14], n = 3; [15], n = 4; [16], n >50
[12] - [15] were obtained by preparative HPLC and are
sharp melting crystalline solids (melting points: [12], 85°;
[13], 105°; [14], 183°). The ^1H and ^{13}C 100 MHz NMR spectra of
compounds [14] and [15] showed no evidence for the formation of
more than one stereoisomer. The 270 MHz methylene ^1H NMR spectrum of [14] showed the expected AB quartet centered at δ = 2.30
ppm in addition to an apparent singlet at δ = 2.40 ppm. The
use of 600 MHz spectrometry ([8]) failed to resolve this singlet
so that the formation of a racemic isomer is possible, although
not certain.

The presence of a methyl group could affect the oligomerization through its presence at the carbanion or the penultimate
site. In order to separate these effects, anion [17], prepared
by addition of 2-(2-pyridyl)propene to lithio-2-ethylpyridine
in THF at -78°C, was reacted with CD_3I. (Eqn. 3)

$$CH_3-CH - CH_2-\overset{\overset{CH_3}{|}}{C} -, \; Li^+ \quad \xrightarrow{CD_3I}{CH_3I} \quad CH_3-CH -CH_2-\overset{\overset{CH_3}{|}}{C} -CH_3(CD_3)$$
$$\quad\quad 2\text{-py} \quad\quad 2\text{-py} \quad\quad\quad\quad\quad\quad\quad\quad 2\text{-py} \quad\quad 2\text{-py}$$

(Eqn. 3)

$$[17] \quad\quad\quad\quad\quad\quad\quad\quad [18]$$

The geminal methyl groups are diastereotopic, and absorb at
1.16 and 1.26 ppm. The corresponding CD_3 substituted product
shows only the absorption at δ = 1.26 ppm so that this alkylation is highly (>90%) stereoselective. This finding is quite
reasonable on the basis of the proposed model. Thus, from
inspection of [10a] and Fig. 1, the presence of an alpha alkyl
substituent is not expected to interfere with the relative
stability of [10a] and [11'a].

The situation is different in the case of a γ substituted
methyl group. Structure [10a] and Fig. 1 both show the rather
severe steric repulsions caused by the substitution of the γ
methine proton by CH_3. The methylation (CD_3I) of trimer anion
[19], however, proceeded with better than 95% stereoselectivity, although the absolute stereochemistry of this reaction is
not known with certainty.

$$CH_3-\overset{\overset{CH_3}{|}}{C} - CH_2-\overset{\overset{CH_3}{|}}{C} - CH_2-\overset{\overset{CH_3}{|}}{C} -, \; Li^+$$
$$\quad 2\text{-py} \quad\quad 2\text{-py} \quad\quad 2\text{-py}$$

$$[19]$$

It is by no means certain, moreover, that the cause of this
stereoselectivity is identical with that observed for [17].

Polymers [16] were prepared in THF at -78° according to
Equation 2 with Li   and Cs as counterions in and without the

presence of the strongly Li ion complexing cryptand 2.1.1.
In all cases, three methyl absorptions could be observed
at δ = 0.28, 0.68, and 1.29 ppm in addition to the methylene
absorptions at 2.30 and 2.44 ppm. The results (Table 4) show
that the differences in stereochemistry between the various
counterions are not nearly as important as observed for the
oligomerization of 2-vinylpyridine.
An unambiguous stereochemical assignment of the alpha
methyl absorptions is difficult to make. On the basis of the
similarity of the observed spectra to the closely similar poly
(alphamethylstyrene) (9, 10, 11), it is tempting to assign the
upfield and downfield absorptions to syndiotactic and isotactic
triads, respectively. This assignment would be supported by the
300 and 600 MHz ^1H spectra of tetramer [14] that suggest predominant (>90%) formation of a racemic dyad.
On the other hand, the similarity of 2-(2-pyridyl)propene
and 2-vinylpyridine would lead one to expect a similar oligomerization and polymerization stereochemistry, and this is
consistent with the formation of single stereoisomers in the
oligomerization of both monomers in the presence of Li ion.
Crystallographic studies on a single crystal of [14] are presently carried out in order to resolve this question.
We wish to acknowledge NSF support of this work under
Grant DMR 76-17781, A01, Polymers Program.

Literature Cited

1. Tien, C.F.; Hogen-Esch, T.E., Macromolecules, 1976, 9, 871;
    Ibid. J. Am. Chem. Soc., 1976, 98, 7109;
    Jenkins, W.L.; Tien, C.F.; Hogen-Esch, T.E., J. Pure Appl.
    Chem., 1979, 51, 139, and references therein.
2. Hogen-Esch, T.E.; Tien, C.F., J. Polym. Sci., 1979, 17, 281.
3. Hogen-Esch, T.E. and Tien, C.F., J. Polym. Sci.-Polymer
    Letters, 1979, 17, 431.
4. Fisher, M. and Szwarc, M., Macromolecules, 1970, 3, 23.
5. Tardi, M.; Rouge, D. and Sigwalt, P., Europ. Polym. J., 1967,
    3, 85.
6. The stereoisomer shown [11'a] corresponds to the chelated
    enantiomer of [11].
7. Unpublished results from this laboratory on solvation of
    alkali salts of the acidic hydrocarbon fluoradene (Hogen-
    Esch, T.E., J. Am. Chem. Soc., 1973, 95, 639) show that
    substituted pyridines coordinate alkali ions considerably
    stronger than esters.
8. We wish to thank Professor Bothner By at Carnegie Mellon
    University for the use of the 600 MHz spectrometer.
9. Brownstein, S.; Bywater, S. and Worsfold, D.J., Makromol.
    Chem., 1961, 48, 127; Lenz, R., Makromol. Chem., 1976, 177,
    (3), 653.
10. Lenz, R.W., J. Macromol. Sci.(Chem.), 1975, A9(6), 945.
11. Elgert, K.F.; Seiler, E.; Puschendorf, G.; Ziemann, W. and
    Cantow, H.J., Makromol. Chem., 1971, 144, 73.
    Elgert, K.F.; Seiler, E., Makromol. Chem., 1971, 145, 95.

RECEIVED February 9, 1981.

Living Anionic Stereospecific Polymerization of 2-Vinylpyridine

ALAIN SOUM and MICHEL FONTANILLE

Laboratoire de Recherches sur les Macromolecules, de l'Universite Paris—Nord (ERA 607), 93430 Villetaneuse, France

Living anionic systems are among the most suitable to study the mechanism of polymerizations. Indeed, the high efficiency of anionic initiators, combined with the good stability of active centers, allows one to study directly the active species by common physico-chemical methods.

Natta et al. have shown (1,2) that stereospecific polymerization of 2-vinyl pyridine (2VP) can be achieved by using de-solvated Grignard reagents as initiators, in solution in hydrocarbon solvents. Although the essential characteristics were not revealed in this work, there is no doubt that an anionic mechanism is operative in such a polymerization.

Recently, Hogen Esch et al. (3,4,5), obtained the stereo-specific oligomerization of 2-VP, by using organolithium deriva-tives as initiators in THF solution. They proposed a stereo-regulating mechanism mainly based on the analysis of the con-figuration of low molecular weight products formed.

Our study concerns the polymerization of 2-VP, initiated by organomagnesium derivatives of the type R Mg R', in hydrocarbon solvents. In order to obtain a detailed knowledge of the stereo-regulating mechanism, we focused this study on the determination of the nature and of the structure of active centers.

1. Initiation of the Polymerization

In spite of their low reactivity towards ethylenic double-bonds, the organomagnesium compounds are able to initiate the polymerization of 2-VP in THF solution (6), but the process is not stereospecific.

We have found that, in hydrocarbon solvents, the activating effect of the pyridine ring upon the reactivity of the double-bond, allows nucleophilic addition from non-solvated R Mg R' derivatives.

Nevertheless, all these compounds cannot be used indiscrim-inently. Indeed, the pyridine ring of the 2-VP molecule is sen-sitive to the attack of nucleophilic reagents, leading to side-

0097–6156/81/0166–0239$07.75/0
© 1981 American Chemical Society

reactions during the initiation step as well as during propagation by attack onto the pVP units of the polymeric chains.

For example, dibutyl magnesium leads to high polymers whose M_n is much higher than that calculated from the initiator concentration. U.V. spectra of living species show the presence of large amount of anionic by-products.

We used the dibenzyl magnesium (Bz_2 Mg) to initiate the polymerization. However this compound is not soluble in hydrocarbon solvents and initiation takes therefore place in heterogeneous phase. The excess of Bz_2 Mg must be filtered off, in order to limit the subsequent attack of the pyridine rings. Nevertheless, as indicated by Figure 1 , side reactions still slowly occur, explaining the appearance of a shoulder in the U.V. spectrum, in the 380 nm region. This shoulder corresponds probably to the formation in the system of the following groups:

The corresponding mechanism was already proposed in a previous paper (6).

The number average molecular weight analysis of the polymers obtained, gives values about twice those calculated from the amount of Bz_2 Mg reacted (Table 1).

Moreover, viscosity measurements on active and deactivated polymers in solution, show that the viscosity of the solution remains roughly constant after deactivation* (Figure 2).

To explain these unexpected results, we assumed a disymmetrical structure of active species,

corresponding to the monofunctionality of the initiator, in spite of the bivalent nature of the cation.

----

*Viscosity measurements were performed in THF solution, after substitution of benzene by THF, in order to prevent a possible aggregation of species in hydrocarbon solvents.

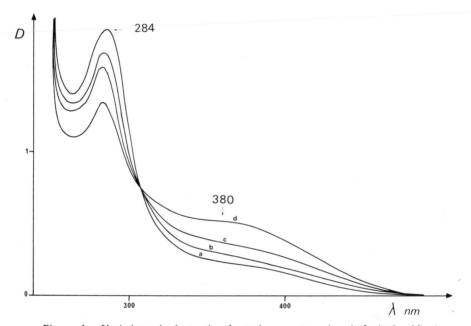

Figure 1. Variation of electronic absorption spectra of poly(2-vinylpyridine) MgBz initiated by dibenzylmagnesium as a function of time: (a) 5 min; (b) 70 h; (c) 400 h; (d) 1000 h.

TABLE 1: MOLECULAR WEIGHTS OF POLY(2-VINYL PYRIDINE) PREPARED FROM ORGANOMAGNESIUM COMPOUNDS AT 25°C IN BENZENE SOLUTION[a].

Sample	Initiator	Initiator conc.[b] $[I] \cdot 10^3$ mole.l^{-1}	Initial monomer conc. $[M_0] \cdot 10^2$ mole.l^{-1}	$\frac{[M_0]}{[I]}$	Theoretical M_n calculated from $[M_0]/[I]$	Experimental M_n
PS 1	s-Bu$_2$Mg	1	17	170	17,800	145,000
PS 2	Bz$_2$Mg	~ 10(solid)	12	12	1,260	3,000
PS 3	Oligomers[c]	0.36	39	1083	113,800	230,000
PS 4	Oligomers[c]	0.54	74	1370	144,000	300,000
PS 5	Oligomers[c]	1.84	80	435	45,700	100,000
PS 6	BPMg	3.6	100	278	29,200	62,000
PS 7	BPMg	0.56	13.6	243	25,500	49,000
PS 8	Oligomers[d]	0.1	1	100	10,500	22,000

a) Yields of polymerizations are quantitative
b) Concentration in carbanions
c) Oligomers prepared from dibenzylmagnesium
d) Oligomers prepared from benzylpicolylmagnesium

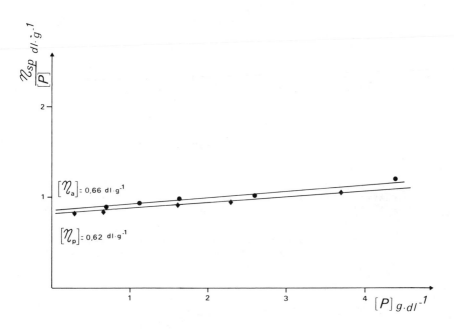

Figure 2. Effect of deactivation on viscosity of a living solution in THF of poly(2-vinylpyridine) MgBz prepared from dibenzylmagnesium: living polymer (●); dead polymer (♦).

In order to demonstrate such an assumption, we hydrolyzed a concentrated solution of living oligomers and we obtained an amount of toluene corresponding to one benzylic carbanion per living chain. The disymetrical structure of the species indicates that the reactivity of the second benzylic carbanion of the initiator is affected by the substitution of the first one. This behavior was already observed by other authors.

Benzyl picolyl magnesium is therefore a model molecule of the living chain and we have prepared such a compound by reacting Bz_2 Mg with α-picoline in methylcyclohexane and recrystallized the product in the form of yellow needles. The disymmetrical structure of this compound was established by ^1H NMR spectroscopy and by V.P.C. analysis of the products after hydrolysis. The corresponding spectrum and assignments will be discussed later.

As indicated by Figure 3 , a solution of BPMg in cyclohexane or in benzene, absorbs in the U.V. region, respectively at 270 and 285 nm, without any shoulder between 300 and 400 nm.

The benzyl picolyl magnesium obtained, as described above, can be used as an initiator and produces living isotactic polymers on which:

- The U.V. spectrum does not present any shoulder between 300 and 400 nm (Figure 4).

- The $\overline{M}_n$ values are in agreement with the ones calculated assuming initiation by the picolyl carbanion only (Table 2).

BPMg is therefore a convenient monofunctional initiator leading to pure living species with a disymmetrical structure, and all the results reported here were obtained with this molecule as an initiator.

2.   Study of the Nature of the Active Centers

In order to determine if the true active species are in a predominant concentration in the system studied, we have carried out a kinetic study of the propagation reaction.

The straight lines in the plot of log $[M_0]/[M]$ vs time, (Figure 5) indicate that the kinetic order is equal to 1 with respect to monomer concentration and that the concentration of active centers remains constant.

The linearity of the plot k = f (C), for concentrations between 5.10^{-4} -6.10^{-5}M (Figure 6), demonstrates the kinetic order 1 with respect to the active centers. The rate constant of the bimolecular propagation reaction is k_p = 3.0 $1.mole^{-1}.s^{-1}$ at 20°C. From experiments performed at different temperatures (Figure 7), it is possible to obtain the activation energy (E_a = 71 KJ.mole^{-1}) and the entropy of activation: Δs^* = -58 u.e. This strongly negative value of Δs^* indicates a very rigid

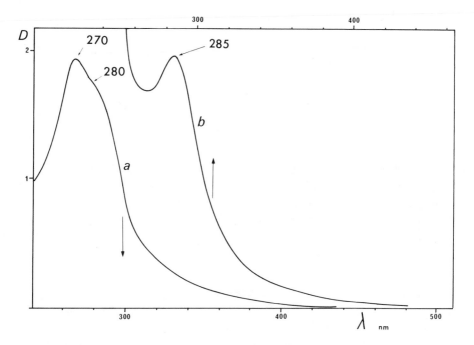

Figure 3. Electronic absorption spectra of benzylpicolylmagnesium: (a) in cyclo-hexane; (b) in benzene.

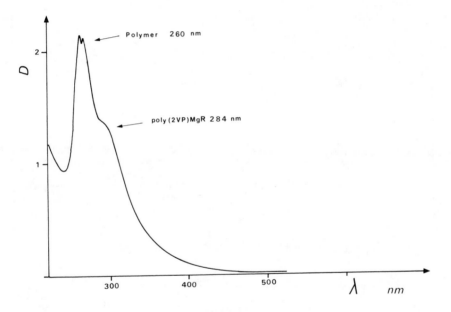

Figure 4. Electronic absorption spectra of poly(2-vinylpyridine) MgBz initiated by benzylpicolylmagnesium in benzene.

TABLE 2: DETERMINATION OF NUMBER AVERAGE MOLECULAR WEIGHTS OF POLY(2-VINYL PYRIDINE) INITIATED BY BENZYL PICOLYL MAGNESIUM IN BENZENE.

Run N°.	[I] Initiator concentration $\times 10^4 \cdot mol.l^{-1}$	$[M_0]$ Initial monomer concentration $\times 10^2 \cdot mol.l^{-1}$	Experimental molecular weights $\overline{M}_n$ exp.	Theoretical molecular weight calculated from $[M_0]/[I]$ $\overline{M}_n$ th.
1	4.8	8.2	18100	20000
2	3.7	8.5	24000	25000
3	1.7	7.1	43100	46000
4	4.8	7.1	155000	170000
5	5.8	7.3	13300	14000
7	1.5	15	105000	115000

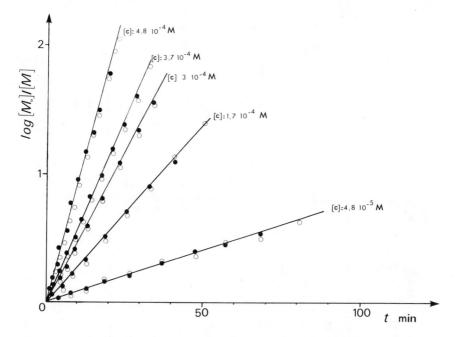

Figure 5. Variation of monomer concentration vs. time for different values of initiator concentration: calculated from appearance of polymer (○); calculated from disappearance of monomer (●).

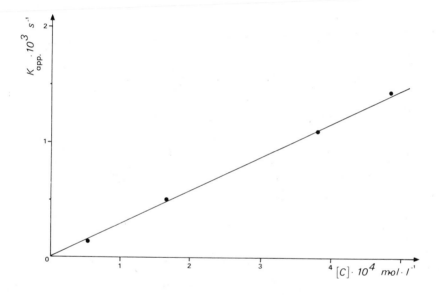

Figure 6.   Variation of k_p *vs. concentration in active centers:* $k_\pm = 3.0 \; L \cdot mol^{-1} \cdot s^{-1}$.

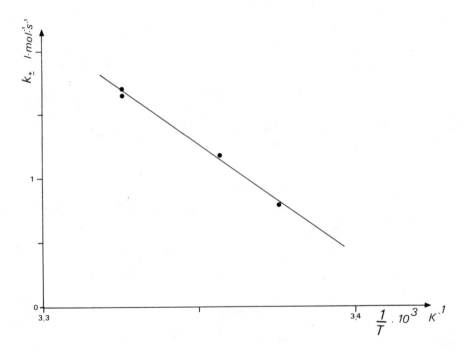

Figure 7. Determination of activation energy and activation entropy of the propagation reaction: $E_a = 71 \pm 4\ KJ \cdot mol^{-1}$; $\Delta s^* = -58\ kJ \cdot K^{-1}$.

transition state, consistent with a strong coordination between active centers and monomer molecules.

From the order 1 with respect to the concentration of active centers and taking into account the agreement between measured and theoretical molecular weights, we can deduce that true active species are in a predominant concentration.

The only uncertainty about the nature of active centers that remains, concerns their aggregation state. In order to measure the aggregation number, we have performed viscometric measurements on the polymer solutions in benzene, before and after deactivation of the active species.

As indicated by Figure 8, there is no significant change in viscosity upon deactivation. This experiment therefore confirms the disymmetrical structure of the active species and provides good proof that the living species are non-aggregated. This surprising non-aggregation of an organomagnesium derivative in an hydrocarbon solvent, will be discussed in a next paragraphy.

3. Study of the Structure of Active Centers

Our attempt to prepare low DP oligomers were unsuccessful, probably due to the initiation step being slightly slower than the propagation step. Accordingly, we chose the benzyl picolyl magnesium (BPMg) as a model molecule of the living chain.

The structure of this compound was studied first by ^{1}H NMR spectroscopy (Figure 9). The assignment of resonance signals is easy for the benzylic ring protons and for the H_6 proton of the pyridine ring but more difficult for other protons. For methylenic protons particularly, it was difficult to assign the corresponding signals to benzyl and picolyl respectively. To resolve this ambiguity we have prepared partially deuterated BPMg from picoline partially deuterated on the methyl group and observed, on the ^{1}H NMR spectrum, the variation of the relative intensity of the 2 signals.

The investigation of the chemical shifts and the comparison wiht the chemical shifts of the corresponding hydrocarbons, indicates that all signals are displaced to higher fields. (Figure 10) gives the values of the corresponding increase of charge.

This shift towards high fields, even for methylenic protons, may appear normal but is quite surprising. Indeed, as reported by Takahashi et al. for the same study performed on picolyllithium in THF solution (8), the change in hybridization of the methylenic carbon from sp^3 in picoline to sp^2 in picolyl Li, produces a shift towards low fields. In our case, we must admit that the 2 carbanions, picolyl and benzyl, are sp^3 hybridized. This result is corroborated by the NMR study of BPMg at low temperature. Indeed, in the case of picolyl Li and ethyl pyridyl Li, Takahashi (7) and Hogen-Esch (5) respectively showed that the partially ethylenic nature of the C_α-C_2 bond leads to the non-equivalency of the two methylenic protons and produces the split-

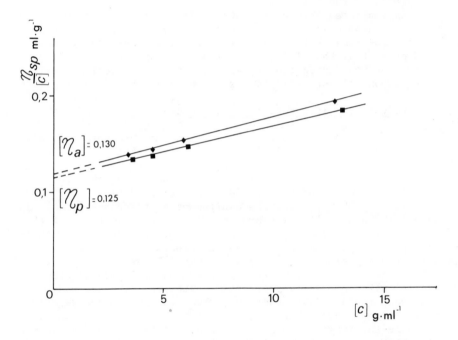

Figure 8. Effect of deactivation on viscosity of living solutions in benzene, of poly(2-vinylpyridine) MgBz prepared from benzylpicolylmagnesium: living polymer (◆); dead polymer (■).

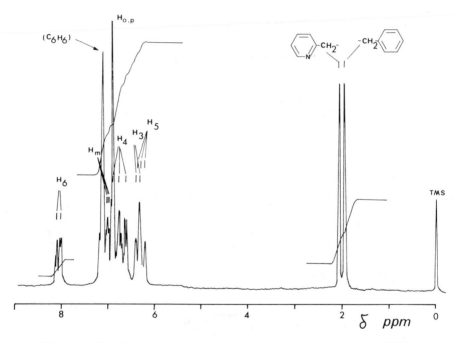

Figure 9.   H-1 NMR spectrum of benzylpicolylmagnesium in C_6D_6 at 28°C.

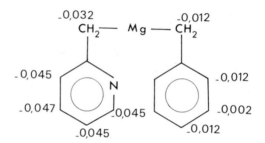

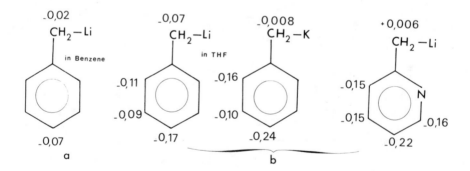

Figure 10. Modification of charge density on carbon atoms of benzylpicolyl-magnesium calculated from H-1 NMR spectrum: (a) Ref. 9; (b) Ref. 7.

ting of the signal at low temperature. The BPMg in toluene solution does not show this behavior, not even at -70°C. Examination on the chemical shifts also indicates a lower ionicity for the Mg-benzyl bond compared to the Mg-picolyl bond. This may explain the difference observed in the nucleophilic reactivity for the 2 groups and also the formation of disym-metrical active centers.

The detailed study of the ^{1}H NMR spectrum of BPMg provides us with further information:

- generally, the signal of the H_6 proton of pyridine and pyridine derivatives, is widened by the quadrupole moment of the nitrogen atom in close proximity. In our system, we observe a very sharp peak.

- moreover, the calculation of increase of charge on protons of the pyridine ring indicates an abnormal increase of the electron density on the H_6 proton compared to other similar organometallic compounds.

We assumed that these phenomena are due to the partial complexation of the nitrogen atom of BPMg, to the magnesium cation. We shall discuss our assumption on the basis of the ^{13}C NMR spectra.

The ^{13}C spectrum of BPMg is given in Figure 11. The most remarkable feature derived from the analysis of this spectrum is the confirmation of the sp^3 hybridization of the two carbanions. Indeed, the chemical shifts of these species are only slightly different of those of the corresponding hydrocarbons, whereas Takahashi for picolyl Li (8) and Waack et al. for benzyl Li (9), observed strong shifts due to the sp^2 nature of carbanions (Figure 12).

Moreover, the calculation of the charge density variation on the different carbon atoms, shows abnormal values for the carbons of the pyridine ring. As previously seen, we attribute this behavior to the coordination of the nitrogen atom to the magnesium atom. If this coordination takes place, it produces an electronic deficiency and a change in the distribution of π electrons.

We have compared the spectrum of picoline with that of picolinium chloride (Figure 13) and observed exactly the same effect as with BPMg (Figure 14).

Nevertheless, the intramolecular coordination of nitrogen to magnesium, does not exclude the possibility of aggregation of this compound in non-polar solvents, through the vacant sites of coordination of the magnesium cation. We have measured the aggregation number of BPMg by cryometry in benzene. As indicated by the results given in Table 3, and whereas we demonstrated previously the lack of aggregation of pVP Mg Bz, the BPMg is aggregated in a dimeric form. This difference in the behavior of the two species, is very significant. It shows that the PB Mg is not a suitable model to explain all the facets of the behavior of

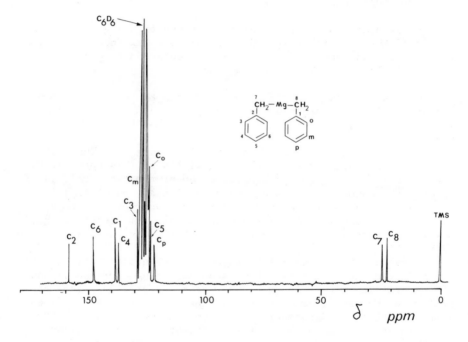

Figure 11.   C-13 NMR spectrum of benzylpicolylmagnesium in C_6D_6 at $20°C$

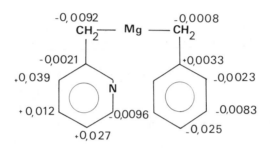

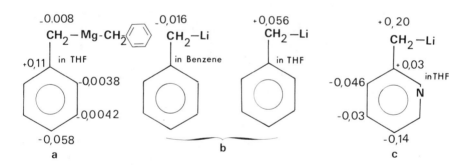

Figure 12. Modification of charge density on carbon atoms of benzylpicolyl-magnesium calculated from C-13 NMR spectrum: (a) Ref. 10; (b) Ref. 9; (c) Ref. 8.

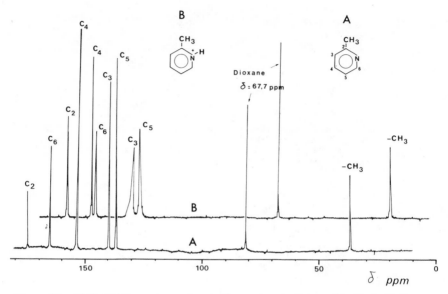

Figure 13.   C-13 NMR spectrum of α-picoline and α-picolinium chloride in D₂O.

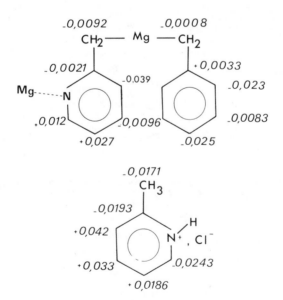

Figure 14.   Modification of charge density on carbon atoms of benzylpicolyl-magnesium and α-picolinium chloride calculated from C-13 NMR.

TABLE 3 : DETERMINATION OF ASSOCIATION NUMBER OF BPMg BY CRYOSCOPIC MEASUREMENTS IN BENZENE SOLUTION.

$\Delta T^\circ K$	Concentration of BPMg calculated from U.V. spectra $\times 10^2$ mol.kg^{-1}	Concentration of BPMg calculated from cryoscopic measurements $\times 10^2$ mol.kg^{-1}	Association number a
0.12	4.31	2.39	1.8
0.07	2.43	1.38	1.7

pVPMgBz. As the main difference in the structure is the unique
pyridine ring in the model molecule, we assume that the penultimate pyridine ring of the pVPMg Bz is responsible for the lack of
aggregation as shown in the Figure 15. The vacant sites on
magnesium allow aggregation of PBMg but are occupied by complexation by the penultimate nitrogen atom. Such an assumption was
already proposed by Sigwalt and colleagues, to explain the
conductimetric behavior of pVP, Na (11).

For the 2 last units of the living chain each possessing 2
asymmetrical carbons, we can propose 4 configurational stereoisomers:

    2 leading to a meso diad

    2 leading to a racemic diad.

In Figure 16 (a,b,c,d) we have represented these structures.

Obviously, among the 4 configurational stereoisomers, none
seems strongly favored. Thus, from the structure of the living
species only, we cannot propose a stereoregulating mechanism and
we searched for an explanation by the mode of complexation of
monomer molecules to the active centers.

We studied this complexation by using as a model of the
active ion-pair, the dicarbazyl Mg, a molecule soluble in hydrocarbons and unable to initiate the polymerization of 2VP. The
preparation of this molecule necessitates the presence of a small
amount of THF. From the ^{1}H NMR spectrum, we determined that 2
THF molecules are complexed with the magnesium cation. Indeed,
the internal chemical shift ($\Delta\delta$THF) of the THF ($\Delta\delta_{THF} = \delta_{-CH_2O^-}$
$-\delta_{-CH_2-}$) in benzene solution is equal to 125.4 Hz, whereas,

$\Delta\delta_{THF} = 133.2$ Hz in the presence of (carbazyl)$_2$ Mg.

If 2VP is added to the system, it decomplexes the THF
($\Delta\delta_{THF} = 126.1$ Hz) and it coordinates with Mg, as indicated by
the shift of the vinylic protons (Figure 17) probably by means of
the pyridinic nitrogen. Hence, the solvating power of the 2VP
molecule is stronger than that of THF.

The study of the stereoregularity of the polymers prepared,
provides also information about the stereoregulating mechanism.
The probability of formation of the different types of sequences,
was determined on the basis of the resonance of the quaternary
carbon of pVP (12). The ^{13}C NMR spectrum performed at 15 MHz
allows one to determine the concentration of triads. The values
summarized in Table 4 do not agree with those expected for
bernouillian statistics. Hence, more than the last unit of the
living chain is involved in the process. In order to obtain more
precise information about the process, it is necessary to measure
the probability of formation of pentads. Such measurements are
possible with spectra performed at 63 MHz (Figure 18). In spite

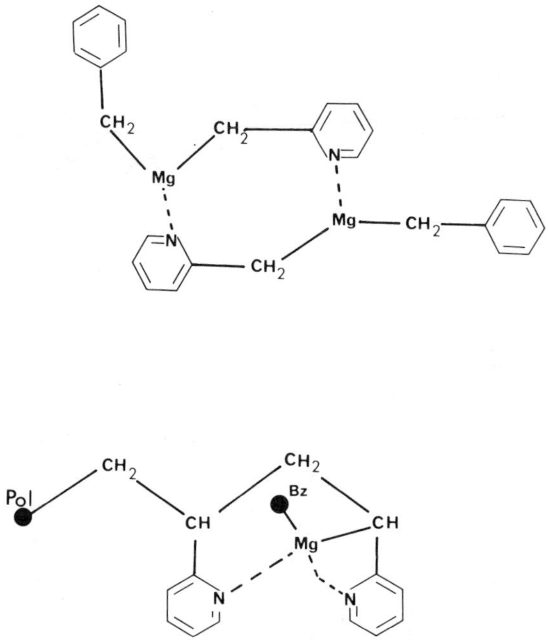

Figure 15.   Interpretation of the difference of aggregation between benzylpicolyl-magnesium and poly(2-vinylpyridine) MgBz.

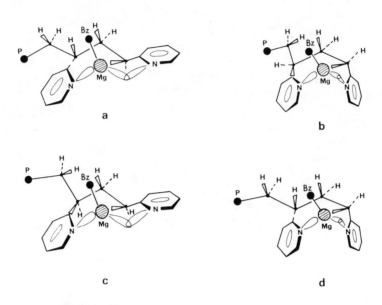

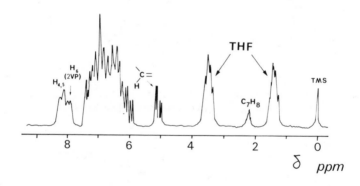

Figure 16.   Structure of diastereomeric ion pairs of poly(2-vinylpyridine) MgBz.

*Figure 17.   H-1 NMR spectrum of an equimolecular solution of dicarbazylmag-
nesium and 2-vinylpyridine in C_7D_8.*

TABLE 4: DETERMINATION OF POLY(2-VINYL PYRIDINE) TACTICITY FROM ^{13}C (RING CARBON C_2) NMR AT 15.09 MHz.

SAMPLES	I (mm)	H (mr)	S (rr)	$\dfrac{4\ IS}{H^2}$
PS$_1$	0.92	0.07	0.01	7.5
PS$_2$	0.89	0.09	0.02	8.9
PS$_3$	0.91	0.07	0.02	14.8
PS$_4$	0.92	0.06	0.02	20.4
PS$_6$	0.88	0.11	0.01	2.9
PS$_7$	0.93	H + S = 0.07		–
PS$_8$	0.88	0.10	0.02	7
PS$_6$[a]	0.875	0.115	0.010	2.8
Poly(vinyl-4 pyridine)	0.27	0.49	0.23	1.03

a) Measured at 62.86 MHz on ring carbon C_2.

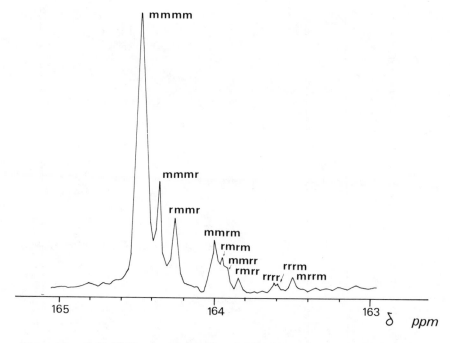

Figure 18.   C-13 NMR spectrum at 62 and 86 MHz of poly(2-vinylpyridine) in CD₃OD/CH₃OH (C₂ ring carbon).

of the relatively low accuracy on the determination of the
degrees of pentads, the results summarized in Table 5 are in
agreement with a Markov 1st order statistics. This result
corroborates the assumption of the participation of the last and
the penultimate units in the stereoregulating process.

These results as a whole, enable us to propose a stereospecific mechanism for the incorporation of the 2VP monomer
molecule into the living growing chain.

The active center is represented in Figure 19 in the
presence of a monomer molecule, which displaces the last pyridine
ring coordinated with magnesium and complexes at vacant sites of
the magnesium cation. In a second step, the 2VP molecule is
incorporated in the growing chain after going through a planar
four centers transition state with simultaneous electron transfer
between. the carbon-metal bond and the vinylic bond, with
retention of configuration as indicated in the following scheme:

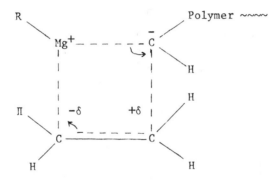

Nevertheless, this transition state can form a priori, either
above the plane defined by the nitrogen, the Mg atom and the
carbanion, or below this plane, according to the position of the
double-bond when the monomer molecule complexes with the Mg
cation.

Considering the possibility of formation of both transition
states, we see unambiguously that incorporation below the plane
is possible going from the "RR" diad, to a mm triad (Figure 19).

On the contrary, the incorporation below the plane starting
from a "RR" diad is difficult whereas the incorporation into a
"SS" diad has a high probability (Figure 20). Therefore,
whatever the configuration of the terminal diad, r or m, steric
hindrance occurs and favors the formation of a new meso diad.
Hence, we obtain predominantly mm triads from this terminal diad
m and rm triads from the terminal diad r.

The formation of rr triads is extremely improbable, thus explaining the very low concentration measured for such triads in
the polymers prepared.

TABLE 5: DETERMINATION OF POLY(2-VINYL PYRIDINE) TACTICITY FROM ^{13}C NMR (RING CARBON C_2) AT 62.86 MHz.

Pentads	mmmm	mmmr	mmom	rmmr	rmom	mmrr	rmrr	mrrm	rrrm	rrrr
Theoretical[a] ratio	0.756	0.110	0.110	0.004	0.008	0.008	0.00057	0.004	0.00057	0.00002
Theoretical[b] ratio	0.770	0.101	0.09	0.004	0.006	0.002	0.016	0.007	0.003	0.0002
Experimental ratio	0.784	0.09	0.08	0.015			0.018	0.008		0.003

Theoretical[a] braces: rmmr + rmom + mmrr = 0.020 ; mrrm + rrrm + rrrr = 0.0046

Theoretical[b] braces: rmmr + rmom + mmrr = 0.012 ; mrrm + rrrm + rrrr = 0.01

a) Calculated from a Bernouillan statistics.

b) Calculated from a Markov 1st order statistics.

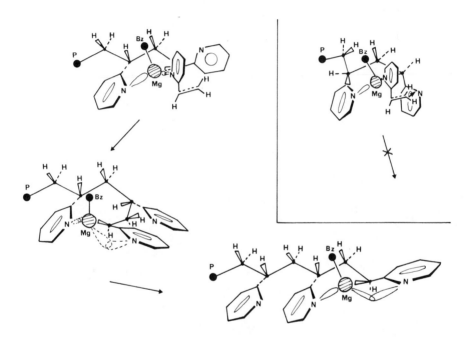

Figure 19.   Representation of the insertion of one 2-vinylpyridine molecule on R–R and S–S ion pairs of poly(2-vinylpyridine) MgBz. Insertion above the plane.

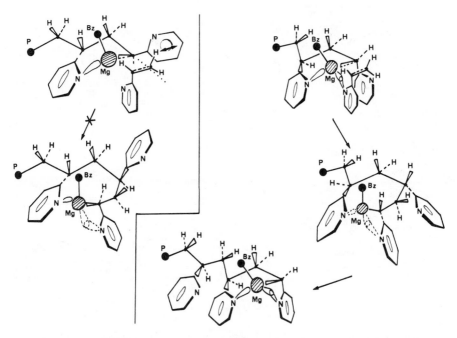

Figure 20. Representation of the insertion of one 2-vinylpyridine molecule on R–R and S–S ion pairs of poly(2-vinylpyridine) MgBz. Insertion under the plane.

Abstract

The anionic polymerization of 2-vinyl pyridine initiated by benzyl-2 pyridylmethylmagnesium (benzyl picolyl magnesium) in hydrocarbon solvents, leads to highly isotactic polymers.

Analysis of multads distribution by ^{13}C NMR, propagation kinetics and study of electron density on active centers model molecules, suggest the stereoregulating mechanism.

Carbanionic active centers are non aggregated and sp^3 hybridizated. Isoregulation is ascribed to a most probable position for the complexation of active centers by the monomer molecules.

Literature Cited

1.   G. Natta, G. Mazzanti, Makromol. Chem., 37, 160 (1960).

2.   G. Natta, G. Mazzanti, J. Polym. Sci., 51, 487 (1961).

3.   C. F. Tien, T. E. Hogen-Esch, Macromolecules, 9, 871 (1976).

4.   C. F. Tien, T. E. Hogen-Esch, J. Am. Chem. Soc., 98, 7109 (1976).

5.   C. F. Tien, T. E. Hogen-Esch, J. Polym. Sci., Polym. Lett. Ed., 16, 297 (1978).

6.   M. Fontanille, P. Sigwalt, A. Soum, J. Polym. Chem. Ed., 15, 659 (1977).

7.   K. Konishi, K. Takahashi, R. Asami, Bull. Chem. Soc. (Japan) 44, 2281 (1971).

8.   K. Konishi, K. Takahashi, Bull. Chem. Soc. (Japan), 50, 2512 (1977).

9.   D. McKeever, R. Waack, J. Org. Chem., 28, 145 (1971).

10.   D. Leibfritz, B. O. Wagner, J. O. Roberts, Leibigs Ann. Chem., 763, 173 (1972).

11.   M. Tardi, D. Rouge, P. Sigwalt, Europ. Polym. J., 3, 85 (1967).

RECEIVED April 13, 1981.

Kinetics and Mechanism of Anionic Polymerization of Lactones

STANISLAW SŁOMKOWSKI and STANISLAW PENCZEK

Polish Academy of Sciences, Center of Molecular and Macromolecular Studies, 90–362 Łodz, Poland

In spite of hundreds of papers and patents devoted to the polymerization of lactones only recently the first data on the rates of elementary reactions became available [1,2].

The earlier and comprehensive kinetic study by Hall [3] gave the rate coefficients for a number of lactones, without however determining the involved rate components, i.e. rate constants for various ionic species simultaneously present, in the studied solutions.

More recently Teyssie' determined the rate constants in the polymerization of ε-caprolactone (εCL) initiated with aluminium alkoxides, believing that the covalent species are the only ones responsible for propagation [4]. For the same monomer Yamashita estimated tentatively rate coefficients of propagation using an anionic initiator [5]. Lenz in his studies of substituted β-propiolactones (βPL) observed peculiar influence of structure on reactivity that can have its origin in the multiplicity of ionic structures involved [6].

Thus, in the present paper we review the available data, pertinent to the kinetics and mechanism of anionic polymerization of lactones and discuss the recent data of our own, giving eventually an access to the rate constants of propagation on macroions and macroion--pairs.

Some unusual monomer properties and major characteristics of polymerizing systems involving lactones. There are at least two features of monomeric lactones that make them particularly attractive for mechanistic studies. Lactones are highly polar compounds (e.g. βPL: D = 46, εCL: D = 40, both at 25⁰) and can react ambidently, namely by O-alkyl or O-acyl

0097–6156/81/0166–0271$05.00/0

ring opening. The latter ability opens an important new direction for studies of copolymerization. Since both βPL and εCL have been shown to give living systems in anionic polymerization [1,4] one can expect that the quantitative understanding of the anionic polymerization of lactones will allow to answer some questions open by studies of the less polar monomers and, thus, less accentuating the novelty of these systems in comparison with hydrocarbons, namely ethylene oxide and propylene sulphide [7,8].

Initiation and Structure of Growing Species

The earlier data on initiators used, and on the mechanism of initiation are comprehensively collected in the Lundberg's review [9]. Unfortunately, the evidence presented till now on the mode of initiation and bond cleavage (O-alkyl or O-acyl) as a function of initiator and monomer structures:

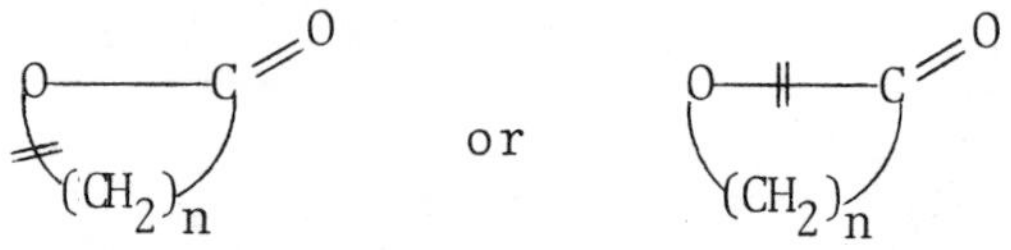

is not, at least in our judgment, sufficiently convincing and this area will have to be reinvestigated by using the modern instrumental methods. Nevertheless, the suggestion of Cherdron [10], that the growing species for polymerization of unsubstituted lactones are alcoholate anions has been widely accepted, β-propiolactones, however, being apparently an exception and propagate with carboxylate anions [11].

This particular behaviour of βPL is probably due to its lower basicity and higher electrophilicity in comparison with εCL (pK_b = 10.06 for βPL and 5.31 for εCL respectively [12]). Thus, even a weaker nucleophile, like carboxylate anion can accomplish initiation of βPL.

Tertiary amines initiate polymerization by zwitterion formation [13,14].

Recently we studied initiation of βPL polymerization with phosphines, including the optically active methyl,n-propyl,phenylphosphine, and established the structure of the betaine-type species directly by ^{31}P-NMR (δ ^{31}P = 28.2 ppm) [15,16].

$$R_3P + \underset{\substack{| \quad | \\ O \!-\! C \diagdown\!\!_O}}{CH_2 \!-\! CH_2} \xrightarrow{k_i} R_3P^+ \!-\! CH_2CH_2C\diagup\!\!\!\!\diagdown\!\!\!\!\!\!_O^- \qquad (1)$$

Zwitterions are stable and can provide a living polymerization, however, at the smaller chain lenght (smaller distance between ions) formation of ylids was noted :

$$R_3P^+ \{CH_2CH_2\overset{O}{\overset{\|}{C}}-O\}_n CH_2CH_2C\diagup\!\!\!\!\diagdown\!\!\!\!\!\!_O^- \longrightarrow$$

$$\qquad\qquad\qquad\qquad\qquad\qquad\qquad (2)$$

$$\longrightarrow R_3P{=}CH{-}CH_2\overset{O}{\overset{\|}{C}}{-}O\{CH_2CH_2\}_{n-1}CH_2CH_2COOH$$

Similar process of decomposition of betaines has been known in case of aromatic amines, but proceeding with elimination of protonated amine [17].

Data on kinetics of initiation are available only for initiation of βPL polymerization with triphenyl-phosphine and methyl,n-propyl,phenylphosphine [16]. The rate constants k_i (cf. eqn (1)) were found at 25^0 to be equal to $2.3 \cdot 10^{-5}$ mole$^{-1} \cdot$l$\cdot$s^{-1} and $5.2 \cdot 10^{-4}$ mole$^{-1} \cdot$l$\cdot$s^{-1} respectively.

Mechanism and Kinetics of Propagation

Kinetics of anionic ring-opening polymerization has hitherto been quantitatively studied and gave for two monomers, namely ethylene oxide [18,19] and propylene sulfide [8,20]. Studies on these systems revealed that the living conditions can be achieved, facilitating quantitative determination of rate constants of propagation on various kinds of ionic growing species.

These and related heterocyclic monomers are usually highly polar and strongly nucleophilic compounds. During polymerization chains containing heteroatoms are formed and they can, as well as monomers themselves, interact with components of ionic growing species. The interaction of the macroion-pairs with the elements of the chains has well been documented for the polymerization of ethylene oxide [7].

Lactones belong to the most polar compounds and constitute a group of monomers of highest dielectric constants (cf. preceding section). Thus, one could expect even more pronounced effects in the polymerization of this class of monomers and due to interactions of monomers themselves with different ionic growing species.

We have recently shown, that the living conditions can be achieved in the polymerization of βPL by using as initiators compounds providing higher proportion of macroions. This favourably decreases the importance of the parasitic side reactions by relatively increasing the overall rate of propagation processes [2]. The following initiators were used:

(where in 1 the crown ether is dibenzo-18-crown-6 ether (DBC) and Φ in 2 denotes C_6H_5).

Apparently similar enhancement of the propagation processes was achieved with cryptated cations [22], although in this case a special care has to be taken not to introduce the zwitterionic initiation.

Data on kinetics of propagation, as it has already been mentioned, are available at present only for polymerization of unsubstituted and substituted βPL [1-3,6], initiated with carboxylates, and for βPL and εCL, initiated with bimetallic oxoalkoxides [4,23].

In all but one of these studies the apparent rate constants of propagation were determined. The observed apparent propagation rate constant (k_p^{app}) is a summ of products of rate constants on various forms of active centers (e.g. free ions, ion-pairs, aggregates) and their fractions in the polymerizing mixture

$$k_p^{app} = \alpha \cdot k_p^- + \beta \cdot k_p^{\mp} + \ldots \qquad (3)$$

where α and β are fractions of free ions and ion-pairs. Any change of lactone structure, of nature of solvent or temperature will influence the rate constants of elementary propagation reactions and position of equilibria between different kinds of active centers.

In our recently published paper on kinetics of elementary reactions in polymerization of βPL, initiated with acetate anion with potassium crowned with dibenzo--18-crown-6 ether (DBCK$^+$) counterion in methylene dichloride, we determined k_p^- and $k_p^{\oplus}$ (where $k_p^{\oplus}$ denotes rate constant of propagation on crowned macroion-pairs) [2]. In the temperature ranging from -20 to +35°C and for starting monomer concentrations 1.0 and 3.0 mole·1^{-1} polymerization was found to be the living process with molecular weight of polymer increasing linearly with monomer conversion.

Conductivity measurements for solutions of living poly-βPL with DBCK$^+$ counterion in CH_2Cl_2/βPL mixture indicated that macroions and macroion-pairs are present in the system. In Fig.1, taken from Ref. 2 the Vant' Hoff's plots are given for dissociation constants of poly-βPL macroion-pairs with DBCK$^+$ counterion (K_D) and similar plot for dissociation constants of $Ph_4B^-DBCK^+$ (K_{D1}). Dissociation of $Ph_4B^-DBCK^+$ was investigated because this salt was further used in the kinetic measurements to shift the equilibrium between macroions and macroion-pairs towards the latter ones.

In Table 1 values of thermodynamic equilibrium parameters and values of K_D and K_{D1} at 25° are given.

T a b l e   1

Dissociation constants and pertinent activation parameters

Solvent	poly-βPL, DBCK$^+$			$(C_6H_4)_4B^-$, DBCK$^+$		
	ΔH_D	ΔS_D	$K_D^{25°}$	ΔH_{D1}	ΔS_{D1}	$K_{D1}^{25°}$
CH_2Cl_2/βPL [βPL]$_0$ = 3.0 mole·1^{-1}	0.5±2	-18±3	$5 \cdot 10^{-5}$	0.6±0.5	-10±2	$2.3 \cdot 10^{-3}$
CH_2Cl_2/βPL [βPL]$_0$ = 1.0 mole·1^{-1}	0.6±0.2	-22±2	$5.6 \cdot 10^{-6}$			

ΔH_D and ΔH_{D1} in kcal·mole^{-1} ; ΔS_D and ΔS_{D1} in cal·mole^{-1}·deg^{-1} ; K_D and K_{D1} in mole·1^{-1}.

By ussing the Fuoss' equation distance (a) between ions constituting the ion-pair was calculated. For [βPL]$_0$ = 1.0 mole·1^{-1} a=3.3 A^0 and for [βPL]$_0$ = 3.0 a=2.6 A^0 , both being independent on temperature.

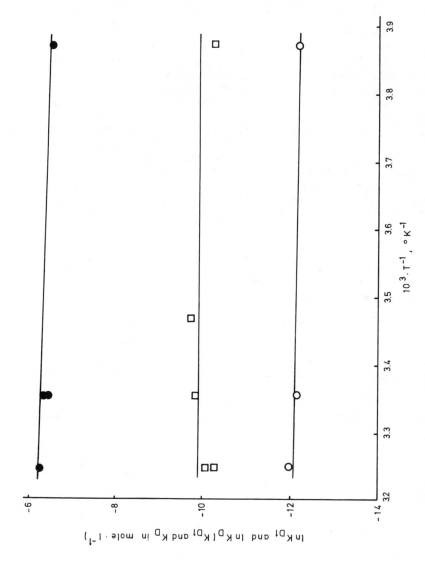

Figure 1. Dependence of ln K_D ([βPL]₀ = 1.0 (○); [βPL]₀ = 3.0 mol · L⁻¹ (□)) and ln K_{D1} ([βPL]₀ = 3.0 mol · L⁻¹ (●)) on reciprocal temperature.

T a b l e 2

Rate constants of propagation : k_p^- and $k_p^{\oplus}$

Temperature	-20	-15	0	10	15	25	35
$10^4 \cdot k_p^-$, mole$^{-1} \cdot$l$\cdot$s^{-1}	1.0	9.5	29	110	400	1500	3700
$10^4 \cdot k_p^{\oplus}$, mole$^{-1} \cdot$l$\cdot$s^{-1}	1.4	9.0	17	19	6	10	40
$10^4 \cdot k_p^-$, mole$^{-1} \cdot$l$\cdot$s^{-1}		490				8000	26000
$10^4 \cdot k_p^{\oplus}$, mole$^{-1} \cdot$l$\cdot$s^{-1}		2.0				18	25

First two rows: $[\beta PL]_0$ = 3.0 mole·l^{-1}

Last two rows: $[\beta PL]_0$ = 1.0 mole·l^{-1}

Propagation rate constants on macroions (k_p^-) and on macroion-pairs ($k_p^\oplus$) were determined from the dependence of k_p^{app} on the fraction of macroions (degree of dissociation α). For systems with only macroions and macroion-pairs participating in equilibrium eqn (3) simplifies to:

$$k_p^{app} = k_p^{\overline{\oplus}} + \alpha(k_p^- - k_p^{\overline{\oplus}}) \qquad (4)$$

Degree of dissociation (α) was calculated by using values of K_D and assuming the quantitative initiation. In some experiments $Ph_4B^-DBCK^+$ was added with purpose to increase fraction of macroion-pairs.

Values of k_p^- and $k_p^\oplus$ for βPL polymerization with starting monomer concentrations 1.0 and 3.0 mole·1^{-1} carried out in temperature region from -20 to 35°C are given in Table 2 (Ref. 2).

From the Arrhenius plots (Fig.2) the activation parameters for propagation on macroions and on macroion-pairs were evaluated. They are given in Table 3.

T a b l e   3

Activation parameters for propagation on macroions and on macroion-pairs in polymerization of βPL with DBCK⁺ counterion.

Solvent	$\Delta H_p^{\ddagger}(-)$	$\Delta S_p^{\ddagger}(-)$	$\Delta H_p^{\ddagger}(\overline{\oplus})$	$\Delta S_p^{\ddagger}(\overline{\oplus})$
	kcal·mole^{-1}	e.u.	kcal·mole^{-1}	e.u.
CH_2Cl_2/βPL $[\beta PL]_0 = 3.0$ mole·1^{-1}	16±1	-10±4		
CH_2Cl_2/βPL $[\beta PL]_0 = 1.0$ mole·1^{-1}	6±1	-40±5	6±1	-52±4

Before discussing reactivities of both kinds of active centers it is necessary to establish the structure of ion-pairs involved in propagation. Independence of $\Delta H_p^{\ddagger}(\overline{\oplus})$ and $\Delta S_p^{\ddagger}(\overline{\oplus})$ on the starting concentration of monomer, being the most polar component of the system, (Table 3) and linearity of Arrhenius plots for $k_p^\oplus$

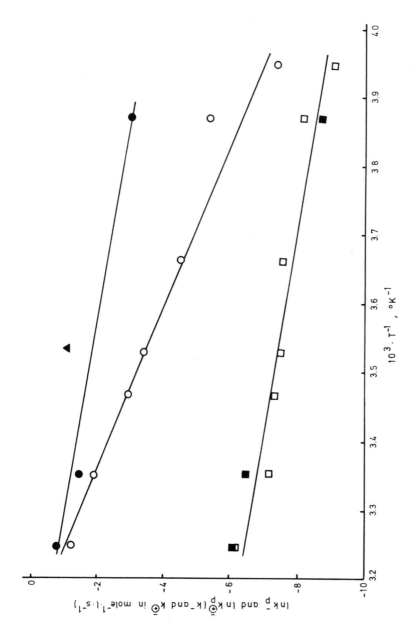

Figure 2. Dependence of propagation rate constants on reciprocal temperature; k_p^- ($[\beta PL]_0 = 1.0$ *mol* · L^{-1} ($\bullet$), $[\beta PL]_0 = 3.0$ *mol* · L^{-1} ($\bigcirc$)); k_p^+ ($[\beta PL]_0 = 1.0$ *mol* L^{-1} ($\blacksquare$), $[\beta PL]_0 = 3.0$ *mol* · L^{-1} ($\square$)). *For* k_p^- ($[\beta PL]_0 = 5 \cdot 10^{-1}$ *mol*$^{-1}$ *and PNP· counterion* ($\blacktriangle$)).

(Fig.2) suggest that only one type of ion-pairs is present in the system.

Enthalpy of dissociation close to zero and entropy of dissociation higher than -25 e.u. indicate that additional solvation of ions upon dissociation is low and enthalpy of this solvation merely balances the electrostatic energy necessary to separate the ions. Apparently, the only moderate increase of solvation of ions accompanying the dissociation is observed because the charges are diffused in both carboxylate anion and DBCK⁺ cation. Diffusion of charge in DBCK⁺ cation means that due to polarization some positive charge is induced in the exterior part of crown ether molecule. Low value of the interionic distance (a≈3 A⁰) indicates that the ion-pairs are the contact ones. Therefore, we postulate for the structure of ion-pairs the disk-like shape of crowned cation with anion approaching the cation perpendicularly to the plane of the disk, and assume that this is the only kind of ion-pairs participating in polymerization. Thus, the other isomer postulated by Hogen Esch and Smid in the polymerization with carbanions [24,25], where K⁺ cation is withdrawn from the immediate proximity of anion, and physically separated from it by elements of the crown ether does not have to be involved in the polymerization of βPL.

Activation parameters for propagation on macroions were found to depend significantly on the starting concentration of βPL, whereas the corresponding activation parameters for propagation on macroion-pairs were found to be virtually independent on the composition of the medium.

Apparently, solvation of ion-pairs is not very much different from solvation of activated complex.

Dependence of $\Delta H_p^{\ddagger}(-)$ and $\Delta S_p^{\ddagger}(-)$ on the starting concentration of βPL, showing a typical compensation phenomenon (Table 3) can only have its origin in the specific solvation of the growing species by molecules of monomer itself. It does not mean, however, that molecules of monomer engaged in the solvation shell are properly oriented for the chemical change. Being properly oriented from the solvation view point these molecules energetically differ from an average molecule in bulk solution and it may be necessary to remove, reorient and/or provide additional energy to these particular molecules to force them to participate in the chemical change (propagation step).

At higher starting concentration of βPL increases the energy needed to desolvate the ground state in order to permit the chemical reaction to proceed. Thus, with increasing the starting concentration of monomer:

$[\beta PL]_0$ from 1.0 mole·1^{-1} to 3.0 mole·1^{-1} the enthalpy of activation increases as much as by 10.0 kcal·mole^{-1}.

Schematic representation of two modes of solvation: namely nonspecific for the macroion-pairs and specific for macroions is given below :

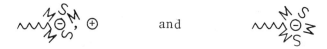

(where S and M denote solvent and monomer molecules respectively).

Thus, we assume, that in case of the macroion-pair solvent and monomer molecules are not oriented in any specific way and disorderly pack the available space around the macroion-pair. On the contrary, macroanion is specifically solvated at least by monomer molecules, due to the strong interaction between a negative charge and highly polar monomer and the further passage to the transition state, being much less polar than the ground state of macroanion, requires higher energy. This is much more apparent at higher monomer concentration.

Comparison of entropies of activation for propagation on macroions and on macroion-pairs leads to conclusions being hand in hand with the analysis of enthalpies. The large negative entropy observed for macroion-pairs (-52.0 cal·mole^{-1}·deg^{-1}) may reflect formation of transition state with more ionic character. Thus, some of the solvent and monomer molecules may even become more firmly oriented than in the corresponding ground state.

Abstract

The quantitative aspects of the elementary reactions involved in the polymerization of lactones are discussed. Some earlier papers are reviewed and analysed together with the recent data of the present authors. It has been shown, that in the polymerization of βPL rate constant of propagation on macroanions (k_p^-) decreases with increasing the starting concentration of monomer ($[\beta PL]_0$), whereas the rate constants of propagation on macroion-pairs ($k_p^{\oplus}$) do not depend on $[\beta PL]_0$. Thus, the ratio $k_p^-/k_p^{\oplus}$ does depend on $[\beta PL]_0$. These observations stem from the high dielectric constant of βPL and its high solvation ability, apparently stronger for ions than for the ion-pairs.

Literature Cited

1. St.Slomkowski and St.Penczek, Macromolecules, 9, 367 (1976).
2. St.Slomkowski and St.Penczek, Macromolecules,in press.
3. H.K.Hall, Jr., Macromolecules, 2, 488 (1969).
4. A.Hamitou, T.Ouhadi, R.Jerome, and P.Teyssie, J.Polym.Sci.,Polym.Chem.Ed., 15 865 (1977).
5. K.Ito, Y.Hashizuka, and Y.Yamashita, Macromolecules 10, 821 (1977).
6. E.Bigdeli and R.W.Lenz, Macromolecules, 11, 493 (1978).
7. K.S.Kazanskii, Chimia i Technologia Vysokomol. Soedin., Ed. N.S.Enikolopov, VINTI, Moscow, 9, 5 (1977).
8. P.Sigwalt and S.Boileau, J.Polym.Sci.: Polymer Symp., 62, 51 (1978).
9. R.D.Lundberg and E.F.Cox in "Ring-Opening Polymerization", Ed. K.C.Frisch and S.L.Reegen, Marcel Dekker, New York and London, 1969, p.266.
10. H.Cherdron, H.Ohse, and F.Korte, Makromol.Chem., 56, 187 (1962).
11. T.Shiota, Y.Goto, and K.Hayashi, J.Appl.Polym.Sci., 11, 753 (1967).
12. T.Tsuda, T.Shimizu, and Y.Yamashita, Kogyo Kagaku Zasshi, 68, 2473 (1965).
13. Y. Etienne and R.Soulas, J.Polym.Sci., C4, 1061 (1963).
14. V.Jaacks and N.Mathes, Makromol.Chem., 131, 295 (1970).
15. T.Biela, L.S.Corley, J.Michalski, St.Slomkowski, St.Penczek, and O.Vogl, Makromol.Chem., in press.
16. L.S.Corley, St.Slomkowski, St.Penczek, T.Biela, nad O.Vogl, Makromol.Chem., in press.
17. Y.Yamashita, K.Ito, and F.Nakakita, Makromol.Chem., 127, 292 (1969).
18. A.A.Solov'yanov and K.S.Kazanskii, Vysokomol. Soedin., A12, 2114 (1970), ibid A14, 1063 (1972).
19. A.Deffieux and S.Boileau, Polymer,18 1047 (1977).
20. P.Hemery, S.Boileau, and P.Sigwalt, Europ.Polym.J., 7, 1581 (1971).
21. D.Kotynska, St.Slomkowski, and St.Penczek, in preparation.
22. A.Deffieux and S.Boileau, Macromolecules, 9, 369 (1976).
23. T.Ouhadi and J.M.Heuschen, J.Macromol.Sci.,-Chem., A9, 1183 (1975).
24. T.E.Hogen Esch and J.Smid, J.Phys.Chem., 79, 233, (1975).
25. U.Takaki, T.E.Hogen Esch, and J.Smid, J.Am.Chem. Soc., 93, 6760 (1971).

RECEIVED March 5, 1981.

Use of Cryptates in Anionic Polymerization of Heterocyclic Compounds

SYLVIE BOILEAU

Laboratoire de Chimie Macromoléculaire associé au CNRS, Collège de France,
11 Place Marcelin–Berthelot, 75231 Paris Cedex 05, France

The anionic polymerization of heterocyclic monomers in
homogeneous media has been relatively less studied than that of
vinyl or dienic monomers. The propagation reaction occurs
through active centers that are generally much more stable than
carbanions, and the polymerization rates are smaller, thus
allowing convenient conductance and kinetic measurements (1,2).
However only few monomers, namely propylene sulfide and ethylene
oxide, which give true living polymers, have been studied in
detail. Difficulties of different types are encountered for
several heterocyclic monomers such as the insolubility of the
resulting polymers, chain transfer to the monomer or polymer or
associations of ion pairs into higher aggregates. Considerable
progress has been made in this field in the last few years by the
use of macrobicyclic ligands (I) discovered by Lehn (3).

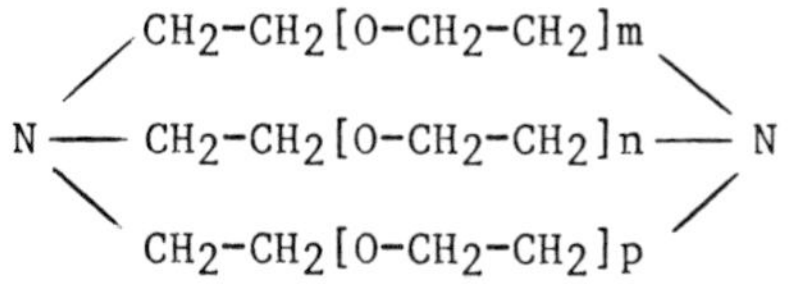

$$N \begin{array}{c} \diagup CH_2-CH_2[O-CH_2-CH_2]m \diagdown \\ \text{—}\ CH_2-CH_2[O-CH_2-CH_2]n\ \text{—}\ N \\ \diagdown CH_2-CH_2[O-CH_2-CH_2]p \diagup \end{array}$$

(I)

for instance m = n = p = 2 designated [222]

These ligands form extremely stable cation inclusion complexes,
called cryptates, in which the cation is completely surrounded by
the ligand and hidden inside the molecular cavity, and this leads
to a considerable increase of the interionic distance in the ion
pairs. It has been shown that such ligands have a marked
activating effect on anionic polymerizations (4,5,6). Moreover,
the aggregates are destroyed and simple kinetic results have been
obtained in the case of propylene sulfide (7,8,9), ethylene oxide
(9,10,11) and cyclosiloxanes (12) polymerizations. Though the

cryptates have been used for the anionic polymerization of other heterocyclic monomers like lactones (4,13,14), cyclic carbonates (4) and lactams (15), we will present the results obtained with propylene sulfide, ethylene oxide and cyclosiloxanes in this paper.

I. Anionic Polymerization of Propylene Sulfide

Propylene sulfide gives polymers with perfectly stable thiolate living ends in ethereal solvents (16). The propagation reaction has been studied in THF (7-9,17-21) and in THP (8,9,21,22,23) with Na^+, Cs^+, Bu_4N^+ and several cryptates as counterions, by dilatometry. It has been shown that even for non cryptated species, the associations of ion pairs do not generally occur in the range of concentrations examined ([C]<10^{-3} mole.l^{-1} in THF and < 2 10^{-4} mole.l^{-1} in THP). The main ionic species are ion pairs in equilibrium with free ions:

$$\sim\sim\sim S^- \ M^+ \ \underset{\longleftarrow}{\overset{K_D}{\longrightarrow}} \ \sim\sim\sim S^- + M^+ \qquad (1)$$

The fraction of free thiolate anions is given by the relation:

$$\frac{[C]\alpha^2}{(1-\alpha)} = K_D \qquad (2)$$

The K_D has been measured directly for Na^+ + [222] as counterion in THF, and calculated from the interionic distance a according to the Fuoss equation for other cryptates, using the Stokes radius values R_s^+ obtained from conductimetric studies of the corresponding tetraphenylborides (24). The value of K_D for Na^+ + [222] in THP was deduced from that found in THF assuming that the interionic distance remains constant in both solvents.

Kinetic measurements were performed at several concentrations [C] of living ends for a given counterion, at a fixed temperature. The values of the apparent propagation rate constant:

$$k_p = \frac{R_p}{[M][C]} \qquad (3)$$

were determined for each experiment (R_p = rate of polymerization, [M] and [C] = monomer and living ends concentrations). According to equilibrium (1), k_p depends on the fraction of free ions:

$$k_p = (1 - \alpha) \ k_{\pm} + \alpha k_- \qquad (4)$$

$k_{\pm}$ and k_- being the propagation constants of the ion pairs and of the free ions, respectively. Examples of plots of k_p versus α are shown in Figure 1, for $Na^+ + [222]$ in THF and in THP, at $-30°C$, giving k_+ and k_-.

The values of k_+ and of k_- found in THF at $-30°C$ are listed in Table 1. The values of k_- may be considered to be the same within the experimental errors. The k_+ values increase rapidly with the size of the counterion. The most surprising fact is that k_+ becomes higher than k_- with all cryptated counterions. A plot of log k_+ versus $1/\underline{a}$ gives approximately a straight line (Fig. 2) and it may be seen that the experimental value for k_- (obtained from kinetics for Na^+) is quite far from the extrapolated value for $\underline{a} = \infty$.

It seems that the reactivities of ion pairs and free ions cannot be readily compared. It may be imagined that a polarization of the CH_2-S bond in the monomer occurs before the ring opening and is induced by the ion pair dipole of the living end. Such a modification should increase with the magnitude of the ion pair dipole and then with the interionic distance $\underline{a}$. This leads to an increase of the interaction with the monomer, together with the reaction rate. It is understandable that the modification of the polarization of the CH_2-S bond may be quite different when the interaction occurs with free ions.

From the results shown in Figure 1, it can be seen that, for $Na^+ + [222]$, the reactivities of cryptated ion pairs and of free ions are nearly the same in THP whereas k_+ is three times higher than k_- in THF. Moreover, k_- is found to be the same in THF and in THP. The plot of log k_+ versus $1/\underline{a}$ for the results found in THP gives a straight line parallel to the line observed in THF. In a general manner, the ion pairs reactivity is lower in THP than in THF for a given counterion because the separation of the charges in the transition state is more difficult in a medium of lower dielectric constant.

II. Anionic Polymerization of Ethylene Oxide

The propagation reaction of ethylene oxide anionic polymerization has been studied in THF at $20°C$ with K^+, $K^+ + [222]$, Cs^+ and $Cs^+ + [TC]$ as counterions (10,11). [TC] is a spheroidal macrotricyclic tetramine hexaether (25) which forms a very stable complex with Cs^+.

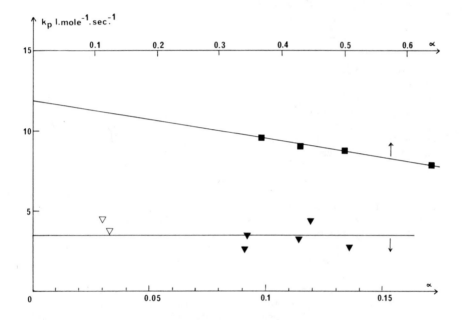

Figure 1. Linear dependence of the apparent bimolecular rate constant of living
polypropylene sulfide propagation on the fraction of free ions α with Na⁺ + [222]
as counterion at −30°C: in THF (■); in THP (▼); with φ₄BNa + [222] in
THP (▽).

TABLE I

Propagation Constants of Free Ions and Ion Pairs for the
Anionic Polymerization of Propylene Sulfide in THF at $-30°C$

Counterion	$10^5 K_D$	$\overset{F}{\underset{\overset{o}{A}}{a_K}}$	$k_{\pm}$	k_-
			$1.mole^{-1}.sec.^{-1}$	
Na^+	$0.0054^{a)}$	3.7	0.0025	3.8
Cs^+	-	4.3	$0.23^{b)}$	-
$Bu_4N^{+c)}$	1.90	6.5	1.7	3.5
$Li^+ + [211]$	2.53	6.7	4.9	2.8
$Na^+ + [221]$	3.35	7.0	8.9	3.0
$Na^+ + [222]$	4.20	7.2	11.9	5.6
$Na^+ + [2_o2_o2_s]$	4.55	7.3	8.3	3.3
$Cs^+ + [322]$	5.50	7.5	8.7	4.7

a) Obtained from kinetic experiments made with and without
   ϕ_4 BNa

b) Extrapolated from the value at $+15°C$ using $\underline{E_+}= 11$ Kcal.mole^{-1}

c) Obtained from kinetic data in ref. 19 using a calculated K_D

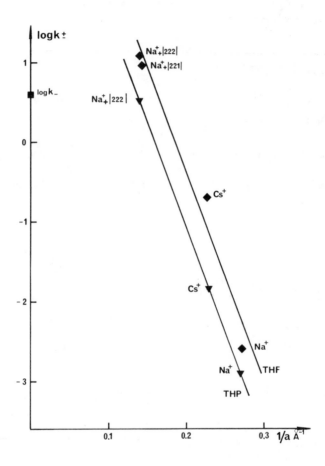

Figure 2. Plot of log $k_{\pm}$ vs. the reciprocal of interionic distance parameter of ion pairs for the anionic polymerization of propylene sulfide at $-30°C$: in THF (♦); in THP (▼).

designated [TC] (II)

Kinetic measurements were made by dilatometry at concentrations in living ends lower than $7 \cdot 10^{-4}$ mole.l^{-1} since associations of ion pairs become too important and cannot be neglected at higher concentrations.

Use of cryptates leads to the formation of a simple equilibrium between cryptated alkoxide ion pairs and free ions. The K_D has been measured directly for $K^+ + [222]$: $K_D = 3 \times 10^{-7}$. This value is in good agreement with that determined from two sets of kinetic experiments made with and without $\phi_4 BK + [222]$ In this case, the plot of k_p versus α gives k_+ and k_- as shown in Figure 3. For $Cs^+ + [TC]$, K_D was calculated from the kinetic results obtained with and without $\phi_4 BCs + [TC]$ assuming that k_- has the same value whatever is the nature of the counterion.

Kinetic experiments have been made with K^+ and Cs^+ as counterions in the absence of cryptand for [C] lower than $1.5 \cdot 10^{-4}$ mole.l^{-1} in order to avoid the associations of ion pairs. The corresponding values of K_D are very low (26), thus k_p can be considered to be equal to k_+.

The results obtained with different counterions are shown in Table II. Free alkoxide ions are about seventy times more reactive than cryptated ion pairs. Cryptated ion pairs are surprisingly slightly less or as reactive as the corresponding non complexed ion pairs within the experimental errors though the charges are more separated. K_D is equal to 3×10^{-7} for $\sim\sim0^-$ $K^+ + [222]$ in THF at 20°C (10) instead of 1.8×10^{-10} for $\sim\sim0^-$ K^+ (26). The former value leads to an interionic distance a equal to 4.6 Å according to the classical Fuoss equation. This value is too small compared to the results obtained for cryptated living polypropylene sulfide (8) and for cryptated tetraphenyl-borides in THF (24). This might mean that either K^+ is not located inside the cavity of the ligand or the oxanion can penetrate into the cavity of the cryptand. This last explanation is consistent with comparative conductivity data made on model compounds (17) as shown in Table III.

The K_D of cryptated potassium phenoxide is higher (about 4 times) for 2,6 dimethylphenoxide compared to 3,5 dimethylphenoxide.

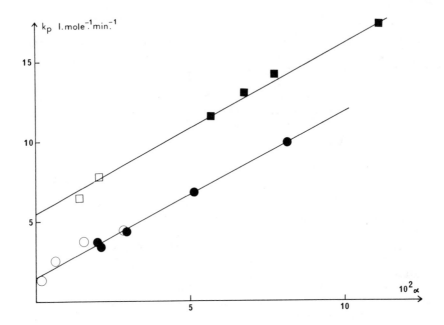

Figure 3.   Linear dependence of the apparent bimolecular rate constant of living polyethylene oxide propagation on the fraction of free ions α with cryptates as counterions in THF at 20°C: K⁺ + [222] (●): with φ₃BK + [222] (○); Cs⁺ + [TC] (■); with φ₃BCs + [TC] (□).

TABLE II

Propagation Constants of Ion Pairs and Free Ions for the
Anionic Polymerization of Ethylene Oxide in THF at 20°C

Counterion	K_D	$k_{\pm}$	k_-
		$1.mole^{-1}.min.^{-1}$	
Na^+	–	~ 0	–
K^+	$1.8 \ 10^{-10}$	~2.9	–
Cs^+	$2.7 \ 10^{-10}$	7.3	–
$Na^+ + [222]$	–	~ 0	–
$K^+ + [222]$	$3 \ 10^{-7}$	1.5	100
$Cs^+ + [TC]$	$1 \ 10^{-6}$ [a]	5.6	100 [b]

[a] Determined in assuming that $k_- = 100 \ 1.mole^{-1}.min^{-1}$

[b] k_- is assumed to be the same as for $K^+ + [222]$.

TABLE III

Conductivity Data for Cryptated Alkoxides and Phenoxides in THF
at 20°C

Salt	K_D	$\frac{a}{Å}$
$\sim\sim\sim CH_2CH_2O^-$ K^+ + [222]	$3 \ 10^{-7}$	4.6
$-O^-K^+$ + [222]	$1.9 \ 10^{-6}$	5.3
$-O^-K^+$ + [222]	$6.8 \ 10^{-6}$	6.0
Theoretical Value	—	5.9

This is due to the methyl groups which hinder the penetration of the oxanion into the cavity of the cryptate when they are located in the 2 and 6 positions. Moreover, there is a good agreement between the theoretical value for the interionic distance and the value found for the cryptated 2,6 dimethylphenoxide.

III. Anionic Polymerization of Cyclosiloxanes

The anionic polymerization of cyclosiloxanes has been extensively studied; despite this, the knowledge of the mechanism of the process is poor. This is due to the fact that the polymerization is of the equilibrium type and besides linear polymers, cyclic structures exist in solution: side reactions which are intra and intermolecular attacks of the chains by living centers occur during the propagation. Moreover, the nature of active propagation species as well as their relative proportion is not well known in the investigated systems (28,29,30). We thought that use of cryptates for the anionic polymerization of cyclosiloxanes might suppress the aggregates and simplify the mechanism (12,23,31).

This study was began with hexamethylcyclotrisiloxane (D_3) as monomer since it has been shown that under certain conditions it polymerizes to give linear polymer of negligible cyclosiloxane content at a much more rapid rate than in the case of octamethylcyclotetrasiloxane (D_4) (29,32). A suitable initiator is butyllithium which can be mixed with D_3 in hydrocarbon solvents to form $Bu-Si(CH_3)_2OLi$ (33). No polymerization occurs even in the presence of excess D_3 until a donor solvent such as THF (29,32), triglyme (29), DME (34), HMPA (34) or DMSO (32) is added; then a reasonably rapid polymerization starts to give near monodisperse polymers.

We have shown that living polydimethylsiloxanes (PDMS) can be prepared by adding the [211] cryptand to benzene solutions of D_3 after reaction with n-butyllithium, at 25°C, as can be seen from the results of Table IV. The propagation reaction is much faster with [211] than with THF.

1. **Anionic Polymerization of D_3.** Kinetics of propagation reaction were followed by dilatometry under high vacuum, at 20°C, at various living end concentrations, with nearly the same value of $[M]_0$ ($\sim$0.5 mole.1^{-1})(12,23). It was established that D_3 disappears according to a first-order law at least up to 80% conversion for each experiment. On plotting $R_p/[M]$ versus $[C]$ (figure 4), a straight line passing through the origin is obtained, showing that the reaction order in active centers is equal to 1. This result can be interpreted either by assuming that only one type of active species is present – presumably cryptated ion pairs – or by assuming that if there are different types of active centers, they might have the same reactivity.

TABLE IV

Polymerization of D_3 Initiated by n-BuLi, in Benzene at Room Temperature

D_3 (g)	BuLi (10^5 mole)	Benzene/THF (v/v)	[211]	Propagation Time (h.)	Yield (%)	$\overline{Mn}$ th. 10^{-5}	$\overline{Mn}$ osm. 10^{-5}
5.1	3.95	27/26	-	7	98	1.27	1.14
6.0	4.27	26/0	11.6	1	100	1.40	1.10

Initiation time: 16 h. at 25°C. Termination reagent: trimethylchlorosilane

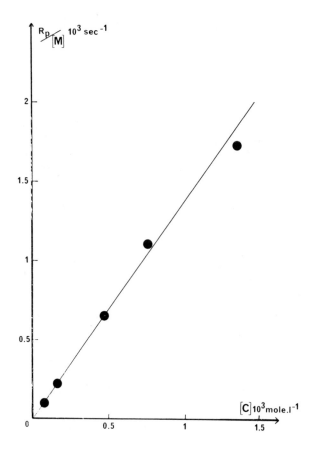

Figure 4.   Plot of $R_p/[M]$ vs. [C] for the anionic polymerization of D_3 with Li^+ + [211] as counterion, in benzene at $20°C$.

In order to elucidate this point, viscosity measurements of
living and deactivated PDMS solutions were performed in toluene,
with Li^+ + [211] as counterion. As no significant change was
observed, it can be deduced that the fraction of aggregates is
negligible ($<$ 1% for [C] = 10^{-3} mole.1^{-1}). Moreover, conductance
measurements made on model silanolates in THF indicate that the
fraction of free ions is very low. In our system (benzene) we
conclude therefore that the contribution to the reactivity from
free ions can be neglected. Thus the main ionic species are
cryptated ion pairs, and

$$k_p = \frac{R_p}{[M] [C]} = k_{\pm} = 1.4 \text{ 1.mole}^{-1}.\text{sec.}^{-1}$$

in benzene at 20°C.

It is well known that cyclic oligomers (D_x) are formed as
by-products through intramolecular attacks of the chains by
active centers, during the propagation. The amounts of consumed
D_3 and formed D_4, D_5 and D_6 were determined by GLC analysis of
the polymerization mixture during the propagation. The results
are shown in figure 5 for [C] = 3.08 10^{-4} mole.1^{-1}. The maximum
yield of polymer ($\sim$90%) is rapidly obtained whereas the
proportion of cyclosiloxanes is very low (weight percentage of D_4
+ D_5 + D_6 $<$ 3%) in the case of Li^+ + [211] as counterion, in
benzene at 20°C.

2. <u>Anionic Polymerization of D_4</u>. It has been recently
shown that addition of cryptands markedly increases the rate of
the bulk polymerization of D_4 initiated by KOH at 160°C (<u>35</u>).
The anionic polymerization of D_4, in toluene, at room temperature, with Li^+ + [211] as counterion appears to follow a different course than in the case of D_3. The rate of polymerization
is much lower and cyclic oligomers formation is more important as
can be seen from the results shown in figure 6 for [C] = 2.76
10^{-3} mole.1^{-1}.

Polymerization of D_4 is represented as follows:

$$\sim\sim Me_2Si \text{ } O^- \text{ } \textcircled{Li}^+ + D_4 \xrightarrow{k_{p4}} \sim\sim Me_2SiO(Me_2SiO)_3Me_2SiO^- \text{ } \textcircled{Li}^+$$

Since it has been shown that the main ionic species are cryptated
ion pairs in the case of Li^+ + [211] in benzene or toluene, k_{p4} =
$k_{\oplus 4}$.

By-products cyclosiloxanes such as D_5, D_6 and D_7 are formed
during the polymerization. Equilibrium between linear and cyclic
oligomers are established as shown below (<u>36</u>):

$$\sim\sim Me_2SiO(Me_2SiO)_{x-1}Me_2SiO^- \text{ } \textcircled{Li}^+ \underset{k_{px}}{\overset{k_{dx}}{\rightleftharpoons}}$$

$$\sim\sim Me_2SiO^- \text{ } \textcircled{Li}^+ + D_x$$

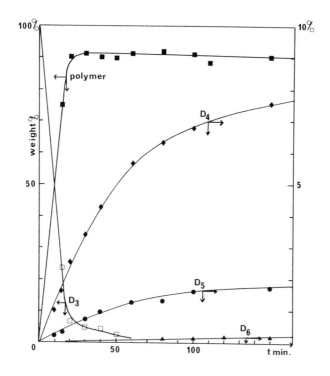

Figure 5. Conversion of D_3 and formation of D_4, D_5, and D_6 as a function of time: $[C] = 3.08 \cdot 10^{-4}$ mol $\cdot$ L^{-1}; $[M]_o = 0.49$ mol $\cdot$ L^{-1}.

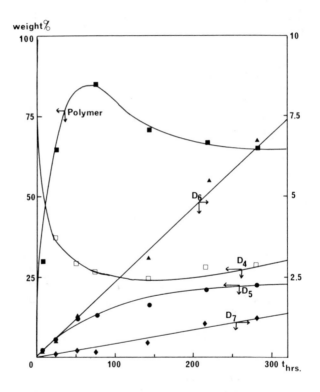

*Figure 6.   Conversion of D_4 and formation of D_5, D_6, and D_7 as a function of time:
[C] = 2.76 10^{-3} mol · L^{-1}; $[M]_o$ = 1.06 mol · L^{-1}.*

with $K_x = k_{dx}/k_{px} = [D_x]eq.$ (x > 4)   (5)

where k_{dx} and k_{px} are the rate constants of formation and of propagation of D_x, respectively, K_x is the molar cyclization equilibrium constant and $[D_x]eq.$ is the concentration of D_x in equilibrium.

The rates of consumption of D_4 and formation of cyclic oligomers D_x are given by kinetic equations (6) and (7) (37):

$$- \frac{d[D_4]}{dt} = (k_{p4} [D_4] - k_{d4}) [C] \qquad (6)$$

$$\frac{d[D_x]}{dt} = (k_{dx} - k_{px} [D_x]) [C] \qquad (7)$$

If the rate of formation of D_4 is assumed to be negligible compared to its rate of polymerization, at the beginning of the reaction, Eqn. (6) becomes:

$$- \frac{d [D_4]}{dt} \sim k_{p4} [D_4][C] \qquad (8)$$

Thus, it is possible to evaluate k_{d4} from Eqn. (5) by determining $[D_4]eq.$ for each experiment.

Concerning the formation of by-product cyclosiloxanes, two types of approximations can be made. The first one consists in assuming that the rate of polymerization of D_x is neglibible compared to its rate of formation at the beginning of the reaction. Thus Eqn. (7) becomes:

$$\frac{d [D_x]}{dt} \sim k_{dx} [C] \qquad (9)$$

and k_{px} can be evaluated from Eqn. (5) by determining $[D_x]eq.$ On the other hand, integration of Eqn. (7) gives:

$$[D_x] = \frac{k_{dx}}{k_{px}} (1 - e^{-k_{px}[C]t}) \qquad (10)$$

if $[D_x]_o = 0$

If the first terms of the series expansion of $\dfrac{1}{e^{k_{px}[C]t}}$ are taken, Eqn. (10) becomes:

$$\frac{1}{[D_x]} \sim \frac{k_{px}}{k_{dx}} + \frac{1}{k_{dx}[C]t} \tag{11}$$

It is thus possible to determine k_{px} and k_{dx} by plotting $\dfrac{1}{[D_x]}$ versus $1/t$.

Both methods have been applied in the case of the anionic polymerization of D_4, in toluene, at room temperature, with Li^+ + [211] as counterion, which give nearly the same results. Moreover, k_{px} and k_{dx} do not vary significantly with [C] as expected since only one type of ionic species is present in the medium. The results are listed in Table V.

A comparison of reactivity of different cyclosiloxanes towards silanolates is shown in Table VI. The results found in the base-catalysed solvolysis of siloxanes in alcohols (38) as well as those observed for the ring opening of cyclosiloxanes by potassium phenyldimethylsilanolate in heptane-dioxane (95:5, v/v) at 30°C (30) are also reported. Our results are in good agreement with those found by Lasocki (38) whereas they do not fit well with the results of Chojnowski (30). It is difficult to comment in detail on the results of these authors because the nature of involved ionic species is not well known. However, while in an alcohol the free anions constitute the active form, in the non-polar medium used by Chojnowski the attacking species have the structure of aggregates. It is reasonable to conceive that our results would better agree with those found in methanol than with those found in heptane since, though we are working in a non-polar medium, the addition of a cryptand suppresses the aggregates and lead to a significant separation between the negative and positive charges in the ion pairs.

3. <u>Molecular weight distributions of PDMS</u>. Gel permeation chromatography of PDMS prepared from D_4 with Li^+ + [211] as counterion, in toluene, shows a bimodal distribution. The ratio of the percentage of low molecular weight polymers over the percentage of high molecular weight PDMS in nearly constant during the course of the polymerization. Its value is rather important: $1/4 < R < 1/3$. Moreover, Mw/Mn values are always equal to 2 for the main peak. This means that intermolecular attacks of the chains by silanolate ends are important and occur at the beginning of the propagation.

GPC diagrams of the PDMS obtained from D_3 with Li^+ + [211] as counterion in benzene show that the proportion of oligomers is very low. However, Mw/Mn values corresponding to high yields of conversion are nearly equal to 2.

Polymers of narrow molecular weight distribution can be obtained using a cryptand the cavity of which is larger than that of the [211] like the [221] or the [222] ligand. It can be

TABLE V

Rate Constants of Formation and of Propagation of D_x in Toluene
at Room Temperature, with Li^+ + [211] as Counterion

D_x	D_4	D_5	D_6	D_7
k_{px}				
$1.mole^{-1}.h^{-1}$	17	6.5	1.2	–
k_{dx}				
$h.^{-1}$	4	0.9	0.05	0.006

TABLE VI

Comparison of the Reactivity of D_X Towards Silanolates

D_X	D_3	D_4	D_5	D_6
k_{px} [a] $l.mole^{-1}.h^{-1}$	5000	17	6.5	1.2
k_{px}/k_{p4}	280	1	0.4	0.07
k_{Dx}/k_{D4} [b]	357	1	0.1	0.06
k_{Dx}/k_{D4} [c]	88	1	1.1	11.8

[a] Found in toluene at 20°C with Li^+ + [211] (31) except for
     D_3 (benzene) (23).

[b] Found in the KOH-catalysed solvolysis of siloxanes in
     methanol (38).

[c] Found in the ring opening of D_X by potassium
     phenyldimethylsilanolate in heptane-dioxane, v/v (95 : 5 at
     30°C (30).

noticed that the rates of propagation decrease on increasing the size of the ligand. Moreover, the cyclic by-product formation is completely inhibited with Li^+ + [221] or Li^+ + [222] as counterions. The results are shown in Table VII.

Conclusion

In conclusion, it has been shown that use of cryptates for the anionic polymerization of heterocyclic monomers leads to a tremendous increase of the rates of polymerization. There are two main causes to the higher reaction rates observed with cryptates. The first one is a suppression of the association between ion pairs in the non polar media, and the second one is the possibility of ion pairs dissociation into free ions in ethereal solvents like THP or THF. By this way, it has been possible to make detailed studies of the propagation reaction for propylene sulfide, ethylene oxide, and cyclosiloxanes.

In the case of a polarizable monomer like propylene sulfide, cryptated ion pairs are not only much more reactive than the corresponding non-complexed ion pairs but they are even more reactive than free ions in THF. For ethylene oxide, the results are different since free alkoxide ions are significantly more reactive than cryptated ion pairs which are themselves slightly less reactive than the corresponding non complexed ion pairs. This anomalous behavior can be explained by a penetration of the oxanion into the cavity of the cryptate.

The advantages of lithium cryptates for the anionic polymerization of cyclosiloxanes are numerous. With Li^+ + [211], only one type of active species is observed in benzene or toluene. Thus, the kinetics of the propagation of the by-products cyclosiloxanes formation can be studied in detail for the first time. The reactivity of cryptated silanolate ion pairs towards the ring opening of D_3 is greatly enhanced compared to that of the other systems used up to now. Moreover, the amount of by-products cyclic compounds is very low when the polymer yield reaches the maximum value. On the other hand, polymers of narrow molecular weight distributions could be obtained using a cryptand the cavity of which is larger than that of the [211] like the [221] or the [222] ligand.

More detailed studies on other monomers are in progress.

Acknowledgements

I would like to thank Drs. P. Hemery, A. Deffieux, H. Dang Ngoc, H. Porte, and Mrs. N. D'Haeyer, who participated in the investigations of the laboratory in this field. I would like to thank also Professor J. M. Lehn of the "Collège de France" for many stimulating discussions.

Literature Cited

1. Bywater, S. Prog. Polym. Sci., 1975, 4, 27.

2. Billingham, N. C. "Developments in Polymerization"; Haward, R. N., Ed.; Applied Science Publishers, 1979, ch. 4.

3. Lehn, J. M. Structure and Bonding, 1973, 16, 1.

4. Lehn, J. M.; Schue, F.; Boileau, S.; Kaempf, B.; Cau, A.; Moinard, J.; Raynal, S. Fr. Patent, 1973, 2 201 304 and 1979, 2 398 079.

5. Boileau, S.; Kaempf, B.; Lehn, J. M.; Schue, F. J. Polym. Sci., (B), 1974, 12, 203.

6. Boileau, S.; Hemery, P.; Kaempf, B.; Schue, F.; Viguier, M. J. Polym. Sci., (B), 1974, 12, 217.

7. Hemery, P.; Boileau, S.; Sigwalt, P.; Kaempf, B. J. Polym. Sci. (B), 1975, 13, 49.

8. Hemery, P.; Boileau, S.; Sigwalt, P. J. Polym. Sci. Polym. Symp., 1975, 52, 189.

9. Sigwalt, P.; Boileau, S. J. Polym. Sci. Polym. Symp., 1978, 62, 51.

10. Deffieux, A.; Boileau, S. Polymer, 1977, 18, 1047.

11. Boileau, S.; Deffieux, A.; Lassalle, D.; Menezes, F.; Vidal, B. Tetrahedron Letters, 1978, 1767.

12. Dang Ngoc, H.; Porte, H.; Hemery, P.; Boileau, S. Internat. Symp. on Macromolecules, Mainz, Germany, 1979, Prep. Vol. I p. 137.

13. Deffieux, A.; Boileau, S. Macromolecules, 1976, 9, 369.

14. Haggiage, J.; Hemery P.; Boileau, S. Communication at the 5th Symposium on Cationic and Other Ionic Polymerizations. Kyoto, Japan, 1980.

15. Sekiguchi, H.; Tsourkas, P.; Coutin, B. Makromolek. Chem., 1977, 178, 2135.

16. Boileau, S.; Champetier, G.; Sigwalt, P. Makromolek. Chem., 1963, 69, 180.

17. Favier, J. C.; Boileau, S.; Sigwalt, P. Eur. Polym. J., 1968, 4, 3.

18. Guerin, P.; Hemery, P.; Boileau, S.; Sigwalt P. Eur. Polym. J., 1971, 7, 953.

19. Tersac, G.; Boileau, S.; Sigwalt, P. Makromolek. Chem., 1971, 149, 153.

20. Guerin, P.; Boileau, S.; Sigwalt, P. Eur. Polym. J., 1971, 7, 1119.

21. Hemery, P.; Boileau, S.; Sigwalt, P. Eur. Polym. J., 1971, 7, 1581.

22. Tersac, G.; Boileau, S.; Sigwalt, P. J. Chim. Phys., 1968, 65, 1141.

23. Dang Ngoc, H. Thèse Doctorat d'Etat, Paris, 1979.

24. Boileau, S.; Hemery, P.; Justice, J. C. J. Solut. Chem., 1975, 4, 873.

25. Graf, E.; Lehn, J. M. J. Am. Chem. Soc., 1975, 97, 5022.

26. Solovyanov, A. A.; Kazanskii, K. S. Vysokomol. Soedin. 1972, A14, 1063.

27. Deffieux, A. Thèse Doctorat d'Etat, Paris, 1978.

28. Chojnowski, J.; Mazurek, M. Makromolek, Chem., 1975, 176, 2999.

29. Zilliox, J. G.; Roovers, J. E. L.; Bywater, S. Macro-molecules, 1975, 8, 573.

30. Mazurek, M.; Chojnowski, J. Makromolek. Chem., 1977, 178, 1005.

31. Porte, H. Thèse Docteur-Ingénieur, Paris, 1980.

32. Lee, C. L.; Frye, C. L.; Johansson, O. K. Polym. Prep., 1969, 10, 1361.

33. Frye, C. L.; Salinger, R. M.; Fearon, F. W. G.; Klosowksi, J. M.; De Young, T. J. Org. Chem., 1970, 35, 1308.

34. Hölle, H. J.; Lehnen, B. R. Eur. Polym. J., 1975, 11, 663.

35. Bargain, M.; Millet, C. Fr. Patent; 1976, 2 353 589.

RECEIVED April 27, 1981.

Anionic Polymerization of Acrolein

Kinetic Study and the Possibility of Block Polymer Synthesis: Mechanisms of Initiation and Propagation Reactions

D. GULINO, Q. T. PHAM, J. GOLÉ, and J. P. PASCAULT

Laboratoire de Materiaux Macromoleculaires, ERA 745, Batiment 403, Institut National Des Sciences Appliquées, 69621 Villeurbanne, France

Acrolein is a very reactive monomer with a high tendency to polymerize. Free-radical and cationic polymerizations lead to insoluble polymers even if the conversions are low. On the other hand, anionic polymerization gives soluble polyacroleins under well-adapted conditions. Because of the difunctional nature of this monomer, the chains of polyacroleins contain different types of units:

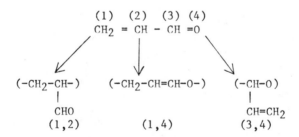

Chains of polyacroleins contain different functional groups. Because of this, these polymers are among the most reactive and may be transformed into a variety of derivatives. Particularly, polyacroleins are radiation-sensitive polymers and can be used to prepare printing plates. However, the elucidation of the structure of the polyacroleins is made extremely difficult by their lack of stability. Indeed, these polymers eventually become insoluble some time after their synthesis.

Several authors (1,2,3,4) have tried to determine the polyacrolein microstructures by chemical analysis.. The results are very often contradictory: only Gole, et al. (5) have shown the presence of (1,4) units. In a previous article (6), we have identified the different types of units using ^1H and ^{13}C NMR. These methods have enabled us to analyze all anionically prepared polyacroleins.

0097–6156/81/0166–0307$05.00/0

Koton, et al. (7,8) have studied the acrolein propagation
initiated by Na-Naphthalene complex or by t-butyllithium in
tetrahydrofuran (THF) at various temperatures. From these results, a mechanism has been deduced, but it ignores the transfer
reactions and the different complexations of the living end.

In this article, we have summarized the main results of our
kinetic study (particularly transfer and propagation rate constants) obtained in THF with Li^+ and Na^+ Naphthalene complex as
initiators. We have also given the microstructure of polyacroleins obtained with the same initiators.

Furthermore, by studying butadiene -acrolein block polymerizations, we have tried to understand the mechanism of the acrolein polymerization.

Experimental Techniques

Synthesis

The usual techniques of the anionic polymerization have been
employed for the solvent purification and the syntheses. The
monomer is distilled three times from calcium hydride and then
once more just before the polymerization.

To synthesize homopolyacroleins, the naphthalene complexes
(Li^+, Na^+, K^+) have been used in tetrahydrofuran (THF). Then,
living oligomers of butadiene, initiated by the naphthalene complex systems were reacted with acrolein to build the block polymers. Other living oligomers of butadiene were first functionalized with ethylene oxide before reacting with acrolein. In this
case, the viscosity is very high and the addition of acrolein is
conducted at room temperature. On the other hand, the same reaction is made at $-40°C$ for the oligobutadienes. The functionalization with acrolein was studied with two different living ends:
the carbanions of butadiene at $-40°C$ and the oxanions of ethylene
oxide at $20°C$ (Table 2). The same concentration of butadiene
units has been employed for the chromatograms of each block polymerization stage and the observed variations between the molecular weight distributions are quantitative.

Molecular Weights

Molecular weight and polydispersity determinations were obtained by GPC with THF as elution solvent with styragel columns
(60, 100, 500, 10^3, 10^4, 10^5 and 3.10^5 Å). Eluted macromolecule
was observed simultaneously by refractive index and by ultraviolet absorption.

For block polymers, the determination of the molecular
weight distribution has been realized on a Waters apparatus with
five columns (three Waters columns: 500 Å and 2 x 100 Å and two
Shodex columns S 802 and S 803.

To easily explain the results of the molecular weight
distributions of the block polymers, it is necessary to
synthesize initiating polymers having molecular weights lower
than 5000. Indeed, the resolution of our apparatus is very high

for molecular weights ranging between 1000 up to 5000.
Therefore, a small shift in the chromatograms after the acrolein
addition can be observed.

^{1}H-NMR Spectroscopy

We have used the VARIAN XL 100 (100 MHz), VARIAN DA 60 IL
(60MHz) and CAMECA (350 MHz) as apparatus. The spectra have been
obtained in deuterated chloroform at room temperature with
tetramethylsilane as an internal reference. We cannot work at
elevated temperatures to reduce the broadening of the bands
because polyacroleins are very sensitive to heat.

Results

Kinetic Results

The kinetics of the anionic polymerization of acrolein has
been studied by dilatometry. The kinetic orders are shown to be
unity for monomer and also for initiator: The living ends are
not associated at the studied concentrations of initiators $\{I_o\}$
in THF with Li^+ and Na^+ as counter-ions ($1.4.10^{-3} < \{I_o\} < 2.10^{-2}$
mole.dm^{-3} for Li^+ and $2.2.10^{-4} < \{I_o\} < 1.4.10^{-3}$ mole.dm^{-3} for
Na^+).

The polymerization rate is greatly modified by the cation
and increases such as : $Li^+ < Na^+ < K^+$. (The polymerization
lasts several days with Li^+, a few hours with Na^+ and only a few
minutes with K^+). Moreover, it is not affected by whether the
initiators is a carbanion or a oxanion.

Terminations of growing chains do not occur. However, numerous transfer reactions to monomer and to polymer do take place
during this polymerization. During our efforts to synthesize
block polymers, these transfer reactions have been easily proved
(Figure 1). The initiator is an α-ω diNa polyisoprene with a
high molecular weight (Mn = 90,000) to which acrolein is added.
After this addition, the refractive index response (RI) shows
chains having molecular weights lower than those of the initiator. And, simultaneously, no shift of the maximum peak of the
curve (Ve -34 counts) corresponding to the polyisoprene initiator
is observed. On the other hand, only the polyacrolein weight
distributions can be seen on the UV responses because the initiator has no absorption at the wavelength λ = 254mμ. On these UV
chromatograms the maximum peak is situated widely down from Mn =
90,000. Numerous chains of homopolyacroleins are synthesized by
transfer reactions to monomer.

Transfer reactions to polymer are easily observed. Indeed,
for the very high yields of the acrolein polymerization, insoluble gels are obtained. With K^+ as counter ion, the polymerization being very fast, this phenomenon cannot be avoided. On the
other hand, with Li^+ and Na^+, no gel is obtained if the yields
remain lower than 70%.

These transfer reactions have no effect on the polymerization rate, but, nevertheless, greatly modify the molecular weight

of the polyacroleins. The measured $\overline{M}n$ is much lower than the theoretical $\overline{M}n$. Moreover, very often, the molecular weight distributions of the polyacroleins are bimodal as shown on the Figure 2. In these conditions, the polymerization rate can be given by the following equation (10):

$$V_p = \frac{-d\{A\}}{dt} = (k_{pr} + 2\, h_m)\, \{A\}\{I_o\}$$

where {A} is the monomer concentration,
$\{I_o\}$ the initial concentration of the initiator,
k_{pr} the propagation rate constant,
h_m the transfer rate constant (transfer to the monomer only),

and $k_p = (k_{pr} + 2\, h_m)$ the polymerization rate constant.

Without detailing the calculations again, the last equation for k_p is obtained after the steady-state approximation for the living ends A^-, has been introduced. A new hypothesis is now presented: $\{A^-\}$ is $<< \{Pn^-\}$, where $\{A^-\}$ is the concentration of the living ends created after transfer to monomer and $\{Pn^-\}$ the concentration of chains having different lengths. Furthermore, the initiation reactions are very fast and do not interfere with the kinetics. Under these conditions:

$$\frac{1}{\overline{DP}} = \frac{\{I_o\}}{2\,\{A_o\}} \cdot \frac{1}{P} + \frac{h_m}{k_p}$$

where $\overline{DP}$ is the experimental degree of polymerization and $P = \dfrac{R}{100}$, R is the poly acrolein yield.

The experimental values for k_p and h_m at different temperatures are summarized in Table 1. In every case h_m is much lower than k_p and $k_p \approx k_{pr}$. Moreover, with temperature, the polymerization rates increase and simultaneously k_p/h_m becomes greater for both cations. Transfer reactions to monomer are more numerous at lower temperatures and also when Li^+ is used: At $-40°C$, the values of the k_p/h_m ratio are 47 for Li^+ and 140 for Na^+. With Na^+, as a cation, the activation energies for the anionic polymerization of acrolein and propylene sulfide (11) are approximately the same. On the other hand, with Li^+, it is impossible to compare the acrolein activation energy with the same monomer or another polar monomer because no result is found in the literature. Moreover, for the acrolein polymerization, $(E_{a,\,Li^+})$ is lower than $(E_{a,\,Na^+})$.

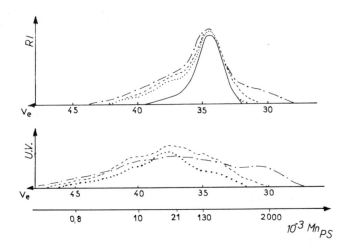

European Polymer Journal

Figure 1. Molecular weight distributions of a polyacrolein initiated by a living polyisoprene in THF at $-38°C$ initiator (———); t = 25 min (· · ·); t = 110 min (– – –); t = 340 min (– · – ·).

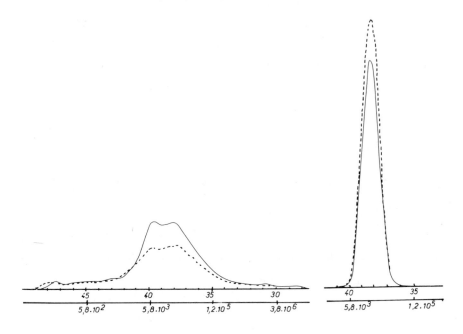

Figure 2. Molecular weight distributions of Polyacrolein $PA_{41}(\overline{M}_{nPS} = 4300)$ and of a polystyrene standard $(\overline{M}_n = 17500)$: refractometric index (———); UV absorption (– – –).

TABLE I

Kinetic Results for the Acrolein Polymerization

	T°C	$k_{P} dm^3 mol^{-1} sec^{-1}$	$h_{m} dm^3 mol^{-1} sec^{-1}$	$E_a, Kcal.mole^{-1}$
Li+	-41.0	$(1.6 \pm 0.2)10^{-4}$	$(3.3 \pm 0.7)10^{-6}$	
	-26.5	$(3.3 \pm 0.5)10^{-4}$	$'3.4 \pm 0.3)10^{-6}$	
	-5.2	$'7.5 \pm 0.7)10^{-4}$	$(6.1 \pm 1)10^{-6}$	6
	+7.6	$(8.7 \pm 0.9)10^{-4}$	$(9.2 \pm 1.4)10^{-6}$	
Na+	-41.9	0.24 ± 0.02	$(1.7 \pm 1.1)10^{-3}$	
	-35.0	0.43 ± 0.03	$'2.9 \pm 0.2)10^{-3}$	11
	-20.3	1.65 ± 0.09	$(7.1 \pm 0.7)10^{-3}$	

Study on Microstructures

Without detailing the ^{1}H and ^{13}C NMR method to analyse the
polyacroleins (12), the main results of the polymer microstructures obtained in THF with Li^+ and Na^+ can be summarized as follows (Figure 3):
- (1,2) units are less observed with polymers initiated
  with Na^+.
- With Li^+ the increase of the molecular weight involves a
  variation of the polyacrolein microstructures. When $\overline{Mn}$
  varies between 1000 up to 2500, the percentage of (1,2)
  units decreases.

Block Polymer Synthesis

Following these results, we wanted to reach two goals:
- Firstly, to prove the possibility of block polymer
  synthesis between the acrolein and a second monomer.
- Secondly, to explain the different mechanisms of
  propagation and transfer and the evolution of the
  microstructures. The experimental conditions are summed
  up in Table 2.

a) Figure 4 shows the GPC chromatograms of a polybutadiene
initiator (Mn = 4200) and that of the corresponding block polymer
PAB_6. A bimodal distribution can be seen simultaneously on the
RI and UV responses of PAB_6. The living polybutadiene reacts
upon acrolein and the new living ends can propagate and build the
block polymer. The maximum peak at 36 counts corresponds to the
block polymer distribution. Small amounts of homopolyacrolein
are present because molecular weights lower than the smallest
chains of "initiator" appear on the final chromatogram. With the
oxanion of ethylene oxide as initiator, a bimodal distribtuion is
also observed (Figure 5).

b) The PAB_{10} chromatogram (Figure 6) also shows a bimodal
distribution. And here, since the molecular weight of the initiator is very low (the resolution of our apparatus is the best in
this range of the molecular weights), a simultaneous shift of the
two distributions above the molecular weight of the initiator can
be noticed. All the living ends of polybutadiene initiate the
acrolein polymerization but this polymerization takes place on
two different living ends and leads to bimodal distribution chromatograms.

c) No variation in the chromatogram is observed when a new
quantity of butadiene monomer is added to the living block
polymer PAB_6 , fifteen minutes after the acrolein addition. The
living ends of polybutadiene which have not reacted upon acrolein
are absent. Moreover, the new living ends of acrolein cannot
initiate a new butadiene polymerization. Likewise, the living
ends of acrolein cannot initiate the polymerization of ethylene
oxide (PAO_7).

It seems important to separate the molecular weight distribution between three fractions using GPC and then to analyse
each fraction. This fractionating method has been applied to the

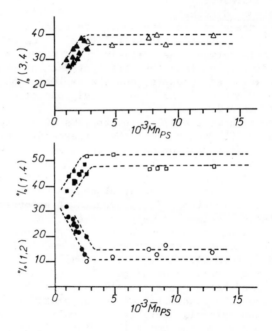

Figure 3.   *Dependence of the microstructures of polyacroleins with the molecular
weights M*n .

TABLE 2

Experimental Conditions for the Different Syntheses of Block Polymers,
(First Monomer: I = Isoprene; B = Butadiene; O = Butadiene Functionalized with Ethylene Oxide)
Na$^+$ as a Cation Except for PAB15 (Li$^+$)

N°	[Bo] mole dm^{-3}	[Ao] mole dm^{-3}	[Io] mole dm^{-3}	[Ao]/[Io]	$\overline{Mn}$ initiator	T°C
PAI3	0.56	1.09	$6.7.10^{-4}$	1627	90,000	-38
PAB6	0.57	0.37	$1.5.10^{-2}$	25	4200	-36
PAB10	0.32	0.37	$2.2.10^{-2}$	17	1600	-40
PAB14	0.6	0.35	$9.4.10^{-2}$	4	690	-49
PAB15	0.6	0.35	$9.4.10^{-2}$	4	690	-46
PAO7	0.89	0.37	$2.5.10^{-2}$	15	3900	+20

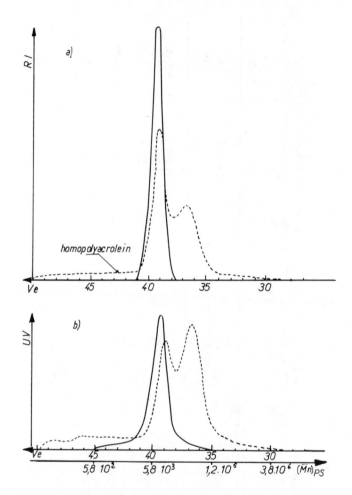

Figure 4. Acrolein initiation by an α–ω di-Na oligobutadiene in THF at −40°C
initiator $\overline{M}_n = 4200$ (———); PAB_6 after the acrolein polymerization (– – –); idem
after a 2nd addition of butadiene: (a) RI response; (b) UV response.

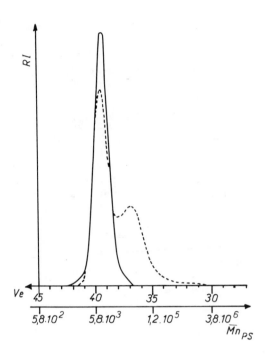

Figure 5. Acrolein initiation by an α—ω di-Na oligobutadiene functionalized with ethylene oxide in THF at + 20°C initiator $\overline{M}_n$ = 3900 (———); PAO_7 after the acrolein polymerization (———); idem after the 2nd addition of ethylene oxide.

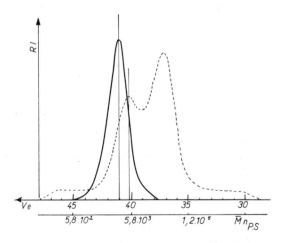

Figure 6. Acrolein initiation by an α—ω di-Na oligobutadiene in THF at −40°C: initiator $\overline{M}_n$ = 1600 (———); PAB_{10} after the acrolein polymerization (———).

transitioned PAB_{14} and PAB_{15} block polymers. The first is initated by the Na^+ Naphthalene complex, the second by the Li^+ Naphthalene complex. The chromatograms of the fractions are shown in the Figure 7 (PAB_{14}) and Figure 8 (PAB_{15}). ^1H NMR (350 MHz) has been used to analyse each fraction (Figure 9).

The polybutadienes used as initiators for the PAB_{14} and PAB_{15} syntheses have a molecular weight $\overline{Mn}$ = 690 but $\overline{Mn}_{ps}$ = 1070 with GPC apparatus. The corresponding block polymers have approximately the same $\overline{Mn}_{ps}$ = 1700. This value corresponds to the theoretical $\overline{Mn}$ = 1500. The polyacrolein blocks are built with 4.5 acrolein units and the corresponding acrolein $\overline{Mn}$ = 250. Moreover, the polydisperisty index increases from 1.2 for the polybutadiene initiators up to 1.4 for the block polymers.

Compared to the NMR spectrum, of the polybutadiene (80% 1,2) initiator, new bands appear on the spectra of block polymers for each fraction (Figure 9). These bands correspond to the acrolein blocks attached to the polybutadiene chains. In particular we can observe the presence of:

    − Aldehydic protons between 9.2 and 9.7 ppm which belong to the (1,2) units ($-CH_2-CH-$) or to the structure − $CH_2-CH-\underline{CHO}$ (this
                              $\underline{C}HO$
last structure is created after the deactivation of the living end (1,4)). No observation of this type of protons is made in high molecular weight polyacroleins.

    − At about 5.2 ppm, the protons of (3,4) unit of acrolein ($-O-\underline{C}H-O-$)
      $\underline{C}H=\underline{C}H_2$
    − Between 3.0 and 4.8 ppm, the protons of (1,4) units of acrolein ($-\underline{C}H_2-CH=CH-O-$).

Some acrolein protons overlap with the polybutadiene bands; for example, the protons ($-CH_2-\underline{C}H=CH-O-$) and ($-\underline{C}H_2-CH-$)
                                                    CHO
which belong to the (1,4) and (1,2) units of acrolein.

Since t−butanol is used as a killing agent, it is believed that some living ends become attached to the $-C(CH_3)_3$ functional groups, which explains the singlet at 1.3 ppm.

Because of the coalescence of the bands belonging to the polyacrolein and polybutadiene blocks, the (1,4) and (3,4) units of acrolein cannot be estimated separately. The results of the acrolein microstructure in the PAB_{14} and PAB_{15} block polymers are summed up in Table 3.

With Li^+ as counter−ion, homopolyacroleins having a $\overline{Mn}$=2500 contain 12% of (1,2) units, those of Mn = 1200, 32% (Figure 3). By extrapolation of the curve (Figure 3) to the $\overline{Mn}$ of polyacrolein blocks ($\overline{Mn}$ = 250) a value ranging between 50% up to 60% for (1,2) units may be found. With Li^+ for PAB_{15} block polymer, Table 3 gives a comparable value since % (1,2) = 50% for the polyacrolein blocks.

This microstructure variation has not been observed with homopolymers synthesized with Na^+. For $\overline{Mn}$ > 3000, % (1,2) = 12%.

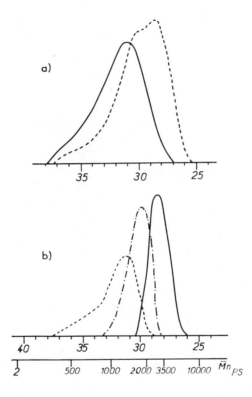

Figure 7. Acrolein initiation by an α–ω di-Na oligobutadiene in THF at −49°C: (a) initiator $\overline{M}_n = 700$ (———); PAB_{14} after the acrolein polymerization (– – –); (b) fractions: PAB_{14} (1) (———); PAB_{14} (2) (– · – ·); PAB_{14} (3) (– – –).

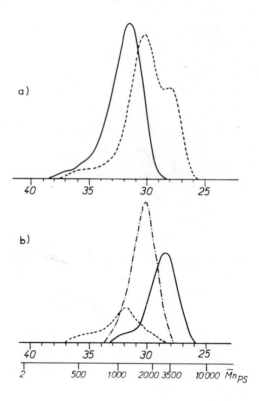

Figure 8. *Acrolein initiation by an α–ω di-Li oligobutadiene in THF at −46°C:*
(a) initiator M_n = 700 (——); PAB$_{15}$ after the acrolein polymerization (– – –);
(b) fractions: PAB$_{15}$ (1) (——); PAB$_{15}$ (2) (– · – ·); PAB$_{15}$ (3) (– – –).

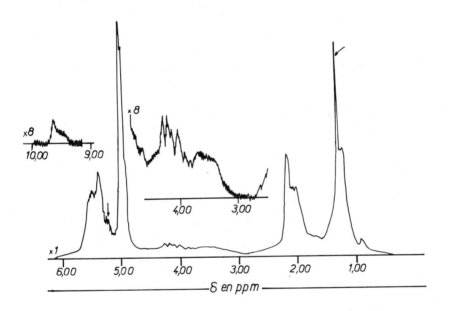

Figure 9.   H-1 NMR spectrum (350 MHz) of the PAB$_{15}$ (2) block polymer.

TABLE 3

Results for the PAB14 and PAB15 Block Polymers

Initiators: Polybutadiene M_n = 700 (M_{nps} – 1000) M_w/M_n = 1.2

N°Fraction $\overline{M}_{nps}$	Weight Percent	% Butadiene	% Acrolein	Microstructures of acrolein blocks k%(1,2) %(3,4 + 1,4)	
PAB14 1700	100	62	38	62	38
(1500)*					
(1) 2500	32	52	48	56	44
(2) 1770	33	69	31	65	35
(3) 1000	35	67	33	65	35
PAB15 1600	100	62	38	50	50
(1500)*					
(1) 2300	33	57	43	51	49
(2) 1730	47	65	25	40	60
(3) 850	20	41	59	59	41

* Theoretical $\overline{M}_n$

Nevertheless, contrary to these high molecular weight polyacroleins, the small acrolein blocks (Mn = 250) initiated with
Na^+ have a high (1,2) percentage. This value is greater than
that obtained for Li^+. Consequently, differences in the acrolein
microstructures exist between blocks synthesized with Li^+ and
Na^+. The termina-tion of the $\sim\sim\overline{CH_2}-\overline{CH}-\overline{CH}-O$ living end results
in the structure

$$-\overset{\beta}{CH_2}-\overset{\alpha}{CH_2}-CHO$$

which corresponds neither to a
(1,2) unit nor to a (1,4) unit. Moreover, H_α and H_β protons belong to the aliphatic NMR bands and involve a % (1,2) over-
estimate. For block polymers, the living end concentration is
very high so that this phenomenon is very intense and perhaps can
explain the differences between the acrolein block microstructures observed with the two counter-ions.

In PAB$_{14}$ block polymer, the weight percentage is approximately same for each fraction. Logically, the highest molecular
weight fraction (1) has the highest acrolein percentage and contains the lowest (1,2)%.

The comparison between the microstructures obtained for the
PAB$_{14}$ and the PAB$_{15}$ fractions is difficult. Indeed, these values
depend on the fractionating quality: the weight percentages of
each fraction are very different between PAB$_{15}$ compared from
PAB$_{14}$. In PAB$_{15}$, the fraction (1) also contains more acrolein
units (43%) than the average value (38%). On the other hand, the
lowest molecular molecular weight fraction (3) has too many
acrolein units (59%). This can be explained by the presence, in
this fraction (3), of homopolyacroleins synthesized by transfer
reactions. A microstructure heterogeneity also exists between
the different PAB$_{15}$ fractions. Here, contrary to PAB$_{14}$, the
fraction (3) contains the highest (1,2) unit percentage (59%)
instead of 50% for the average value.

Discussion and Conclusion

Acrolein polymerization given (1,2), (3,4) and (1,4) units.
To explain the formation of all these units, the propagation
mechanism must involves two different types of living ends: the
(3,4) oxanion living end (Figure 10) and the (1,4) delocalized
living end (Figure 11).

The (1,4) living end propagation corresponds to the
alkylation of an enolate. The studies of these reactions (13,14)
show (Figure 11) that the α attack (O-alkylation) is favored,
compared with the γ attack (C-alkylation) when cation and (1,4)
living end are loosely paired. In these conditions, the (1,4)
unit formation is favored with respect to (1,2) unit formation
(Figure 11).

The living end attacks on the C$_3$ monomer carbon (Figure 10)
involves the formation of (3,4) living ends. The structure of

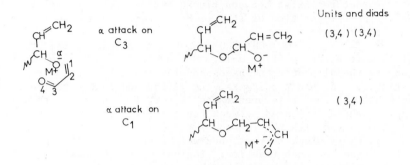

Figure 10.   Propagation by the (3,4) living ends.

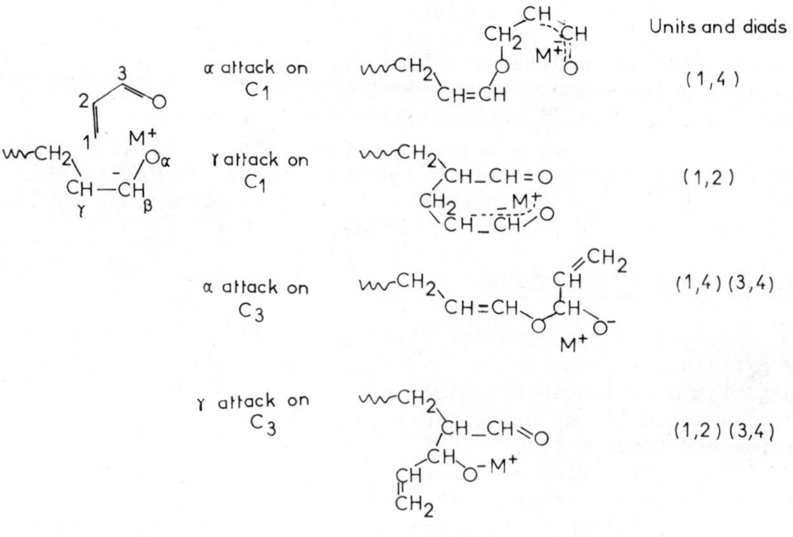

Figure 11.   Propagation by the delocalized (1,4) living ends.

these living ends is determined before a new monomer addition, contrary to (1,4) growing ends. The attack either on C_1 or on C_3 acrolein carbons depends on living end- monomer complexing state before the propagation. The acrolein has the possibilities to coordinate with the cation either by only its carbonyl bond (attack on C_3) or by both double bonds (attack on C_1).

Results of microstructures show (1,2) units are mainly synthesized at the beginning of polymerizations. The (1,2) % decreases with the $\overline{Mn}$ increase and becomes constant around 12%. At the beginning of polymerizations, the living end-cation complexation is very strong and then diminishes during the propagation. Numerous oxygen atoms present along a living polyacrolein chain can solvate and attract the cation. For these conditions, the cation is moved away from the living end. It must be noticed (15) that a similar phenomenon has been shown on methyl methacrylate polymerizations. The monomer units located at the ends of the growing chains may also complex with the cation, thus, interfering in the mechanism.

Copolymers with only short acrolein blocks can by synthesized. In these conditions, the formation of homopolyacroleins can be avoided (differences between the PAI_3 and the PAB_{14} chromatograms Figures 1 and 7). In fact, it would better to define this approach as functionalization rather than block polymer synthesis.

According to the nature of initiator, the respective living end percentages varies. With Na^+ as counter ion, the (1,4) living ends are favored. This is shown by the differences of microstructures observed for the fractions (1) of PAB_{14} and PAB_{15} (likewise with fraction (2)).

Moreover, the size of the polyacrolein blocks depends on the ratio of the propagation rate constant k_{pr} to the transfer rate constant h_m (k_{pr}/h_m). This ratio may depend on the nature but also on the complexation of the living ends. Indeed, an evolution of the complexation state exists for the (1,4) living end and may involve a variation of (k_{pr}/h_m) (1,4) during the propagation. The previous mechanism can explain the bimodal weight distributions obtained for both homopolyacroleins and block polymers.

In this block polymer study, the initiation reactions are favored rather than the propagation reactions because of the small $\{A_o\}/\{I_o\}$ ratio. But for high molecular weight polyacroleins, this effect diminishes because lower initiator concentrations are used. Furthermore, with Na^+ as counter ion, the (1,2) % decreases very fast with Mn while, with Li^+, the same phenomenon is slow and can be easily observed. Indeed, with the latter cation, the transfer reactions are more numerous than with Na^+. Consequently, during the reaction time, new chains begin to grow. This involves the formation of new (1,2) units.

Literature Cited

1. R. C. Schultz, G. Wagner, W. Kern, J. Polym. Sci., C 16, 989
    (1967).

2. I. V. Andreyeva, M. U. Koton, V. N. Artermeva, Yu.N.Sazanov,
    G. N. Fedorava, Vysokomol. Soedin, A 18, 1707 (1976).

3. P. Thivollet, J. Gole, J. Polym. Sci., 11, 1615 (1973).

4. I. Rashkov, I. M. Panayotov; Makromol. Chem., 151, 275
    (1972).

5. H. Calvayrac, P. Thivollet, J. Gole, J. Polym. Sci., 11,
    1631 (1973).

6. D. Gulino, J. P. Pascault, J. Gole, unpublished results.

7. I. V. Andreyeva, M. U. Koton, Yu.V. Medvedev, Vysokomol.
    Soedin, A 15, 852 (1973).

8. I. V. Andreyeva, Yu.V. Medvedev, Vysokomol. Soedin, A 11,
    2244 (1969).

9. D. Gulino, J. Gole, J. P. Pascault, Europ. Polym. J., 15,
    469 (1979).

10. D. Gulino, J. Gole, J. P. Pascault, Makromol. Chem., 182, to
    be published, 1981.

11. P. Hemery, Thesis AO 12869 (1976).

12. D. Gulino, J. P. Pascault, Q. T. Pham, Makromol. Chem., in
    preparation.

13. F. Guibe, P. Sarthou, G. Bram, Tetrahedron, 30, 3139 (1974).

14. A. L. Kurtz, A. Mathias, I. P. Beletskaya, O. A. Reutor,
    Tetrahedron Letters, 32, 3037 (1971).

15. W. Fowells, C. Schuerch, F. A. Bovey, F. P. Hood, J. Amer.
    Chem. Soc., 89, 1396 (1967).

RECEIVED April 27, 1981.

Studies on the Anionic Polymerization of Methyl Methacrylate Initiated with Butyllithium in Toluene by Using Perdeuterated Monomer

KOICHI HATADA, TATSUKI KITAYAMA, KINZO FUMIKAWA, KOJI OHTA, and HEIMEI YUKI

Department of Chemistry, Faculty of Engineering Science, Osaka University, Toyonaka, Osaka 560, Japan

Many papers have been published on the polymerization of methyl methacrylate (MMA) with alkyllithium. It was reported that in the initiation step an attack of alkyllithium on monomer occurred in the first few seconds and took place on both olefinic and carbonyl double bonds and a considerable amount of oligomer was formed in the early stage of the polymerization (1, 2, 3, 4, 5). Kawabata and Tsuruta (5) studied the reaction mode of butyllithium (BuLi) with MMA in the initiation step. Glusker and his coworkers postulated the pseudo-cyclization of chain end as a cause of oligomer formation (6). However, it is still unresolved how the butyl group of the initiator is incorporated into the polymer chain, and how the oligomer terminates its further growth.

In this article perdeuterated MMA was polymerized with un-deuterated BuLi and the resultant polymer and oligomer were studied for the initiator fragment by using ^{1}H NMR spectroscopy (7a). The undeuterated monomer was also polymerized with BuLi in toluene at -78°C and chromatography and each fraction was analyzed by ^{1}H NMR and mass spectroscopies (7b). The formation of butane in this polymerization reaction was also studied in some detail (7c).

Experimental

Perdeuterated MMA was prepared in 32% yield from acetone cyanohydrin-d$_7$ and methanol-d$_4$ according to the method of Crawford (8). Bp 100 ~ 101°C. The isotopic purity was found more than 99.7% by NMR using precision coaxial tubing method (9). Undeuterated MMA was obtained from commercial source. Each monomer was stirred over calcium dihydride overnight and vacuum distilled. The distillate was stirred over fresh calcium dihydride on the vacuum line for 5 hr and redistilled at room temperature.

Butyl isopropenyl ketone was prepared by the reaction of methacryloyl chloride with BuLi in ether at -78°C. The crude product was purified by the fractional distillation under reduced nitrogen pressure. Bp 56.5°C(14mmHg). n$_D^{20}$=1.4363. Anal. Found:

0097–6156/81/0166–0327$05.00/0

C, 75.92%, H, 11.18%. Calcd for $C_8H_{14}O$: C, 76.14%; H, 11.18%.
The ^{1}H NMR spectrum (10% CCl_4 solution) showed peaks at 0.92ppm
(triplet, 3.0 H) assigned to the methyl protons of butyl group,
at 1.42ppm (multiplet, 4.1 H) assigned to the protons of the two
methylenes adjacent and next but one to the methyl in butyl group,
at 1.83ppm (singlet, 3.0 H) assigned to the α-methyl protons, at
2.60ppm (triplet, 2.0 H) assigned to the protons of methylene
group adjacent to the carbonyl group and at 5.64 and 5.84ppm (two
multiplets, 2.0 H) assigned to the β-protons of vinyl group.

Polymerization was carried out in a sealed glass ampoule and
terminated by the addition of a small amount of methanol. The
reaction mixture was poured into a large amount of methanol to
precipitate the polymer formed. After standing it overnight the
precipitate was collected by filtration, washed with methanol and
dried in vacuo at 60°C. The combined filtrates were evaporated
to dryness under reduced pressure and the residue was redissolved
in benzene to remove a small amount of insoluble material. The
methanol-soluble oligomer was freeze-dried from the benzene solution.

In the polymerization designed to detect the amount of lithium methoxide formed, the reaction was terminated with a small
amount of acetic acid, and the reaction mixture was analyzed for
methanol by gas-liquid chromatography. It was assumed that the
methoxide formed during the polymerization was converted quantitatively to methanol (10).

The polymerization was also terminated with CH_3OH or CH_3OD in
the case to detect butane. From the volatile fraction distilled
out from the reaction mixture on a vacuum line, low-boiling compounds were collected by redistillation and subjected to the analysis by combined gas chromatography-mass spectrometry. Butane
(BuH) and monodeuterated butane, $CH_3CH_2CH_2CH_2D$ (BuD), employed as
authentic samples, were obtained by the reaction of BuLi with
CH_3OH and CH_3OD, respectively.

The fractionation of MMA oligomer was carried out by using a
JASCO model FLC-A10 high speed liquid chromatograph with polystyrene gel column. Chloroform was employed as an eluent. The
gel was prepared and kindly given by Dr. Tanaka of Tokyo University of Agriculture and Technology. The sample oligomer was separated as a heptane-soluble fraction from a methanol-soluble oligomer.

Depolymerizations of polymer and oligomer were performed by
heating   them at 300 ∿ 350°C under vacuum. The products were
collected in the reservoir cooled in liquid nitrogen bath and
analyzed by gas-liquid chromatography.

The ^{1}H NMR spectra of the polymer and oligomer were measured
in nitrobenzene-d_5 at 110°C on JNM-FX100 Fourier transform NMR
spectrometer(JEOL) being operated at 100MHz. The hydrogen content
of sample polymer or oligomer was determined from the relative
intensity of the signal of interest to the signal due to the
remaining protons in the nitrobenzene-d_5. The absolute intensity

of the latter signal was measured by precision coaxial tubing
method (9). Spin-lattice relaxation time, T_1, was measured by the
inversion-recovery Fourier Transform method.

Field desorption and electron impact ionization mass spectra
were taken by using a Matsuda type double focussing spectrometer
(11) equipped with silicon emitter (12) and a JMS-01SG-2 mass
spectrometer(JEOL), respectively.

Number average molecular weight was determined in benzene
solution by Hitachi 117 vapor pressure osmometer(VPO) at 42°C.

Results

1. Polymerization of MMA-d8 with BuLi in Toluene. Polymerization of MMA-d8 was carried out in toluene with BuLi at -78°C.
The results and the ^{1}H NMR spectra of the resultant polymer and
oligomer are shown in Table 1 and Figure 1, respectively. The

Table I

Polymerization of Perdeuterated MMA with BuLi in Toluene
at -78°C for 5hr (7a)

| | Yield | | $\overline{M_n} \times 10^{-3}$ | |
	(g)	(mmol)[b]	VPO	NMR
Polymer	0.295	0.011	26.9	26.8
Oligomer	0.271	0.249	1.09	1.63

[a] Monomer 5.2 mmol, BuLi 0.50 mmol, toluene 10ml.
[b] Calculated from yield and $\overline{M_n}$ measured by VPO.

(Polymer Bulletin)

triplet at 0.79ppm should be due to the methyl protons in the
butyl group, which originally came from the initiator. The singlet
at 2.57ppm was assigned to the terminal methine proton

$$(\text{\small$\sim\sim\sim$}CD_2\text{-}\underset{\underset{COOCD_3}{|}}{\overset{\overset{CD_3}{|}}{C}}\text{-H}\quad),$$ which was introduced from methanol in the
termination reaction. The signal disappeared when the polymeriza-
tion was terminated by CH_3OD instead of CH_3OH. The polymer and
oligomer amounted to 0.011 and 0.249mmol, respectively, based on
the yield and the $\overline{M_n}$. On the other hand, the contents of the
methine proton in the polymer and oligomer were measured to be
0.011 and 0.167mmol, from which the molecular weights of the polymer and oligomer were calculated to be 26,800 and 1,630 respectively. Each molecule was assumed to contain one terminal methine
proton. The molecular weight of the polymer thus obtained was
consistent with that measured by VPO, but the oligomer showed different molecular weights by these two methods (Table I). The
results indicate that most of the polymer chain was still living
at the end of the reaction, and the spontaneous termination through
the formation of cyclic ketone structure did not occur at the chain
end of the polymer.

It was reported that the oligomer chain was essentially inactive at the later stage of polymerization, owing to the pseudo-

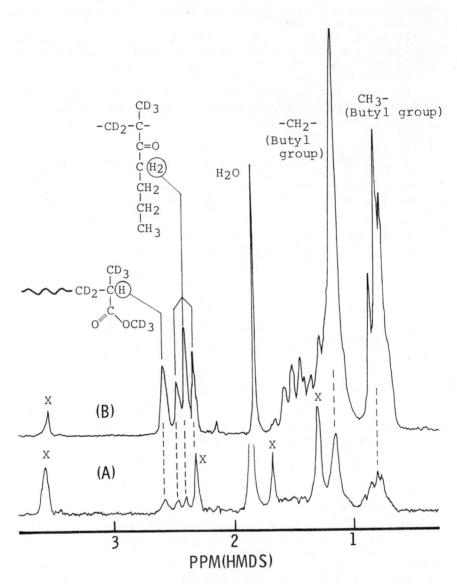

Figure 1.   H-1 NMR spectra of the polymer (A) and oligomer (B) of MMA-d₈ prepared in toluene with BuLi at −78°C (7a). The signals labelled X arose from the remaining protons in the monomer unit of the chain.

cyclization (6). The difference between the molecular weights of the oligomer measured by NMR and VPO shows that about 30% of the oligomer molecules did not contain a terminal methine proton (Table II), indicating the possibility of termination by true

Table II

Analysis of Butyl Group and Terminal Methine Proton in the Polymer and Oligomer of MMA-d_8 Prepared in Toluene with BuLi at $-78°C$(7a)

	C_4H_9-(total)		C_4H_9CO-		Terminal methine	
	mmol	mol/mol[a]	mmol	mol/mol[a]	mmol	mol/mol[a]
Polymer	0.0209	1.9	0.0107	1.0	0.011	1.0
Oligomer	0.424	1.7	0.182	0.7	0.167	0.7

[a] Represents the number of butyl group or terminal methine proton per a molecule of polymer or oligomer.

cyclization with a release of methoxide. However, the remaining 70% of oligomer were still reactive to the proton of methanol.

Intensity measurement of the methyl proton signal at 0.79ppm indicated that the polymer and oligomer had 1.9 and 1.7 butyl group in each molecule, respectively, (Table II). The butyl group should be incorporated not only in the initiation step but also through the reaction of BuLi with carbonyl group. The triplet at 2.40ppm was assigned to the protons of the methylene group adjacent to carbonyl group ($-COCH_2CH_2CH_3$). The resonance turned out a singlet by the irradiation at 1.4ppm, where a weak multiplet appeared, indicating that the latter resonance was ascribed to the methylene group after next to the carbonyl group. The contents of butyl carbonyl group in the polymer and oligomer were determined to be 1.0 and 0.7 per a molecule, respectively. Consequently, each one molecule of the polymer and oligomer had one terminal butyl group which was introduced in the initiation step.

In the polymerization of MMA by BuLi the initiator was reported to react first with many more carbonyl groups than with vinyl double bonds at low temperatures (3). Therefore, the butyl carbonyl group must be incorporated through the copolymerization of MMA with butyl isopropenyl ketone formed by the former reaction.

$$CH_2=C\begin{matrix}CH_3\\|\\C=O\\|\\OCH_3\end{matrix} \xrightarrow{BuLi} CH_2=C\begin{matrix}CH_3\\|\\Bu-C-OLi\\|\\OCH_3\end{matrix} \longrightarrow CH_2=C\begin{matrix}CH_3\\|\\C=O\\|\\Bu\end{matrix} + LiOCH_3$$

Now it will be necessary to elucidate the location of the butyl isopropenyl ketone unit in the polymer chain. The spin-lattice relaxation time, T_1, of the protons in the polymer and oligomer was measured. The T_1 of methylene protons adjacent to the carbonyl group was nearly the same level as the T_1 of methyl protons in the terminal butyl group or terminal methine proton (Table III) but much longer than the T_1 of the protons in interior sequences of polymer (13). These indicate that the butyl carbonyl group in the polymer or oligomer locates at or near to the forefront or the end of the chain.

Table III
[1]H Spin-Lattice Relaxation Times T_1's in Second of the Protons
in Butyl Group and the Terminal Methine Proton
in Nitrobenzene-d_5 at 110°C (7a)

	$\underline{CH}_3CH_2CH_2CH_2-$	$CH_3CH_2CH_2\underline{CH}_2CO-$	Terminal methine
Polymer[a]	3.9	1.8	5.1
Oligomer[a]	5.4	2.1	6.1

[a] Prepared in toluene by BuLi at -78°C.

(Polymer Bulletin)

The results of the polymerization terminated 10min after the
initiation are shown in Table IV. As compared with the results of
the polymerization for 5hr, the yield and molecular weight of the
polymer and oligomer were lower, but the number of butyl carbonyl
group in one polymer molecule was almost constant regardless of the
polymerization time. This may indicate that the butyl isopropenyl
ketone unit in the polymer molecule locates at or near to the
forefront of the chain.

Table IV
Polymerization of MMA-d_8 with BuLi in Toluene
at -78°C for 10min[a]

	Yield (g)	$\overline{M}_n$ $\times 10^{-3}$	C_4H_9-(total) (mol/mol)[b]	C_4H_9CO- (mol/mol)[b]	Terminal methine (mol/mol)[b]
Polymer	0.045	14.5	2.0	1.0	1.0
Oligomer	0.204	0.84	1.7	0.7	0.6

[a] Monomer 5.2mmol, toluene 5ml.
[b] Represents the number of butyl group or terminal methine
proton per a molecule of polymer or oligomer.

2. Fractionation and Structure Determination of MMA Oligomers
(7b). MMA was polymerized in toluene with 10mole% of BuLi at -78°C
for 24hr, and the methanol-soluble oligomer obtained was separated
into the soluble and insoluble parts in heptane. In order to elucidate the location of butyl isopropenyl ketone unit more clearly,
the heptane-soluble oligomer was fractionated by liquid chromatograph and five fractions were collected. By the inspection of the
^{1}H NMR and mass spectra these fractions were found to be a tertiary
alcohol(A), dimer(B, n=1), trimer(B, n=2), tetramer(B, n=3), and
pentamer(B, n=4) of the following structures.

$$
\begin{array}{ll}
\text{(A)} \quad
\begin{array}{c}
\quad\quad CH_3 \\
CH_2 = \overset{|}{C} \\
C_4H_9-\overset{|}{C}-OH \\
\overset{|}{C}_4H_9
\end{array}
, &
\text{(B)} \quad
\begin{array}{c}
\quad\quad CH_3 \quad\quad\quad CH_3 \\
C_4H_9\text{--}(CH_2-\overset{|}{C}\text{---})_n CH_2-\overset{|}{CH} \\
\quad\quad\overset{|}{C}=O \quad\quad\quad\overset{|}{C}=O \\
\quad\quad OCH_3 \quad\quad\quad C_4H_9
\end{array}
\quad n = 1, 2, 3, 4.
\end{array}
$$

The existence of the alcohol shows that the butyl isopropenyl
ketone formed was again attacked by BuLi. The field desorption
mass spectra of the trimer, tetramer and pentamer showed only their
own M^+, $(M+1)^+$ and $(M+2)^+$ peaks, and the mass numbers of M^+ peaks

$$CH_2=\overset{\overset{\displaystyle CH_3}{|}}{\underset{\underset{\displaystyle C_4H_9}{|}}{\underset{C=O}{|}}} \quad\xrightarrow{\text{BuLi}}\quad CH_2=\overset{\overset{\displaystyle CH_3}{|}}{\underset{\underset{\displaystyle C_4H_9}{|}}{\underset{C_4H_9-C-OLi}{}}}$$

were 384, 484 and 584, respectively. On the other hand, in the electron impact ionization mass spectrum of the tetramer (Figure 2) there appeared peaks of the mass numbers 127, 158, 258, and 358, which corresponded to the following ions, C and D (n=0, 1, and 2), respectively.

$$(C)\quad \left[\ CH_2-\overset{\overset{\displaystyle CH_3}{|}}{\underset{\underset{\displaystyle C_4H_9}{|}}{\underset{C=O}{|}}}H\ \right]^{+} \qquad (D)\quad \left[\ C_4H_9\!-\!\!\left(CH_2-\overset{\overset{\displaystyle CH_3}{|}}{\underset{\underset{\displaystyle OCH_3}{|}}{\underset{C=O}{|}}}\right)_{\!n}\!\!-CH_2-\overset{\overset{\displaystyle CH_3}{|}}{\underset{\underset{\displaystyle OCH_3}{|}}{\underset{C=O}{|}}} + H\ \right]^{+}\quad \begin{array}{l}n=0,1,\\ \text{and } 2.\end{array}$$

The trimer showed the peaks of 127, 158 and 258, while the dimer showed the peaks of 127 and 158. The results strongly indicate that the ketone unit locates at the chain end of the oligomer, which could not be determined by ^{1}H NMR spectroscopy. Then, it is reasonable to assume that the oligomer molecule containing a terminal methine proton has one butyl isopropenyl ketone unit at the chain end.

3. Formation of Lithium Methoxide and Butane in the polymerization of MMA with BuLi(7c,14).

Undeuterated MMA was polymerized in toluene with BuLi and the reaction was terminated by a small amount of acetic acid. The reaction mixture was analyzed for methanol and butane by gas-liquid chromatography (Table V). The amount of

Table V
Formation of Lithium Methoxide and Butane in the
Polymerization of MMA with BuLi in Toluene[a]

BuLi (mmol)	Polymerization Temp.(°C)	Time	CH$_3$OLi (%)[b]	Butane (%)[b]
1.0	−78	10min	51.6	9.8
1.0	−78	2hr	53.0	9.4
1.0	−78	12hr	51.0	7.8
1.0	−40	2hr	55.8	5.3
0.5[c]	−78	10min	17.0	—
0.5[d]	−78	10min	2.9	—

[a] MMA 10mmol, toluene 10ml.
[b] Percentage based on the BuLi used.
[c] 1,1-Diphenylhexyllithium was used instead of BuLi.
[d] Reaction products of equimolar amounts of butyl isopropenyl ketone and BuLi were used instead of BuLi.

methanol corresponds to that of lithium methoxide which was produced in the polymerization reaction. The lithium methoxide amounted to 51.6% of BuLi used 10min after the initiation at −78°C and the amount was almost constant during the prolonged polymerization, but slightly increased in the polymerization at −40°C.

When 1,1-diphenylhexyllithium or the reaction products of

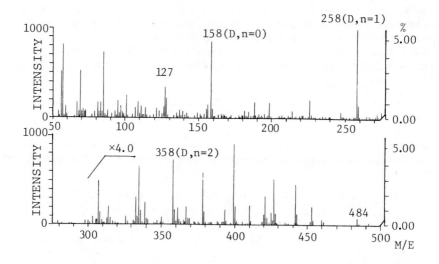

Figure 2.   Electron ionization MS of tetramer (7b).

equimolar amounts of butyl isopropenyl ketone and BuLi were used as initiator, the formation of lithium methoxide was greatly decreased. The isotacticities of the polymers prepared by these initiator were higher than that of the polymer obtained with BuLi. The amounts of oligomers were decreased as compared with that of the oligomer formed in the polymerization with BuLi (Table VI). The results

Table VI
Polymerization of MMA in Toluene at -78°C
with Several Initiators[a]

| Initiator | Polymer | | | | Oligomer (%) |
| | Yield (%) | Tacticity(%) | | | |
		I	H	S	
BuLi	65.3	68	19	13	33.8
DPHLi[b]	82.9	86	10	4	17.1
BuLi/Ketone[c]	72.2	79	14	7	11.0

[a] Monomer 10mmol, initiator 0.5mmol, toluene 10ml.
[b] 1,1-Diphenylhexyllithium.
[c] BuLi was reacted with an equimolar amount of butyl iso-propenyl ketone in toluene at 0°C for 10min, and then the mixture was used as initiator.

indicate that the lithium methoxide formed in the reaction mixture contributes to the formation of syndiotactic sequences in the polymer chain, as previously proposed for the polymerizations with BuLi of MMA (3) and ethyl methacrylate (15). In order to clarify the origin of the butane formed in the polymerization, the poly-merizations of MMA and MMA-d_8 with BuLi were terminated by CH_3OD and CH_3OH, respectively, and the resulting volatile fractions were analyzed by combined gas-liquid chromatography-mass spectrometry (7c). The amount of butane formed in the polymerization for 10min. at -78°C was about 10%, and slightly decreased with an increase in the polymerization time. It also decreased to about 5% at -40°C. In Figure 3 are illustrated the mass spectra of the butane frac-tions obtained 10min after the initiation in the polymerization of MMA terminated by CH_3OD (MMA—CH_3OD system) and the polymerization of MMA-d_8 terminated by CH_3OH (MMA-d_8—CH_3OH system). The spectra of BuH and BuD are also shown for the reference. From the inspection of these spectra it is clear that the butane formed in MMA—CH_3OD system mainly consists of BuD and that formed in MMA-d_8—CH_3OH system is mainly BuH. The results mean that the butane formed in both systems was originated from a small amount of BuLi remaining unreacted in the polymerization mixture.

$$BuLi + CH_3OD \longrightarrow BuD + LiOCH_3$$
$$BuLi + CH_3OH \longrightarrow BuH + LiOCH_3$$

In the mass spectra of the butane formed in the polymerization of MMA for 12hr (Figure 4) the intensities of the peaks at m/e=58 and 56 are much increased as compared with those in the spectra of the butane obtained in 10min (Figure 3). The increasing intensity of the peak at m/e=58 indicate that the proton abstraction by the remaining BuLi occurs gradually during the polymerization.

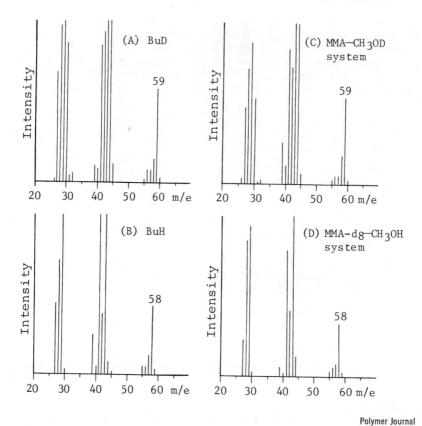

Figure 3.   MS of butane formed when the polymerization of methyl methacrylate by BuLi in toluene was terminated by adding a small amount of methanol 10 min after the initiation (7c).

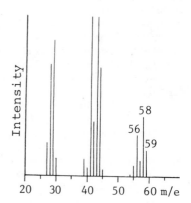

Figure 4.   MS of the butane obtained 12 h after the initiation in the polymerization of MMA terminated by CH₃OD (7c).

$$BuLi \; + \; CH_2=\overset{\underset{\displaystyle C=O}{|}}{\underset{\displaystyle OCH_3}{|}}\overset{CH_3}{C} \; \longrightarrow \; BuH \; + \; CH_2=\overset{\underset{\displaystyle C=O}{|}}{\underset{\displaystyle OCH_3}{|}}\overset{CH_2Li}{C}$$

On the other hand, the intensity enhancement of the peak at $m/e=56$ probably means that the prolonged polymerization time causes the formation of butene from the remaining BuLi, although the mechanism is not clear at present.

4. Depolymerization of the Polymer and Oligomer of MMA (14).
A sample of the polymer prepared in toluene with BuLi at -78°C was heated to 350°C under vacuum. Volatile materials were obtained in 86% yield. More than 90% of the materials were the MMA monomer. In the gas-chromatogram there appeared three small peaks at longer retention times than that of MMA. Among these peaks the one after next to the MMA peak was found to correspond to butyl isopropenyl ketone by using authentic sample, and g.l.c-mass spectrometry. The number average molecular weight of the PMMA sample was 24,000, and the amount of butyl isopropenyl ketone unit in this polymer could be calculated to be 0.42mol% of MMA unit if a polymer molecule was assumed to contain one ketone unit. The amount of the ketone formed in the depolymerization reaction was 0.3mol% of MMA formed and corresponded to about 70% of the theoretical value. The depolymerization of the oligomer also gave low-boiling products in 53% yield, which contained about 9% of butyl isopropenyl ketone. This is the direct evidence for the existence of the ketone unit in the polymer and oligomer chains.

Discussion

In the previous communication(7a) we postulated the initiation mechanism involving the attack of BuLi on the vinyl double bond of the butyl isopropenyl ketone formed. However, from the results mentioned above we will propose the following mechanism for the polymerization of MMA with BuLi in toluene, which is slightly different from the previous one. On mixing the monomer with BuLi the initiator reacts with both olefinic and carbonyl double bonds of the monomer. The attack on the olefinic double bond produces the MMA anions, which add the monomer to form the growing chains(E).

$$BuLi + CH_2=\overset{\underset{\displaystyle C=O}{|}}{\underset{\displaystyle OCH_3}{|}}\overset{CH_3}{C} \longrightarrow BuCH_2-\overset{\underset{\displaystyle C=O}{|}}{\underset{\displaystyle OCH_3}{|}}\overset{CH_3}{\underset{\displaystyle}{C}}{}^-Li^+ \longrightarrow BuCH_2-\overset{\underset{\displaystyle C=O}{|}}{\underset{\displaystyle OCH_3}{|}}\overset{CH_3}{C}\sim\sim\sim CH_2-\overset{\underset{\displaystyle C=O}{|}}{\underset{\displaystyle OCH_3}{|}}\overset{CH_3}{\underset{\displaystyle}{C}}{}^-Li^+$$

(E)

The attack of the initiator on the carbonyl group gives butyl isopropenyl ketone with a release of lithium methoxide, and the ketone molecule formed soon attacks most of the growing chains owing to its higher reactivity than that of MMA.

$$BuLi \; + \; CH_2=\overset{\underset{\displaystyle C=O}{|}}{\underset{\displaystyle OCH_3}{|}}\overset{CH_3}{C} \; \longrightarrow \; CH_2=\overset{\underset{\displaystyle C=O}{|}}{\underset{\displaystyle Bu}{|}}\overset{CH_3}{C} \; + \; LiOCH_3$$

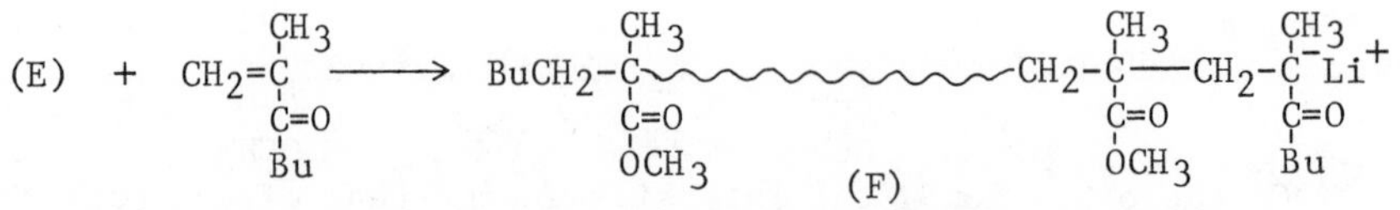

$$(E) \; + \; CH_2{=}\underset{\underset{Bu}{\overset{|}{C{=}O}}}{\overset{CH_3}{\overset{|}{C}}} \longrightarrow BuCH_2{-}\underset{\underset{OCH_3}{\overset{|}{C{=}O}}}{\overset{CH_3}{\overset{|}{C}}}\wwave\wwave CH_2{-}\underset{\underset{OCH_3}{\overset{|}{C{=}O}}}{\overset{CH_3}{\overset{|}{C}}}{-}CH_2{-}\underset{\underset{Bu}{\overset{|}{C{=}O}}}{\overset{CH_3}{\overset{|}{C}}}{}^{-}Li^{+}$$

(F)

The resulting anions are less reactive than the MMA anions and add the monomer very slowly, but once attacked by MMA they turn out to add the monomer rapidly to form high molecular weight polymers. Most of the chains ending in the ketone unit remain unreacted during the polymerization and are recovered as methanol-soluble oligomer after the polymerization and only a small fraction can be transformed to high molecular chain, which contained one ketone unit near to its forefront. The transformation of the oligomer chain into the high polymer chain was also suggested in the polymerization of ethyl methacrylate in toluene by BuLi (15).

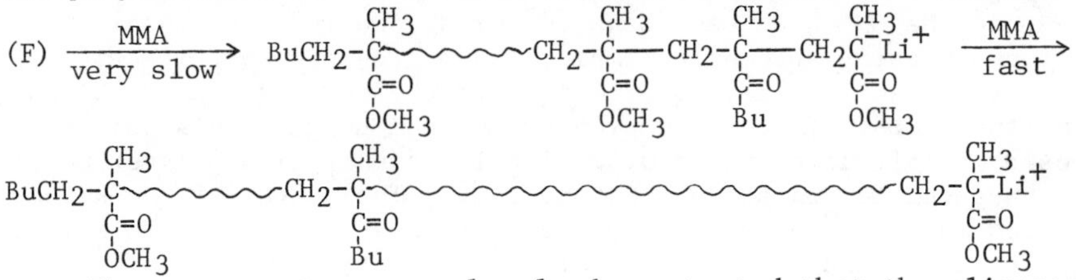

$$(F) \; \xrightarrow[\text{very slow}]{MMA} \; BuCH_2{-}\underset{\underset{OCH_3}{\overset{|}{C{=}O}}}{\overset{CH_3}{\overset{|}{C}}}\wwave CH_2{-}\underset{\underset{OCH_3}{\overset{|}{C{=}O}}}{\overset{CH_3}{\overset{|}{C}}}{-}CH_2{-}\underset{\underset{Bu}{\overset{|}{C{=}O}}}{\overset{CH_3}{\overset{|}{C}}}{-}CH_2\underset{\underset{OCH_3}{\overset{|}{C{=}O}}}{\overset{CH_3}{\overset{|}{C}}}{}^{-}Li^{+} \; \xrightarrow{\substack{MMA \\ fast}}$$

$$BuCH_2{-}\underset{\underset{OCH_3}{\overset{|}{C{=}O}}}{\overset{CH_3}{\overset{|}{C}}}\wwave CH_2{-}\underset{\underset{Bu}{\overset{|}{C{=}O}}}{\overset{CH_3}{\overset{|}{C}}}\wwave CH_2{-}\underset{\underset{OCH_3}{\overset{|}{C{=}O}}}{\overset{CH_3}{\overset{|}{C}}}{}^{-}Li^{+}$$

The mass spectroscopy clearly demonstrated that the oligomer molecule contained one butyl isopropenyl ketone unit at the chain end. The possibility that the ketone unit locates at the chain end of the polymer may be rejected because the low molecular polymer obtained in the 10min polymerization also contained one ketone unit similarly to the high polymer.

There is also some posibility that the butyl isopropenyl ketone formed was again attacked by BuLi to produce an anion as follows:

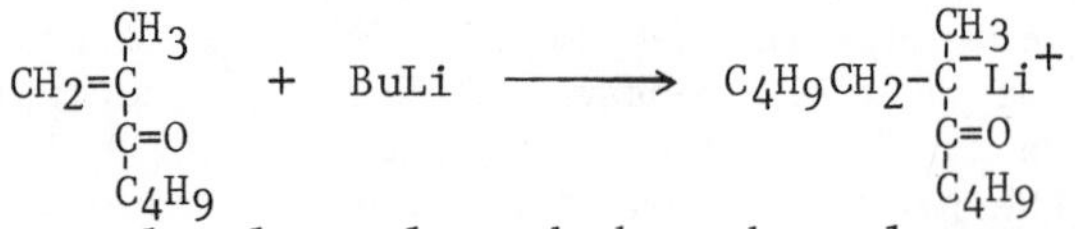

$$CH_2{=}\underset{\underset{C_4H_9}{\overset{|}{C{=}O}}}{\overset{CH_3}{\overset{|}{C}}} \; + \; BuLi \longrightarrow C_4H_9CH_2{-}\underset{\underset{C_4H_9}{\overset{|}{C{=}O}}}{\overset{CH_3}{\overset{|}{C}}}{}^{-}Li^{+}$$

It can not be completely neglected that the polymer of following structure is formed by the attack of the monomer on this species.

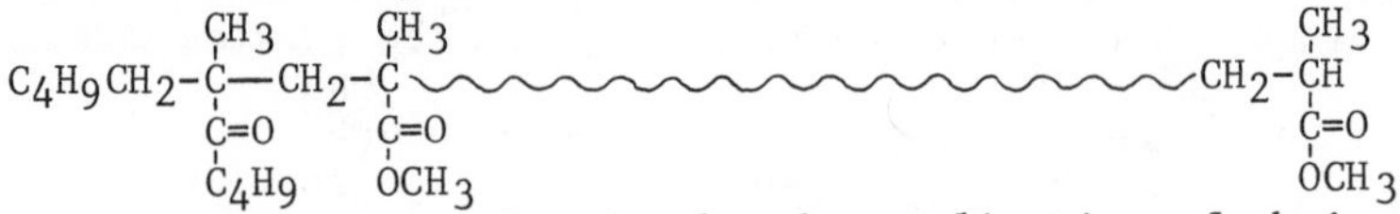

$$C_4H_9CH_2{-}\underset{\underset{C_4H_9}{\overset{|}{C{=}O}}}{\overset{CH_3}{\overset{|}{C}}}{-}CH_2{-}\underset{\underset{OCH_3}{\overset{|}{C{=}O}}}{\overset{CH_3}{\overset{|}{C}}}\wwave\wwave CH_2{-}\underset{\underset{OCH_3}{\overset{|}{C{=}O}}}{\overset{CH_3}{\overset{|}{CH}}}$$

The spontaneous termination by the cyclization of chain end seems to occur only in the growing chain of short length in the polymerization, although the reason is not clear at present.

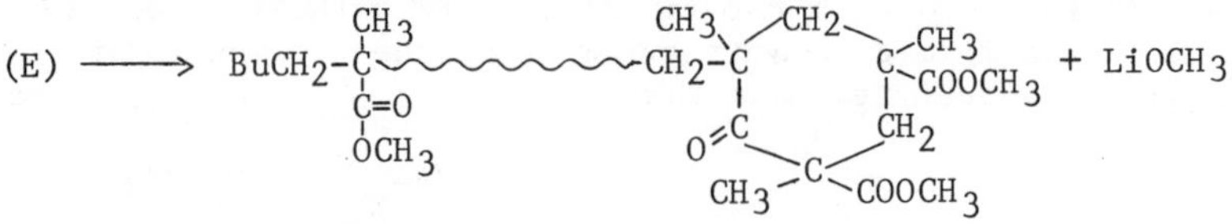

$$(E) \longrightarrow BuCH_2{-}\underset{\underset{OCH_3}{\overset{|}{C{=}O}}}{\overset{CH_3}{\overset{|}{C}}}\wwave CH_2{-}\overset{CH_3}{\underset{}{C}}\big\langle\ldots\big\rangle \; + \; LiOCH_3$$

In the polymerization of MMA with BuLi it was reported that all the BuLi added could not be accounted for if each polymer molecule was assumed to contain one butyl group (1). As mentioned already the polymer and oligomer prepared at −78°C had some two butyl groups in each molecule and about half of them was incorporated through the carbonyl attack by BuLi. The amount of butyl group summed to 0.44mmol (Table 2), and was very close to the amount of the BuLi used (0.5mmol). The difference (0.06mmol) nearly corresponded to the amount of butane (0.05mmol) formed in the polymerization, and accordingly, all the BuLi used could be completely accounted for as shown in Table VII.

Table VII

Fate of the Initiator in the Polymerization of MMA in Toluene with BuLi at −78°C

	Account (%)[a]
Butane and butene	9.8
BuCH$_2$-C with CH$_3$ and COOCH$_3$ groups { Oligomer	48.4
Polymer	2.0
BuCO- { Oligomer	36.4
Polymer	2.1
	Total 98.7

[a]Percentage based on the BuLi used.

The formation of butane in the polymerization of MMA with BuLi was reported by Kawabata and Tsuruta (5), and by Tsvetanov et al (16). They proposed the metalation of the monomer with BuLi. It was also reported that the BuLi disappeared instantaneously on mixing MMA and the initiator (1). However, the results in this work clearly show that most of butane was originated from unreacted BuLi when the polymerization was terminated, especially short time after the initiation. The existence of unreacted BuLi was also observed by Vankerckhoven and Van Beylen in the polymerization of methacrylonitrile in toluene at −78°C. It was suggested that the remaining unreacted BuLi molecules are engaged in very stable and inactive mixed associated particles (17). This may be true in our polymerization. If the polymerization time is prolonged, the proton abstraction by the remaining BuLi from the monomer comes to occur to some extent. The abstraction may occur on the α-methyl group in the monomer.

It is well known that a small amount of lithium methoxide was formed at the initial stage of the polymerization (3). In the polymerization in tolune at −78°C the lithium methoxide formed at the initial stage amounted to 51.6% of the initiator used, and this was nearly equal to the sum of the amounts of the butyl carbonyl group in the polymer and oligomer, and of oligomer containing no terminal methine proton (Table VIII). This indicates that the oligomer molecule containing no terminal methine proton should have cyclized to form a cyclic ketone structure at the chain end with a release of lithium methoxide.

Table VIII
Lithium Methoxide Formed in the Polymerization of MMA
in Toluene by BuLi at -78°C[a]

	Amount formed(%)[b]
LiOCH$_3$	51.6
C$_4$H$_9$CO- incorporated in the polymer and oligomer	38.5[c]
Oligomer containing no terminal methine proton	15.0[c]

[a] Monomer 5.2mmol, BuLi 0.50mmol, toluene 5ml.
[b] Percentage on the basis of the BuLi used.
[c] The data refer to the polymerization of perdeuterated MMA.

Acknowledgement

The authors express their sincere thanks to Professor H.
Matsuda and Dr. T. Matsuo, College of General Education, Osaka
University and Dr. I. Katakuse and Dr. S. Kusumoto, Faculty of
Science, Osaka University for the measurements of mass spectra.
They are also indebted to Mr. T. Ochi and Mr. M. Nagakura for their
help in laboratory work, and to Mrs. F. Yano for her clerical
assistance in the preparation of this manuscript.

Literature Cited

1. Cottam, B. J.; Wiles, D. M.; Bywater, S. Can. J. Chem., 1963, 41, 1905.
2. Glusker, D. L.; Stiles, E.; Yoncoskie, B. J. Polym. Sci., 1961, 49, 297.
3. Wiles, D. M.; Bywater, S. Trans. Faraday Soc., 1965, 61, 150.
4. Goode, W. E.; Owens, F. H.; Myers, W. L. J. Polym. Sci., 1960, 47, 75.
5. Kawabata, N.; Tsuruta, T. Makromol. Chem., 1965, 86, 231.
6. Glusker, D. L.; Lysloff, I.; Stiles, E. J. Polym. Sci., 1961, 49, 315.
7. (a) Hatada, K.; Kitayama, T.; Fujikawa, K.; Ohta, K.; Yuki, H. Polymer Bulletin, 1978, 1, 103., (b) Hatada, K.; Kitayama, T.; Nagakura, M.; Yuki, H. Polymer Bulletin, 1980, 2, 125., (c) Hatada, K.; Kitayama, T.; Yuki, H. Polymer J., 1980, 12, 535.
8. Crawford, J. W. C. B. P. 405,699, 1934.
9. Hatada, K.; Terawaki, Y.; Okuda, H. Org. Magn. Res., 1977, 9, 518.
10. Wiles, D. M.; Bywater, S. J. Phys. Chem., 1964, 68, 1983.
11. Matsuda, H. Atomic Masses Fundam, Constants, 1976, 5, 185.
12. Matsuo, T.; Matsuda, H.; Katakuse, I. Anal. Chem., 1979, 51, 69.
13. Hatada, K.; Kitayama, T.; Okamoto, Y.; Ohta, K.; Umemura, Y.; Yuki, H. Makmorol. Chem., 1978, 179, 485.
14. Hatada, K.; Kitayama, T.; Yuki, H. unpublished data.

15. Hatada, K.; Kitayama, T.; Sugino, H.; Umemura, Y.; Furomoto, M.; Yuki, H. Polymer J., 1979, 11, 989.
16. Tsvetanov, Ch. B.; Petrova, D. T.; Li, Pham H.; Panayotov, I. M. Europ. Polym. J., 1978, 14, 25.
17. Vankerckhoven, H.; Van Beylen, M. Europ. Polym. J., 1978, 14, 189.

RECEIVED April 20, 1981.

Synthesis of Copolymers Using Ionic Techniques

D. H. RICHARDS

Propellants, Explosives, and Rocket Motor Establishment, Ministry of Defense, Waltham Abbey, Essex EN9 1BP, U.K.

In recent years increasing effort has been devoted to the synthesis of novel copolymers and block copolymers in order to develop materials with predictable and controllable properties. Such materials include thermoplastic elastomers, polymers of predetermined functionality, and polymers which are vital ingredients in the formulation of polymer alloys and blends. In this paper ways of synthesizing polymers and copolymers possessing ionic groups at specified points along their backbones are considered and some of the properties of such materials are described. Because work in this area of synthesis is continuing, this paper is necessarily a combination of well established and half-established facts interlaced with considerable speculation. It is hoped that it is made clear into which category any given topic falls.

The specific theme dealt with here is the development of methods of synthesizing polymers and copolymers possessing specifically sited quaternary ammonium bonds. The basic polymerization techniques employed are either anionic or cationic and these are considered separately.

Anionic Techniques

We have recently shown[1] that living anionic polybutadiene reacts with pyridine in tetrahydrofuran (THF) to yield the 4-adduct (equation 1), and that this product can be re-aromatized by refluxing in a suitable solvent (equation 2) to

$$\sim\sim\sim But^- \ Li^+ \ + \ \langle \bigcirc \rangle N \ \rightarrow \ \sim\sim\sim But \ \overset{H}{\diagup}\diagdown N^- \ Li^+ \quad (1)$$

form polymer with a terminal pyridine. It has subsequently been

$$\sim\sim\sim But \ \overset{H}{\diagup}\diagdown N^- \ Li^+ \ \overset{\Delta}{\longrightarrow} \ \sim\sim\sim But \ \overline{\langle\bigcirc\rangle} N \ + \ LiH \quad (2)$$

shown that although the 4-adduct is the exclusive product in THF, a mixture of the 4- and the 2-adduct is formed in hydrocarbon

solvents.[2] More importantly from a synthetic viewpoint it was
also found that, although adduct formation appeared to be quantitative, the re-aromatization stage was subject to some side-
reaction which reduced the overall yield to a maximum of about
70%.   Nevertheless, the yields were high enough to demonstrate
the dramatic effects which occur when hydrocarbon polymers possessing such tertiary amine groups are quaternized.

Low molecular weight polybutadienes with terminal pyridine
groups, like their hydrogen terminated counterparts, are mobile
liquids.   However, when quaternized in bulk with benzyl bromide
(equation 3) they are transformed into solid materials.   This

$$\sim\sim\sim But \overline{\bigcirc} N \ + \ \bigcirc - CH_2Br \ \rightarrow \ \sim\sim\sim But \overline{\bigcirc} N^+ - CH_2 - \bigcirc \quad (3)$$
$$Br^-$$

phenomenon is ascribed to association of the highly ionic pyridinium salts into micelles in the presence of the non-polar matrix (Figure 1).   This results in a dramatic increase in the apparent molecular weight and, when polybutadiene terminated with
pyridine at both ends is used, the micelles formed on quaternization act as ionic physical crosslinks to yield a polymeric network, but one which is capable of being broken by heat or by solvent action.

These results indicate that if polydienes and similar polymers can be prepared quantitatively with tertiary amine terminal
groups, then they can be combined with other halogen functional
polymers using established techniques to create interesting new
block copolymer systems.   For example, consider the reaction between telechelic pyridine terminated polybutadiene and monofunctional bromine terminated polystyrene (equation 4) -the latter
has been prepared in 95% yield.[3,4] The product would be an ABA

$$2 \ \sim\sim\sim CH_2 - CH \ Br \ + \ N\bigcirc - But\sim\sim\sim\sim But \overline{\bigcirc} N$$

$$\sim\sim\sim CH_2 - CH - N^+\bigcirc - But\sim\sim\sim\sim But \overline{\bigcirc} N^+ - CH - CH_2 \sim\sim\sim \quad (4)$$
$$Br^- \qquad\qquad\qquad\qquad Br^-$$

block copolymer which differs from the conventional SBS thermoplastic elastomer by the fact that the linking bond is ionic and
therefore incompatible with either polymeric component.   The material should therefore separate into three phases in contrast to
the two phase nature of the SBS block copolymer and such behavior
ought to confer certain superior properties on the product.   For
example, the added interaction of the ionic groups should result

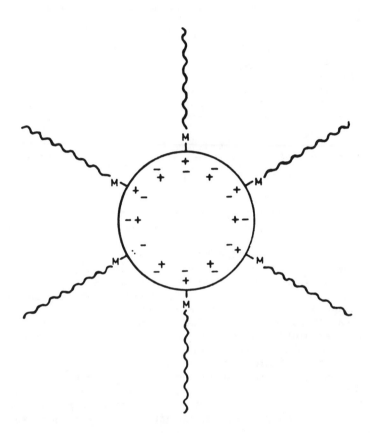

Figure 1. Schematic of ionic association of polybutadiene possessing terminal quaternary ammonium salts.

in increased strength and in its retention to higher temperatures. The product should also exhibit interesting annealing
properties since the maximum phase separation requires a high degree of polymer ordering and its attainment should therefore be
time dependent.

The successful synthesis of such materials requires that,
initially, telechelic tertiary amine terminated polybutadiene be
prepared quantitatively. Alternative routes are therefore being
sought, two of which are outlined in this paper.

The first approach is to polymerize small amounts of 4-vinyl
pyridine on to the ends of anionic living polybutadiene, mono- or
difunctional, to produce what are essentially AB or ABA block copolymers (equation 5). Materials possessing values of n typically averaging about 3 have been prepared and shown to produce
solids when quaternized with benzyl bromide. The result of

$$\sim\sim\sim PBd^- \ Li^+ \ + \ n \ CH = CH_2 \ \xrightarrow[(2) \ H^+]{(1) \ ^P OZ} \ \sim\sim\sim PBd \ (-CH_2 - CH)_n \ H \quad (5)$$

replacing this halide with polystyrene bromide in the quaternization process will not be the well defined ABA system so far
discussed, but a multi graft thermoplastic elastomer. Nevertheless, since Fetters has shown that similar but non-polar graft
SBS systems possess properties superior to the triblock polymer,[5]
the goal seems an attractive one to aim for.

In the second approach an α-chloroalkyl-ω-tertiary amine is
generated and simultaneously vaporized by reacting the hydrochlo-
ride with a potassium hydroxide melt. The vapor after drying is
passed into a solution of living polybutadiene until decoloriza-
tion is complete. Initial experiments indicate that metathetical
reaction of the butadiene anion with the halide ligand takes
place to form the tertiary amine terminated polymer before appreciable self-quaternization of the reagent can occur, and that
this reaction is quantitative. However, further experimentation
is required to confirm this, and no attempt has yet been made to
quaternize the product.

Cationic Techniques

It has now become well established that living cationic
polyTHF can be prepared under carefully controlled conditions
which ensure that no appreciable transfer reactions occur.[6,7]
The propagating species in this system is an oxonium ion (I)
associated with the anion of a strong acid such as HPF_6 or $HSbF_6$.
The oxonium ion is the product of coordination of a THF molecule
and, in principle, this species could be replaced by reaction
with a stronger nucleophile. Amines are suitable candidates for

$$\sim\sim\sim O(CH_2)_4 - \overset{+}{O}\square \quad PF_6^-$$

I

this role, and the remainder of this paper is devoted to a
description of the reactions of this family of compounds with
living polyTHF.

Tertiary Amines

The reaction of pyridine with living polyTHF at -10°C has
been examined in detail by gpc and 'H nmr, and the formation of
polymer pyridinium salts (equation 6) has been shown to be rapid
and quantitative.[8] The salt is stable and does not exchange with
any excess oxonium ions present in solution. Interestingly,

$$\sim\sim\sim O(CH_2)_4 - \overset{+}{O}\square \; PF_6^- + \bigcirc_N \rightarrow \sim\sim\sim O(CH_2)_4 - \overset{+}{N}\bigcirc PF_6^- + \; \overset{\square}{O} \quad (6)$$

II

neither II nor the polymer pyridinium bromide (prepared by
exchange from II and excess LiBr) exhibits the dramatic increase
in viscosity demonstrated by its polybutadiene analogue. This
behavioral difference has been explained in terms of solvation of
the ion pairs by the polyether reducing the degree of solvation
of interaction and hence of association (Figure 2), and is
characteristic of all the polymer-quaternary ammonium salts
prepared.

A number of other heterocyclics have been similarly studied
and shown by 'H nmr, to produce quaternary ammonium salts with
living polyTHF.[9] Moreover, their rates of reaction are a direct
function of their basicities, the following order of reactivities
being observed: ethyl pyridine > pyridine > isoquinoline >
quinoline > acridine. Aliphatic tertiary amines also react in
the same way; the order of reactivities was found to be
triethylame > tributylamine > diethylaniline. In all cases
studied, the quaternary ammonium salt once formed did not
exchange with any excess oxonium ions.

Experiments have also confirmed that certain tertiary
diamines act as efficient linking agents with living polyTHF, and
both tetramethylethylene diamine (TMEDA) and 4,4'-bipyridyl have
been used in this manner. The reaction may be carried out
directly with stoichiometric quantities of reagents to yield with
monofunctionally polyTHF a polymer possessing two ionic groups
centrally placed along the polymer backbone. Alternatively, the
reaction may be carried out in two stages; the monofunctionally
terminated polymer being first prepared by reaction of the living

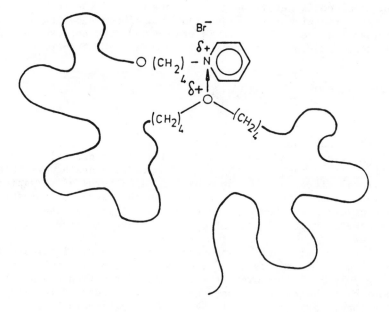

Figure 2. Self-solvation of polytetrahydrofuran possessing terminal quaternary ammonium salts.

polyTHF with excess diamine, and then isolated and subsequently reacted with fresh living polyTHF. The latter procedure has the advantage that the two ionic groups can be placed at any predetermined position along the polymer backbone by suitable selection of the chain length of the living polyTHF used in the two reactions.

Perhaps not surprisingly, pyrazine, (N◯N), has been shown to act only as a monofunctional reagent. Evidently the creation of a quaternary nitrogen deactivates the other nitrogen in the para position to a point where it is not capable of replacing the solvating THF coordinated to the second growing polyTHF chain. There is some evidence, however, that this species does participate in the solvation of the terminal oxonium ion to stabilize it and thereby retard its rate of propagation.

Lastly, living polyTHF has been shown to react rapidly and efficiently with the pendant pyridine units of an AB poly(styrene-b-4-vinylpyridine) which had been prepared anionically. The resulting graft copolymer possessed distinctly different solubility characteristics from its parent block copolymer and its composition and hence its properties could be closely controlled by suitable selection of the relative reagent concentrations.

Secondary Amines

Reactions analogous to that outlined with tertiary amines may be carried out with other amines. The reaction of secondary amines with polyTHF should result in tertiary amines terminated polymers (equation 7), and experimentation involving gpc and ^{1}H nmr analysis has broadly confirmed this expectation. The reaction with diethylamine, dicyclohexylamine and piperidine

$$\sim\!\!\sim\!\!O(CH_2)_4 - O^+ \Big| \quad PF_6^- + R_1R_2NH \rightarrow \sim\!\!\sim\!\!O(CH_2)_4 - NR_1R_2 \cdot HPF_6 + \Big|_O$$

$$+ \; OH^- \qquad (7)$$

$$\sim\!\!\sim\!\!O(CH_2)_4 - N \; R_1R_2$$

showed rapid and total conversion of living polyTHF to the appropriate tertiary amine while reaction with diphenylamine was incomplete and no reaction was observed with carbazole. Evidently the basicities of the last two reagents are insufficient to allow complete conversion. These, however, may be increased by converting the amines to the alkali metal salts prior to reaction (equation 8). All five reagents reacted rapidly and quantitatively with living polyTHF by use of this procedure, and the method has the further advantage that the tertiary amine is formed directly. The results are summarized in Table 1.

Table I.   Routes for quantitative reaction of some secondary amines with living polytetrahydrofuran.

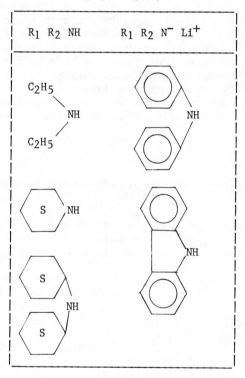

$R_1R_2NH + nBuLi \rightarrow n\ BuH + R_1R_2NLi$

$$\left|\ \substack{\sim\sim O(CH_2)_4 - \overset{+}{O}\boxed{}\ PF_6^-}\right.\qquad(8)$$

$$\downarrow\ \sim\sim O(CH_2)_4 - NR_1R_2 + Li\ PF_6$$

Of course the tertiary amine terminated polymer thus produced is capable of being quaternized by alkyl halides and this has been confirmed experimentally. Moreover, it can be reacted with further living polyTHF to yield a polymer possessing a quaternary ammonium salt moiety at a point along the chain controlled by the relative molecular weights of the two living THF polymers reacted in the system, and again experiments have shown this to occur.

Primary Amines

It is evident that reactions similar to those carried out on secondary amines should be capable of application to primary amines to yield polymers with terminal secondary amine groups. In this instance, however, excess amine may have to be employed to minimize the possibility of disubstitution by the living polymer. Collaborative work in this area is currently being carried out at Waltham Abbey and the University of Montellier[11] and therefore, the results quoted here are in the nature of a progress report only.

It is already apparent that rapid and quantitative reaction can be achieved with butylamine, cyclohexylamine and aniline directly, and that the product can be neutralized to prepare the secondary amine (equation 9, for aniline). Moreover, evidence

$$\sim\sim O(CH_2)_4 - \overset{+}{O}\boxed{}\ PF_6^- + NH_2\!\!\bigcirc \rightarrow \sim\sim O(CH_2)_4 - NH\!\!\bigcirc\cdot HPF_6 + \boxed{}O$$

$$\downarrow OH^- \qquad\qquad (9)$$

$$\sim\sim O(CH_2)_4 - NH\!\!\bigcirc$$

has been obtained that the aniline terminated polymer may be reacted with further living polyTHF to create the polymer tertiary amine III. No attempt has been made as yet, however, to convert III into a star shaped quaternary ammonium polymer by reaction with another aliquot of living polyTHF.

$$\sim\sim O(CH_2)_4 - N - (CH_2)_4\ O\sim\sim$$

$$\underset{\text{III}}{\bigcirc}$$

Ammonia and Hydrazine

Little work has as yet been carried out with either of these reagents, although the principle of reacting ammonia (or sodamide) with living polyTHF to produce a polymer possessing a terminal primary amine group has been established experimentally. If all the reaction sequences so far described are applied sequentially to this material then, steric factors allowing, a four stranded polymer star in the form of a quaternary ammonium salt should be synthesized. Similar multistrand polymers could be prepared from hydrazine.

The techniques described here are capable of numerous adaptations to produce new closely defined polymers, whether homo, block or graft. The major restriction in their application is the requirement of a living cationic polymerization system in which transfer reactions are absent on which to carry out the reactions. This currently limits the techniques to ring opening systems characterized by THF. Nevertheless, considerable scope exists for extending these principles to other Lewis bases – for example, triphenyl phosphine has been used as a reagent to determine the number and nature of propagating centers in mixed cyclic ether systems because of its quantitative reaction with living oxonium ions.[12]

Literature Cited

1. Burgess, F. J., Cunliffe, A. V. and Richards, D. H., Eur. Pol. J. (1978), 14, 509.
2. Cunliffe, A. V., Richards, D. H. and Thompson, D., unpublished results.
3. Burgess, F. J., Cunliffe, A. V., MacCallum, J. R. and Richards, D. H., Polymer (1977), 18, 719.
4. Burgess, F. J., Cunliffe, A. V., MacCallum, J. R. and Richards, D. H., Polymer (1977), 18, 726.
5. Bauer, B. J. and Fetters, L. J., Rubber Chem. Tech., (1978), 51, 406.
6. Croucher, T. G. and Wetton, R. E., Polymer (1976), 17, 205.
7. Burgess, F. J., Cunliffe, A. V., Richards, D. H. and Thompson, D., Polymer (1978), 19, 339.
8. Cunliffe, A. V., Hartley, D. B., Kingston, S. M., Richards, D. H. and Thompson D., Polymer, in press.
9. Cunliffe, A. V., Hartley, D. B., Richards, D. H. and Robertson, F., Polymer, in press.
10. Hartley, D. B., Hayes, M. S. and Richards, D. H., unpublished results.
11. Cohen, P., Hartley, D. B., Richards, D. H. and Schue, F., unpublished results.
12. Brzezinska, K., Chwalkowska, W., Kubisa, P., Matyjaszowski and Penczek S., Makromol Chem., (1977), 178, 2491.

RECEIVED April 20, 1981.

Anionic Polymerization of Triphenylmethyl Methacrylate

YOSHIO OKAMOTO, KOJI OHTA, KOICHI HATADA, and
HEIMEI YUKI

Department of Chemistry, Faculty of Engineering Science,
Osaka University, Toyonaka, Osaka 560, Japan

Stereospecific polymerization of various methacrylates has been extensively studied (1). Among many, triphenylmethyl methacrylate (TrMA) is one of the most interesting monomers and brings about some unusual results as follows: This monomer forms only

highly isotactic polymers with butyllithium (BuLi) not only in toluene but in tetrahydrofuran (THF) (2,3), and even with radical initiator it gives a polymer rich in isotacticity (2,3,4,5). This unique nature of TrMA has been ascribed to the bulky triphenylmethyl group. The large ester group also affects greatly its reactivity in the copolymerization with other methacrylates, such as methyl methacrylate (MMA) or α-methylbenzyl methacrylate (MBMA) with BuLi (6,7,8). Some of the copolymers of (S)-MBMA and TrMA shows large positive optical rotations, which are opposite in sign to the rotation of the homopolymer of (S)-MBMA. This positive rotation is attributed to the isotactic sequence of TrMA units which has helical conformation spiraled in single screw sense (9, 10, 11).

Recently, we found that the polymerization of TrMA with chiral anionic catalysts gave an optically active polymer the chirality of which is caused by helicity (12). This is the first example of optically active vinyl polymer the activity of which arises only from the helicity. This article describes the detailed results of the polymerization of TrMA by chiral anionic catalysts in addition to a brief review on our earlier studies described above.

Polymerization of TrMA with BuLi (2, 3)

Table I shows the results of the polymerization of TrMA with BuLi. Both in toluene and THF highly isotactic polymers were

obtained. The radical polymerization initiated by AIBN also af-
forded a polymer rich in isotacticity. These are unusual results
because most methacrylates form polymers rich in syndiotacticity
with BuLi in THF and with radical catalysts (1). TrMA did not
polymerize with phenylmagnesium bromide, while most methacrylates
do (1).

Table I
Polymerization of TrMA by BuLi[a]

Solvent	Temp. (°C)	Time (hr)	Yield (%)	Tacticity,%		
				I	H	S
Toluene	−78	24	21	96	2	2
Toluene	0	24	58	93	4	3
Toluene[b]	60	144	92	64	22	14
THF	−78	2.5	56	96	3	1
THF	0	24	98	81	13	6

[a] TrMA, 5mmol; solvent, 10ml; TrMA/BuLi=20.
[b] Catalyst, AIBN.

Copolymerization of TrMA and MMA with BuLi (6, 7)

Copolymer composition curves in the copolymerization of MMA
(M_1) and TrMA (M_2) with BuLi at -78°C are shown in Figure 1. From
these curves the monomer reactivity ratios were determined to be
$r_1 = 6.28$ and $r_2 = 0.13$ in toluene and $r_1 = 0.62$ and $r_2 = 0.62$ in THF.
In THF both the monomers showed similar reactivity. The compositional and configurational analyses of the copolymers indicated
that the copolymerization approximately follows the terminal model
in this solvent.

The tacticity of a copolymer can be calculated in terms of
conditional parameters, P_{ij}: P_{12} and P_{21}, for the sequence distribution and configurational parameters, σ_{ij}: σ_{11}, σ_{22}, σ_{12}, and
σ_{21}. The P_{ij} is the probability of finding the M_j unit as the
next neighbor of the M_i unit and can be calculated from monomer
reactivity ratios, r_1 and r_2 values (13). The σ_{ij} is the probability of generating a meso dyad when a new monomer unit M_j is
formed at the M_i end of a growing chain. The σ_{11} and σ_{22} are
obtainable from the dyad or triad contents of MMA and TrMA homopolymers, and σ_{12} and σ_{21} can be estimated by a trial and error
method by comparing the calculated triad contents with experimental
values. The latters were obtained for the poly(methyl methacrylate) (PMMA) derived from poly(MMA-co-TrMA). The observed and
calculated triad tacticities are shown in Figure 2. The curves
were obtained by using $r_1 = r_2 = 0.62$, $\sigma_{11} = 0.25$, and $\sigma_{22} = 0.94$ and
assuming $\sigma_{12} = \sigma_{21} = 0.25$. Good agreement was obtained between the
experimental and calculated data, when the MMA content in the
copolymer was greater than 20%, indicating that the copolymerization approximately follows the terminal model in THF. The probability of forming an isotactic placement between MMA anion and
TrMA or TrMA anion and MMA is much lower than that between TrMA
anion and TrMA.

Contrary to the copolymerization in THF, the copolymerization in toluene proceeded abnormally, i.e. it formed an isotactic homo-polymer of MMA and a copolymer with low stereoregularity. The copolymerization of an equimolar mixture of MMA and TrMA was carried out in toluene with BuLi at -78°C, and the polymer obtained was treated in methanolic hydrochloric acid in order to hydrolyze selectively the TrMA units into methacrylic acid units. Then the resulting polymer was separated into two fractions with methanol, as shown in Table II. The methanol insoluble fraction was found

Table II
Fractionation of Copoly(Methyl Methacrylate-Methacrylic Acid)
Obtained by Hydrolysis of Copoly(Methyl Methacrylate-
Trityl Methacrylate)

	Weight,mg	$\dfrac{[MMA]}{[MAA]}$[a]	Tacticity,%		
			I	H	S
Hydrolyzed polymer	94	6.4	74	17	9
MeOH insoluble part	54	∞	83	11	6
MeOH soluble part	40	2.2	56	31	13

[a] MAA: Methacrylic acid.

to be a homopolymer of MMA the isotacticity of which was higher than that of PMMA obtained under the same conditions of polymeri-zation, and the soluble fraction was the copolymer of lower ste-reoregularity. The highly isotactic growing chain of MMA does not seem to allow the incorporation of bulky TrMA. This is one of the reasons why TrMA showed the extremely low reactivity in the co-polymerization in toluene.

Copolymerization of TrMA and MBMA with BuLi (8, 9, 10, 11)

Figure 3 shows the copolymer composition curves observed in the copolymerization of TrMA (M_2) with optically pure (S)- and racemic (RS)-MBMA (M_1). In toluene (S)- and (RS)-MBMA showed different reactivity; $r_1 = 8.55$, $r_2 = 0.005$ in the former case and $r_1 = 4.30$, $r_2 = 0.03$ in the latter. The MBMA homopolymer having a high isotacticity (86%) and the copolymer with lower isotacticity again formed in the former copolymerization similarly as in the copolymerization of TrMA and MMA, while the formation of the MBMA homopolymer was not observed in the copolymerization with racemic MBMA and the copolymer obtained showed low stereoregularity. The isotacticity of poly[(RS)-MBMA] prepared with BuLi in toluene was 56% and considerably lower than that (78%) of poly[(S)-MBMA] obtained under the same conditions (14). This indicates that in the polymerization of (RS)-MBMA, which is practically the copoly-merization of an equimolar amount of (R)- and (S)-MBMA, the inter-action between the optical antipodes, one the growing chain end and the other the incoming monomer, is unfavorable for isotactic addition. In the copolymerization of (RS)-MBMA and TrMA, even when the chain is composed of MBMA units, it is less isotactic and

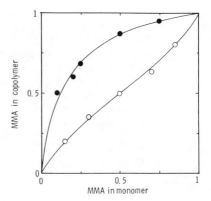

Figure 1. Copolymerization of TrMA-
(M_2) and MMA(M_1) with BuLi at $-78°C$
in toluene (●) ($r_1 = 6.28$; $r_2 = 0.13$) and
in THF (○) ($r_1 = 0.62$; $r_2 = 0.62$).

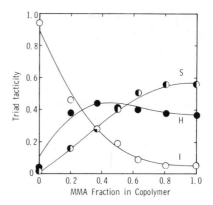

Figure 2. Triad tacticity of poly(MMA–
co-TrMA) obtained with BuLi in THF
at $-78°C$.

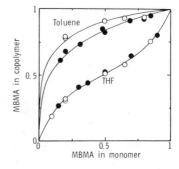

Figure 3. Copolymerization of TrMA-
(M_2) with (S)- or (RS)-MBMA in toluene
and THF with BuLi at $-78°C$: $M_1 =$
(S)-MBMA (○); $M_1 =$ (RS)-MBMA
(●).

the chain end has a labile conformation; therefore, TrMA is able to add to this end to form a copolymer of low stereoregularity.

In THF, however, no difference in the monomer reactivity ratios was observed between the (S)-MBMA-TrMA and (RS)-MBMA-TrMA systems, and the ratios ($r_1 = 0.39$ and $r_2 = 0.33$) showed similar reactivity of MBMA (M_1) and TrMA (M_2). The copolymerization seemed to proceed without termination and chain transfer reactions. An abnormal optical property was observed in some of the copolymers of (S)-MBMA and TrMA. Table III shows the tacticity and optical data of the copolymers which were obtained in various polymer yields from the monomer mixtures of a constant molar ratio, $[M_1]_0/[M_2]_0 = 1/4$. The (S)-MBMA content in the copolymers decreased

Table III

Copolymerization of (S)-MBMA(M_1) and TrMA(M_2) with BuLi in THF at $-78°C^a$

No.	$\dfrac{[M_2]}{[M_1]+[M_2]}$	Yield (wt%)	$\dfrac{m_2}{m_1+m_2}^b$	$\bar{M}_n^c$ ($\times 10^3$)	Tacticity,%			$[\alpha]_D^{20\,d}$	
					I	H	S	obsd.	calcd.
1	0.00	93	0.00	6.8	12	28	60	-103.8	-103.8
2	0.80	13	0.69	4.0	33	38	29	-34.1	-21.8
3	0.80	22	0.68	7.3	37	36	27	-33.2	-22.0
4	0.80	30	0.68	8.6	37	36	27	-31.2	-21.9
5	0.80	50	0.70	8.6	41	34	25	-19.4	-20.2
6	0.80	62	0.72	12.8	44	33	23	+5.5	-18.7
7	0.80	72	0.75	11.9	49	31	20	+39.0	-17.1
8	0.80	87	0.78	25.0	55	28	17	+59.1	-14.9
9	0.80	95	0.80	19.7	58	26	16	+56.0	-13.4
10	1.00	56	1.00	7.4	96	3	1	0.0	0.0

^{a} $[M_1]_0 + [M_2]_0 = 20$mmol; BuLi, 0.22-0.30mmol; THF, 20ml.
^{b} Copolymer composition. ^{c} By VPO. ^{d} In toluene.

and the isotacticity increased with the increase of polymer yield. The initial copolymers (No. 2-4) showed negative optical rotations, the sign of which was the same as that of poly[(S)-MBMA], and the absolute values were higher than those calculated from the weight fractions of the optically active monomer units in the copolymers. However, the copolymers obtained in higher yields showed large positive rotations which were opposite in sign to that of poly-[(S)-MBMA]. The chains of these copolymers should have a tapered structure with respect to the sequence distribution of the monomer units, namely the length of TrMA sequence is short at the initially formed end of the polymer chain and increases gradually to the other end. Figure 4 illustrates the change in the length of (S)-MBMA and TrMA sequences in the increments of the growing chain which are newly formed with increasing conversion. The curves were estimated from the polymer yield and the monomer reactivity ratios $r_1 = 0.39$ and $r_2 = 0.33$ using the intetgated copolymer composition equation of Meyer and Lowry (15). The length of the(S)-MBMA sequence is close to unity over the whole polymer chain. Therefore, the copolymers are composed of the TrMA sequences

interconnected with (S)-MBMA unit. The average length of TrMA
sequence in the initially formed end is sbout 2.3, but increases
steeply above 60% conversion. The isotacticity of copolymer also
increased with increasing conversion. This increase in the isotacticity is attributed to the highly isotactic configuration of
the long TrMA sequences formed with the progress of copolymerization. The long TrMA sequences are considered to take rigid helical
conformation. When TrMA adds to the (S)-MBMA end in the copolymerization, the steric interaction between the chiral end and the
incoming monomer may be large enough to force some chiral addition
of TrMA, and the following isotactic addition of TrMA results in
the formation of a helix in one screw sense. This helical sequences must cause the large positive optical rotations of the
copolymers obtained at higher conversions.

Figure 5 illustrates the CD spectra of the copolymers and
poly[(S)-MBMA]. The CD data were correlated well with the optical
rotations of the copolymers. The CD spectrum of copolymer 2 in
Table III showed a negative band at 230 nm due to the TrMA units
besides a band at 211 nm mainly due to the (S)-MBMA units. Furthermore, a band at 250-280 nm is positive, similar to that of
poly[(S)-MBMA]. This suggests that TrMA units in this copolymer
contribute slightly to the optical activity of the copolymer in
the same sign as that of the (S)-MBMA homopolymer. On the other
hand, the CD spectrum of copolymer 9 showed large positive bands
at 208 and 232 nm, the wave lengths of which correspond to those
of the ultraviolet absorptions of poly[TrMA]. These strong CD
bands should be closely associated with the largely positive optical rotation of copolymer 9.

Polymerization of TrMA by Optically Active Catalysts (12, 16)

Several attempts were made to prepare optically active vinyl
polymers, $\{CH_2-CXY\}_n$, which have chirality in the polymer chain
(17-23). However, all the polymers obtained showed no or only
small rotation which arose from the chiral terminal groups. It
has been pointed out that even if an isotactic polymer composed of
units of a single configuration is formed in the process of the
polymerization, the polymer can not be optically active because
the polymer becomes pseudo asymmetric (24, 25).

However, there will be a possibility of obtaining an optically
active polymer when the polymerization proceeds by forming a tight
helix of polymer chian and the helix is sufficiently stable in
solution (26). This concept was first demonstrated with poly(tert-
butyl isocyanide) [$C=N-C(CH_3)_3]_n$ (27). In this case, the polymer
was resolved into two fractions of positive and negative optical
rotations. This is not a vinyl polymer. The results of the copolymerization of (S)-MBMA and TrMA prompted us to investigate the
possibility of forming an optically active homopolymer of TrMA.

TrMA was polymerized with two different types of chiral catalysts, lithium (R)-N-(1-phenylethyl)anilide (LiAn) and (-)-spar-

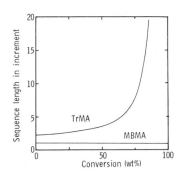

Figure 4. The change of (S)-MBMA and TrMA sequence lengths that are newly formed at growing end.

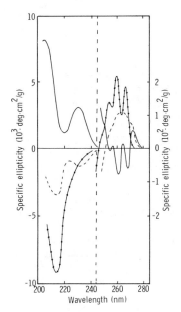

Figure 5. CD spectra of poly[(S)-MBMA] (· · ·), Copolymer 2 (– – –), and Copolymer 9 (———) in Table III.

teine-BuLi complex (Sp-BuLi). The results of the polymerization
are summarized in Table IV. Optically active polymers were obtained with both the catalysts. When LiAn was used, the polymers

LiAn   Sp-BuLi

of negative rotation were formed both in toluene and THF and most
were soluble in THF.

Table IV

Polymerization of TrMA with LiAn and Sp-BuLi at -78°C[a]

No.	Catalyst	Solvent	Time hr	Yield %	THF-soluble polymer				THF-insol. polymer Yield,%
					Yield %	$\overline{DP}$	$[\alpha]_D^{20}$ (deg) THF	Tol.	
11	LiAn	Toluene	0.7	50	50	—	-57	-75	0
12	LiAn	Toluene	2.9	73	73	62	-70	-85	0.5
13	LiAn	THF	2.3	93	93	21	-82	-104	0
14	Sp-BuLi	Toluene	2.0[b]	100	100	42	+262	+363	0
15	Sp-BuLi	Toluene	1.0	19	19	—	+67	+93	0
16	Sp-BuLi	Toluene	1.0	25	25	35	+137	+193	0
17	Sp-BuLi	Toluene	2.0	41	41	48	+190	+258	0
18	Sp-BuLi	Toluene	3.5	55	54	58	+254	c	1
19	Sp-BuLi	Toluene	5.2	90	18	43	+245	c	72[d]
20	Sp-BuLi[e]	Toluene	26.0	100	3	—	+19	—	97[f]
21	Sp-BuLi	THF	2.4	100	100	21	+7	+10	0

[a] [TrMA]/[Catalyst]=20, Solvent/TrMA=20 (ml/g).   [b] Temp., -40°C.
[c] A part of the polymer was insoluble in toluene.   [d] $\overline{DP}$ = 74.
[e] [TrMA]/[BuLi] = 50.   [f] $\overline{DP}$ = 210.

The specific rotation $[\alpha]_D^{20}$ of the polymers obtained with
Sp-BuLi in toluene increased to positive direction with the polymer
yield and reached above +250° (in THF). The polymers of high molecular weights were insoluble in common organic solvents. In
general, the reaction mixture was homogeneous in appearance during
the polymerization at low temperature, although it formed gel at
the final stage of the reaction. The polymer obtained in THF
showed a very small rotation.

The change of optical rotation of the reaction mixture during
the polymerization in toluene was directly followed in a 1.0-cm
optical cell at -78°C (Figure 6). When LiAn was used as catalyst,
the rotation became more negative as the polymerization proceeded
and reached a constant value after about 3 hr. During this time
most monomer was consumed. From the final value the specific

rotation $[\alpha]_D^{-78}$ of the polymer was estimated to be -110° assuming complete polymerization.

When the polymerization was initiated with Sp-BuLi in toluene at -78°C, the rotation became slightly negative initially and then turned to the positive direction, and increased noticeably. The reaction rate was slower than that by LiAn and it took about 15 hr to complete the reaction. The $[\alpha]_D^{-78}$ of the polymer at complete reaction was estimated to be about +520°. At -40°C, the reaction was completed in about 30 min and the final optical rotation was smaller than that observed at -78°C.

All PMMA derived from poly(TrMA) were almost fully isotactic and showed almost no optical rotation.

The polymers obtained in toluene and THF with LiAn seem to show similar specific rotation at the same polymer yield. This suggests that the nature of the ion pair at growing chain end is not important because most anionic polymerizations by alkyl-Li are greatly affected by the polarity of the solvent. The anilino group attached to the chain end does not seem to be the main source of the optical rotation, which is too large to arise only from end groups. The large optical rotation must be attributed to the helicity of the isotactic poly(TrMA). The rotation, however, was lower than that of the polymer prepared with Sp-BuLi, although the DP of the polymers were not too different. The first addition of the monomer to LiAn forms two diasteromers one of which must be in excess and the subsequent addition of monomer proceeds mainly with retention of configuration, because the polymer produced appeared to be fully isotactic. The polymer chain may be either completely left- or right-handed helix. The rotation of the polymer corresponds to the rotation of the excess enantiomeric polymer. In these process the large triphenylmethyl group plays a very important role in stabilizing the helix.

The polymer obtained with Sp-BuLi in toluene exhibited a very large optical rotation, which seemed to increase with DP of the polymer. The specific rotation at -78°C became over +500°, although the polymer having such a high rotation was insoluble after precipitation at room temperature. The first addition of the monomer to BuLi in toluene probably forms one of two enantiomers in excess because of the presence of (-)-sparteine. The further addition of the monomer to one of the enantiomers may proceed with retention of configuration resulting in the formation of a polymer helix having a positive optical rotation. The isomer which is the optical antipode of the above enantiomer may add some monomers in the same configuration, but the addition may be more or less restricted by the presence of (-)-sparteine and, at a certain stage, the growing end will change its structure to the form suitable for the production of the polymer of positive rotation. Therefore, the content of the previaling helix may increase along with the progress of the polymerization. (-)-Sparteine did not work as effective asymmetric ligand in THF. The polar solvent coordinates the lithium cation and must weaken the effect of (-)-sparteine.

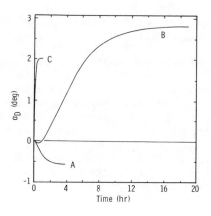

Figure 6.   Polymerization of TrMA with LiAn at −78°C (A), Sp-BuLi at −78°C (B), and Sp-BuLi at −40°C (C) in toluene in a 1.0-cm cell (toluene, 3 mL; TrMA 0.15 g; [TrMA]/[Li] = 20).

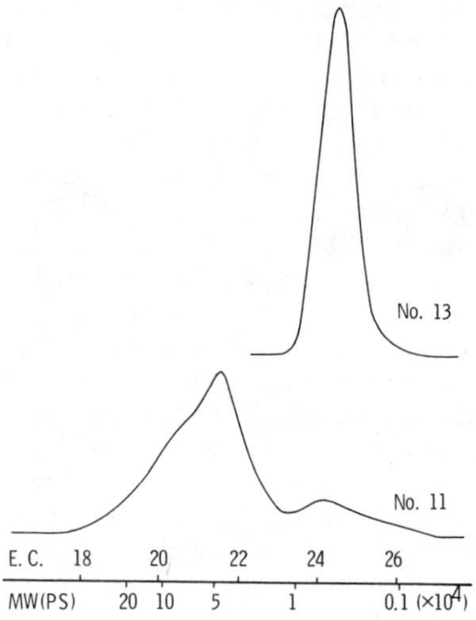

Figure 7.   GPC curves of poly(TrMA) (No. 1 and 3 in Table IV) obtained with LiAn at −78°C.

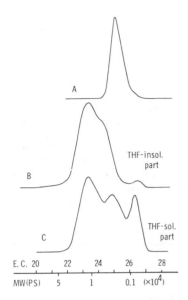

Figure 8. GPC curves of PMMA derived from poly(TrMA) (A = No. 21, B and C = No. 19 in Table IV) obtained with Sp-BuLi in THF and toluene at −78°C.

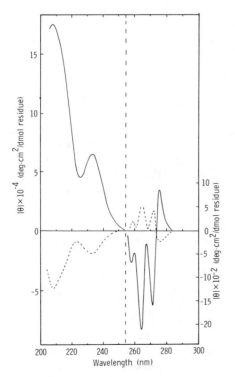

Figure 9. CD spectra of poly(TrMA) in THF: No. 14 in Table IV, [α]$_D^{20}$ + 262° (THF) (———); No. 13 in Table IV, [α]$_D^{20}$ −82° (THF) (− − −).

Figure 7 illustrates the GPC curves of poly(TrMA) obtained by LiAn in toluene (A) and THF (B); Figure 8 is the GPC curves of the PMMA derived from the poly(TrMA) obtained with Sp-BuLi. Both polymers obtained in THF with LiAn and Sp-BuLi showed rather narrow distributions and initiator efficiency was about 100%. The polymerization in toluene was complicated and a large portion of the initiators was consumed for the formation of oligomers.

The specific rotation of polymer No. 14 in THF and toluene did not change after 24 hr at room temperature and even after heating for 30 min at 50°C. The CD spectra of the poly(TrMA) obtained with Sp-BuLi and LiAn catalysts are shown in Figure 9. Both spectra were almost mirror images of one another as expected from their opposite rotations. The relative intensities of the peaks were nearly proportional to the ratio of the specific rotations. The peaks at 208 and 232 nm may be assigned to the absorptions of the phenyl and ester groups, respectively. The spectral pattern is very similar to that of the copolymer 9 of TrMA with a small amount of (S)-α-methylbenzyl methacrylate illustrated in Figure 5. This also indicates that the large positive rotations of the copolymers 7-9 in Table III are attributed to the helical conformation of isotactic TrMA units preferential in one screw sense.

The polymer of high molecular weight in the solid stage exhibited high crystallinity under a polarized microscope and insoluble in common organic solvents. When the polymer with high optical rotation was used as stationary phase or sorbent for the chromatographic resolution of racemic compounds, it showed the ability of resolution for many kinds of compounds, such as alcohols, amines, esters, and even hydrocarbons (28).

Acknowledgment

The authors are pleased to acknowledge the experimental contributions of Y. Kikuchi, T. Niinomi, S. Nakashima, K. Suzuki, and I. Okamoto. A part of this work was supported by a Grant-in-aid for Scientific Research from the Ministry of Education.

Literature Cited

1. Yuki, H.; Hatada, K. Adv. Polym. Sci., 1979, 31, 1.
2. Yuki, H.; Hatada, K.; Kikuchi, Y.; Niinomi, T. J. Polym. Sci. B, 1968, 6, 753.
3. Yuki, H.; Hatada, K.; Niinomi, T.; Kikuchi, Y. Polymer J., 1970, 1, 36.
4. Niezette, J.; Desreux, V. Makromol. Chem., 1971, 149, 177.
5. Niezette, J.; Hadjichristidis, V.; Desreux, V. Makromol. Chem., 1976, 177, 2069.
6. Yuki, H.; Okamoto, Y.; Ohta, K.; Hatada, K. J. Polym. Sci. Polym. Chem. Ed., 1975, 13, 1161.
7. Okamoto, Y.; Nakashima, S.; Ohta, K.; Hatada, K.; Yuki, H. J. Polym. Sci. Polym. Lett. Ed., 1975, 13, 273.

8. Ohta, K.; Okamoto, Y.; Hatada, K.; Yuki, H. J. Polym. Sci. Polym. Chem. Ed., 1979, 17, 2917.
9. Yuki, H.; Ohta, K.; Okamoto, Y.; Hatada, K. J. Polym. Sci. Polym. Lett. Ed., 1977, 15, 589.
10. Ohta, K.; Hatada, K.; Okamoto, Y.; Yuki, H. J. Polym. Sci. Polym. Lett. Ed., 1978, 16, 545.
11. Yuki, H.; Ohta, K. Polymer Preprints, 1979, 20, 747.
12. Okamoto, Y.; Suzuki, K.; Ohta, K.; Hatada, K.; Yuki, H. J. Am. Chem. Soc., 1979, 101, 4763.
13. Ito, K.; Yamashita, Y. J. Polym. Sci. Part A, 1965, 3, 2165.
14. Yuki, H.; Ohta, K.; Ono, K.; Murahashi, S. J. Polym. Sci. A-1, 1968, 6, 829.
15. Meyer, V. E.; Lowry, G. G. J. Polym. Sci., A, 1965, 3, 2843.
16. Okamoto, Y.; Suzuki, K.; Yuki, H. J. Polym. Sci. Polym. Chem. Ed., in press.
17. Marvel, C. S.; Frank, R. L.; Prill, E. J. Am. Chem. Soc. 1943, 65, 1647.
18. Marvel, C. S.; Overberger, C. G. J. Am. Chem. Soc., 1946, 68, 2106.
19. Overberger, C. G., Palmer, L. C. J. Am. Chem. Soc., 1956, 78, 666.
20. Murahashi, S.; Nozakura, S.; Takeuchi, S. Bull. Chem. Soc. Jpn., 1960, 33, 658.
21. Fray, G. I.; Robinson, R. Tetrahedron, 1962, 18, 261.
22. Pino, P.; Ciardelli, F.; Lorenzi, G. P. J. Polym. Sci. Part C, 1963, 4, 21.
23. Braun, D.; Kern, W. J. Polym. Sci. Part C, 1963, 4, 197.
24. Frisch, H. L.; Schuerch, C.; Szwarc, M. J. Polym. Sci., 1953, 11, 559.
25. Pino, P. Adv. Polym. Sci., 1965, 4, 393.
26. Millich, F.; Baker, G. K. Macromolecules, 1969, 2, 122.
27. Nolte, R. J. M.; Van Beijnen, A. J. M.; Drenth, W. J. Am. Chem. Soc., 1974, 96, 5932.
28. Okamoto, Y.; Okamoto, I.; Yuki, H. to be published.

RECEIVED April 6, 1981.

C-13 NMR Study of the Propagating Live End in the Butyllithium Polymerization of 1,3-Butadiene

ADEL F. HALASA—Kuwait Institute of Scientific Research, Safat, Kuwait

V. D. MOCHEL—The Firestone Tire and Rubber Co., Central Research Laboratories, Akron, OH 44317

G. FRAENKEL—The Ohio State University, Department of Chemistry, Columbus, OH 43210

Several workers ($\underline{1},\underline{2},\underline{3},\underline{4}$) have used ^{1}H nmr to study the propagating chain end in the polymerization of 1,3-butadiene (1,3 BD) with a butyllithium initiator. They concluded that the poly(butadienyl) lithium chain end is virtually 100 percent 1,4 with no 1,2 structures, even though 1,2 units are incorporated in the chain. The lithium is σ bonded to the α carbon, and there is no evidence of a π-allyl type of delocalized bonding involving the γ carbon. However, the presence of vinyl in-chain units was taken as evidence for the presence of an undetectable amount of the π bonded chain ends in equilibrium with the σ bonded chain ends. Glaze and coworkers ($\underline{3}$) further suggested that the stereochemical course of allyllithium reactions may depend on the aggregation of the reactive species.

Bywater and coworkers ($\underline{5},\underline{6}$) employed ^{13}C nmr to identify the the shift positions and to estimate the charge distribution over the various carbon positions in several organoalkali metal model compounds of butadiene and isoprene. They concluded that all of the compounds are delocalized ionic compounds in the solvents, C_6D_6 and several ethers.

In some organolithium compounds the [13]C-[7]Li spin-spin coupling can be observed in the ^{13}C nmr spectrum, but in many cases, quadrupolar coupling of the [7]Li nucleus broadens and obscures the coupling. The [6]Li isotope has a much smaller quadrupole moment than [7]Li, so organolithium compounds prepared with [6]Li would offer a much better chance to observe the [13]C-[6]Li coupling under conditions of slow exchange. In this paper experiments were carried out using n-Bu[6]Li and t-Bu[6]Li as the initiators. Also, butadiene monomer enriched with ^{13}C was used to improve further the prospects of gaining insights into the type of bonding at the live ends in butadiene polymerization.

0097–6156/81/0166–0367$05.25/0

Variable temperature measurements which provide information about the state of aggregation of the chain ends will also be presented.

Experimental

Samples of 90% ^{13}C-enriched 1,3 butadiene-$^{13}C_1$ were supplied by Merck and Co. The same company also prepared a sample with 90% ^{13}C enrichment in the 2 position, i.e., 1,3-BD-$^{13}C_2$. The n-Bu^{6}Li and the T-Bu^{6}Li samples were prepared for us by Org Met. Chemicals Co. following the procedure developed by one of us (GF).

The ^{13}C-enriched butadiene samples were transferred and sealed in nmr tubes, along with the solvent and initiator, under a vacuum of about 10^{-5} torr. They were held at liquid nitrogen temperatures before warming to the reaction temperature. The non-enriched butadiene monomer was syringed from a hydrocarbon solution and injected through a rubber septum into an nmr tube containing solvent and initiator at liquid nitrogen temperature. The tube was then sealed under vacuum.

The ^{13}C nmr spectra were obtained with a JEOL PFT100 nmr spectrometer operating at 25.15 MHz for ^{13}C nuclei in the FT mode with noise decoupling unless otherwise noted. T_1 measurements were carried out to establish proper operation conditions for quantitative results. The longest T_1 for the oligomers was determined to be 5.3 s. All spectra were run with a repetition time of 5.3 s and a tip angle of 70° (19.2 μs pulse width). Most of the spectra were also run in a gated decoupling mode in which the Nuclear Overhauser Effect (NOE) was eliminated.

Results and Discussion

The poly(butadienyl) lithium chain ends can assume the following forms in which the lithium is σ bonded to the α or the γ carbon:

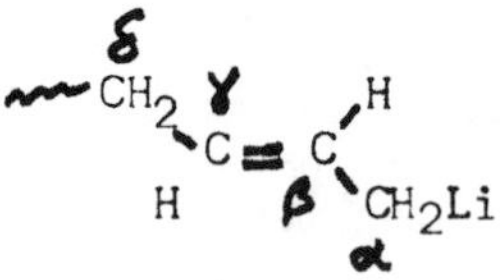

I (trans)

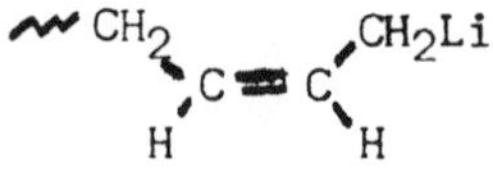

II (cis)

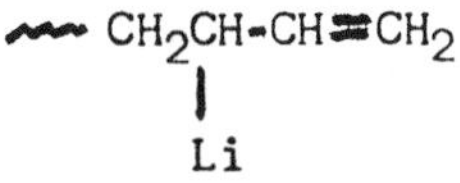

$$\sim\!\!\sim CH_2CH\text{-}CH\!=\!CH_2$$

$$|$$
$$Li$$

III (1,2)

In addition, the chain ends can experience a delocalized π-allyl type bonding with lithium as shown below.

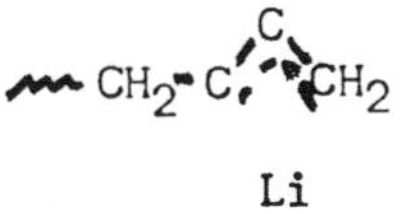

$$\sim\!\!\sim CH_2\text{-}C\overset{C}{\underset{}{\diagup\!\!\diagdown}}CH_2$$

Li

IV (π-allyl)

The FT ^{13}C nmr technique is ideally suited for following the course of a polymerization if the rate is sufficiently slow for good spectra to be obtained. Valuable kinetic and thermodynamic data can quite readily be obtained in this way.

Figure 1 shows the early stages of the polymerization of 6.0 M 1,3 BD with 1.0 M n-BuLi in cyclohexane. Spectrum 1a was taken before any reaction had taken place. The two peaks at 117.26 and at 138.33 ppm arise from carbons 1 and 2, respectively, of the unreacted BD. The aliphatic portion of the spectrum shows the four peaks of the unreacted n-BuLi and the strong cyclohexane peak at 27.6 ppm. The three peaks marked "h" are due to hexane, the solvent for the n-BuLi, and the small peak at 29.4 ppm is an impurity in the solvent.

Figure 1b is the spectrum taken 0.6 hr after warming the sample to 26°C, the reaction temperature. Small resonances, due to oligomer, start to appear. The weak, broad resonance, at 100 ppm, is due to the γ carbon in agreement with the assignment made by Bywater, et al., (5) with model compounds. This peak is known to be associated with the live end because it disappears when the chains are terminated with CH_3OH. A gated decoupling technique in which the C-H coupling is restored shows that the resonance at 100 ppm is due to a methine carbon by its doublet splitting.

The shape of the C-1 carbon resonance of unreacted n-BuLi at 11.7 ppm, in Figure 1b, is important. The formation of the shoulder on the right side is believed to be due to a change in association and might be reflecting a series of equilibria between the hexameric and monomeric species. Figure 2, an expansion of this resonance, reveals there are at least three additional peaks, thus suggesting that the intermediate stages of association are being observed.

Figure 1c shows that after 1.8 hours, most of the n-BuLi has reacted. The three equal intensity peaks at 14.46, 23.39, and 32.71 ppm are due to reacted butyl groups at the start of each chain. The remaining peaks are due to polymer showing both

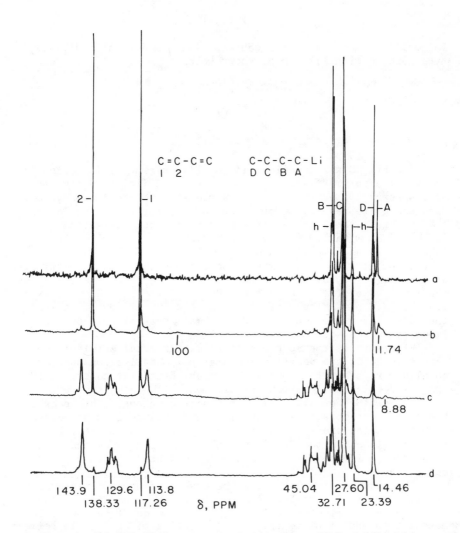

Figure 1. $^{13}C \{^1H\}$ *NMR spectra of polymerization at 26°C of 6.0M 1,3-buta-diene with 1.0M n-BuLi in cyclohexane: (a) 0 h; (b) 0.6 h; (c) 1.8 h; (d) 4.2 h.*

1,4 and 1,2 structures. At this stage of the reaction, the chains are composed of sixty percent 1,2 units and forty percent 1,4 units.

In Figure 1d, (4.2 hours) only a small amount of unreacted BD monomer remains. After 5 hours, all of the BD monomer has reacted and no further changes are observed in the spectrum.

If the sample is heated to 50°C, the broad resonance at 100 ppm sharpens considerably. This behavior is indicative of either an association change or an exchange process.

If a sample with live ends and with an excess of n-BuLi is heated to 50°C or above, three new resonances appear at 154.7, 77.7 and 61.8 ppm, as shown in Figure 3a. This is a partial spectrum of a sample of 7.5 mmoles of 1,3 BD which was reacted with neat n-Bu^{6}Li at room temperature and then heated to 70°C a short time. The ^{6}Li isotope was used to reduce broadening from the larger quadrupole moment of the ^{7}Li nucleus. The peaks at 143.9 and 113.82 ppm are due to in-chain vinyl and the resonance at 129.64 ppm is due to in-chain 1,4 units. This indicates a higher DP than would be expected from the high concentration of initiator.

Figure 3b is the same sample after heating approximately 24 hours at 70°C. It is seen that the γ carbon resonance of the live end at 100 ppm appears to shift to 79.2 ppm. The distinctive doublet pattern might suggest equal amounts of trans and cis units in the live end, but it is not possible to determine this from the aliphatic portion because of overlapping of peaks. the doublet might be revealing equal states of aggregation. The β carbon resonance at 138.6 ppm also decreases upon heating and appears to shift to 154.7 ppm. The new resonance at 61.8 ppm seems to be the result of a change in the aliphatic part of the spectrum, but this again is not observable because of the overlapping of peaks in that area. Figure 3c is a gated decoupling spectrum showing by the doublet splitting of all three peaks that new resonances are due to methine carbons. It is suggested that the three new resonances observed in Figures 3a and 3b are the result of metalation, as will be discussed further below.

Because of the overlapping of peaks in the aliphatic region, it has not been possible to identify the carbon bonded to the lithium. Therefore, experiments were carried out using ninety percent enriched 1,3 BD-^{13}C$_1$ monomer. Figure 4 follows the course of the polymerization of 1.4 M ^{13}C-enriched BD in neat n-Bu^{6}Li at 26°C. Figure 4a shows, after ten minutes, development of two peaks in the aliphatic region in addition to the taller unreacted n-BuLi peaks. Only C$_1$ and C$_4$ peaks will be prominent. The peak at 22.86 ppm is somewhat broadened as might be expected for a carbon bonded to a lithium in an associated state. The other peak at 32.42 ppm is undoubtedly the C$_4$ (δ) carbon of a trans unit in a 1:1 adduct. The small peak at 28.15 ppm would be the δ carbon peak of the corresponding cis unit. The selective ^{13}C enrichment allows the observation and the

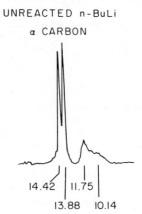

UNREACTED n-BuLi

α CARBON

14.42   11.75

13.88   10.14

Figure 2. Expansion of C-1 resonance
of Figure 1c.

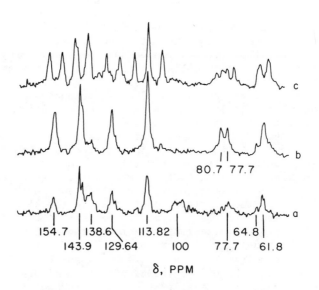

80.7 77.7

154.7   138.6   113.82

143.9   129.64   100   77.7   61.8

64.8

δ, PPM

Figure 3.   ¹³C {¹H} NMR spectra of live-end butadiene heated to 70°C: (a) 1 h;
(b) 24 h; (c) gated decoupling.

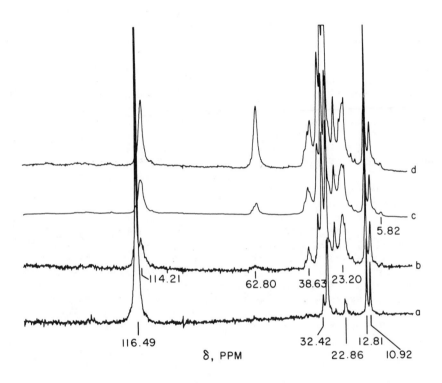

Figure 4. ^{13}C {^1H} NMR spectra of ^{13}C-enriched 1,3-butadiene-$^{13}C_1$ polymerized in neat n-Bu^{6}Li at 26°C: (a) 10 min; (b) 65 min; (c) 310 min; (d) 23 h.

positive identification of both the α and the β carbon peaks of
the live end. Without the enrichment these resonances are quite
obscure and in doubt. The intense peak at 116.49 ppm is the C_1
carbon of the unreacted butadiene monomer.

Figure 4b shows the development of other peaks which are
responding to sequence-related information in the growing chain.
The peak at 114.21 ppm is due to the methylene olefin carbon of
a 1,2 unit in the chain. Of special significance is the resonance peak at 62.80 ppm which is the same as that observed in
Figure 3 as a result of heating at 70°C. Thus, even at room
temperature, some "metalation" is taking place in the presence
of excess n-BuLi.

Figure 4c shows the further increase in this peak at 310
min and indicates that all of the BD monomer has reacted. Figure
4d is the spectrum after 23 hours showing the large increase in
the peak at 62.80 ppm and a somewhat subtle decrease in the co-
valently σ-bonded peak assigned to the carbon carrying the
lithium at 23.20 ppm.

This trend continues upon heating at 50°C or above. After
2 hours at 70°C, the peak at 23.20 ppm and the one at 32.42 ppm
have all but vanished, while the peak at 62.80 ppm has increased
considerably. This is shown in Figure 5. The number above each
of the peaks B-I gives the change in the percentage of the total
aliphatic resonance of that peak from the start of the appearance of the peak at 62.80 ppm until the final spectrum. Thus,
27 percent of the aliphatic resonance is now located at 62.80
ppm. Peak H, the α carbon resonance (the area bounded by the two
lines), decreased by 16 percent, and peaks E and G (the β carbon
peaks) had a combined decrease of about 14 percent. The two
peaks labeled "I" due to unreacted n-BuLi also decreased by about
6 percent. Peaks C and F both gained roughly 6 percent.

These data strongly suggest that lithiation has taken place
even at 26°C in the live ends as illustrated below:

$$C\text{-}C\text{-}C\text{-}CLi + \sim C\text{-}C\text{≡}C\text{-}C\text{-}Li \longrightarrow$$
(Excess)

$$C\text{-}C\text{-}C\text{-}C \quad + \sim C\text{-}C\text{≡}C\text{-}C\text{-}Li$$
$$\qquad\qquad\qquad\qquad\quad |$$
$$\qquad V \qquad\qquad Li$$

$$+ \sim C\text{-}C\text{≡}C\text{-}C\text{-}Li$$
$$\qquad\quad | \qquad |$$
$$\qquad\quad Li \qquad Li$$

VI

The fact that the metalated peak at 62.80 ppm is much larger than the decrease observed in the α carbon peak requires a metalated species such as VI in addition to V. Keeping in mind that the in-chain 1,2 unit content (Peak A) did not change and the resonance at 62.80 ppm is from methine carbons (Figure 3c), structures V and VI seem the most likely. This would require both the α and the δ carbons bonded to Li to resonate at the same position. This is probably possible if structures V and VI have an ionic character. Lithiation of in-chain 1,4 units cannot be ruled out as an additional possibility.

Figure 6 shows the progression of the polymerization at 26°C of 4.5 M BD-^{13}C$_1$ with 6.2 M n-Bu^{6}Li in cyclopentane (CP). Only the aliphatic portion of the ^{13}C spectrum is shown. The time of reaction in minutes is given at the right of each spectrum. At zero time only the solvent and the n-Bu^{6}Li peaks are observed. At 20 minutes three additional peaks appear. The peaks at 33.78 ppm and 23.54 ppm are due to the trans δ and α carbons, respectively. The small peak at 22.57 ppm is probably the cis α peak, but the corresponding δ peak is not clearly discernible. At 30 minutes, however, the cis δ peak is clearly visible at 28.15 ppm. As the reaction proceeds, the α carbon peaks broaden and are shifted slightly downfield (to 24.51 ppm). At the same time, the remaining unreacted n-Bu^{6}Li α carbon peak shifts upfield (12.28 to ~10 ppm). This is probably reflecting the changes in the states of association of the live ends and of the unreacted BuLi. After about 60 minutes in-chain 1,2 units start to appear in the olefinic region (not shown), and therefore resonances from the methylene carbons become evidenced in Figure 6b. In-chain 1,4 units also are observed.

Figure 7a shows the entire spectrum of the sample of Figure 6 after 23 hr. reaction at 26°C. It is interesting that even in solution and with only a slight excess of unreacted n-BuLi, an easily observable amount of metalation has taken place as shown by the peak at 63.63 ppm.

Figure 7b is the spectrum of the same sample after opening the tube in a dry box and adding a chelating diamine, dipiperidyl ethane (DPE). We have found that this polar modifier can be used with n-BuLi and 1,3 BD to give polybutadiene with high 1,2 addition content. More complete details of the polymerization mechanism with the use of this modifier will be given in a later publication. The observed effect of the DPE in Figure 7b is the disappearance of the α carbon resonance at 24.95 ppm and the reinforcement of several peaks, but especially of one at 34.99 ppm. This strongly suggests a change from the predominantly σ-bonded character of the α carbon in CP to a delocalized, π-allyl character in the presence of DPE in agreement with the work of Bywater and Worsfold[6].

Figure 8a is the ^{13}C nmr spectrum of a predominantly 1:1 adduct of 1.4 M 1,3-BD-^{13}C$_1$ and 1.16 M t-Bu^{6}Li in CP before all of the monomer reacted. The peak at 117.21 ppm is the C$_1$ carbon

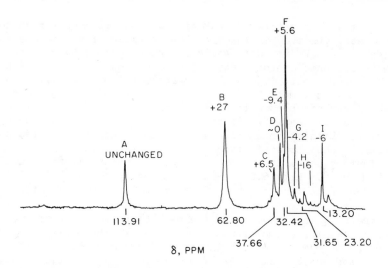

Figure 5. ^{13}C {^1H} NMR spectrum of sample in Figure 4 heated at 70°C for 2 h.
Numbers above the peaks show the change in the percentage of the total aliphatic
resonance for each peak from the start of the appearance of Peak B.

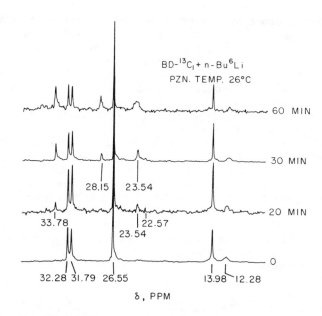

Figure 6a. ^{13}C {^1H} NMR spectra showing the progress of polymerization of
4.5M butadiene-$^{13}C_1$ with 6.2M n-Bu^{6}Li in cyclopentane at 26°C. Aliphatic portion
only. Times of reaction are given at the right side.

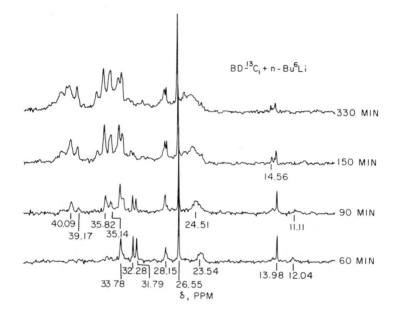

$BD-^{13}C_1 + n-Bu^6Li$

330 MIN

150 MIN

14.56

90 MIN

40.09
39.17
35.82
35.14
24.51
11.11

60 MIN

33.78
32.28
31.79
28.15
26.55
23.54
13.98
12.04

δ, PPM

Figure 6b. Continuation of Figure 6a.

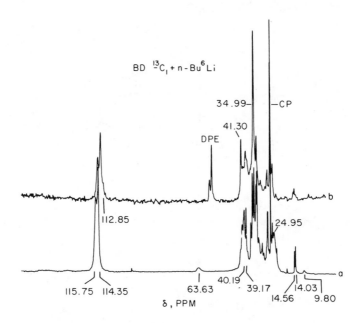

Figure 7.   Complete spectrum of sample of Figure 6 after 23-h reaction (a); same sample after adding DPE (b).

of the unreacted BD. The upper trace is the spectrum after complete reaction. The peak at 114.59 ppm shows there are some in-chain 1,2 units. The peaks at 48.78 and 42.32 are the δ carbons of trans and cis units, respectively in agreement with the assignments of Bywater and Worsfold[6]. The integrated intensities of these two peaks indicate a trans/cis ratio of 3/1 in agreement with the ^{1}H nmr work of Glaze and coworkers[3]. The meaning of the two shoulder peaks at 50.04 and 48.05 is not clear at this point, but they might be due to the methylene carbon of in-chain 1,2 units in an oligomer of DP2 or 3 as illustrated below:

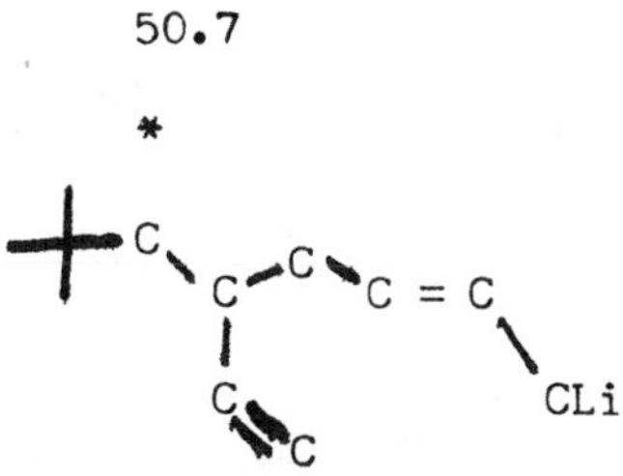

50.7

VII

51.2

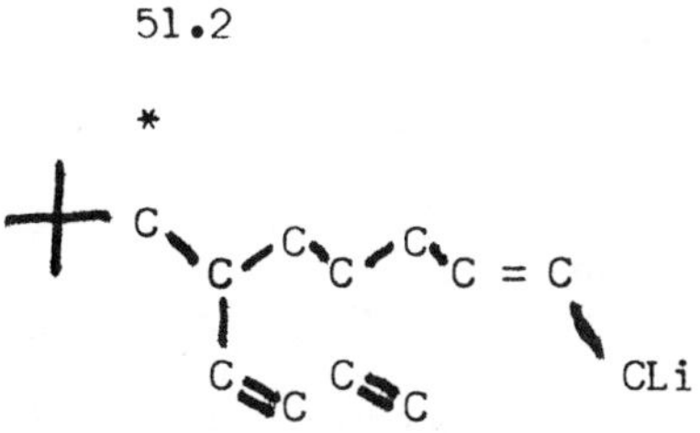

VIII

Empirically determined chemical shift additivity parameters have been determined[7] for diene-type polymers. The shift contribution of a quaternary carbon which is α to the carbon in question was not determined by those authors. However, using their additivity parameters and the shift positions of the δ carbons in Figure 8, a value of +15.4 ppm can be estimated for the contribution of a neighboring (α) quaternary carbon. Using this value, the shift positions of the * carbons in structures VII and VIII are calculated as shown. If the first 1,2 unit were on the chain in a 2,1 manner, the methylene carbon resonance would be at a considerably higher field, but it would be difficult to estimate its position with any certainty because the β quaternary effect

is not known. Thus it seems reasonable that one or more of the
satellite peaks of the 1,4 peaks might arise from 1,2 methylene
carbons as shown in VII and VIII. It is not certain what the
multiple peaks near 35 ppm represent, but they probably arise
from ^{13}C-^{13}C spin-spin coupling in oligomers with a DP of 2 or
more. The sharp peak at 29.70 is due to the methyl carbons of
the reacted t-butyl group.

The α carbon peak at 23.59 does not show splitting due to
coupling with ^6Li as might be expected for a σ bond. A coupling
between ^{13}C and ^7Li is observed[5] in t-BuLi and s-BuLi but not in
n-BuLi. It has been suggested[6] that rapid exchange processes
would eliminate it in the butadiene live end even if it existed.

Previous authors[3,8] have suggested that the stereochemistry
of the polymer chain may depend upon the changing state of aggregation of the reactive species. Variable temperature nmr
measurements of the live ends would be expected to shed light
on the states of aggregation and any exchange processes associated with them. Figure 9 shows the changes that take place in
the aliphatic part of the spectrum of the sample of Figure 8 as
the temperature is varied from 26° to -50°C. Both the α and the
δ carbon peaks are broadened as the temperature is lowered, but
the broadening is considerably more extensive in the α carbon
resonance. Although the strong CP resonance obscures the
carbon area, several new peaks become distinguishable and it
appears that we might be observing exchange between several
states of aggregation. Part of the increased broadening could
be caused by ^6Li-^{13}C coupling which is averaged out at the higher temperatures. Fraenkel and coworkers[2] have studied the
dynamic behavior and the states of aggregation of propyllithium
with ^{13}C and ^6Li nmr and found somewhat similar behavior.

The δ carbon resonances (48.88 and 42.08 ppm) are interesting in that the upfield (right side) shoulder peaks remain
sharp at low temperatures. We would conclude that these two
peaks are due to the methylene carbons of the in-chain 1,2 units
as discussed above, whereas the broadened peaks are the δ carbons
of the live end 1,4 units. The cis δ carbon (42.08) is especially revealing in that at least three resonances are discernible,
especially at -10 and -20°C.

Although there was no evidence in oligomers with normal BD
that Li formed a localized σ bond at the γ carbon as illustrated
in III, two experiments with 90 percent ^{13}C enrichment in BD-$^{13}C_2$
were carried out. The γ carbon is "transparent", of course, in
the samples with C_1 enrichment. Figure 10 is the spectrum of
the reaction product at 5°C of 1.0 M t-Bu^6Li and 1.02 M BD-$^{13}C_2$
in benzene. Spectrum a was taken 4 minutes after warming to
reaction temperature. The peak at 137.94 ppm is the C_2 carbon
of the unreacted BD. In trace b, which was taken 23 minutes
after warming to reaction temperature, this peak has disappeared.
The trans β and the cis β peaks are clearly shown at 144.00 and
140.17 ppm, and are in a ratio of 3/1, respectively. The

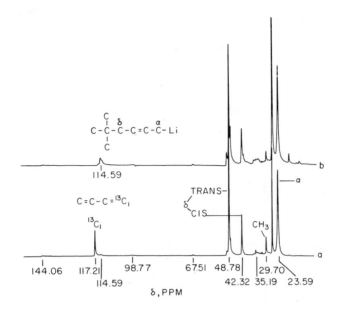

Figure 8.   ¹³C {¹H} NMR spectra of product of 1.4M 1,3-butadiene-¹³C₁ with 1.16M t-Bu⁶Li in cyclopentane: (a) before complete reaction; (b) after complete reaction.

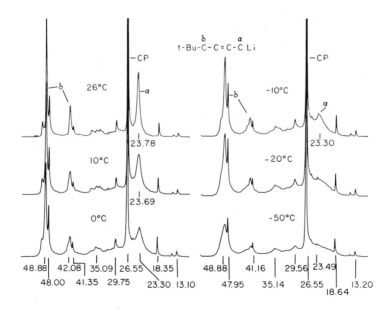

Figure 9.   Variable-temperature measurements on expanded aliphatic portion of sample in Figure 8.

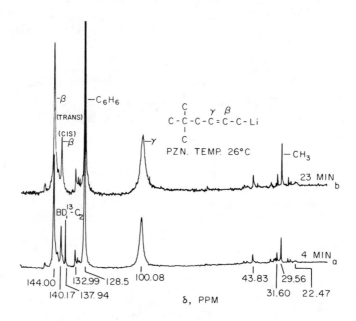

Figure 10. ¹³C {¹H} *NMR spectra of reaction product at 5°C of 1.02M buta-diene-*¹³*C₂ and 1.0M t-Bu⁶Li in benzene: (a) 4 min; (b) 23 min.*

slightly asymmetric **ϒ** carbon appears at 100.08 ppm. The product
is predominantly 1:1 adduct, but the small peak at 43.83 ppm indicates there is a small amount of in-chain 1,2 units. The large
benzene peak obscures part of the in-chain 1,4 unit peaks. The
important feature of these spectra is that there is no evidence
of any appreciable amount of localized C-Li sigma bond at the **ϒ**
carbon as illustrated in structure III. This resonance would
appear in the aliphatic region. All of the major peaks in that
region can be accounted for as in-chain 1,2 methine (43.83),
reacted t-Bu (29.56 and 31.60) or natural abundance 1,4 **α** carbon
(22.47).

Figure 11a shows the spectrum of the reaction product of
2.25 M BD-$^{13}C_2$ and 1.16 M t-Bu^{6}Li in cyclopentane. This should
have a DP of 2. It was prepared in stages by first allowing it
to react at -30°C for about 3 hours and then warming to -10°C for
about 1 hour. The reaction was then quickly completed at 26°C.
There is a fair amount of in-chain 1,2 units as evidenced by the
methine carbons in the aliphatic region. The 1,4 olefinic carbons are also clearly visible in this spectrum. Again, there is
no evidence of a C-Li **σ** bond at the **ϒ** carbon. The **ϒ** carbon of
the 1,4 unit at 97.80 has a noticeable hump on the upfield side.
Figure 11b is the spectrum of the oligomer of BD-$^{13}C_1$ from
Figures 8 and 9. The combination of these two figures should
allow the complete assignment of the early part of the BD chain.

Figure 12 shows the effect of temperature upon the **ϒ** carbon
resonance of Figure 11 over the range +50 to -50°C. These spectra
quite dramatically show that exchange is taking place and possibly
three different species are observed. Below 50°C the resonances
start to broaden, but exchange is rapid enough to average the
resonances. Below 10°C we observe a slowing down of the exchange
process to the extent that three new resonances are observed.
Below -30°C no additional changes are noted. It is curious that
one of the resonances is very broad while the other two remain
quite sharp. This broadening could be due to ^{13}C-^{6}Li coupling
and/or quadrupolar broadening from the ^{6}Li nuclei. ^{13}C nmr experiments with simultaneous ^{1}H and ^{6}Li decoupling are being done
at high field strengths to clarify this. If an interaction with
^{6}Li is shown in this resonance, this would be an extremely important finding because it would show that Li is bound in some
way at the **ϒ** carbon, possibly in different states of aggregation.
It is interesting that the **ϒ** carbon resonance is more sensitive
to the exchange processes than the **α** carbon resonance shown in
Figure 9.

Conclusions

^{13}C nmr has been used to follow the polymerization of 1,3
butadiene with either n-BuLi or t-BuLi. The disappearance of
unreacted monomer and the growth of the initial polymer peaks
are readily followed. Kinetic and thermodynamic data can be
obtained.

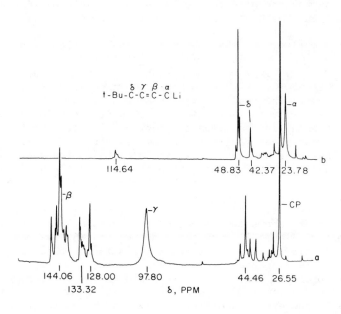

Figure 11. ¹³C {¹H} *NMR spectrum of reaction product of 2.25M butadiene-*
¹³C₂ *and 1.16M t-Bu⁶Li in cyclopentane (a); spectrum of sample of Figure 8 (b).*

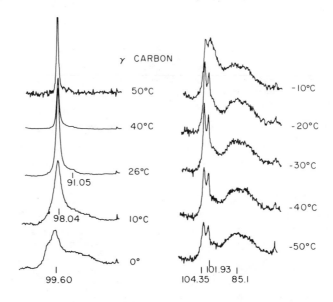

Figure 12. Variable-temperature measurement of γ-carbon resonance of sample from Figure 11a; + 50 to − 50°C.

Experiments with 90 percent ^{13}C enrichment in the 1 or 2 position of 1,3 butadiene confirm earlier work that the live end is predominantly a 1,4 unit with a trans/cis ratio of 3/1. The Li is bound to the α carbon of the 1,4 butadiene unit in what appears to be a highly localized σ bond. However, the presence of partial ionic character in the bond cannot be ruled out. There is no evidence of Li being σ bonded to the γ carbon. When a chelating diamine such as dipiperidyl ethane is added to the live cement, a drastic change takes place in the spectrum which suggests complete conversion to a delocalized ionic bonding.

In the presence of excess n-BuLi, metalation of the live end unit is thought to occur when the sample is heated. With the enriched butadiene samples it was found that metalation takes place even at room temperature and in solution.

Variable temperature measurements of the ^{13}C enriched samples show that exchange processes are operating and different states of aggregation are observed. It is not clear at this point if inter- or intra-aggregate exchange is dominating.

Acknowledgement

We are grateful to the Firestone Tire and Rubber Company for permission to publish this paper.

Abstract

[13]C nmr was used to monitor the polymerization of 1,3 butadiene with either n-BuLi or t-BuLi. Experiments with 90 percent ^{13}C enrichment in the 1 or the 2 position of 1,3 butadiene provide detailed insights into the structure and the bonding of the live ends. The live end is a 1,4 unit with a trans/cis ratio of 3/1. The Li is bound to the α carbon of the 1,4 unit in what appears to be a highly localized bond. There is no evidence of Li being σ bonded to the γ carbon. Metalation is observed even at room temperature and in solution when an excess of n-BuLi is present. Variable temperature measurements show that exchange processes are operating and different states of aggregation are observed.

Literature Cited

1. M. Morton, R. D. Sanderson and R. Sakata, Polymer Letters, 9, 61 (1971).

2. M. Morton, R. D. Sanderson and R. Sakata, Macromolecules, 6, 181 (1973).

3. W. H. Glaze, J. E. Hanciak, M. L. Moore and J. Chaudhuri, J. Organometallic Chem., 44, 39 (1972).

4. S. Bywater, D. J. Worsfold and G. Hollingsworth, Macro-molecules, 5, 389 (1972).

5. S. Bywater, P. Lachance and D. J. Worsfold, J. Phys. Chem., 79, 2148 (1975).

6. S. Bywater and D. J. Worsfold, J. Organometallic Chem., 159, 229 (1978).

7. A. L. Segre, M. Delfini, F. Conti and A. Boicelli, Polymer 16, 338 (1975).

8. H. S. Makowski, M. Lynn and A. N. Bogard, J. Macromolecular Sci., Chem. 2, 665 (1968).

9. G. Fraenkel, M. Henrichs, J. M. Hewitt, B. M Su and M. J. Geckle, J. Amer. Chem. Soc., In Press.

RECEIVED February 4, 1981.

Anionic Polymerization:
Some Commercial Applications

H. L. HSIEH and R. C. FARRAR—Phillips Petroleum Co., Bartlesville, OK 74004

K. UDIPI—Monsanto Plastic & Resins Co., Indian Orchard, MA 01151

The phenomenal growth in commercial production of polymers by anionic polymerization can be attributed to the unprecedented control the process provides over the polymer properties. This control is most extensive in organolithium initiated polymerizations and includes polymer composition, microstructure, molecular weight, molecular weight distribution, choice of functional end groups and even monomer sequence distribution in copolymers. Furthermore, a judicious choice of process conditions affords termination and transfer free polymerization which leads to very efficient methods of block polymer synthesis.

Commercial polymers from organolithium initiator systems cover a wide range, and Phillips commitment in this field on a world-wide basis is illustrated in Table I.

TABLE I

ORGANOLITHIUM BASED SOLUTION POLYMERIZATION PLANTS
(Phillips)

Plant	Date of Commission	Annual Capacity Metric Tons
Borger, USA[a]	December, 1962	48,000
Calatrava, Spain[b]	August, 1966	105,000
PACL, Australia[a]	December, 1966	23,000
Negromex, Mexico[b]	August, 1967	55,000
Petrochim, Belgium[b]	January, 1968	60,000
JEC, Japan[c]	April, 1969	32,000
ANIC, Italy[c]	September, 1972	25,000

a. Wholly owned
b. Joint venture
c. Licensee

Work on organometal initiator systems for solution polymerization increased significantly with the discovery of the titanium based initiators for olefin polymerization by Karl Ziegler.

0097–6156/81/0166–0389$05.00/0

These efforts coupled with the much earlier work on sodium and
lithium initiated polymerizations led to an appreciation of the
stereospecificity of the alkyllithium initiators for diene polymerization both industrially and academically. Polymerization of
isoprene to a high cis polyisoprene with butyllithium is well
known and the details have been well documented.[1-7] Control over
polybutadiene structure has also been demonstrated.[8] This report
attempts to survey the unique features of anionic polymerization
with an emphasis on the chemistry and its commercial applications
and is not intended as a comprehensive review.

Polydienes

     In alkyllithium initiated, solution polymerization of dienes,
some polymerization conditions affect the configurations more than
others. In general, the stereochemistry of polybutadiene and
polyisoprene respond to the same variables.[9] Thus, solvent has a
profound influence on the stereochemistry of polydienes when
initiated with alkyllithium. Polymerization of isoprene in non-
polar solvents results largely in cis-unsaturation (70-90 percent)
whereas in the case of butadiene, the polymer exhibits about equal
amounts of cis- and trans-unsaturation. Aromatic solvents such as
toluene tend to increase the 1,2 or 3,4 linkages. Polymers prepared in the presence of active polar compounds such as ethers,
tertiary amines or sulfides show increased 1,2 (or 3,4 in the case
of isoprene) and trans unsaturation.[4, 10-13] It appears that the
solvent influences the ionic character of the propagating ion
pair which in turn determines the stereochemistry.
     Figures 1 and 2 show the dependence of polymer microstructure
on the molecular weight of the polymer and therefore on the
initial initiator concentration. The polymerization temperature
also has an effect on the microstructure as can be seen in Figure
3 for polybutadiene. The overall heat activation energy leading
to 1,2 addition is greater than that leading to 1,4 addition.[9, 12]
In summary, the stereochemistry of polymerization of butadiene
and isoprene is sensitive to initiator level, polymerization temperature and solvent. The initiator structure (i.e., organic
moiety of the initiator), the monomer concentration and conversion
have essentially no effect on polymer microstructure.
     Firestone and Shell started commercial production of cis-
polyisoprene by the anionic process in the 1950's but these plants
are no longer in operation now. About the same time, Phillips
started the manufacture of polybutadienes by the anionic route
and ever since, there has been a steady growth in their use, particularly in the tire-tread as well as tire-carcass formulations.
These solution polybutadienes, generally, have low vinyl contents
but recently, Phillips has found some interesting applications
for medium vinyl polybutadienes as well.[14] Polybutadienes with
50-55 percent vinyl contents behave like emulsion polymerized
SBR in tire tread formulations and exhibit very similar tread

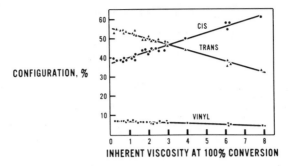

Journal of Polymer Science

Figure 1. *Microstructure of polybutadiene initiated with alkyllithium in cyclohexane at 50°C (9).*

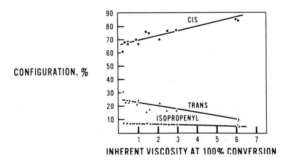

Journal of Polymer Science

Figure 2. *Microstructure of polyisoprene initiated with alkyllithium in cyclohexane at 50°C (9).*

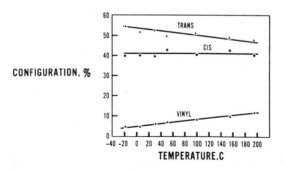

Journal of Polymer Science

Figure 3. *The effect of temperature on microstructure of polybutadiene prepared in cyclohexane (9).*

wear, wet skid resistance, low heat build-up, etc. A three-way blend of 45 percent vinyl polybutadiene with SBR and cis-polybutadiene (30/35/35) was even better in that respect than a 65/35 blend of SBR and cis-polybutadiene. The medium vinyl polybutadienes were also found to be effective as partial or complete replacement for SBR in a variety of non-tire applications.

Polydienes prepared with enough initiator to give a molecular weight of 250,000 will have a desirable combination of vulcanizate properties, but also will have high cold flow, high compounded Mooney viscosity and poor processability. Increasing the molecular weight does lower the cold flow, but greatly decreases the processability. Lowering the molecular weight improves the processability but leads to intolerable cold flow problems. The solution to these problems is usually found by altering the molecular weight distribution and/or introducing branching in the polymer molecule.

There are several ways to broaden the molecular weight distribution. An alkyllithium can be picked which will give a slow rate of initiation compared to the rate of polymerization.[6] Generally, this results in a relatively minor broadening as shown by the GPC curve in Figures 4A and 4B. An initiator of limited solubility can be used so that it slowly dissolves and initiates polymerization throughout the process. However, a more controllable and practical procedure is continuous initiator addition to a batch process.[15] By programming the rate of addition, a variety of molecular weight distributions can be obtained. A very common technique for broadening the molecular weight distribution is continuous polymerization (see Figure 5). Here, all reagents are added continuously to the reactor and the product is constantly withdrawn at the same rate. The molecular weight distribution of polymer from such a system is quite broad with both very low and very high molecular weight species. Processability of these broad molecular weight products is improved by the presence of low molecular weight material, but cold flow still can be unacceptably high.

Cold flow can be most effectively decreased and performance of the final product improved by branching the polymer molecule. A variety of techniques have been used to accomplish this. Branching comonomers such as divinylbenzene can be used.[16] The amount needed to branch the polymer adequately is generally so low that the reagent can hardly be detected in the final product. Additives which cause metalation of the polymer chains can be included in the polymerization mixture or added after polymerization to create new growth sites along the polymer chain.[17] Or, polymer can be heat soaked by increasing the residence time. While these methods result in random branching with very little control over placement or extent of branching in any particular molecule, the products can be quite useful commercially.

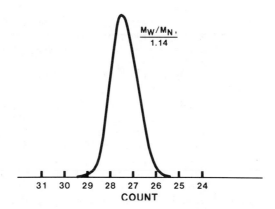

Figure 4A. Molecular weight distribution of essentially monodisperse polymer.

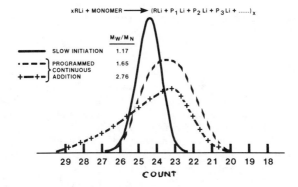

Figure 4B. Molecular weight distribution of polymer from slow initiation or programmed continuous initiator addition.

Linear polymers can be branched by employing a polyfunctional coupling agent at the end of polymerization. This technique has been used to prepare model polymers to allow rheological studies[18],[19] of specific types of branching and commercially for some rubbers produced by Phillips. By controlling the functionality of the reagent, it is possible to prepare nearly pure trichain, tetrachain, star- and comb-shaped branched polymers. (see Figure 6). Use of a polyfunctional initiator and branching comonomer in conjunction with terminal coupling will result in both branching and molecular weight broadening.

Introduction of one or two long chain branches into a polydiene molecule to form tri- or tetrachain molecules leads to profound changes in rheological behavior. It has been found[20] that in polybutadiene, at low molecular weights, the Newtonian viscosity is decreased relative to a linear polymer of same molecular weight (see Figure 7). At molecular weights exceeding 60,000 (trichain) or 100,000 (tetrachain), the Newtonian viscosity rises rapidly above the corresponding value for a linear polybutadiene. Figure 8 shows the relationship between viscosity (η) and molecular weight (M_w) at higher shear rate (s = 20 sec^{-1}). At this shear rate, the viscosity of the branched polymers is uniformly lower than that of the linear samples of identical molecular weight. In other words, the non-Newtonian behavior of branched polymers becomes rapidly more pronounced at higher molecular weight. Long chain branched polymer has higher resistance to flow at low shear rate (i.e., low cold flow) and more flow at moderate and high shear rate (i.e., better processing) than the corresponding linear polymer.

Diene-Styrene Copolymers

In the copolymerization of butadiene or isoprene and styrene, the reactivity ratios are influenced by the type of solvent used.[21],[22],[23] Typical conversion curves of a 75/25 butadiene/ styrene copolymerization in various hydrocarbon solvents are shown in Figure 9. It is noticed that the solvents change only the overall rates but the general shapes of the curves remain similar. As one analyzes samples at various conversions (Figure 10) it is observed that the styrene contents are initially lower than in the monomer charge and gradually increase until the inflection points of the conversion curves of Figure 9 are attained and thereafter increase very rapidly. Furthermore, in the analysis of the samples by oxidative degradation,[24] polystyrene segments are recovered only after the inflection points are reached and the same increase thereafter (see Figure 10). Since the final polymers are essentially homogeneous in composition and in molecular weight, it follows that the process has resulted in a tapered (or graded) block copolymer, one segment being a butadiene-styrene copolymer and the other, a polystyrene block with an overall B/S-S type structure. These results seem to be in

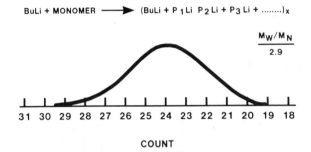

Figure 5. *Molecular weight distribution of polymer from a continuous polymerization.*

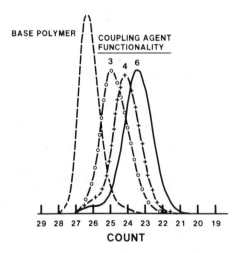

Figure 6. *Molecular weight distribution of polymer terminally coupled with polyhalide coupling agents (28).*

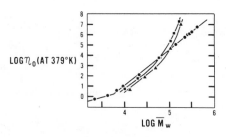

LOG η_0 (AT 379°K)

LOG $\overline{M}_w$

Figure 7. Dependence of Newtonian
viscosity on molecular weight: linear (●);
trichain (■); tetrachain (▲) (20).

Journal of Polymer Science

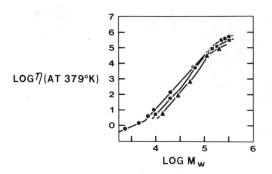

LOG η (AT 379°K)

LOG M$_w$

Journal of Polymer Science

Figure 8. Viscosity vs. molecular weight at shear rate 20 sec^{-1}: linear (●); trichain (■); tetrachain (▲) (20).

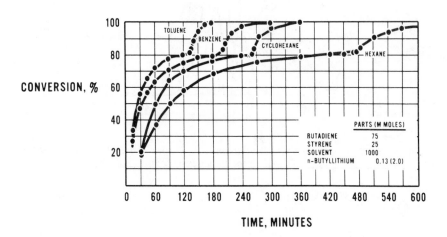

Syracuse University Press

Figure 9. Polymerization of butadiene-styrene in different solvents at 50°C (39).

contrast to what is generally known about anionic homopolymerization of styrene, i.e., styrene homopolymerizes faster than butadiene and yet, in a mixture, butadiene polymerizes first. However, if one examines the cross propagation rates in hydrocarbon solvents,[7, 25-27] the reason for this behavior is evident.

$$\sim\sim\sim B^{\ominus} + B \xrightarrow{\ k_{bb}\ } \sim\sim\ BB^{\ominus}$$

$$\sim\sim\sim B^{\ominus} + S \xrightarrow{\ k_{bs}\ } \sim\sim\ BS^{\ominus}$$

$$\sim\sim\sim S^{\ominus} + S \xrightarrow{\ k_{ss}\ } \sim\sim\ SS^{\ominus}$$

$$\sim\sim\sim S^{\ominus} + B \xrightarrow{\ k_{sb}\ } \sim\sim\ SB^{\ominus}$$

$$r_s = \frac{k_{ss}}{k_{sb}} \qquad\qquad r_b = \frac{k_{bb}}{k_{bs}}$$

Where r_b is about 50 times larger than r_s. The inversion phenomenon, as one would expect from a kinetic point of view, is independent of polymerization temperature and is shown in Figure 11.

Only a limited number of monomer pairs form block copolymers in this manner. Examples are conjugated dienes and vinyl aromatics that have similar Q-e values. The nature of the anionic initiator, i.e., the ionic character of the carbon-metal bond plays an important role in both the amount and sequence of block formation. For instance, when potassium or cesium initiators are used, styrene polymerizes first as can be seen in Figure 12.

A tapered block copolymer containing 75 percent butadiene and 25 percent styrene, marketed as Solprene 1205, was the first solution copolymer produced commercially by Phillips in 1962. This polymer has outstanding extrusion characteristics, low water absorption, low ash and good electrical properties. As a result, it is useful in such diverse applications as wire and cable coverings, show soles, floor tiles, etc.

It was also discovered at Phillips[28] that the four rate constants discussed above can be altered by the addition of small amounts of an ether or a tertiary amine resulting in reduction or elimination of the block formation. Figures 13 and 14 illustrate the effect of diethyl ether on the rate of copolymerization and on the incorporation of styrene in the copolymer. Indeed, random copolymers of butadiene and styrene or isoprene and styrene can be prepared by using alkyllithium as initiator in the presence of small amounts of an ether or a tertiary amine. Use of these polar randomizers also increases the vinyl unsaturation in the copolymer. Butadiene-styrene random copolymers can also be prepared by a very slow and continuous addition of monomers[29] or by an incremental addition of butadiene to a styrene-rich monomer mixture during polymerization. These two

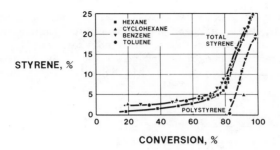

Syracuse University Press

Figure 10. Copolymerization of styrene from butadiene–styrene (75/25) at 50°C (39).

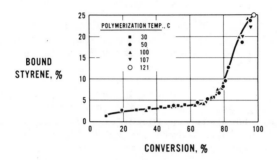

Syracuse University Press

Figure 11. Styrene incorporation from butadiene-styrene (75/25) in cyclohexane solutions at 30–121°C (39).

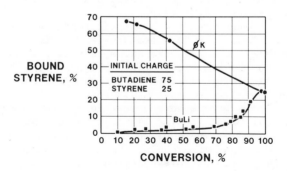

Syracuse University Press

Figure 12. Styrene incorporation (39).

methods, unlike the ether or tertiary amine complexing systems, would produce copolymers of low vinyl unsaturation ($\sim < 10\%$). Constant composition copolymers can also be produced by continuous polymerization.

There is yet another general method to prepare random copolymer. As stated earlier, when one uses potassium, rubidium or cesium initiator, styrene polymerizes first, to give a S/B-B type of tapered block polymer. But when one mixes an alkyllithium with a potassium compound such as potassium t-butoxide, quite a different system is obtained.[30-32]

Alkyllithium compounds as well as polymer-lithium associate not only with themselves but also with other alkalimetal alkyls and alkoxides. In a polymerization initiated with combinations of alkyllithiums and alkalimetal alkoxides, dynamic tautomeric equilibria between carbon-metal bonds and oxygen-metal bonds exist and lead to propagation centers having the characteristics of both metals, usually somewhere in between. This way, one can prepare copolymers of various randomness and various vinyl unsaturation. This reaction is quite general as one can also use sodium, rubidium or cesium compounds to get different effects.

The solution random copolymer generally contains about 32 percent cis-, 41 percent trans- and 27 percent vinyl-unsaturation compared to 8 per cent cis-, 74 percent trans- and 18 percent vinyl-unsaturation in emulsion copolymer of the same monomer composition. The principal effect of slightly higher vinyl unsaturation in solution copolymer is a small increase in the glass transition temperature (-58 C versus -62 C for the emulsion copolymer). However, both solution and emulsion polymerized copolymers exhibit satisfactory low temperature performance for general uses.

Inherently, solution copolymers have very narrow molecular weight distributions. Typical molecular weight distribution curves for the solution and emulsion copolymers are shown in Figure 15. The low molecular weight component in the emulsion copolymer helps it process better, but has an undesirable influence on hysteresis. The solution copolymer on the other hand is characterized by a sharp peak with only a small amount of higher molecular weight fraction. This feature leads to improved hysteresis properties and gives better abrasion resistance, but also results in more difficult processing. The solution copolymers, like polydienes, are also known to exhibit cold flow and as in the case of dienes, the solution to these two problems is found by altering the molecular weight distribution and/or branching of the polymer molecule. Branching of random copolymers can be accomplished as mentioned earlier for polybutadienes. In addition, Phillips has developed a series of multichelic initiators which are ether-free, hydrocarbon soluble, multiple-lithium initiators to produce both branching and molecular weight broadening in many of our random copolymers both low and medium vinyl. These initiators are based upon the reaction of an alkyllithium such as sec-butyllithium and a multifunctional vinyl compound such as divinylbenzene,[33] diisopropenylbenzene,[34] trivinylphos-

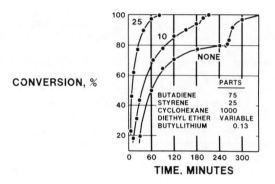

Figure 13.   Effect of amount of diethyl ether on rate of copolymerization at
50°C (16).

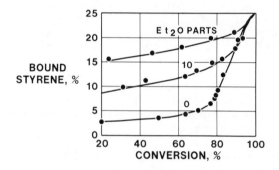

Figure 14.   Effect of amount of diethyl ether styrene incorporation at 50°C (16).

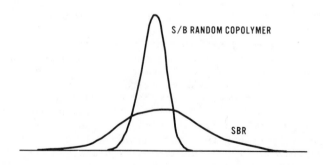

Figure 15.   Molecular weight distribution of anionically polymerized styrene-
butadiene random copolymer and emulsion polymerized SBR.

phine,[35],[36] or tetravinylsilane.[35],[36] Functionality of these initiators can be varied to provide more or less branching and/or molecular weight broadening in these polymers.

Solution random copolymers prepared by the above procedures have performed well in tire-tread formulations.[37] They require about 20 percent less accelerator (see Table II) as compared to an emulsion SBR and give higher compounded Mooney, lower heat build-up, increased resilience and better retread abrasion index.

TABLE II

EVALUATION OF ALKYLLITHIUM INITIATED BUTADIENE-STYRENE

RANDOM COPOLYMER (SOLPRENE® 1204) IN COMPOUNDED STOCK

	Solprene 1204	SBR 1500
Rubber	100	100
ISAF Black	60	60
Highly Aromatic Oil	25	25
Sulfur	2.0	2.0
Accelerator	1.1	1.3
Processing:		
Compounded Mooney, ML-4	44	36
Scorch at 280°F	17	17
Extrusion at 250°F		
grams/minute	115	105
Rating (12 is perfect)	17	17
Mill Banding	Good	Good
Laboratory Tests*		
300% Modulus, psi	1120	1080
Tensile, psi	3100	3050
Elongation, %	630	600
Heat Build-up, °F	64	69
Resilience, %	59	52
Blowout, minutes	4	4
Shore A Hardness	59	59
Retread Tire Test Performance		
Abrasion Index	114	100

*Compounds cured 30 minutes at 307°F.

So far the discussion was focused on copolymers derived from a mixture of styrene and a diene. In view of the "living" nature of organolithium polymerization, it is also possible to synthesize block polymers in which the sequence and length of the blocks are controlled by incremental (or sequential) addition of monomers.[38-41] This general method of preparing block polymers is readily adaptable to commercial production, and, indeed, a number of block copolymers are manufactured this way. Those that have received the most attention in recent years are the diene-styrene two-phase

systems with two distinctly different glass transition temperatures. These polymers have three or more blocks and are characterized by their high raw strength, complete solubility in common organic solvents and thermoplasticity. They are referred to
as "thermoplastic elastomers" because they behave as vulcanized
elastomers at room temperature and yet can be processed as thermo-
plastics at elevated temperatures. The simplest form of this
diene based elastomer is the linear triblock polymer SBS or SIS.
Where S represents polystyrene, B, polybutadiene and I, polyisoprene. The other variations are known as radial block polymers
and have a controlled number of branches of equal length and composition, $(SB)_n$- X where X is a coupler residue and n can be 3,
4, 5 or higher. Details of radial block polymers and their preparation are described elsewhere.[42] Teleblock polymers can also
be made, using multifunctional initiators in a single step polymerization of a butadiene-styrene mixture. Some of the multichelic initiators mentioned earlier yield high green strength if
the initiator preparation is properly conducted. By varying the
monomer ratios, it is also possible to vary the properties from
elastomeric to plastic. Thus, certain styrene rich block polymers
of butadiene and styrene with rather sophisticated chain architectures are finding increasing applications as clear resins of
medium impact strength.[43] Production of elastomeric radial block
polymers began in October, 1967, and at present there are several
in the market that belong to the radial block polymer family.

Elastomeric block polymers of styrene and butadiene or isoprene and their products of hydrogenation are finding increasing
use in a variety of fields.[44] Linear and radial block polymers
are used extensively in injection molded rubber goods, footwear,
pressure sensitive and hot melt adhesives and in mechanical rubber
goods such as hose, tubing, cove base, toys, drug sundries, rubber
bands, stoppers, erasers, etc.

The outstanding feature of diene-styrene block polymers in
pressure sensitive adhesive formulations is their excellent
resistance to adhesive creep. It is stated[45] that these adhesives
have established new performance highs for shear holding when
applied to selected metal and plastic surfaces, while maintaining
a high degree of tack. Radial block polymers have an advantage[46,47] over linear polymers in that at equal molecular weights,
the radial polymers exhibit lower melt and solution viscosities.
This is important because it allows the use of radial polymers of
higher molecular weights with a corresponding improvement in shear
resistance. Isoprene-styrene block polymers have an advantage
over butadiene-styrene polymers since in an oxidative atmosphere,
polyisoprene degrades by chain scission rather than gelling by
crosslinking resulting in better tack retention.

Styrene-butadiene block polymers also find applications in
blends with polystyrene and ABS plastics. When minor amounts of
rubbery SBS or $(SB)_n$- X block polymers are blended with high
molecular weight polystyrene such that the polystyrene forms the

continuum, impact resistance improves with increasing styrene
block length. For good results, the styrene block lengths should
be of the order of 20,000 as shown in a study[48] where butadiene-
styrene diblock polymers (25% styrene) were blended with polystyrene (1:3) and cured with 0.1% dicumyl peroxide. When triblock
or branched multiblock polymers are used in blends with polystyrene, it is not necessary to crosslink the rubbery domains[49,50]
although peroxide treatment does produce some additional improvement in impact and tensile strength. Another use of butadiene-
styrene block copolymers is in blends with ABS. When ABS scrap is
reprocessed, there results a loss in toughness which may be restored by the addition of small amounts ($\sim$5%) of a block polymer.[50]
Interestingly, SB diblocks are as effective as SBS or $(SB)_n$- X
multiblock polymers in this application. It must be remembered
that the butadiene-styrene diblock polymers are also used as the
source of elastomeric components in the polymerization of certain
types of ABS resins. Although very little information is available on this application, it is conceivable that the diblock polymers are sometimes preferred over polybutadienes because of the
advantages in the processes and/or properties of the final products.

Blends of butadiene-styrene block polymers with polyolefins,
particularly polypropylene are mentioned in literature[51] to improve the impact strength of the latter. Since similar improve-
ments can be realized from the use of polyolefin block polymers,
the blends have not gained much recognition. However, butadiene-
styrene radial teleblock polymers are blended into polyethylene
film, to increase the tear resistance and tensile impact.[52]

Butadiene-styrene block polymers with functional end groups
such as carboxy terminated polymers have been developed[53] by
Phillips as impact modifiers in sheet molding compounds based on
fiber reinforced unsaturated polyester and styrene. These blends
on curing, exhibit a good balance of mechanical properties,
improved impact strength, low shrinkage, no sink, excellent surface, and improved pigmentability and paintability.

A relatively new development which promises to gain in importance in the future is the modification of asphalt by butadiene-styrene block polymers.[44] The block polymers help reduce
the low temperature brittleness and impart resistance to flow at
elevated temperatures. Applications in mastics, automobile body
undercoatings and waterproofing materials such as high quality
roofing membranes are envisaged.

Anionic polymerization has also been used to make telechelic
polymers (Greek *telos*, end, and *chele*, claw), i.e., polymers with
reactive terminal groups.[54,55] We coined the term, telechelic
in 1957 and it has been accepted ever since in technical as well
as patent literature. Liquid carboxy- and hydroxy telechelic
polybutadienes initiated with difunctional organolithium initiators are commercially produced since 1962. Some of the physical
properties,[56-58] production details[59] and uses as in solid rockets

of these polymers have been described.[60] We also developed[55] solid, telechelic elastomers of which the polymer chain ends incorporate into the vulcanizate network and thus obtain superior physical properties. Mercapto-, hydroxy- and aziridinyl telechelic elastomers when cured with peroxide or sulfur accelerator recipes exhibited improved stress-strain and dynamic properties in comparison to those of the controls. The mercapto- and aziridinyl telechelic butadiene-styrene copolymers also showed outstanding properties in tread formulations. However, these polymers are not produced commercially.

Selectively hydrogenated random and block copolymers of vinyl aromatic monomers and dienes are used as viscosity index improvers in multigrade lubricating oils.[61-64] The block copolymers are of the SB type where S is most commonly styrene and B is butadiene or isoprene. Hydrogenation of the diene component imparts resistance to oxidative degradation. Crystallinity is undesirable in these applications and it is prevented by control of microstructure and a random incorporation of styrene. Since the viscosity of a heavy lubricating oil stock decreases with increasing temperature more rapidly than that of a lighter stock, such a (viscosity index improving) polymer is added to the light oil to match the viscosity of a heavier oil at high temperature. The modified oil though thickened must still be more fluid than the heavy oil at low temperatures. In other words, the desired property of a VI improver is a relative viscosity which remains constant or decreases as the temperature falls.

An outstanding property of these polymers is their shear stability. The sonic shear stability tests[65] indicate that these polymers are superior to some of the currently used polymers of ethylene-propylene or methacrylate type. The excellent stability of the hydrogenated diene-styrene polymers is attributed to their relatively low molecular weight and narrow distribution consistent with the established theory of shear degradation of polymers.[66] The most recent developments in this field are block polymer VI improvers with dispersancy properties, built into the molecule by chemical modification of the rubber block.[67,68]

In conclusion, it is evident from the above discussion that anionic polymerization has emerged from a laboratory curiosity to an important industrial process in a relatively short span of time. Currently, over a million tons of polymers are produced by the anionic route in about twenty manufacturing plants around the world. We at Phillips are quite proud of being one of the pioneers along with Firestone and Shell in harnessing this new technology to commercial applications. The fact that our polymers find such wide ranging applications from tire treads to injection molded blood filters and from lubricant additives to solid rocket binders bears ready testimony to this.

Acknowledgements

The authors wish to express their thanks to many of our colleagues at Phillips Petroleum Company who contributed so much to the success of this work.

Abstract

Anionic polymerizations initiated with alkyllithium compounds enable us to prepare homopolymers as well as copolymers from diene and vinylaromatic monomers. These polymerization systems are unique in that they have precise control over such polymer properties as composition, microstructure, molecular weight, molecular weight distribution, choice of functional end groups and even copolymer monomer sequence distribution. Attempts have been made in this paper to survey these salient features with respect to their chemistry and commercial applications.

Literature Cited

1. Stavely, F. W., et al., Ind. Eng. Chem., 48, 778 (1956).
2. Alliger, G., Willis, J. M., Smith, W. A., and Allen, J. J., Rubber World, 134, 549 (1956).
3. Hsieh, H. L., and Tobolsky, A. V., J. Polymer Science, 25, 245 (1957).
4. Hsieh, H. L., Kelley, D. J., and Tobolsky, A. V., J. Polymer Sci., 26, 240 (1957).
5. Hsieh, H. L., and Glaze, W. H., Rubber Chem. and Tech., 43, 22 (1970).
6. Bywater, S., Adv. Polymer Sci., 4, 66 (1965).
7. Forman, L. E., in J. P. Kennedy and E. Törnqvist, eds., Polymer Chemistry of Synthetic Elastomers, Wiley-InterScience, New York, Part II, 491.
8. Short, J. N., Thornton, V., and Kraus, G., International Synthetic Rubber Symposium, London, 1967.
9. Hsieh, H. L., J. Polymer Sci., A3, 153 (1965).
10. Morita, H., and Tobolsky, A. V., J. Am. Chem. Soc., 79, 5853, (1957).
11. Tobolsky, A. V., and Rogers, C. E., J. Polymer Sci., 40, 73 (1959).
12. Stearns, R. S., and Forman, L. E., ibid., 41, 381 (1959).
13. Kuntz, I., and Gerber, A., ibid., 42, 299 (1960).
14. Railsback, H. E., and Stumpe, N. A., Rubber Age, 107, 12, 27 (1975).
15. Farrar, R. C., IUPAC Meeting, Rio de Janerio, 1974.
16. Zelinski, R. P., and Hsieh, H. L., (to Phillips Petroleum Company), U. S. Patent 3,280,084.
17. Halasa, A. F., ACS Polymer Reprints, 13, No. 2, 678 (1972).
18. Zelinski, R. P., and Wofford, C. F., J. Polymer Sci., A3, 93 (1965).
19. Uraneck, C. A., and Short, J. N., Rubber Chem. and Tech., 41, 1375 (1968).

20. Kraus, G., and Gruver, J. T., J. Polymer Sci., A3, 105 (1965).
21. Phillips Petroleum Company, British Patent 895,980.
22. Zelinski, R. P., (to Phillips Petroleum Company), U. S. Patent 2,975,160.
23. Cooper, R. N., (to Phillips Petroleum Company), U. S. Patent 3,030,346.
24. Kolthoff, I. M., Lee, T. S., and Carr, C. W., J. Polymer Sci., 1, 429 (1946).
25. Morton, M., and Ells, F. R., ibid., 61,25 (1962).
26. Smid, J., and Szwarc, M., ibid., 61, 31 (1962).
27. Johnson, A. F., and Worsfold, D. J., Makromol. Chem., 85, 273 (1965).
28. Zelinski, R. P., (to Phillips Petroleum Company), U. S. Patent 2,975,160.
29. Short, J. N., (to Phillips Petroleum Company), U. S. Patent 3,094,512.
30. Hsieh, H. L., and Wofford, C. F., J. Polymer Sci., A1, 7, 461 (1969).
31. Hsieh, H. L., ibid., A1, 8, 533 (1970).
32. Wofford, C. F., (to Phillips Petroleum Company), U. S. Patent 3,294,768.
33. Farrar, R. C., (to Phillips Petroleum Company), U. S. Patent 3,652,516
34. Farrar, R. C., (to Phillips Petroleum Company), U. S. Patent 3,734,973.
35. Farrar, R. C., (to Phillips Petroleum Company), U. S. Patent 3,784,637.
36. Farrar, R. C., (to Phillips Petroleum Company), U. S. Patent 3,624,057.
37. Hanmer, R. S., and Railsback, H. E., Paper presented at the Am. Chem. Soc. Rubber Division Meeting, Detroit, May 1964.
38. Zelinski, R. P., and Childers, C. W., Rubber Chem. and Tech., 41, 161 (1968).
39. Hsieh, H. L., in "Block and Graft Copolymers", Burke, J. J., and Weiss, V., Eds., Syracuse University Press, 1973, Ch. 3, p. 51.
40. Legge, N. R., Holden, G., Davison, S., and De LaMare, H. E., in "Applied Polymer Science", Gaven, J. K., and Tess, R. W., Eds., American Chemical Society, Washington, D.C., 1975, Ch. 29, p. 396.
41. Morton, M., in "Block Polymers", Aggarwal, S. L., Ed., Plenum Press, New York, 1970, p. 1.
42. Hsieh, H. L., Rubber Chem. and Tech., 49, 5, 1305 (1976).
43. Fodor, L. M., Kitchen, A. G., and Biard, C. C., in "New Industrial Polymers", Deanin, R. D., Ed., ACS Symposium Series 4, 1974, Chap. IV.

44. Kraus, G., and Hall, D. S., Paper presented at the Midland Macromolecular Institute, Midland, Michigan, August 20-24, 1979.

45. Goken, G. L., Paper presented at Am. Chem. Soc. Rubber Division Meeting, Atlanta, Georgia, 1979.
46. Marrs, O. L., Zelinski, R. P., and Doss, R. C., J. Elastomers and Plastics, 6, 246 (1976).
47. Marrs, O. L., Naylor, F. E., and Edmonds, L. O., in "Recent Advances in Adhesion", Lee, L. H., Ed., Gordon and Breach Science Publishers, 1973, p. 213.
48. Childers, C. W., Kraus, G., Gruver, J. T., and Clark, E., in "Colloidal and Morphological Behavior of Block and Graft Polymers", Molau, G., Ed., Plenum Press, New York, 1971, pp. 193-207.
49. Durst, R. R., Griffith, R. M., Urbanic, A. J., and vanEssen, W. J., Amer. Chem. Soc., Organ. Coatings and Plastics Div. Preprints 34 (2), 320 (1974).
50. Burr, R. H., "Modification of ABS", Phillips Petroleum Company Bulletin 825-P.
51. Bull, A. L., and Holden, G., Rubber Chem. and Tech., 49, 1351 (1976).
52. Burr, R. H., "Solprene Plastomer Blends in HDPE Film", Phillips Petroleum Company Bulletin 818-P.
53. South, Jr., A., Paper presented at Soc. of Plastics Engineers Meeting, Detroit, Michigan, November 6-8, 1979.
54. Uraneck, C. A., Hsieh, H. L., and Buck, O. G., J. Polymer Sci., 46, 535 (1960).
55. Uraneck, C. A., Hsieh, H. L., and Sonnenfeld, R. J., J. Applied Polymer Sci., 13, 149 (1969).
56. Crouch, W. W., Rubber Plastics Age, 42, 276 (1961).
57. French, D. M., Casey, A. W., Collins, C. T., and Kirchner, P., Am. Chem. Soc. Meeting, 1966, Polymer Preprints, 7, 447 (1966).
58. Screaton, R. M., and Seemann, R. W., Am. Chem. Soc. Meeting, 1967, Polymer Preprints, 8, 1379 (1967).
59. Wentz, C. A., and Hopper, E. E., Ind. Eng. Chem. Res. Devel., 6, 209 (1967).
60. Sayles, D. C., Rubber World, 153, No. 2, 89 (1965).
61. Schiff, S., Johnson, M. M., and Streets, W. L., (to Phillips Petroleum Company), U. S. Patent 3,554,911.
62. St. Clair, D. J., and Evans, D. D., (to Shell Oil Company), U. S. Patent 3,772,196.
63. Anderson, W. S., U. S. Patent 3,763,044.
64. Eckert, R. J. A., and Heemskerk, J. (to Shell Oil Company), U. S. Patent 3,752,767.
65. Johnson, T. W., and O'Shaughnessy, M. T., in "The Relationship Between Engine Oil Viscosity and Engine Performance", Stewart, R. M., and Selby, T. W., Eds., Soc. Auto. Engrs. Publication SAE-SP-Y16, 1977, pp. 57-69.
66. Bueche, F., J. Applied Polymer Sci., 4, 101 (1960).
67. Trepka, W. J., (to Phillips Petroleum Company), U. S. Patent 4,145,298.
68. Kiovski, T. W., (to Shell Oil Company), U. S. Patent 4,033,888.

RECEIVED February 19, 1981.

Industrial Applications of Anionic Polymerization: Past, Present, and Future

ADEL F. HALASA[']

Central Research Laboratories, The Firestone Tire and Rubber Co., Akron, OH 44717

The first report on anionic polymerization appeared in the patent literature in 1910-1911. Matthews and Strange (1) in 1910 and later Harries (2) in 1911, described the preparation of polyisoprene using sodium and potassium as initiators. They mentioned the use of lithium as a possible initiator for this polymerization, but   there seems to be no description of the polymer made from it. Similarly, Schlenk et al (3) described the polymerization of 1,3 butadiene using sodium dust. The modern day concept of anionic polymerization was not appreciated by earlier researchers and it was not until 1920 that Staudinger (4) recognized the chain character of the addition of polymerization.

The new "non-termination" concept that laid the foundation for living polymerization was actually put forth by Ziegler and co-workers (5,6,7). They have conducted an extensive investigation on the addition of alkali metal and its organic derivatives, particularly lithium and alkyl-lithium compounds with butadiene, isoprene, piperlyene and 2,3 dimethyl-butadiene in polar media. The products described by Ziegler and his coworkers were resinous and rubber-like substances resembling polybutadiene made by sodium. The polymeric materials prepared by Ziegler and previous workers were low in molecular weight and contained a large percentage of 2,3 addition products, in the case of polyisoprene, and 1,2 addition products, in the case of polybutadiene. These rubbery materials have a high glass-transition temperature, which gave them the resinous-like behaviour described by Ziegler.

It was not until 1955 that a process was developed for the synthesis of an elastomer closely resembling natural rubber and the technological methods for the use of this elastomer in the compound were disclosed. This

' Current address: Kuwait Institute for Scientific Research, P.O. Box 24885, Safat, Kuwait.

0097–6156/81/0166–0409$05.00/0

process was developed in the Firestone research laboratories under contract to the United States Department of Defense and the Office of Synthetic Rubber, Federal Facilities Corporation (9). This work appeared in the literature in detail at the Annual Meeting of the American Chemical Society (10). The detailed descriptions published later laid the foundations for future work on anionic polymerization (11). This opened many areas of investigation on this subject in several laboratories around the United States and the world.

In his early work Szwarc (12,13) rediscovered the living nature of anionic polymerization and   demonstrated its potentials. He clearly showed that a living chain of polystyrene would accept another monomer such as isoprene and form a block co-polymer of polystyrene-polyisoprene. The nature of the living polymer was revealed by the step addition of monomers in these experiments. Szwarc demonstrated that the chain ends resume their growth whenever a monomer is added. If the monomer was different from the one previously used, a block copolymer resulted (14,15,16). This,indeed, was the most versatile technique for the preparation of block copolymers and will be discussed in more detail later.

Much of the early work on anionic polymerization was spent on the development of synthetic rubber having the same microstructure as natural rubber,which limited the investigation to the polymerization of isoprene thus neglecting the early work of Szwarc on block copolymers. After many years of research on the use of synthetic polyisoprene made by anionic initiators in the compounds, it was concluded that physical properties such as stress induced crystallization, green-strength and wind-up tack were much inferior to natural rubber. So, in the early 1960's, the program on synthetic polyisoprene and various modifications to improve the above-mentioned properties came to a standstill. Publications by Brock and Hackathorn (17) have shed some light on the lack of these fundamental properties found in polyisoprene made by lithium. In their work, they concluded that even though polyisoprene made by lithium has 94% Cis 1,4 microstructure, the sequence distribution of head-tail arrangement of the isoprenyl units are low and lack uniformity in packing of the unit cell.

Meanwhile, development of coordination catalyst was proceeding full scale. The polyisoprene prepared using this coordination catalyst ($TiCl_4$: AlR_3) proved to be more suitable in physical properties than the one made by lithium metal or organolithium compounds in hydrocarbon media. The Ziegler polyisoprene, as it was called, has greater stereoregularity and stress-induced crystallization properties than polyisoprene made by the alkyl lithium catalyst. However, the Ziegler catalyst failed to copolymerize vinyl aromatic and conjugated diene to yield block copolymers. This deficiency in the Ziegler catalyst to produce block copolymers and the abilities of anionic initiators to produce it kept the interest in anionic initiators active in many industrial laboratories. This interest in anionic research in these laboratories paid off handsomely in the areas of block and random copolymers. In this review major emphasis will be focused on the major products from both homo and block copolymers currently being manufactured by anionic technique and future trends in this area.

HOMOPOLYMERIZATION

As of this date, there is no lithium or alkyl-lithium catalyzed polyisoprene manufactured by the leading synthetic rubber producers in the industrial nations. However, there are several rubber producers who manufacture alkyl-lithium catalyzed synthetic polybutadiene and commercialize it under trade names like "Diene Rubber"(Firestone) "Soleprene"(Phillips Petroleum), "Tufdene"(Ashai KASA Japan). In the early stage of development of alkyl-lithium catalyzed poly-butadiene it was felt that a narrow molecular distribution was needed to give it the excellent wear properties of polybutadiene. However, it was found later that its narrow molecular distribution, coupled with the purity of the rubber, made it the choice rubber to be used in the reinforcement of plastics, such as high impact polystyrene. Till the present time, polybutadiene made by alkyl-lithium catalyst is,for many chemical and technological reasons, still the undisputed rubber in the reinforced plastics applications industries.

The unique feature about anionic polymerization of diene to produce homopolymer was that the microstructure of the homopolymer could be altered and changed at will to produce unique physical and chemical properties. These microstructural changes can be introduced before, after or during the polymerization. For example, chelating diamines, such as tetramethyl ethylene and diamine (TMEDA) (18), with the alkyl-lithium catalyst have been used to produce polymer with 80% 1,2 addition products, while the use of dipiperidine ethane (DPE),with same catalyst has produced polybutadiene with 100% 1,2 addition product.

This new development in the microstructural architecture of polybutadiene has opened the door for the preparation of various block copolymers made from the same monomer. For example, one can use this concept to prepare various polybutadiene rubbers in which the chain segment contains various glass transition temperatures, depending on its microstructural arrangements. Similarly, manipulating the polymerization temperature using the same modifier and

catalyst, one can alter the microstructure of the same polybutadiene chain. For example, raising the polymerization
temperature of n-BuLi TMEDA catalyzed 1,3 - butadiene from
50°C to 100°C causes a decrease in the 1,2 content of the
polybutadiene from 80% to 40%, which, when translated into
physical properties,means that the glass transition temperature was changed from -20°C to -55°C. A more dramatic
change in the 1,2 content is illustrated when n-BuLi DPE
is used as catalyst for 1,3-butadiene polymerization. Ad-
justing the polymerization temperature from 5° to 50° causes
a drop in 1,2 content from 100% to 70%. This means that
the glass transition temperature changes from $+5^{\circ}$C to -30°C,
a rather dramatic change in the cold temperature properties
of this rubber (19).

        This versatility gives the synthetic polymer chemist
a tool to tailor the chain of the polybutadiene to fit
the desired application. It can be used to manipulate
the low temperature properties important in rubber
goods and tire applications. Moreover, the chemical
reaction that chemists want to perform on these rubbers
can be altered, depending on the reactivity of the 1,2
polybutadiene double bonds fraction versus that of 1,4
polybutadiene. The work of Halasa and co-workers (21)
illustrates this point. These workers hydrogenated medium
vinyl polybutadiene (1,2 content varied between 60-60%
1,2 adduct) to produce thermoplastic elastomers based
on polybutadiene, but taking advantage of the versatility
of anionic polymerization by changing the microstructure at
the end of the polymerization. The data in Table I illustrate these findings.

TABLE I

Block Copolymer of Poly(Bd)
of 1,4 Poly(Bd) - 1,2

$\overline{Mn}$ B-1,4	$\overline{Mn}$ B-1,2	Tg$^{\circ}$C	Tm$^{\circ}$C	Tensile(psi)	Elong %
94,000	294,000	-52	None.	120	100
162,000	306,000	-55	None.	140	120
58,000	396,000	-40	None.	250	130

        The physical properties of this block copolymer
are improved by hydrogenation using cobalt-copralactam
reduced by tri-isobutyl alumimium. This homogeneous
catalyst reduces the 1,2 double bond at a rate four times
faster than it reduces the 1,4 double bond. The saturated
1,4 double bond, yields polyethylene block and while
saturating the 1,2 double bond, yields an elastomer with

a low glass transition temperature of -25°C. The results of saturating block copolymer of 1,4 and 1,2 units mentioned in Table I are shown in Table II.

It can be seen in Table II that a substantial improvement in tensile strength has occurred. This improvement has been attributed to some unique features in the polymer chain introduced via hydrogenation. One outstanding feature of this block copolymer is that the hard segment is made of crystalline polyethylene. The molecular weight of this hard segment is much greater than that of the styrene 1,3-butadiene-styrene in which the hard segment molecular weight is only 15,000-20,000. The physical properties shown in Table II illustrate that these polymers are diblock, not triblock, and that the physical properties are due to the crystalizable polyethylene chain segments.

It is interesting to note that soft segments in the diblock copolymer shown above is the hydrogenated 1,2 polybutadiene. The reason for the rubber characteristic of polybutadiene 1,2 is the presence of the chiral carbon carrying the vinyl units. This assymmetric carbon is not altered by hydrogenation since the vinyl group is on the side chain of polymers. Therefore, the final product is heterotactic rubbery poly-1-butene, which has a glass transition temperature of -25°C.

One would expect that addition filler of similar structure would reinforce the hard segments. It was found that the addition of thermoplastic polyolefins of high melting point did improve the physical properties. The data in Table III illustrate that the addition of polypropylene showed an improvement in physical properties similar to the commercially available SBS triblock copolymer. This TPE, shown in Table III, shows outstanding physical properties. This unique block copolymer can only be made using the anionic polymerization technique.

TABLE II

Hydrogenated B-1,4 - B-1,2

$\overline{Mn}$ B-1,4	$\overline{Mn}$ B-1,2	%H$_2$*	Tg$^{\circ}$C	Tm$^{\circ}$C	Tensile(psi)	Elong %
94,000	94,000	100	-25	89	625	950
162,000	306,000	100	-29	101	1106	770
58,000	396,000	100	-27	100	403	573

* Hydrogenation can be varied (60-100%)

TABLE III

Blend of 30% IPP* with the hydrogenated
diblock copolymers of
$B-1,4H_2 - B-1,2H_2$

$\overline{Mn}$ B-1,4	$\overline{Mn}$ B-1,2	% H_2	$Tg^\circ C$	$Tm^\circ C$	Tensile Psig.	Elong. %
94,000	294,000	100	-25	89	4250	760
162,000	306,000	100	-29	101	5320	450
58,000	396,000	100	-27	100	2240	460
Kraton 1101	-	-	-60	-	3300	1200

* A broad transition was noted. It was assigned to the
glass transition of the isotactic polypropylene.

A block copolymer of hydrogenated $B-1,4H_2-$
$B-1,2H_2$ gave the best physical properties irrespective
of the soft segments' molecular weight. These phenomena
appear to be inconsistent with the usual thermoplastic
elastomer, especially the styrene-butadiene type. The
comparison between the SBS and these block copolymers
may not be accurate since the strength of these diblock
elastomers is due to the crystallizable polyethylene
chain,while the SBS use is due to the hard polystyrene
chain.

The fundamental concept one has to address oneself
to is the enhanced physical properties present in this
diblock and the nature of interaction that takes place in
these polymers. Perhaps examination of the microstructure
of the parent block polymer B-1,4-B1,2 before hydrogenation
and the resulting block polymer after hydrogenation may
be of interest for the understanding of these forces. The
B-1,4 block contains 13-15% 1,2: 32-30% Cis 1,4;55-50% trans
1,4. Hydrogenation of the B-1,4 block gives polyethylene,
and hydrogenation of the B-1,2 block give amorphous poly-
butane-1. The diblock copolymer has Tm of polyethylene
(+103 to 115°C; pure polyethylene is 118-125°C). The
hydrogenation of the B-1,2 units usually gives a rubber
segment of polybutane-1. However, the random distribution
of the amorphous polybutane-1 in first block, the polyethylene formed from the B-1,4 block chains, are postulated
to form a microphase that allows the rubbery second block
polybutane-1, resulting from $B-1,2H_2$, to be trapped in the
chain folding of the crystallizable polyethylene. This
type of chain folding appears to form hard-soft repeating
units similar to triblock copolymer. If this explanation
is plausible, then the addition of pure polyethylene would
reinforce these types of block copolymers. This indeed
was found to be the case. The addition of 30% polyethylene

to the hydrogenated polymers (Table II) increased the tensile strength to maintain an acceptable ultimate elongation, as shown in Table IV.

TABLE IV

Block copolymers of Hydrogenated B-1,4H$_2$-B-1, 2H$_2$ filled with high density polyethylene

B-1,4	B-1,4-B-1,2	%PE	Ult.Tensile(psi)	Elong. %
94,000	294,000	30	1120	407
162,000	306,000	30	2130	540
58,000	396,000	30	1880	360

A thermoplastic elastomer similar to the above structures was made by utilizing conjugated 1,3 di-olefins that can be polymerized anionically. The work of Halasa and co-workers(25) illustrate the point. These workers polymerized 1,3-butadiene and isoprene to produce a diblock copolymer of poly(butadiene)-poly(isoprene) (see Table V).

These copolymers were made by anionically poly-merizing 1,3-butadiene with n-Buli followed by the addition of isoprene to the live cement. The molecular weight was varied in the 1,4 poly(bd) block to produce the maximum physical properties. The content of the Bd/isoprene in the copolymer was varied 30/70. Similarly, (Table VI) the molecular weight of the diblock was kept constant at 60/40 Bd isoprene ratio, while the molecular weight of the individual block was varied. In Tables V and VI the physical properties of the diblock of the conjugated diene rubber showed elastomeric properties typical of that of the uncrossed elastomer.

TABLE V

Block Copolymer of Polybutadiene-Isoprene

Bd/Isoprene	No. Blocks.	Total M̄n.	Poly(Bd) of (1Block)	Poly(Isoprene) of (1Block)
30/70	2	100,000	30,000	70,000
	3	50,000	7,500	25,000
	3	100,000	15,000	70,000
	3	150,000	22,000	105,000

TABLE VI

Diblock Copolymer of Poly(butadiene)-Poly(isoprene).

Bd/Isoprene	No. Blocks.	Total Mn.	Poly(Bd)of (1 Block)	Poly(isoprene) of (1 Block)
60/40	2	100,000	60,000	40,000
	3	100,000	30,000	40,000
	5	100,000	20,000	20,000
	7	100,000	15,000	13,300
	11	100,000	10,000	8,000

The block copolymers shown in both Table V and VI were hydrogenated. The B-14 block produced polyethylene and the polyisoprene block produced ethylene propylene alternating copolymer. The physical properties of this co-polymer, composed of crystalline polyethylene block and a soft elastomeric segment made of an EPR block, is tabulated in Table VII. The data in this table illustrate the fact that a diblock of hydrogenated polybutadiene-polyisoprene gave excellent physical properties. This is a further illustration of the new concept of soft chain inter-penetrating the crystallizable polyethylene chain via chain folding.

TABLE VII

Hydrogenated Diblock of
Poly(Butadiene)-Poly(Isoprene)
to give Polyethylene-propylene rubber.

PE/PER	MODULUS 100%	300%	Tensile	Elongation at Break
30/70	210	270	280	390
	283	489	600	470
	250	350	415	375
	266	373	495	520
60/40	550	725	1650	680
	550	750	2000	760
	650	825	2025	750

The hydrogenated copolymers of poly(butadiene-poly(isoprene) have been milled with 30% isotactic poly-propylene and high density polyethylene. These thermo-plastic elastomers (TPR's) showed excellent physical properties (illustrated in Table VIII).

TABLE VIII

Hydrogenated diblock Poly(butadiene)-Poly(isoprene)
filled with isotactic polypropylene
and high density polyethylene.

Mn poly(Bd)	Mn poly(I)	Total Mn Diblock.	%Polyolefin	Tensile (psi)	% Elong.
60,000	40,000	100,000	30 PE	2250	620
30,000	70,000	100,000	30 PE	2560	400
20,000	80,000	100,000	30 PP	2210	450
10,000	90,000	100,000	30 PP	2140	600

The data from this table illustrate the semi-compatibility of the phase between isotactic polypropylene and the high density polyethylene with block copolymer without gross interference in the domain structure or the crystalline phases that exist in these TPR's.

If one can take advantage of the microstructure changes caused by anionic polymerization produced by variation in the quantities of polar modifiers and the temperature at different stages of the reaction, one can produce the proper polymers with the desired microstructure. After this tailoring takes place in microstructural changes, the polymer is subjected to hydrogenation reaction; the resulting thermoplastic polymer can be compounded to give the proper physical properties. This is usually done by adding a plastic of high melting point, such as isotactic polypropylene, to form phase segregated domain of hard crystalline segments and rubbery soft segments, which leads to reinforced thermoplastic elastomers (TPD's) as illustrated below (Table IX).

TABLE IX

Hydrogenated Poly(bd)-Poly(Bd 1,2)

Mn B-1,4	Mn B-1,2	Mn B-1,4 – B-1,2	%H$_2$	Tg	Tensile	% Elong.
94,000	200,000	294,000	100	-25	625	950
162,000	144,000	306,000	100	-30	1106	370
58,000	338,000	396,000	100	-27	403	473

The interesting phenomenon that seems to be evident in the above table is that the molecular weight of the hard-crystalline segments appears to be the controlling factor in the physical properties. For example, the polyethylene segment, resulting from the hydrogenation of B-1,4/Mn 58,000, had a lower tensile strength than the sample with the highest poly-

ethylene, resulting from the hydrogenation of the same polymer
with a higher molecular weight. Similarly, many variations
can be produced on these copolymers to improve their physical
properties.

Commercial methods for preparing block copolymers
of styrene-butadiene-styrene utilize cyclohexane or toluene
as a solvent since the polystyryl lithium is insoluble in
straight chain alphatic solvents. Usually the 1,3-butadiene
is added to the polystyryl lithium to produce the diblock
styrene-butadienyl lithium. At this point in the reaction
two processes are employed.

One process utilizes a difunctional joining agent
such as methylene dichloride which gives an SBBS and hence
the total $\overline{Mn}$ of the block copolymer and the $\overline{Mn}$ of the middle
block are constant, since half of the required 1,3 butadiene
and all the styrene were added to make the diblock.

The second process utilizes the two stage method in
which half of the styrene added at the beginning of the reaction followed by all the 1,3-butadiene and then the remaining
half of styrene is added. All these polymeriz-
ation processes are done in cyclohexane since homopolystyrene
with or without lithium terminated is insoluble in all straight
chain or branched hydrocarbon solvents such as heptane, hexane
petroleum ethers or the branched derivatives.

The recent development of using hexane as a solvent
for the preparation of SBS copolymers has been attempted even
though the polystyryl lithium is insoluble in this media, (24).
Polystyryl lithium was dispersed in hexane using 1% of SBS
rubber as a dispersing agent. The block copolymer tried as
a dispersing agent retained the colloidal properties of poly-
styrenyl lithium till the addition of the butadiene monomer.
In this process the styrene-butadienyl lithium becomes soluble
in hexane. Surprisingly, the triblock SBS made by this process
had a rather narrow molecular distribution. The physical properties of SBS made by the dispersion method described above
had properties similar to SBS made in cyclohexane on all homo-
genous processes.

The hydrogenation of the centre block of SBS copolymer
produced oxidation stable thermoplastic elastomer. This product was commercialized by the Shell Development Company under
the trade name of Kraton G. The field of thermoplastic elastomers based on styrene, 1-3-butadiene or isoprene has expanded
so much in the last 10 years that the synthetic rubber chemist
produced more of these polymers than the market could handle.
However, the anionically prepared thermoplastic system is still
the leader in this field, since it produced the best TPR's
with the best physical properties. These TPR's can accommodate
more filler, which reduces the cost. For example, the SBS
Kraton type copolymer varies the monomer of the middle block
to produce polyisoprene at various combinations, then ,followed

by hydrogenation to produce EPR material, have lead to styrene-ethylene propylene-styrene type thermoplastic elastomers with unique physical properties and it is less susceptible to oxidation. These types of thermoplastic elastomers can be filled with polyolefins, which reduces the cost and enhances the physical properties.

Similarly, using their own variation, Halasa and co-workers made polybutadiene 100% 1,2 microstructure. The elastomer was further converted to amorphous poly(butene-1) rubber by hydrogenators. This type of elastomer is difficult to produce by simple polymerization of butene-1 or butene-2 by the Ziegler type of catalyst. However, utilizing anionic polymerization of various conjugated 1,3-olefins and then saturating the double bond by hydrogenation produces another useful and novel elastomer. As a further illustration of this point, the polymerizations of piperylene using organo-lithium compounds and polar modifier DPE(dipiperidine ethane) gave polypiperylene in which the 1-methyl is on the terminal vinyl double bond. Hydrogenation of this elastomer gave the poly (1-pentane) with low glass transition temperature. This type of transformation has opened the field for the synthetic polymer chemist to use his microstructural architecture in the laboratory and to prepare a large variety of poly-diene with 1,4 or 1,2 microstructure with phenyl or alkyl substituent and to produce, with hydrogenation, a new class of elastomers, i.e. plastic or thermoplastic elastomers.

Anionic polymerization of conjugated dienes and olefins retains its lithium on the chain ends as being active moities and capable of propagating additional monomer. This distinguishing feature has an advantage over other methods of polymerization such as radical, cationic and Ziegler polymerization. Many attempts have been made to prepare block copolymers by the above methods, but they were not successful in preparing the clear characterized block copolymer produced by anionic technique.

BLOCK COPOLYMERS OF HIGH VINYL SOFT SEGMENTS

The microstructure of the soft block can be varied to produce SBS in which the middle block is from 100% 1,2 to 45% or a mixture of varied composition. The work of Halasa and co-workers (25) is presented in Table X to illustrate this point.

Furthermore, the microstructure of the middle segment can be altered by sequential addition to produce medium single polybutadiene (see Table XI).

TABLE X

High Vinyl Poly(Bd)(SBS) Block Copolymer

Composition IR			Copolymer Composition			Phys. Prop.	
						Tensile	%
Styrene % ($\overline{Mn}$)	1,2 %	Homo Sty.	SB ($\overline{Mn}$)	% SBS ($\overline{Mn}$)		(Psi)	Elong
31.4 (9,000)	99.6	3.0	(64,000)	96.5		3157	397
				(86,000)			
26.7(20,000)	98.7	6.7	(174,000)	88.9		3100	530
				(227,000)			
28.5(15,000)	98.5	9.0	(183,000)	91.0		2900	450
				(222,500)			

TABLE XI

Medium Vinyl Poly(Bd)   (SBS)   Block Copolymer

Composition IR			Copolymer Composition			Phys. Prop.	
						Tensile	%
Styrene % ($\overline{Mn}$)	1,2	% Homo.Sty.	SB ($\overline{Mn}$)	%SBS($\overline{Mn}$)		(Psi)	Elong
32.9 (15,000)	52.7	2.7	81,000	97.3		2111	900
				(97,000)			
35.6 (15,000)	54.5	2.9	101,000	97.2		4000	735
				(122,000)			
27.7 (23,000)	56.8	2.7	155,000	97.4		3267	890
				(183,000)			

When the above polymers were compounded, without adding reinforcing filler but using extendar aromatic oil, the physical properties did drop, illustrating the close compatibility of aromatic oil and the high and medium vinyl solubility   paratometer (Table XII).

TABLE XII

High Vinyl Poly(Bd) SBS Block Copolymer Compound with 30% Aromatic Oil.

Composition (IR)			Copolymer Composition			Oil Extended Poly	
						Tensile	%
Styrene % ($\overline{Mn}$)	1,2	% Homo Sty.	% SBS ($\overline{Mn}$)	% SBS ($\overline{Mn}$)		(Psi)	Elong
31.4 (15,000)	99.6	3.0	(64,000)	96.5		589	580
				(86,000)			
26.7 (20,000)	98.7	6.7	(179,000)	88.9		923	500
				(227,000)			
28.5 (15,000)	98.5	9.0	(183,000)	91.0		850	683
				(222,000)			

Further illustrations can be seen when the diblock copolymers of high and medium vinyl are made in hexane solvent. This is illustrated by the fact that 1,2-polybutadiene is made first. Then the styrene monomer is added to form $B_{12}S$ diblock. The SB_{12} diblock in which the styrene block is made first in cyclohexane, followed by the dipiperidise ethane modifier to the styreyl lithium then the butadiene to form this block coplymer. The reader may see the difference as illustrated in the following tables (XIII - XIV) as the physical properties vary from $B_{12}S$ to SB_{12}.

TABLE XIII

Block Copolymer Styrene-Poly(Bd) High Vinyl

Composition (IR)		Polymer Composition		Physical Prop.	
%S (M̄n)	% 1,2	% Homo.Sty. (M̄n)	% SB (M̄n)	Tensile (Psi)	% Elong
24.6	99.7	11.0 (7,500)	89.0 (72,500)	1400	583
27.5	100	5.0 (9,700)	95.0 (40,300)	583	600
30.7	93.8	1.2 (13,700)	98.8 (42,900)	725	400

This diblock of high vinyl styrene copolymer was made in hexane but the vinyl poly(Bd) was made initially, followed by the styrene. The table below shows that these polymers were superior in the physical properties to the above diblock SB_{12}.

TABLE XIV

High Vinyl Poly(Bd) Styrene B_{12}-S

Composition					Physical Prop.	
%S (M̄n)	% 1,2	% B-S (M̄n)	Tg °C	Tm °C	Tensile (Psi)	% Elong
36.5(29,000)	98.5	(112,000)	-3.5	115	704	160
28.1 (7,000)	98.5	(86,000)	-4	115	734	207
34.1(15,000)	99.2	(184,000)	-2.5	113	1220	437
26.6(19,000)	98.9	(184,000)	-2	114	1232	633

The $B_{12}S$ block copolymers were subjected to the hydrogenation reaction in which only the 1,2 poly(Bd) were hydrogenated to improve its physical properties. These are illustrated in Table XV.

TABLE XV

Hydrogenated Diblock Copolymer of Styrene-Poly(Bd) (1,2)

Composition IR %S ($\overline{Mn}$)	% 1,2	% H_2	Tg °C		B-1,2H-S($\overline{Mn}$)	Tensile (Psi)	% Elong
36.5(29,000)	98.5	91.0	-23	113	(112,000)	706	340
28.1 (7,000)	98.5	95	-24	104	(86,000)	400	480
34.1(15,000)	98.5	96	-19	102	(184,000)	1025	NB
26.6(19,000)	98.9	95	-20	104	(184,000)	3686	650

RANDOM COPOLYMERS OF STYRENE/BUTADIENE

The preparation of random copolymer styrene/butad-
iene has been achieved by using organo-lithium compound as
initiators. The ability to use living anionic polymerization
gave us the versatility to vary the styrene content at will.
For example, polymers containing as low as 18% styrene and
as high as 33% styrene have been made. These copolymers are
made in a random fashion by a process patented by Firestone
without increasing the vinyl structure of the polybutadiene
content under 15%. Such a process has been able to produce
what is known in the industry as solution SBR. The physical
properties of solution SBR were equivalent to, and in some
cases even better than, emulsion SBR. Living anionic polymerization can be used to prepare such a polymer, because
one. can produce: (a) variation in the microstructure; (b) the
sequence distribution of styrene; (c) molecular weight distribution,and (d) branching.

The last property is related to the processing of
the rubber in the tire making equipment. By using organolithium compound in this case, it was possible to maintain
a vinyl content not greater than 18%, but to produce a polybutadiene styrene copolymer that has random block styrene
and without the use of polar modifiers, which normally will
increase the 1,2 content. This copolymer, when compounded in
the tread recipe, as shown in the Table XVI, gave properties
that are actually equivalent to that of emulsion SBR and in
some cases even better. This is particularly true in the
properties of the Young modulus index, which showed between
-38 to -54°C; the Stanley London Skid Resistant, in which the
control is 100, shows that 110-115 was obtained.

The breaking index was also interesting. Solution
SBRS containing 18 to 33% styrene showed a wearing index of
between 100 and 105% increase in wear, as compared to an
emulsion SBR of 100. This suggests that the amount of styrene
increased accordingly, because living polymerization was used
and because the styrene can be added at different increments
during polymerization. The fact that one. can make this type
of polymer free of gel at high purity gives the solution SBR

TABLE XVI

Tread Stock Recipe, Laboratory Evaluation

Polymer	Emulsion SBR	Solution Butadiene/Styrene Copolymer				Polar* modif.
		18% Styr.	23% Styr.	28% Styr.	32% Styr.	
Stress-strain properties, Cure 23 min. at 300°F 300% Modulus (lbs./sq. in.)	1000	1050	1000	1100	1125	850
Tensile strength (lbs./sq.in)	2750	2800	2750	2800	2875	2500
Ultimate elongation (%)	600	590	650	620	630	660
Shore A hardness	54	56	58	59	57	54
Processing rating			All good			
Running Temp. (°F)	290	280	294	284	280	280
YOUNG's modulus index (°C)	-40	-54	-44	-37	-38	-43
STANLEY-LONDON [15] Wet skid resistance (C.F. 0.39)	110.2	100	102.4	107.7	105.1	112.2

* Butadiene/styrene copolymer containing 18% styrene prepared in hexane solution with n-BuLi modified with diethylene glycol dimethyl ether (diglyme).

an advantage over the emulsion SBR. Since emulsion SBR contains a significant amount of soap catalyst residue, and in some cases a high concentration of gel, it was not accepted in the plastic application, and indeed, an alkylithium system can be used to prepare a more narrow molecular weight linear polymer of good colour, free of gel and high purity. This type of polymer has been used, not only in the tire application field, but also in the plastic industry.

The use of coupling agents and the fact that the polymer end has a lithium on it are rather interesting. This lithium can be reacted with different additives to functionally terminate the polymer chain. For example, ethylene oxide and carbon dioxide, as well as various electron activated groups, such as groups with various functionality, have been added to the polymer chain to terminate it. Thus, the rheology of the polymer can be adjusted to fit varying processing requirements. The broad molecular weight distribution, for example, has been quite useful in producing unsaturated rubbers that can be processed on conventional tire equipment.

For example, a branched polymer of polybutadiene lithium or styrene butadiene lithium has been made by coupling with silicon tetrachloride, dichloromethane, carbon tetrachloride and chloroform. The work of Halasa and coworkers (26 and 27) has shown that even secondary butylchloride can be reacted with the live lithium chain. This has produced what is known as a pre-radical polymerization in non-aqueous solvents. This work has shown that the allylic radical at the end of the chain can only be coupled to produce a very broad, molecular weight with a bi-model distribution of molecular weight of what was initially a narrow molecular weight polymer. This area needs more investigation to show the utility of using live lithium with monohalogenated compounds, such as secondary butyl-chloride, to produce polymers that can have a radical at the end of the polymer chain. It gives the utility to add the monomers that would normally be attacked by the organolithium compound species, but will only copolymerize with free radical initiators. Such a versatility of living polymerization is unique and cannot be found in other systems.

SUMMARY

In summary, one can say that the versatility of living polymerization makes it a very useful and unique technique. Synthetic polymer chemists can utilize the lithium at the end of the chain to add polymer-like conjugated monomers in order to tailor make the polymer chain, or they can use chelating diamines to change the microstructure of the polymer chain. They also can run hydrogenation to produce thermoplastic elastomers. They can transform anionically made polymer by adding halogens to produce vinyl chloride copolymers or by reacting it with carbene reaction to saturate the polymer chain to give dihalocarbene. They can cyclise the vinyl portion with cationic catalyst to give a copolymer of linear chains with cyclic rings. They can use it to prepare ladder polymers. Obviously this review cannot adequately accommodate all the areas for which anionic polymerization can be used, either in preparation or leading to the preparation of unique polymers. It needs much more extensive studies are needed. This author believes it is time for people who are working in anionic polymerization to sit down and write a comprehensive review. Previous reviews in this area have been very sketchy, concentrating on one small area that neglects to take into consideration all the work done on anionic polymerization by competent researchers. This tends to expand certain authors' ideas without giving proper consideration to others who have published extensively. A review which looks at the whole area critically, comprehensively, and informatively would enable young synthetic polymer chemists to benefit from the experience of others. I believe it is the duty of young scientists to do this, one who does not yet have rigid ideas and one who has not taken any school of thought as his guiding light. He should be one who can look at the work of Szwarc, Morton, Bywater , Shue, Halasa, Fetters and various others who have worked and published in this field of anionic polymerization and write a comprehensive review in an unbiased fashion.

Acknowledgment

The author acknowledges the help and suggestions of various colleagues at Firestone - J. Hall, D. Cadson, J. Spewiak, Don Schultz, Tai Cheng, Pete Bathea, J. Ozomiak, V. Mochel, S. Futamura, G. Hamad, L. Vescelius and G.G. Bohn. Many thanks to the Firestone management for permission to publish this work.

Literature Cited

1- F.E. Matthews and E.H. Strange. British Pat. 24790 (1910)

2- C. Harries, Ann., 383 184 (1911); U.S. Pat. 1,058,056 (To Bayer & Co. 1913).

3- W. Schlenk, J. Appenrodt, A. Michael and A.Thal, Chem.Ber., 47, 473 (1914).

4- H. Standinger, Chem.Ber., 53 1073 (1920)

5- K. Zeigler and K. Bahr, Chem. Ber., 61, 253 (1928)

6- K. Zeigler, H. Colonius and O. Schater, Ann. Chem. 473 36 (1929).

7- K. Zeigler and O. Schater, Ann. Chem, 479 150 (1930).

8- K. Zeigler, L. Jakob, H. Wolltham and A. Wenz, Ann. Chem. 511, 64(1934).

9- Private Communication to the Office of Synthetic Rubber
    of the Federal Facilities Corporation; Chem. Eng.News
    33 (Aug. 5, 1955).

10- F.W. Stavely, F.C. Forster, J.L. Binder and L.E. Forman
    Ind. Eng.Chem., 418 778 (1956), paper presented to the
    Div. Rubber Chem., Am. Chem. Soc. Philadelphia, Pa. Nov. 55.

11- F.C. Forster and J.L. Binder, Advances in Chem. Series No.19
    Am. Chem. Soc. P-26 (1957).

12- M. Szwarc, Nature, 178 1168 (1956).

13- M. Szwarc, M. Levy and R. Milkovich, J. Am. Chem. Soc.,
    78, 2656 (1956).

14- M. Szwarc, Advance Chem. Phys., 2 147 (1959)

15- M. Szwarc, Makromol. Chem., 35 132 (1960).

16- M. Szwarc, Proc. Roy. Soc(London) Ser.A. 279 (1964).

17- M.J. Brock and J. Hackathron, Rubber Chem. Techn. 40,590(1967)

18- A.W. Langer, Jr. Trans. N.Y. Acad.Sci. 27 741 (1965).

19- T.A. Antkowiak, A.E. Oberster, A.F. Halasa, J. Polymer
    Sci., A-1 10 1319 (1972)

20- A.F. Halasa F.G. Lohr & J.E. Hall, J. Polymer Letters(In press)

21- A.F. Halasa, D.W. Carlson Private communication.

22- A.F. Halasa, R. Cohn,Polymer Reprint, Div. of Polymer Chem.
    ACS 22 No. 1 (1980).

23- A.F. Halasa private communication U.S. Patent Applications.

24- A.F. Halasa, J.E. Hall, D.W. Carlson, Firestone Library
    report 23.

25- A.F. Halasa, G. Hamed, J.E. Hall, L. Vescelius and
    S. Futamura Private Communication (1979).

26- D.N. Schulz and A.F. Halasa, J Polymer Sci. A-1 1S
    2401 (1977).

27- A.F. Halasa and D.N. Schulz, U.S. Patent 3,862100.

RECEIVED March 5, 1981.

Functionally Terminal Polymers via Anionic Methods

D. N. SCHULZ[1] and J. C. SANDA

The Firestone Tire and Rubber Co., Central Research Laboratories, Akron, OH 44317

B. G. WILLOUGHBY

Rubber and Plastics Research Association of Great Britain, Shrewsburg, U.K.

This paper reviews recent developments in the synthesis of telechelic and semi-telechelic polymers via anionic methods. The two anionic approaches (electrophilic termination and functional initiation) to the synthesis of these materials are discussed. The advantages of anionic methods are noted. Furthermore, the special benefits of the use of protected functional initiators and polymers are highlighted. Besides the usual advantages of the anionic methods, the protected functional initiator approach is a high yield and gel-free procedure that allows the attachment of reactive and/or mixed functionalities to polymer chain ends.

Functionally terminal polymers are valuable material intermediates. The di- and polyfunctional varieties (telechelic polymers) have found theoretical (e.g., model network) and commercial (e.g., liquid rubber) applications (1,2). On the other hand, macromolecules with a functional group at one chain end (semitelechelic polymers) have been used to prepare novel macromolecular monomers (Macromers), as well as block and graft copolymers (3-8).

Telechelic and semitelechelic polymers have been synthesized by a variety of free radical, anionic and, most recently, cationic techniques (1,2,9-12). The advantages of anionic procedures include functional purity, versatility of substitution, and monodispersity and control of polymer molecular weights. This

[1] To whom correspondence should be addressed. Current address: Corporate Research Laboratory, Exxon Research and Engineering Co., Linden, NJ 07036.

paper features recent developments in the synthesis of
functionally terminated polymers via anionic methods.

Experimental

Synthesis of Functional Terminal Polymers. Mono-
and di- terminal -OH and -NH$_2$ polydienes were prepared
according to our published procedures (13,14). Star
-OH polydienes were synthesized analogously
except that ethyl 6-lithiohexyl acetaldehyde acetal
was used as the initiator and trichloromethylsilane as
the joining agent.

α ,ω -Polystyrene diols were prepared according
to modifications of these methods. For example, to a
styrene/toluene blend (21.4% W/W), free of adventi-
tions impurities, at -26°C was added 205 ml of 0.787M
ethyl 6-lithiohexyl acetaldehyde acetal in diethyl
ether. Polymerization was continued for 6.5 hours and
then an excess of ethylene oxide was introduced at
-15°C. There was no increase in viscosity after the
ethylene oxide addition. The ethylene oxide reaction
was quenched with methanol after 18.5 hours. Catalyst
residues were removed by washing with water. The batch
was acidified with aqueous dichloroacetic acid and
refluxed until there was no change in the -OH region
(2.75-2.9γ) and the C-O-C region (8.8-9.0γ) of the
infrared spectrum of the reaction mixture. Nitrogen
entrainment was used to remove volatile by-products
of the hydrolysis. The reaction mixture was then
basified with 10% KOH and refluxed for an additional
3 hours to insure that none of the polymer end groups
had become contaminated with dichloroacetate ester
groups (from the dichloroacetic acid catalyst). After
neutralization, the product was concentrated on a
Rotovac and dried. The OH equivalent weight of the
polymer was 610 g/mol and its Tg (DTA) was +7°C.

Chain Extension of α -ω-Polystyrene Diols. A two-
stage chain extension of the α -ω -polystyrene diols
was accomplished by carboxylation of the diols with
succinic anhydride followed by chain extension with a
diepoxide. The succinic anhydride reaction was carried out 120-130°C under nitrogen. The reaction was
monitored by changes in the carbonyl bands at 1715
and 1740 cm^{-1} in the infrared spectra of the reaction
mixtures. The resulting dicarboxylic acid polymers
were chain-extended in bulk at 130°C for 9 hours with
Dow's DER diepoxide, equivalent weight =171, using
bis(3,5-diisopropylsalicylato)Cr (III) as the catalyst.

Liquid Rubber Cure of Tristar Polybutadienes.
Tristar -OH terminated polybutadienes were converted
to three dimensional networks in a diisocyanate liquid

rubber compound. The recipe (in phr) was 100 polymer; 50 ISAF carbon black; 15 Dutrex 916 oil; 1 Ethyl 702 antioxidant; Isonate 1432 (MDI) was used at several NCO/OH ratios. The samples were cured for 45 min. at 138°C.

Results and Discussion. There are basically two approaches to the preparation of telechelic and semi-telechelic polymers by anionic procedures. One method involves terminating living anionic polymers with suitable electrophiles; another technique utilizes functionally substituted anionic initiators.

There are many electrophiles which not only terminate living polymer chains but also produce end-group substitution. For example, macromolecules with hydroxyl, carboxyl, thiol, or chlorine termini can be prepared by reacting living polymers with such compounds as epoxides, aldehydes, ketones, carbon dioxide, anhydrides, cyclic sulfides, disulfides, or chlorine (15-23). However, primary and secondary amino-substituted polymers are not available by terminations with 1° or 2° amines because living polymers react with such functionalities (1). Yet, tert-amines can be introduced to chain ends by use of p-N-N-di-methylamino-benzaldehyde as the terminating agent (24).

In principle, mono-, di-, and polyfunctional terminal polymers are all available by electrophilic termination of living polymer chains. For example, difunctional polymers can be prepared by the use of dilithium initiation, followed by ditermination (25-35). However, the strict end-use requirements (e.g., linear chain extension) for difunctional materials are especially demanding.

Dilithium initiators must be soluble, stable, and purely difunctional. Many early dilithium initiators (25,29) failed all or some of these criteria (36). More recent workers have overcome one or more of these difficulties (27,28,30-35). In particular, Morton and Fetters (31,32) have produced quality telechelics by use of a soluble, stable oligomeric initiator (I) from lithium metal and 2,4-hexadiene.

α,ω-Polyisoprene diols prepared via this initiator showed a $\bar{f}(OH)$ of 1.9-2.0. The functionality of these polymers was further tested by linear chain extension with TDI to form a polyurethane with a fourteen-fold increase in molecular weight compared with the molecular weight of the diol prepolymer.

Another problem associated with the treatment of dilithium polymers with some electrophilic agents is the formation of an intractable anionic association or gel. The degree of association is related to the charge density of the electrophile and increases in the following order:

$$H_2O \; < \; CO_2 \; < \; \overset{S}{\triangle} \; < \; \overset{O}{\triangle}$$

Of course, the practical result of gel formation is an inability to mix the reagents (10, 32, 37, 38). However, it has recently been reported that this tendency to form gel can be greatly reduced by use of special solvents with solubility parameters of ≤ 7.2; e.g., 2,2,4-trimethyl pentane, or isopentane. Presumably, the use of such poor solvents for the polymers leads to mixable microscopic rather than macroscopic gels (39).

The second approach to the synthesis of telechelic polymers involves the use of functionally substituted anionic initiators. An example of a functional organolithium initiator is p-lithiophenoxide (40). Unfortunately, p-lithiophenoxide shows low solubility even in polar solvents like THF. It is difficult to control polymer molecular weights with such a heterogeneous initiator (41). Trepka (42) has improved the solubility of p-lithiophenoxide by adding alkyl groups to the ring. Nevertheless, only moderate yields of the alkylated initiator have been reported. Hirshfield (41) has described an analogous catalyst, $Bu_4NOC_6H_4Li$, that is soluble in THF. Yet, no yields for its preparation have been given.

Uraneck and Smith (43) recently have described that the product of 2,4-pentadiene-1-ol and sec-butyllithium can be used to synthesize polymers with hydroxyl end groups. However, some of the sec-butyllithium ($\sim$13%) remains unreacted.

To circumvent the low reactivity, low solubility and/or low yield problems associated with the previously mentioned hydroxyl functional initiators, we (13, 14) have built protective hydroxyl functionality into organolithium molecules. Specifically, alkyllithium initiators containing hydroxyl protecting groups have been prepared. These protected functional initiators contain acetals, i.e., tetrahydropyranyl (II) or α-ethoxyethyl ether (III) moieties.

$$Li-R-\overset{\overset{\displaystyle OC_2H_5}{|}}{OCHCH_3}$$

(II)

$$Li-R-O\text{—}\hexagon$$

(III)

The mixed acetal groups are stable to both organo-lithium reagents and living lithium polymers. Such initiators are preparable in high yields (84-92%) and are soluble in diethyl ether or benzene and insoluble in hexane (13, 44, 49).

Initiators (II) or (III) have been used to pre-pare mono-, di-, and even star acetal terminated polybutadienes, or polystyrenes

(IV - VII), where A = $-O$—$\hexagon$ or - $\overset{\overset{\displaystyle OC_2H_5}{|}}{OCHCH_3}$

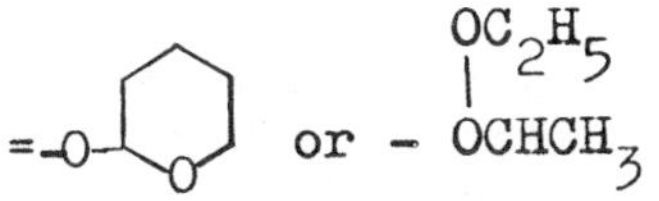

(Eq. 1-5)(13,44)

A-R-Li + M (St,Bd,etc.) $\longrightarrow$ A$\sim\sim$Li   (1)

A$\sim\sim$Li + MeOH $\longrightarrow$ A$\sim\sim$H   (2)

(IV)

A$\sim\sim$Li + $\triangle$(O) $\longrightarrow$ A$\sim\sim$CH$_2$CH$_2$OH   (3)

(V)

A$\sim\sim$Li + Cl-$\overset{\overset{\displaystyle CH_3}{|}}{\underset{\underset{\displaystyle CH_3}{|}}{Si}}$-Cl $\longrightarrow$ A$\sim\sim$$\overset{\overset{\displaystyle CH_3}{|}}{\underset{\underset{\displaystyle CH_3}{|}}{Si}}$$\sim\sim$A   (4)

(VI)

A$\sim\sim$Li + Cl-$\overset{\overset{\displaystyle CH_3}{|}}{\underset{\underset{\displaystyle Cl}{|}}{Si}}$-Cl $\longrightarrow$ A$\sim\sim$$\overset{\overset{\displaystyle CH_3}{|}}{\underset{\underset{\displaystyle A}{|}}{Si}}$$\sim\sim$A   (5)

(VII)

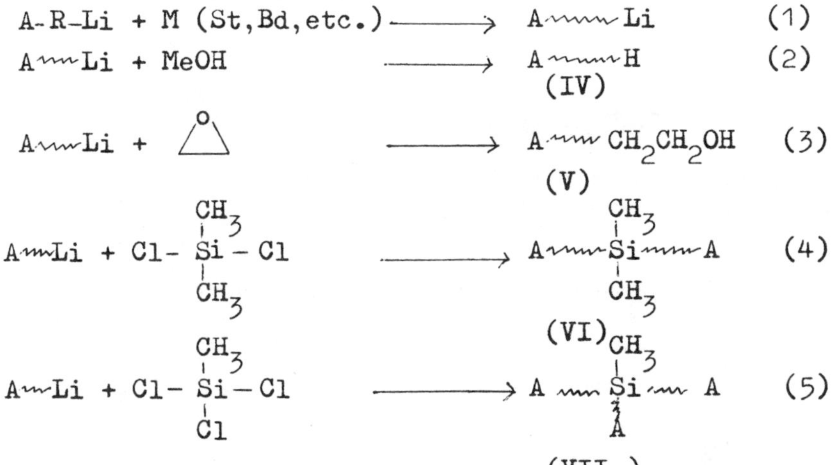

Polymerizations with initiators (II) and (III) are characterized by good molecular weight control and monodispersity of polymer molecular weights (Table I). Also, it is possible to prepare polymers with the same or different end groups in this manner.

TABLE I (13)

ACETAL TERMINATED POLYBUTADIENES

Initiator	$Ms \times 10^{-3}$	$\overline{Mn}(VPO) \times 10^{-3}$	$\overline{Mw}/\overline{Mn}$
II	3.2	3.7	1.06
II	5.8	5.7	1.05
II	2.0	1.9	1.05
III	3.5	5.3	1.06
III	15.7	14.3	1.08

Hydrolysis procedures have been developed for the removal of the acetal protecting groups without accompanying crosslinking of polydiene backbones. Dilute solution hydrolyses are preferred over bulk methods (45). (Eq. 6-9).

$$A \sim\!\!\sim\!\!\sim H \; \underset{}{\overset{H+}{\rightleftharpoons}} \; HO \sim\!\!\sim\!\!\sim H \qquad\qquad (6)$$

(IV)   (VIII)

$$A \sim\!\!\sim CH_2CH_2OH \; \underset{}{\overset{H+}{\rightleftharpoons}} \; HO \sim\!\!\sim CH_2CH_2OH \qquad (7)$$

(V)   (IX)

$$A \sim\!\!\sim \underset{\underset{CH_3}{|}}{\overset{\overset{CH_3}{|}}{Si}} \sim\!\!\sim A \; \underset{}{\overset{H+}{\rightleftharpoons}} \; HO \sim\!\!\sim \underset{\underset{CH_3}{|}}{\overset{\overset{CH_3}{|}}{Si}} \sim\!\!\sim OH \qquad (8)$$

(VI)   (X)

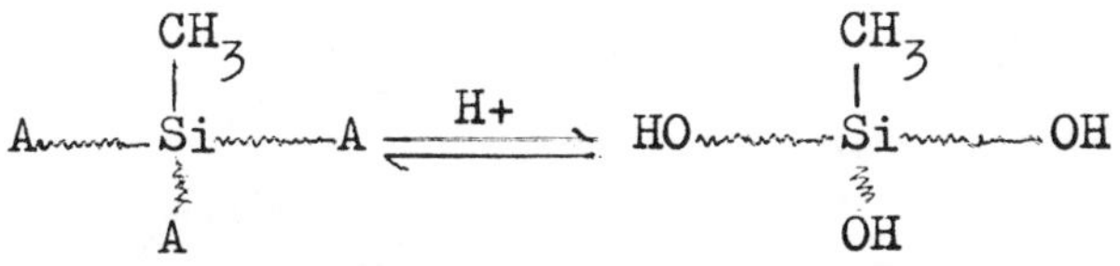

(9)

(VII) (XI)

In addition, hydroxyl polymers prepared by use of lithium alkyl acetal initiators have shown a high degree of functional purity (Table II). The functionality data for XI is a bit low, in part, because a linear GPC calibration was used to calculate $\overline{M}n$ (GPC). It should also be noted that Equations 1-9 proceed in the absence of anionic association or gel.

TABLE II

HYDROXYL FUNCTIONALITY DATA (13)

Polymer Type	$\overline{M}n$ (GPC) g/mol x10^{-3}	OH(Tit) meq/g	$\overline{f}$(OH)[a] Exp.	$\overline{f}$(OH) Theory
V	6.2	0.14	0.87	1.0
V	5.7[b]	0.18	1.02	1.0
V	2.4	0.41	0.99	1.0
VIII	6.2	0.15	0.93	1.0
IX	2.3	0.83	1.93	2.0
IX	2.5	0.70	1.76	2.0
IX	6.0	0.31	1.85	2.0
X	4.0	0.46	1.86	2.0
X	4.9	0.39	1.90	2.0
X	7.2	0.28	2.03	2.0
X	7.3	0.28[c]	2.04	2.0
XI	4.3	0.58	2.49	3.0

a $\overline{f}$(OH)=$\overline{M}n$ (g/mol) x OH (meq/g) x 10^{-3} eq/meq

b $\overline{M}n$ (VPO)

c OH (infrared data)

Journal of Polymer Science

To further test the functional purity of hydroxyl
polymers prepared by the protected initiator and
polymer route, two stage chain extensions of two polystyrene diols were carried out. The first stage involved conversion of the diols to acids with succinic
anhydride; the second stage involved chain extension
with a diepoxide (Table III). If one assumes an overall conversion of 97-98%, a $\overline{DP}$ of 19 requires a functionality $\sim$ 1.93-1.95 based upon step-growth polymerization theory (47). By comparison, a free radically
prepared dicarboxylic acid, Thiokol's HC434, when
chain extended with the same diepoxide, produced
polymers with $\overline{DP}$'s of only 4-6 (48).

TABLE III
TWO STAGE CHAIN EXTENSION OF PS-DIOLS[a]

Polymer	Anhydride	Epoxide	$\overline{DP}$ [e]
PS-Diol-1[b]	Succinic	DER-332[d]	16.2
PS-Diol-2[c]	Succinic	DER-332[d]	18.7

[a] 1st stage carboxylation with succinic anhydride;
2nd stage bulk chain extension of dicarboxyl
polymer with diepoxide, 130°C for 9 hours.

[b] Eq. wt. = 610

[c] Eq. wt. = 369

[d] Eq. wt. = 171

[e] $\overline{DP} = \dfrac{2\,Mn}{M_A + M_B}$

Tristar polybutadienes prepared by the interme-
diacy of lithium acetal initiators were also converted
to three dimensional networks in a liquid rubber
formulation using a diisocyanate curing agent. Table
IV shows normal stress-strain properties for liquid
rubber networks at various star branch $\overline{Mn}$'s. It can
be seen that as the branch $\overline{Mn}$ increases to 2920, there
is a general increase in the quality of the network.
Interestingly, the star polymer network with a star
branch $\overline{Mn}$ of 2920 ($\overline{Mc}$=5840) exhibits mechanical properties in the range of a conventional sulfur vulcani-
zate with a $\overline{Mc}$ of about 6000-8000.

TABLE IV
TRI-STAR POLYMER CURES[a]

Av. Branch $\overline{M}n$	830	1370	2920
Normal Stress-Strain Prop.			
100% Mod., MPa(psi)	5.34	5.17	2.90
	(775)	(750)	(420)
300% Mod., MPa(psi)	-	-	12.5
			(1820)
Tensile Str.,MPa (psi)	10.9	12.9	17.4
	(1575)	(1875)	(2530)
Ult. Elong., %	180	170	380

[a] Recipe (in phr) was 100 polymer; 50 ISAF carbon black; 15 Dutrex 916; 1 Ethyl 702 antioxidant; Isonate 143L (MDI), NCO/OH = 1.25.

Other workers have recognized the benefit of using our protected hydroxyl initiators. For example, Teyssie has successfully used (II) to prepare narrow MWD terminated polysiloxanes with good $\overline{M}n$ control and high functional purity, (Eq. 10, 11) (Table V)(49).

$$A-R-Li+n \left[Si(CH_3)_2 -O\right]_3 \xrightarrow[(2)(CH_3)_3SiCl]{(1)\varnothing H}$$

$$A-R\left[Si(CH_3)_2O\right]_{3n}Si(CH_3)_3 \qquad (10)$$

(XII)

$$A-R\left[Si(CH_3)_2-O\right]Si(CH_3)_3 \; \underset{\longleftarrow}{\overset{H+,H_2O}{\rightleftharpoons}}$$

$$HO-R\left[Si(CH_3)_2-O\right]Si(CH_3)_2 \qquad (11)$$

(XIII)

TABLE V
SEMITELECHELIC POLYSILOXANES (49)

$\overline{M}n$ (Calc)	$\overline{M}n$ (VPO)	$\overline{M}w/\overline{M}n$	Exp. #OH/Chain
3,100	3,200	1.10	0.97
5,060	5,000	1.12	1.04
9,100	8,500	1.16	1.08

Macromolecules

Other telechelics of interest are amine-ended polymers. tert-Amine terminated polymers have been prepared by use of the tert-amine functional initiator, 3-N,N-dimethylpropyllithium (50,51). However, primary amine-ended polymers have required the use of protecting groups because of the mutual reactivity of primary amines and living lithiomacromolecules.

We have found that the primary amine moiety can be protected from organolithium reagents and polymers by use of the N,N-bis(trimethylsilyl)amine, $-N(TMS)_2$, protecting group. In turn, the $-N(TMS)_2$ substituted polymers can be converted to primary amine polymers by simple hydrolysis (Eqs. 12-14) (14,52).

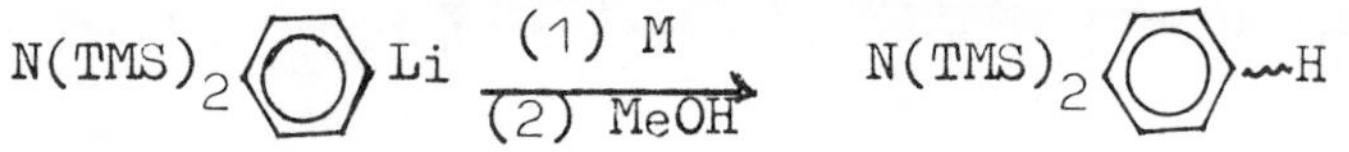

$$N(TMS)_2 \langle \bigcirc \rangle Li \xrightarrow[(2)\ MeOH]{(1)\ M} N(TMS)_2 \langle \bigcirc \rangle \sim H \quad (12)$$

(XIV)

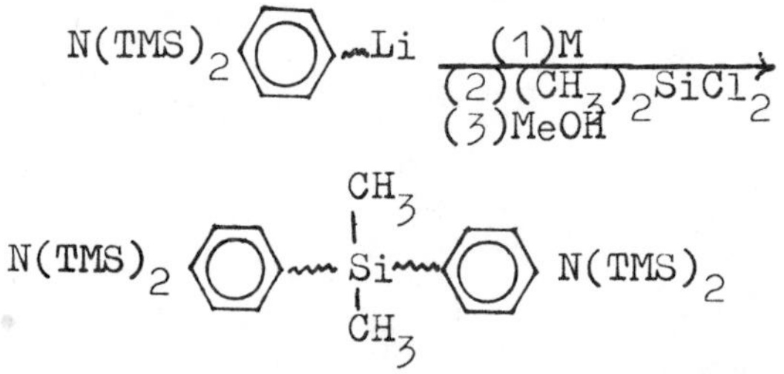

$$N(TMS)_2 \langle \bigcirc \rangle \sim Li \xrightarrow[(3)MeOH]{(1)M \atop (2)(CH_3)_2SiCl_2}$$

$$N(TMS)_2 \langle \bigcirc \rangle \sim \underset{\underset{CH_3}{|}}{\overset{\overset{CH_3}{|}}{Si}} \sim \langle \bigcirc \rangle N(TMS)_2 \quad (13)$$

(XV)

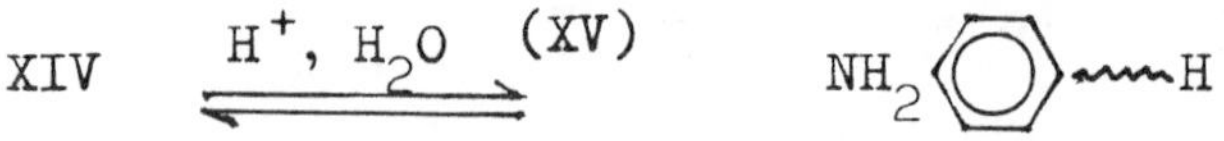

$$XIV \xrightarrow{H^+,\ H_2O} NH_2 \langle \bigcirc \rangle \sim H \quad (14)$$

(XVI)

$$XV \xrightarrow{H^+,\ H_2O} NH_2 \langle \bigcirc \rangle \sim \underset{\underset{CH_3}{|}}{\overset{\overset{CH_3}{|}}{Si}} \sim \langle \bigcirc \rangle NH_2 \quad (15)$$

(XVII)

The protected amine and primary amine ended polymers show the good molecular weight control and narrow MWD's one expects of anionic polymers. In addition, the hydrolysis of the protecting group does not disrupt polydiene backbones. Such procedures are also gel free. The polymers formed exhibit functionalities approaching the theoretical values, especially for low molecular weight (high amine content) polymers (Table VI), (14, 52).

TABLE VI (14)

HYDROLYSIS OF N(TMS)$_2$ POLYMERS

Mn (GPC) x 10^{-3} (Before Hydrolysis)	Mn (GPC) x 10^{-3} (After Hydrolysis)	NH$_2$, meq/g Exptl.	NH$_2$, meq/g Theory
-	7.6	0.13	0.13
7.50	7.3	0.13	0.14
14.0	13.5	0.051	0.074
34.0	34.2	0.035	0.050
60.0	59.5	0.025	0.032

Recently, Teyssie (53) has extended the concept of protected functional initiators and polymers to the preparation and use of the protected carboxyl initiator (XVIII).

(XVIII)

The low efficiency of this initiator limits its utility for polymerization. However, it has been shown to be highly efficacious for grafting onto and substitution of halogenated polymer substrates (53).

Summary: This paper has reviewed recent developments in the synthesis of telechelic and semitelechelic polymers via anionic methods. The two anionic approaches (electrophilic termination and functional initiation) to the synthesis of these materials were discussed. The advantages of anionic methods (e.g.,

functional purity, control of polymer $\overline{M}n$, monodispersity of polymer MWD, as well as versatility of sub-
stitition and backbone architectures) were noted.
Furthermore, the special benefits of the use of protected functional initiators and polymers were mentioned. Besides the usual advantages of the anionic
methods, the protected functional initiator approach
is a high yield and gel-free procedure that allows the
attachment of reactive and/or mixed functionalities
to polymer chain ends.

Acknowledgments

The author wishes to thank Dr. A. F. Halasa for
his inspiration. We are especially grateful to the
Firestone Tire & Rubber Company for permission to
publish this paper.

Literature Cited

1. French, D. M., Rubber Chem. Technol., 1969, 42, 71.

2. Schnecko, H.; Degler, G.; Dongowski, H.; Caspary,
    R.; Anger, G.; Ng, T.S., Ang. Makromol. Chem.,
    1978, 70, 9.

3. Milkovich, R.; Chiang, M.; U.S. Patent 3,786,116
    (Jan. 15, 1974).

4a. Yamashita, Y.; Hane, T., J. Polym. Sci., A-1,
    1973, 11, 425.

4b. Yamashita, Y.; Hane, T., J. Polym. Sci., A-1,
    1973, 11, 823.

5. Hsieh, H., Am. Chem. Soc., Div. Polym. Chem.,
    Polym. Prepr., 1976, 17 (1), 200.

6. Se Kenda, J. J., Macromol. Sci., Chem., 1972, 6,
    1145.

7. Ambrose, R. J.; Hergenrother, W. L., J. Polym.
    Sci., Polym. Symposium, 1977, 60, 15.

8. Perly, B.; Douz, A.; Gollot, B., Makromol. Chem.,
    1976, 177, 2569.

9a. Reed, S.F., J. Polym. Sci. A-1,1971,9,2029.
9b. Reed, S.F.,J. Polym. Sci., A-1, 1971,9,2147.
9c. Reed, S.F., J. Polym. Sci, A-1, 1972,10,649.
9d. Reed, S.F., J. Polym. Sci.,A-1, 1972, 10, 2025.
9e. Reed, S. F., J. Polym. Sci., A-1, 1972, 10, 2493.
10. Reed, S.F. J. Polym. Sci., A-1, 1972, 10, 1187.
11. Kennedy, J.P.; Smith, R.A., Am. Chem. Soc., Div.
    Polym. Chem., Polym. Prepr.,1979,20 (2),316.

12. Fehervari, A.; Kennedy, J.P.; Tudos, F.,Am. Chem. Soc., Div. Polym. Chem., Polym. Prepr.,1979,20,(2), 320.
13. Schulz, D.N.; Halasa, A.F.; Oberster, A.E., J.Polym. Sci., A-1, 1974, 12, 153.
14. Schulz, D. N.; Halasa, A.F.,J. Polym. Sci., A-1, 1977, 15, 2401.
15. Brody, H.; Richards, D.H.; Swarc M.; Chem. & Ind. (London), 1958, 1473.
16. Goldberg,F.J., U.S. Patent 3,055,952 (Sept.25,1962)
17. B.F. Goodrich Co.,British Patent 964,259 (July 22, 1964).
18. Rempp, P.; Loucheux, M.H., Bull. Soc. Chem. France, 1958, 1497.
19. Uraneck, C.A.; Short, J.N.; Zelinski, R.P.,U.S. Patent 3,135,716 (June 2, 1964).
20. Uraneck, C.A.; Short, J.N.; Zelinski, R.P.; Hsieh, H.L.,West German Patent 1,169,674 (May 6,1964).
21. Richards, D.H., J. Polym. Sci. Polym. Let.,1968,6, 417.
22. Cleary, J.W., U.S. Patent 3,048,568 (Aug. 7,1962).
23. Phillips Petroleum Co., British Patent 906,315, (Sept. 19, 1962).
24. Zelinski, R.P.; Hsieh, H.L.; Strobel, C.W., U.S. Patent 3,109,871 (Nov. 5,1963).
25. Eberly, K.C., U.S. Patent 2,947,793 (Aug. 2,1960).
26. Hofmann, N.L.; Pritchett, E.G., British Patent 1,029,451 (May 11, 1966).
27. Fetters, L. J.; Morton, M., Macromolecules, 1969, 2, 453.
28. Karoly, G., Am. Chem. Soc., Div. Polym. Chem., Polym. Prepr., 1969, 10, 837.
29. Lithium Corporation of America., Product Bulletins Nos. 190-194.
30. Foss, R.P.; Jacobson, H.W., Sharkey, W. H., Macromolecules, 1977, 10, 287.
31. Fetters, L. J., U.S. Patent 3,848,008 (Nov.12, 1974).
32. Morton, M.; Fetters, L.J.; Inonata,J.; Rubio, D.C.; Young, R.N.,Rubber Chem. Technol., 1976,49,303.

33. Sigwalt, P.; Guyot, P.; Fontaville, M.; Vaison, J.P., Ger. Offen. 2623344 (Dec. 16, 1976).
34. Sigwalt, P.; Fontaville, M., French Pat. 7517716 (Appl. June 6, 1975).
35. Tung, L. H.; Lo G.Y.S., Blyer, D.E., Macromolecules, 1978, 11, 616.
36. G. Jalics, presented at the 7th Akron Chemists-Chemical Engineers -Technician Affiliate Symposium, Akron, Ohio, October 20, 1977.

37. Brody, H.; Richards, D. H.; Swarc, M., Chem. &
    Ind. (London), 1958, 1473.
38. Richards, D. H., J. Polym. Sci. Polym. Let., 1968,
    6, 417.
39. Komatsu, K.; Nishioka, A.; Ohshima, N.; Takahashi,
    M.; Hara, H., U. S. Patent 4,083,834 (April 11,
    1978).
40. Hirshfield, S. M., NASA Tech. Brief, 71-10184,
    1971; NASA Reports NPO-10998, 1976.
41. Fletcher, J. C.; Hirshfield, S. M., U.S. Patent
    3,755,283 (August 28, 1973).
42. Trepka, W. J., U.S. Patent 3,439,047 (April 15,
    1969).
43. Uraneck, C. A.; Smith, R.L., U.S. Patent 3,993,854
    (Nov. 23, 1976).
44. Halasa, A. F.; Schulz, D. N., U.S. Patent 3,862,
    100 (Jan. 21, 1975).
45. Schulz, D.N.; Anderson, J.N., U.S. Patent 3,954,
    885 (May 4, 1976).
46. Halasa, A.F.; Schulz, D.N., U.S. Patent 4,052,
    370 (Oct. 4, 1977).
47. Billmeyer,Jr., F.W. "Textbook of Polymer Science",
    Interscience, New York, NY. 1962, Ch. 8.
48. Willoughby, B. G., unpublished results.
49. Lefebvre, P.M.; Jerome R.; Teyssie, P., Macro-
    molecules, 1977, 10, 871.
50. Eisenbach, C. D.; Schnecko, H.; Kern, W., Eur.
    Polym. J., 1975, 11, 699.
51. Eisenbach, C. D., Schnecko, H.; Kern, W.,
    Makromol. Chem., 1975, 176, 1587.
52. Schulz, D. N.; Halasa, A. F.; U.S. Patent 4,015,
    061 (March 29, 1977).
53. Boze, G.; Jerome, R.; Teyssie, P., Makromol.
    Chem., 1978, 179, 1383.

RECEIVED February 24, 1981.

Present View of the Anionic Polymerization of Methyl Methacrylate and Related Esters in Polar Solvents

AXEL H. E. MÜLLER

Institut für Physikalische Chemie, Universität Mainz,
D-6500 Mainz, West Germany

A review is given on the kinetics of the anionic polymerization of methyl methacrylate and tert.-butyl
methacrylate in tetrahydrofuran and 1,2-dimethoxy-
ethane, including major results of the author's laboratory. The Arrhenius plots for the propagation
reaction are linear and independent of the counterion
(i.e. Na^+, Cs^+). The results are discussed assuming
the active centre to be a contact ion pair with an
enolate-like anion, the counterion thus exhibiting
little influence on the reactivity of the carbanion.
The kinetics of the termination reaction seem to be
very complex and still are not fully understood. From
the kinetic results it is very likely that side reactions occuring in the initiation process are responsible for chain termination during polymerization.

The nature of the active species in the anionic polymerization of non-polar monomers, e. g. styrene, has been disclosed
to a high degree. The kinetic measurements showed, that the polymerization proceeds in an ideal way, without side-reactions,
and that the active species exist in the form of free ions,
solvent-sparated and contact ion pairs, which are in a dynamic
equilibrium (1-4). For these three species the rate constants
and activation parameters (including the activation volumes), as
well as the rate constants and equilibrium constants of interconversion have been determined (4-7). Moreover, it could be
shown by many different methods (e. g. conductivity and spectroscopic methods) that the concept of solvent-separated ion pairs
can be applied to many ionic compounds in non-aqueous polar
solvents (8).

On the other hand, the anionic polymerization of polar
monomers appears to be very complicated. This is due to the functional groups which give rise to physical interactions (e.g. association), as well as chemical side reactions (e. g. termination

and transfer), especially in non-polar solvents. In the case of the anionic polymerization of methyl methacrylate (MMA) in toluene or toluene/THF mixtures this was demonstrated by broad and multi-modal molecular weight distributions, the existence of high amounts of oligomeric side-products and complex kinetics which still are not fully understood (for reviews see 9,1o,11).

Since the discovery of Fox et al. (12) that in the anionic polymerization of MMA the tacticity of the resulting polymer is fully controlled by the experimental conditions, many attempts have been made to elucidate structures of the active species res-ponsible for the tacticities obtained under different conditions. However, this can only be achieved, if a complete analysis of the polymerization process is performed by investigating the kinetics of polymerization in different systems, as has been done for styrene.

The side reactions complicating the kinetics in non-polar solvents are decreased to a high extent, when polar solvents are used instead of non-polar ones. Wenger (13) reported that PMMA-Li remained living in tetrahydrofuran (THF) at $-78^{\circ}C$ for more than 4o h, which was proven by the polymerization of new monomer added to the living solution. By a comparison of the number-average molecular weights of the polymer before and after monomer addition Graham et al. (14) showed that under similar conditions no new chains are formed from transfer, when new monomer is added, so that the polymerization is initiated by living PMMA-Li. Roig et al. (15) investigated the molecular weight distributions of PMMAs prepared in THF and 1,2-dimethoxyethane (DME), using biphenyl-sodium as the initiator and Wenger's technique of slow destilla-tion of the monomer to the initiator solution. The molecular weight distributions reported are cery narrow ($1.o2 = \overline{M}_w/\overline{M}_n = 1.o6$). Under the same experimental conditions Guzmán and Bello (16) reported bimodal distributions. This was attributed to an equili-brium between two active species (anion and ion pair) with slow interconversion. However, from the rates of dissociation known from other systems, this explanation is not probable (6,69).

Fowells et al. (17) reported conductivity measurements of PMMA-Li in THF. The dissociation constant calculated is very low ($K_d = 2 \cdot 1o^{-8}$ mol/l at $-78^{\circ}C$). However, the conductivity at only one concentration was measured. Figueruelo (18) reported dissoci-ation constants being smaller by two orders of magnitude ($K_d = 4.4 \cdot 1o^{-1o}$ for PMMA-Li and $2.1 \cdot 1o^{-1o}$ for PMMA-K at $-78^{\circ}C$). Löhr and Schulz (19,2o) reported values of $3.5 \cdot 1o^{-9}$ for PMMA-Na and $2 \cdot 1o^{-9}$ for PMMA-Cs at $-78^{\circ}C$). Besides the appreciable discrepan-cies, all values are much smaller than the corresponding ones of the polystyryl salts (2,21), indicating that anions participate in the propagation to a small amount only.

Kinetics of the propagation reaction in the polymerization of MMA

It was not until 1973 that results of kinetic measurements were
reported, which had been carried out with MMA in a polar solvent.
This may be due to the experimental difficulties arising from the
fact that the kinetics cannot be monitored by the decrease of the
u.v. absorption of the monomer, in combination with the very high
rates of polymerization (at $-8o°C$ and a concentration of living
ends of $1o^{-3}$ mol/l, the half-life of monomer addition to ion pairs
is 5 s; at ambient temperature, only o.1 s). Using the flow tube
technique, which had been applied before to the polymerization of
styrene, Löhr and Schulz (19,2o) and Mita et al. (22) investigated
the kinetics of the polymerization of MMA in THF, using Na^+ and
Cs^+ as counterions. Both groups showed that at low temperatures
(T = $-75°$ C) the reaction proceeds in an ideal way: initiation is
fast as compared with propagation, the propagation reaction follows first-order kinetics with respect to monomer concentration
(indicating the absence of termination reactions) and the number-
average degree of polymerization is proportional to conversion
(indicating the absence of transfer reactions). As a consequence,
Löhr and Schulz obtained narrow molecular weight distributions.
Despite the low dissociation constants, from the dependence of the
rate constants on the concentration of living ends it was clearly
demonstrated that both anions and ion pairs participate in the
propagation reaction, i. e. al least two different kinds of propagating species have to be accounted for.
     The kinetic measurements of both groups suffer from the fact
that, when using Na^+ as the counterion, only bifunctional initiators were used (i. e. naphthalene sodium and oligo-α-methylstyryl-
sodium). As was shown by Warzelhan et al. (23,24), bifunctionally
growing chains can form intramolecular associates, which differ
from the monofunctionally growing species in rate constant of propagation as well as in the tacticity of the resulting polymer (see
below). Intermolecular associates have not been detected in the
range of living end concentrations used ($1o^{-4} \leqslant c^* \leqslant 1o^{-3}$ mol/l).
     Schulz and coworkers (25,26) investigated the propagation ki-
netics over a wide temperature range ($-1oo°C \leqslant T \leqslant +2o°C$) using
monofunctional initiators, i. e. cumylcaesium and monofunctional
benzyl-oligo-α-methylstyrylsodium. The main interest in these investigations was the study of the structure of the ion pairs involved, therfore $Na(B\emptyset_4)$ and $Cs(\emptyset_3BCN)$, resprectively, were used
as common ion salts to suppress the dissociation to free ions and
to reduce the probability of triple ion formation. As the use of
the flow tube reactor is unfavourable at very low temperatures,
reactions below $-5o°$ C were conducted in a stirred tank reactor
allowing kinetic measurements of reactions with half-lives of 2 s
and longer by taking samples periodically (27). Thus, molecular
weight distribution, tacticity, activity of chain ends and other
interesting properties of the resulting polymer can be determined
as a function of conversion.

Although an appreciable amount of termination is found at
elevated temperatures, rate constants can be calculated from the
initial slope of the first-order time-conversion curve. The concentration of living ends is calculated from the linear plot of
the number-average degree of polymerization vs. conversion,which
still remains linear when termination occurs, since the total
number of chains remains unaltered, provided nor intermolecular
termination (grafting) nor transfer occurs.

The Arrhenius plots of the resulting propagation rate constants for ion pairs are linear for both Na^+ and Cs^+ as the counterions (cf. Figure 1), thus rendering no evidence for the existence of more than one kind of ion pair. Moreover, within experimental error, the straight lines are identical for both counterions. These results are in strong contrast to the kinetics of the
polymerization of styrene (4,28)(cf. Figure 2). The curved Arrhenius plots in this case are due to a temperature-dependent equilibrium between contact and solvent-separated ion pairs. The un-
differentiated behaviour in the MMA polymerization was explained
assuming the counterion to be solvated intramolecularly by the
penultimate or antepenultimate ester group of the polymer chain
(29) (M^+: Na^+ or Cs^+):

I   II

This model had been proposed by Fowells et al. (17) on the basis
of NMR measurements of partially deuterated polymers prepared in
the system THF/Li^+, and it was applied by Fisher and Szwarc (3o)
and Sigwalt and coworkers (31,32)to the polymerization of 2-vinyl
pyridine.

If the proposed structure was true, then a solvent of higher
solvating power, such as DME, should decrease the stability of the
intramolecular complex. Using sodium as the counterion part of
the active centres should exist in the form of externally solvated
contact ion pairs or even solvent-separated ion pairs. With cae-
sium as the counterion there should be little change, because
this cation is only poorly solvated by both solvents and consequently the possibility of solvent-separated ion pairs to be found
should be extremely small.

The kinetic results obtained by Kraft et al. (33)(cf. Fig.3)
show, that the rate constants are higher and the activation parameters are smaller than those obtained in THF. It is remarkable,
that the difference between the two counterions again is within

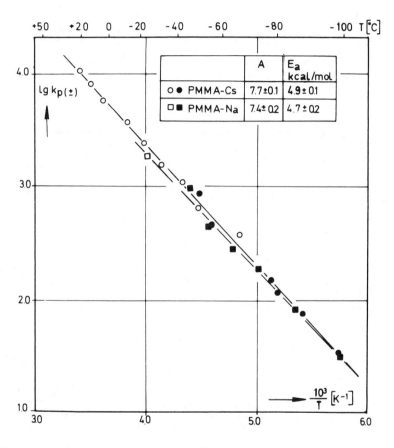

Figure 1. Arrhenius plot of the propagation rate constants in the anionic poly-merization of methyl methacrylate in THF using Na^+ and Cs^+ as the counterions (25)

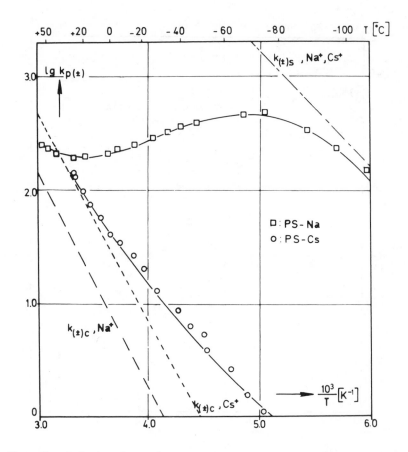

*Figure 2. Arrhenius plot of the propagation rate constants in the anionic poly-
merization of styrene in THF using Na⁺ and Cs⁺ as the counterions (4, 28)*

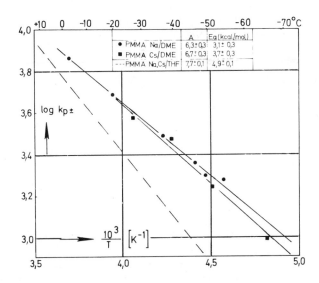

Makromolekulare Chemie, Rapid Communications

*Figure 3.   Arrhenius plot of the propagation rate constants in the anionic poly-
merization of MMA in DME using Na⁺ and Cs⁺ as the counterions (33)*

experimental error. The latter result clearly cannot be explained
by the hypothesis of intramolecular solvation. A model which assumes contact ion pairs for both counterions and both solvents is
neither suitable, since in this case there should be a distinct
difference between Na^+ and Cs^+ in the ion pair propagation rate
constants (a factor of 4 is observed in the polymerization of
styrene, cf. Figure 2). As the rate constants are approximately
equal (in both solvents), it was concluded that part or all of the
active species exist in a special form of contact ion pair, in
which the counterion is less close to the α-carbon atom, thus exhibiting less influence on its reactivity. In the most extreme
case the counterion is supposed to be separated from the carbanion
by a solvent molecule, still being in contact with the enolate
oxygen atom, however. The proposed "true" structure of the ion pair
is somewhere inbetween III and IV, depending on counterion, solvent
and temperature. Part of the ion pairs may exist as structure I
or II, especially in THF or less solvating solvents.

(S: solvent)

III   IV

Eizner and Erussalimsky (34) reported quantummechanical calculations on the free methyl acrylate anion and its lithium salt,
showing that in the anion the negative charge density at the carbonyl oxygen is higher than at the α-carbon atom; in the salt both
charges are equal. However, these calculations generally suffer
from the difficulty of taking solvent effects into account.

A considerable charge density at the carbonyl oxygen is also
indicated by C-13 NMR measurements with the metalated "monomer"
(i.e. methyl isobutyrate) and dimer in THF, which were reported
by Vancea and Bywater (35). They showed that the bond between the
α-carbon and the carbonyl carbon exhibits a distinct double bond
character.

The model suggested above is in accordance with the very low
dissociation constants measured. The corresponding interionic distances of only 3 Å (calculated by means of the Fuoss equation (67))
are easily understood taking into account a coordination of the
counterion with two centres of negative charge or even with the
enolate anion alone, exhibiting a high negative charge density.
The small interionic distances also reflect the absence of solvent-

separated ion pairs, the contribution of which leads to values of
6.4 Å in the case of polystyryl-Na.

Conductivity measurements of Müller at al. ($\underline{36}$) on Cs($\emptyset_3$BCN)
in THF and mixtures of THF with methyl isobutyrate (which serves
as a model for the ester groups of the polymer) resulted lower
dissociation constants in the mixture than in pure THF, as is to
be expected from the lower dielectric constant of methyl isobutyrate. Thus, no specific solvation of Cs$^+$ by the polymer ester
groups could be shown, although a chelate-like effect could enhance the solvating power of the penultimate ester group of the
active species.

Divalent counterions: Kinetic measurements using mono- and
bifunctional initiators and Ba^{++} as the counterion in THF were reported by Mathis and François ($\underline{37}$), who applied adiabatic calorimetry. At -7o$°$C no termination is found and conversion follows
first order with respect to monomer concentration. The rate constants do not depend on the concentration of living ends, indicating the absence of free anions. The rate constants are smaller by
a factor of 2o as compared with those measured with monovalent
counterions. However, they are smaller by a factor of 3 only,
compared with those calculated for chains which are intramolecularly associated (Na$^+$, counterion). The activation energy for
PMMA$_2$Ba in THF is equal to that for monovalent counterions, but
the frequency exponent is smaller by about 1.5 units, reflecting
the fact that the transition state for the dianionic ion pair has
higher steric requirements.

Methacrylates other than methyl: The kinetics of the polymerization of tert.-butyl methacrylate (TBMA) in THF was studied
by Müller ($\underline{38}$) with Na$^+$ and Cs$^+$ as the counterions (cf. Figure 4).
Rate constants in this system are distinctly lower, this being due
to the high activation energy rather than to the frequency exponent, which is higher than in the MMA polymerization. This result
is in contrast to the expectation of a higher steric factor caused
by the bulkiness of the ester group. Rate constants in this system
are strongly dependent on the counterion (Cs$^+$ results rate constants which are four to ten times higher than the corresponding
values for Na$^+$). It was concluded, that in this system the counterion is closer to the carbanion, because of the positive inductive
effect of the tert.-butyl group, increasing the negative charge
density.

The copolymerization of MMA with a large number of alkyl
methacrylates has been studied by Bevington et al. ($\underline{39}$), Ito et
al. ($\underline{4o}$), and Yuki and coworkers ($\underline{41-44}$). The copolymerization parameters yield information on the reactivity of the monomer as a
function of the ester group. A rough correlation between the r_1
values and Taft's σ^* parameter was given by Yuki, whereas there
seems to be no correlation between the r_1 values and Taft's steric
parameter E_s. The ρ^* value reported is cá. 1.1 for butyllithium as

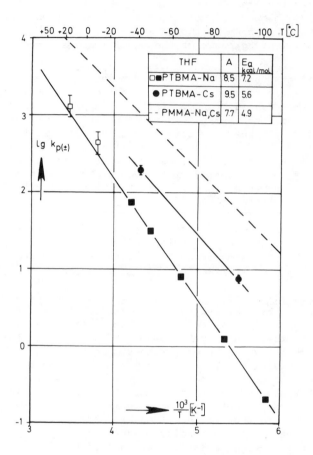

*Figure 4. Arrhenius plot of the propagation rate constants in the anionic poly-
merization of t-butyl methacrylate in THF using Na⁺ and Cs⁺ as the counterions*

the initiator in both THF and toluene as the solvents at $-78°C$.
A combination of these results with those of homopolymerization
kinetics allows the relative reactivity of the active centres to
be estimated. In the case of tert.-butyl methacrylate it was shown
that the tert.-butyl group not only decreases the reactivity of
the monomer but also that of the polymer (38).

Tacticities of the polymers

The tacticities of PMMA samples prepared in different solvents
using different counterions show a distinct tendency concerning
the polarity of the solvent and the solvatability of the counterion.

Table I: Tacticities of PMMA samples obtained in DME and THF,
using mono- and bifunctional initiators and Na^+ and Cs^+ as the
counterions. I,H,S: isotactic (mm), heterotactic (mr/rm) and syndiotactic (rr) triads; m,r: meso and racemic dyads; $\rho = 2mr/H$:
persistence ratio ($\rho = 1$ for Bernoullian statistics).

Counterion/ solvent	T $[°C]$	I	H	S	m	r	ρ	ref.
Na^+/DME	-55	0.01	0.22	0.78	0.12	0.88	1.00	33
Cs^+/DME	-65	0.02	0.37	0.60	0.21	0.79	0.89	33
Na^+/THF	-61	0.04	0.36	0.58	0.23	0.77	0.92	24
Cs^+/THF	-66	0.05	0.53	0.42	0.315	0.685	0.81	29
Na^+/THF bifunctional	-75	0.39	0.41	0.20	0.595	0.405	1.18	24

Table I shows the tacticities of PMMA samples which were prepared in counterion/solvent systems which so far have been investigated kinetically and in which the contribution of free anions
was suppressed by the addition of a common ion salt. It is seen,
that the degree of isotacticity as well as the persistence ratio
decrease with decreasing solvation, exhibiting the following order:
  Na/DME < Cs/DME ≈ Na/THF < Cs/THF.
This indicates that a close contact of the carbanion with the
counterion favours isotactic placements as well as short sequence
length (corresponding to persistence ratios below 1). In the system Cs/THF the marked non-Bernoullian behaviour can be described
by Markovian statistics rather than the Coleman-Fox model, i.e.
the penultimate monomer unit influences the stereochemistry of the
monomer addition (29). This effect can be interpreted by decreasing external solvation (III,IV) and increasing intramolecular
solvation (I,II).
  Intramolecular chain end associates partly formed in the polymerization by bifunctional chains and Na^+ as the counterion in THF

V

(Warzelhan et al. (23,24) ascribed structure V to these associates)
lead to distinctly different tacticities than the monofunctionally
growing species (cf. Table I). These tacticities are intermediate
between those produced in polar solvents by. monofunctional initiators and those produced in non-polar solvents, where association of the growing ends with methoxide or with the monomer has
been proposed.

There are many tacticity measurements on methyl and various
alkyl methacrylates reported in the literature in which the polymerization medium is varied from highly solvating solvents (e. g.
NH_3, HMPA, THF/cryptands) to poorly solvating ones (THP, dioxane,
toluene). A review was given recently by Yuki and Hatada (45).
However, it is difficult to compare the results obtained by different research groups, as in many cases anions are present to an
unknown extent. Often bifunctional initiators were used, leading
to unknown amounts of intramolecular associates. Intermolecular
association may also occur at higher initiator concentration or in
solvents of lower solvating power.

Kinetic investigations on the termination reactions

Side reactions are observed when working at higher temperatures than $-75°C$, or when using very low concentrations of living
ends, as is neccesary for the preparation of polymers with high
molecular weight. The common features of these side reactions are:
(i) a downward curvature of the linear first-order plot of mono-
     mer concentration vs. time

$$\ln([M]_o / [M]) = k_p c^* t,$$

indicating a decrease in the concentration of living ends c^*,
due to chain termination;
(ii) a broadening of the molecular weight distribution. A low mo-
     lecular weight tailing is found in the GPC eluograms. The
     number-average degree of polymerization, however, still is
     proportional to conversion, as the total number of chains
     is not decreased.

Based on these effects it was seen qualitatively that the relative amount of termination depends on the solvating power of the
solvent. Least termination is found in DME, more in THF and even
more in tetrahydropyran (THP) ($\underline{46}$). However, as long as the absolute termination rate constants are not known, it cannot be ex-
cluded that these differences just arise from the dependence of
the propagation rate constant on the solvent. From the abovemen-
tioned effects and the similarity of the propagation rate constants
it can be concluded that in THF the termination rate constant is
higher for Na^+ than for Cs^+ as the counterion ($\underline{47}$).

Using $\underline{tert.}$-butyl methacrylate as the monomer, no termination
has been detected even at ambient temperature ($\underline{38}$).

Evidently the rate, and perhaps the mechanism of termination
is influenced by the structure of the ion pair. Therefore the
study of the termination reaction may give some indirect insight
into the structure of the active species.

Mainly three termination reactions have been discussed in the
literature. They were first proposed by Schreiber ($\underline{48}$):

(i) a reaction of the living end with the carbonyl group of the
monomer resulting a vinyl ketone ("monomer termination"):

$$
\sim\sim\!CH_2\!-\!\underset{\underset{OCH_3}{\overset{|}{C=O}}}{\overset{CH_3}{\underset{|}{\overset{|}{C}}}}
\;+\;
CH_2\!=\!\underset{\underset{OCH_3}{\overset{|}{C=O}}}{\overset{CH_3}{\underset{|}{\overset{|}{C}}}}
\;\longrightarrow\;
\sim\sim\!CH_2\!-\!\underset{\underset{OCH_3}{\overset{|}{C=O}}}{\overset{CH_3}{\underset{|}{\overset{|}{C}}}}\!-\!\overset{\overset{\textstyle CH_3}{\underset{|}{C=CH_2}}}{\underset{\underset{OCH_3}{\overset{|}{C=O}}}{\overset{|}{}}}
\;+\; OCH_3^-
$$

(ii) reaction of the living end with the ester group of another
polymer resulting in chain coupling ("intermolecular polymer
termination")

$$
\sim\!\!\overset{CH_3}{\underset{\underset{OCH_3}{\overset{|}{C=O}}}{\underset{|}{\overset{|}{CH_2\!=\!C}}}}
\;+\;
\underset{}{O\!-\!\overset{O}{\underset{}{\overset{\|}{C}}}\!-\!\overset{CH_2}{\underset{}{\overset{|}{C}}}\!-\!CH_3}
\;\longrightarrow\;
\sim\!\!\overset{H_3C\;\;O\;\;CH_2}{CH_2\!-\!\underset{\underset{OCH_3}{\overset{|}{C=O}}}{\overset{|}{C}}\!-\!\overset{\|}{C}\!-\!\overset{|}{C}\!-\!CH_3}
\;+\; OCH_3^-
$$

(iii) reaction of the living end with the antepenultimate ester
group of its own polymer chain resulting a cyclic ß-ketoester structure ("intramolecular polymer termination";
"backbiting"):

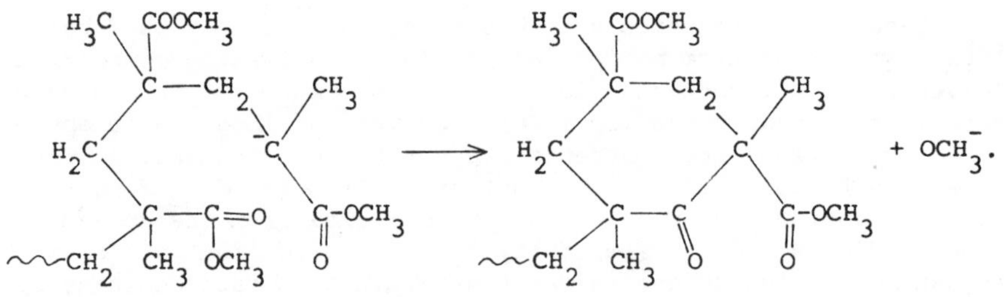

All three reactions produce methoxide as a by-product. Wiles and Bywater ([49],[5o]) and Mita et al. ([22]) showed, that nearly all methoxide is produced in the initial stage of the polymerization. As there is no polymer present at that time, the methoxide must be generated by the attack of the initiator onto the carbonyl group of the monomer. The resulting vinyl ketone may add another initiator molecule at the carbonyl group, if the initiator/monomer ratio is high enough and the initiator is not too bulky. The resulting carbinol was detected by Schreiber ([48]), using phenyllithium in diethylether, Kawabata and Tsuruta ([51]), using butyllithium in hexane and THF, and Hatada ([52],[53],[54]), using butyllithium in toluene. The vinyl ketone may as well add "normally", i. e. with its vinyl group, to the initiator or to a propagating chain end. The latter was postulated by Bullock et al. ([68]) and by Hatada et al. ([53],[54]) based on ESR and NMR results, respectively.

The involvement of the carbonyl group is shown indirectly by the lack of termination in the polymerization of tert.-butyl methacrylate ([38]). The positive inductive effect of the tert.-butyl group decreases the carbonyl activity and hence the termination rate constant.

The methoxide formed in any of the reactions mentioned above is able to initiate the polymerization of MMA in THF, in the contrast to non-polar solvents, as was shown by Müller et al. ([47]). The initiation rate, however, is very slow, so that this effect only gains importance when all the chains started initially have been terminated before total monomer conversion. The newly initiated living end again can suffer from termination, producing methoxide, etc., resulting in a transfer reaction.

The formation of butane, detected by gas chromatography, when initiating with butyllithium in hexane and THF was attributed by Kawabata and Tsuruta ([51]) to the metallation of the monomer by the initiator, leading to an allylic anion:

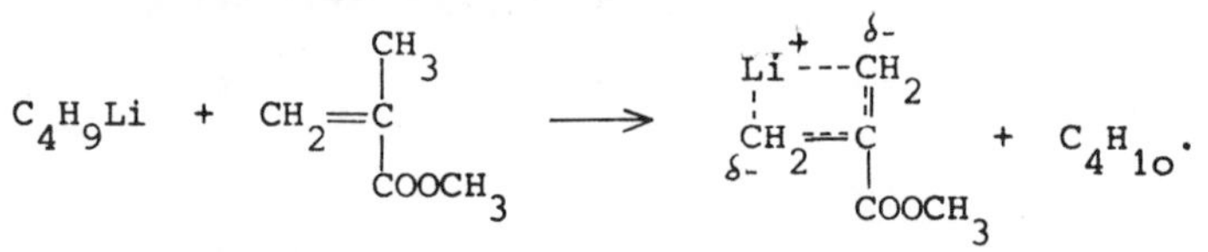

Tsvetanov et al. ([55]) found evidence for this allylic anion from IR spectroscopic measurements on living oligomers in THF. However, this anion may be reactive enough to initiate the polymerization. Thus, this is not a termination reaction. The butane formation was ascribed to unreacted butyllithium by Hatada et al. (54) based on experiments with deuterated initiator.

Up to now there has been no evidence for the existence of an intermolecular termination reaction under normal conditions. Even at room temperature Rempp et al. (56) did not detect a grafting reaction. Living polystyrylsodium, which is much more reactive than PMMA-Na is, attacked the ester groups of a preformed polymer very slowly at room temperature. Neither did Gerner et al. (46) find high molecular weight tailings in the GPC eluograms using Na^+ as the counterion in THF up to $0^{\circ}C$. Trekoval (57) reported a grafting reaction leading to one branch per 10^4 monomer units after 18 h at $+ 20^{\circ}C$ using butyllithium in toluene. As the rate of polymerization is very much higher, this cannot account for termination during polymerization.

By its characteristic carbonyl IR absorption band the product of intramolecular termination, i. e. the cyclic ß-ketoester was detected by Goode et al. (58), Glusker et al. (59), and Owens et al. (60) as one of the products of the oligomerization of MMA with phenylmagnesium bromide or fluorenyllithium in toluene. The same compound was found by Lochman et al. (61) when using metalated methylisobutyrate as the initiator in a THF/pentane mixture at high initiator/monomer ratios.

Allen et al. (62) measured a first-order decrease of the maximum of the UV spectrum of PMMA-Na in toluene/THF mixtures at $+18^{\circ}C$. The same was found by Löhr (63) for PMMA-Cs in THF. The first-order decrease can be explained by intramolecular polymer termination. However, the rate constants measured are too low to explain the termination during polymerization.

Assuming monomer termination as the main termination reaction, Löhr et al. (64) estimated the termination rate constant from the molecular weight distribution of the polymers. No direct evidence for the termination mechanism assumed was given, however.

Direct kinetic methods for the determination of the termination rate constant during polymerization: As a direct method to determine the number of living ends during polymerization, Warzelhan et al. (26) applied the method of labelling the polymers still active with tritiated acetic acid (CH_3COOT). This method had been introduced by Glusker et al. (65). As the design of the reactor used makes it possible to draw samples and terminate them with CH_3COOT every 2 s, it was possible to monitor the decrease of the concentration of living ends during polymerization (40 to 95% conversion). Figure 5 shows, that the decrease of the living end concentration is first order with respect to c^* and that it is not dependent on the actual monomer concentration. This indicates, that polymer termination ("backbiting") is the predominant termination reaction.

This result appears reasonable, because the carbonyl activity of saturated esters, i. e. the polymer, usually is higher than

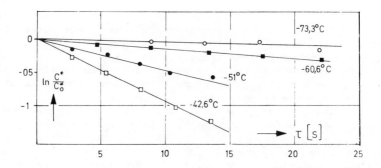

Figure 5. First-order plot of the decrease of living ends in the anionic polymerization of MMA in THF using Na⁺ as the counterion (26)

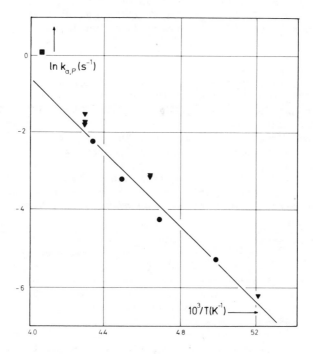

Figure 6. Arrhenius plot of the termination rate constants in the anionic polymerization of MMA in THF using Na⁺ as the counterion (26, 66)

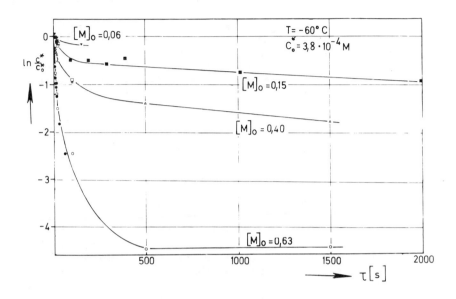

Figure 7.   *First-order plot of the decrease of living ends during and after polymerization of MMA in THF, counterion Na⁺. Dependence on initial monomer
concentration.*

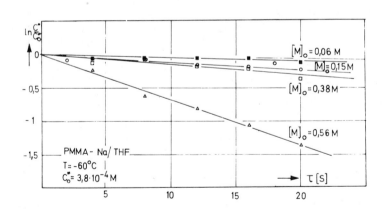

Figure 8.   *Dependence of the initial decrease of living ends on the initial monomer
concentration in the anionic polymerization of MMA in THF using Na⁺ as the
counterion*

that of unsaturated ones, i. e. the monomer. Moreover, the precursor species is identical with one of the possible propagating species (II).

Mathis and François (37) followed the termination reaction of PMMA$_2$Ba in THF by adding monomer at certain time intervals and measuring the first-order time-conversion curve for monomer addition, the slope of which is proportional to c^*. The first-order decrease of living end concentration, which was independent of the concentration of added monomer, was also attributed to intramolecular termination.

Both groups obtained straight Arrhenius plots for the termination rate constants (cf. Figure 6).

Using the labelling method, Gerner (66) tried to verify the termination mechanism by measuring the concentration of active species for a longer period, i. e. during polymerization and after total monomer conversion. The measurements revealed, that the rate of termination sharply decreases at longer times, when little or no monomer is left (cf. Figure 7). In some experiments, the living end concentration even remained constant after some time. The difference between the initial and the final living end concentration is roughly proportional to the initial monomer concentration $[M]_o$; this is also true for the initial rate of termination (cf. Figure 8). It is remarkable, that there is no dependence on the actual monomer concentration, or on the initial concentration of living ends.

Up to now it has not been possible to prove a mechanism which is able to explain all these effects. Neither intramolecular termination alone, nor a combination of intramolecular termination and monomer termination fit the data. However, they can be fitted by a hypothetical kinetic scheme which assumes that a slow reaction takes place between the living end and a deactivating species which is produced in the initial stage of the reaction. The dependence of the rate of termination on the initial monomer concentration can only be explained on the basis of a side reaction occuring in the initiation process. To verify this model, however, the deactivating species has to be identified by analyzing the oligomers produced in the initiation step.

Literature Cited

1. Barnikol, W.K.R.; Schulz, G.V. Z. Physik. Chem. (Frankfurt am Main) 1965, 47, 89
2. Shimomura, T.; Tölle, K.J.; Smid, J.; Szwarc, M. J. Am. Chem. Soc. 1967, 89, 796
3. Szwarc, M. "Carbanions, Living Polymers, and Electron Transfer Processes"; Interscience, New York, NY, 1968
4. Böhm, L.L.; Chmelir, M.; Löhr, G.; Schmitt, B.J.; Schulz, G.V. Adv. Polym. Sci. 1972, 9, 1
5. Schmitt, B.J.; Schulz, G.V. Eur. Polym. J. 1965, 11, 119

6. Böhm, L.L.; Löhr, G.; Schulz, G.V. Ber. Bunsenges. 1974, 78, 1o65
7. Bunge, M.; Höcker, H.; Schulz, G.V.; Makromol. Chem. 1979, 18o, 2637
8. Szwarc, M.; Ed. "Ions and Ion Pairs in Organic Reactions", Wiley-Interscience, New York, NY, Vol. 1: 1972; Vol. 2: 1974
9. Wiles, D.M., in: "Structure and Mechanisms in Vinyl Polymerization", T. Tsuruta and K.F. O'Driscoll, Eds; Marcel Dekker, New York, NY, 1969; p. 233
1o. Szwarc, M., in ref. 8, Vol. 2, p. 4o1
11. Bovey, F.A. "High Resolution NMR of Macromolecules", Academic Press, New York, NY, 1972; p. 65 and 146
12. Fox, T.G.; Garrett, B.G.; Goode, W.E.; Gratch, S.; Kincaid, J.F.; Spell, A.; Stroupe, J.D. J. Am. Chem. Soc. 1958, 8o, 1768
13. Wenger, F. Chem. Ind. (London) 1959, 1o94
14. Graham, R.K.; Dunkelberger, D.L.; Cohn, E.S. J. Polym. Sci. 196o, 42, 5o1
15. Roig, A.; Figueruelo, J.E., Llano, E. J. Polym. Sci. B 1965 3, 171
16. Guzmán, G.M.; Bello, A. Makromol. Chem. 1967, 1o7, 46
17. Fowells, W.; Schuerch, C.; Bovey, F.A.; Hood, F.P.; J. Am. Chem. Soc. 1967, 89, 1396
18. Figueruelo, J.E. Makromol. Chem. 197o, 131, 63
19. Löhr, G.; Schulz, G.V.; Makromol. Chem. 1973, 172, 137
2o. Löhr, G.; Schulz, G.V.; Eur. Polym. J. 1974, 1o, 121
21. Chmelir, M.; Schulz, G.V. Ber. Bunsenges. 1971, 75, 83o
22. Mita, I.; Watanabe, Y.; Akatsu, T.; Kambe, H. Polymer J. 1973, 4, 271
23. Warzelhan, V.; Schulz, G.V. Makromol. Chem. 1976, 177, 2185
24. Warzelhan, V.; Höcker, H.; Schulz, G.V. Makromol. Chem. 198o, 181, 149
25. Kraft, R.; Müller, A.H.E.; Warzelhan, V.; Höcker, H.; Schulz, G.V. Macromolecules 1978, 11, 1o93
26. Warzelhan, V.; Höcker, H.; Schulz, G.V. Makromol. Chem. 1978 179, 2221
27. Warzelhan, V.; Löhr, G.; Höcker, H.; Schulz, G.V.; Makromol. Chem. 1978, 179, 2211
28. Löhr, G.; Schulz, G.V.; Eur. Polym. J. 1975, 11, 259
29. Müller, A.H.E.; Höcker, H.; Schulz, G.V. Macromolecules 1977, 1o, 1o86
3o. Fisher, M.; Szwarc, M. Macromolecules 197o, 3, 23
31. Tardi, M.; Sigwalt, P. Eur. Polym. J. 1973, 9, 1369
32. Honnoré, D.; Favier, J.C.; Sigwalt, P.; Fontanille, M. Eur. Polym. J. 1974, 1o, 425
33. Kraft, R.; Müller, A.H.E.; Höcker, H.; Schulz, G.V. Makromol. Chem. Rapid Commun. 198o, 1, 363
34. Eizner, Yu.Ye.; Erussalimsky, B.L.; Eur. Polym. J. 1976, 12, 59

35. Vancea, L.; Bywater, S. IUPAC 26th International Symposium on Macromolecules Mainz 1979, preprints p. 14o
36. Müller, A.H.E., Schmitt, H.-J.; Löhr, G.; Schulz, G.V.; in preparation
37. Mathis, C.; François, B. 1st European Discussion Meeting on "New Developments in Ionic Polymerization", Strasbourg 1978, preprints p. 75
38. Müller, A.H.E.; Makromol. Chem. 1981, 182, in press
39. Bevington, J.C.; Harris, D.O.; Rankin, F.S. Eur. Polym. J. 197o, 6, 725
4o. Ito, K.; Sugie, T.; Yamashita, Y. Makromol.Chem.1969,125,291
41. Yuki, H.; Okamoto, Y.; Hatada, K. Polymer J. 1974, 6, 573
42. Yuki, H.; Okamoto, Y.; Ohta, K.; Hatada, K. J. Polym. Sci., Polym. Chem. Ed. 1975, 13, 1161
43. Yuki, H.; Hatada, K.; Ohta, K.; Okamoto, Y. J. Macromol. Sci., Chem. 1975, 9, 983
44. Yuki, H.; Okamoto, Y.; Shimada, Y.; Ohta, K.; Hatada, K. J. Polym. Sci, Polym. Chem. Ed. 1979, 17, 1215
45. Yuki, H.; Hatada, K. Adv. Polym. Sci. 1979, 31, 1
46. Gerner, F.J.; Kraft, R.; Müller, A.H.E.; in preparation
47. Müller, A.H.E.; Warzelhan, V.; Höcker, H.; Löhr, G.; Schulz, G.V.; IUPAC International Symposium on Macromolecules Dublin 1977, preprints p. 31
48. Schreiber, H.; Makromol. Chem. 196o, 36, 86
49. Wiles, D.M.; Bywater, S. Chem. Ind. (London) 1963, 12o9
5o. Wiles, D.M.; Bywater, S. Trans. Faraday Soc. 1965, 61, 15o
51. Kawabata, N.; Tsuruta, T. Makromol. Chem. 1965, 86, 231
52. Hatada, K.; Kitayama, T.; Fujikawa, K; Ohta, K; Yuki, H. Polymer Bull. 1978, 1, 1o3
53. Hatada, K.; Kitayama, T.; Nagakura, M.; Yuki, H. Polymer Bull.
54. Hatada, K.; Kitayama, T., Fujikawa, K.; Ohta, K.; Yuki, H.; this book
55. Tsvetanov, Ch.B.; Petrova, D.T.; Pham H. Li; Panayotov, I.M. Eur. Polym. J. 1978, 14, 25
56. Rempp, P.; Volkov, V.I.; Parrod, J.; Sadron, Ch. Bull. Soc. Chim. France 196o, 919
57. Trekoval, J.; Kratocvil, P. J. Polym. Sci. A-1, 1972, 1o, 1391
58. Goode, W.E.; Owens, F.H.; Myers, W.L. J. Polym. Sci. 196o, 47, 75
59. Glusker, D.L.; Lysloff, I.; Stiles, E. J. Polym. Sci. 1961, 49, 315
6o. Owens, F.H.; Myers, W.L.; Zimmerman, F.E. J. Org. Chem. 1961 1961, 26, 2288
61. Lochmann, L.; Rovova, M; Petranek, J.; Lim, D. J. Polym. Sci. A-1, 1974, 12, 2295
62. Allen, P.E.M.; Chaplin, R.P.; Jordan, D.O. Eur. Polym. J. 1972, 8, 271
63. Löhr, G.; unpublished results
64. Löhr, G.; Müller, A.H.E.; Warzelhan, V.; Schulz, G.V. Makromol. Chem. 1974, 175, 497

65. Glusker, D.L.; Stiles, E.; Yonkoskie, B. J. Polym. Sci. 1961, 49, 297
66. Gerner, F.J.; PhD Thesis, Mainz 1981
67. Fuoss, R.M. J. Am. Chem. Soc. 1958, 8o, 5o59
68. Bullock, A.T.; Cameron, G.G.; Elsom, J.M. Eur. Polym. J. 1977, 13, 751
69. Persoons, A.; this book

RECEIVED March 2, 1981.

Anionic Polymerization of Isoprene by the Complexes Oligoisoprenyllithium/Tertiary Polyamines in Cyclohexane. I. Kinetic Study

SYLVIANE DUMAS, JOSEPH SLEDZ, and FRANCOIS SCHUÉ

Laboratoire de Chimie Macromoléculaire, Université des Sciences et Techniques du Languedoc, Place Bataillon, 34060 Montpellier Cedex, France

The influence of two tertiary polyamines on the a-
nionic polymerization of isoprene in cyclohexane
has been studied. TMEDA (N,N,N',N'-tetramethyl-
ethylenediamine) and PMDT (pentamethyldiethylene-
triamine) can increase or decrease the propagation
rate, depending on the concentration range. The
results are discussed on the basis of the nature
of the living species.

The kinetic of the anionic polymerization of dienes and
styrene has been intensively investigated by many workers (1-10)
mainly by BYWATER et al (1-6, 8, 10), but in absence of complexing agents like amines.

Recently several studies have focused on the nature of the active species in the polymerization of ethylene and conjugated
dienes initiated by the chelate of butyllithium and N,N,N',N'-
tetramethylethylenediamine (TMEDA).

LANGER (11) has postulated the presence of a 1:1 chelate and attributed its reactivity in polymerizing ethylene and butadiene
to the absence of association of the metal alkyl and the increased polarization of the metal alkyl bond.

Recent studies, however, suggest a more complicated mechanism.

0097–6156/81/0166–0463$05.00/0

HAY et al ($\underline{12}$, $\underline{13}$) showed that the addition of TMEDA to butyllithium (BuLi) produces a remarkable increase in reactivity to-
ward the polymerization of butadiene. This higher reactivity is
attributed to the absence of association of alkyllithium species
and the presence according to the maximum rate of polymerization, of a separated ion pair during the polymerization.
ERUSSALIMSKY et al($\underline{14}$) described the anionic polymerization of
isoprene in the presence of TMEDA. They showed that the tertiary
diamine causes a significant increase of the polymerization rate
and of the content of 3,4-links in the polymers formed, a plateau
being reached for r = TMEDA/living species = 4.
More recently ERUSSALIMSKY et al ($\underline{15}$) investigated the polymerization of 2,3-dimethylbutadiene induced by oligo-2,3 dimethyl-
butadienyllithium/TMEDA. Contrary to butadiene and isoprene, catalytic amounts of TMEDA decrease the propagation rate of dime-
thylbutadiene.
HELARY and FONTANILLE ($\underline{16}$) have studied the system polystyrylli-
thium/TMEDA in cyclohexane solution. The behaviour depends on
the concentration range : for high concentration of living ends
(8.3 x 10^{-3}mole.1^{-1}) the addition of TMEDA increases the overall
reactivity of the system, at low concentration
(9.2 x 10^{-4}mole.1^{-1}) a slight decrease of the reactivity is noted
upon addition of TMEDA.
So, it seems that there exists a discrepancy about the mechanism
of polymerization. Until this is resolved the proposed mechanism
of the reaction must be in doubt. Therefore, we have realized a
systematic study (UV, NMR, kinetic) of the influence of several
factors determining the reactivity of active centres in the case
of the polymerization of isoprene ($\underline{17}$, $\underline{18}$, $\underline{19}$).
The purpose of this paper is to present some kinetic results re-
lative to the propagation reaction in the polymerization of isoprene by oligoisoprenyllithium complexed with TMEDA or PMDT (pen-
tamethyldiethylenetriamine) in cyclohexane.

Kinetic study of the propagation reaction by ultra violet spectroscopy.

The initiation step is followed by ultra violet spectroscopy
at 273 nm (polyisoprenyllithium absorption). The following experimental conditions are choosen:

    temperature : 18 °C
    t-BuLi   = 4 x 10^{-4} mole.1^{-1}
    isoprene   = 7 x 10^{-4} mole.1^{-1}

After 4 or 5 hours the reaction is complete and a sigmoïdal living ends formation curve is observed (1). The degree of polymerization of the oligoisoprenyllithium equals about ten. By
addition of catalytic amounts of complexing agent (TMEDA or PMDT)
the remarkable increase of the initiation rate is not measurable
by ultra violet spectroscopy. Also, only the propagation step was
studied.

We have followed the propagation rate of the polymerization of
isoprene initiated by cyclohexane solutions of the complexes oli-
goisoprenyllithium/TMEDA and oligoisoprenyllithium/PMDT in two
very different ranges of living ends concentration. In each case
we have determined the order of the reaction with respect to
the living ends concentration.

The monomer consumption is followed at 255 nm (ε isoprene = 39)
and the propagating specie is an oligoisoprenyllithium which
polymerization degree equals ten. The isoprene consumption in
presence of TMEDA or PMDT is characterized by two straight lines
owing for the fact that the complexes polyisoprenyllithium/che-
lating agents present a very high stability (figure 1) ; the
propagation rate v_p = $-d[M]$ /dt is given by the slope of the
lines.

Variation of v_p in function of r = amine/living ends

   Complexation with TMEDA.   The kinetic of the polymerization has been studied by varying the concentration of living
ends and the molar ratio r = $\left[\text{TMEDA}\right]$ / $\left[\text{living ends}\right]$.
For "low" and constant concentrations of polyisoprenyllithium
(3.5×10^{-4} mole.1^{-1}) TMEDA decreases the propagation rate by
varying the ratio r from 0 to 0.5 ; for higher values of r the
tertiarydiamine does not affect the polymerization rate, at least
no sensible variation is observed (figure 2). In order to follow
the propagation rate as a function of the concentration of TMEDA,
we have standardized v_p with respect to the monomer concentration
and plotted $v_R = v_p / \left[\text{M}\right]$ versus r (figure 2). $\left[\text{M}\right]$ represents the
monomer concentration.
For "high" and constant concentrations of polyisoprenyllithium
(5×10^{-3} mole.1^{-1}), TMEDA affects drastically the propagation
rate by varying the ratio r from 0 to 1.0 (figure 3). However a
low increase of the rate is observed for r values less than 0.5.

   Complexation with PMDT.   The study developed above has been
repeated using a tertiary triamine (PMDT).
For "low" concentration of polyisoprenyllithium
(4.5×10^{-4} mole.1^{-1}) PMDT decreases the propagation rate by varying the molar ratio r = $\left[\text{PMDT}\right]$ / $\left[\text{living ends}\right]$ from 0 to 1
(figure 4) ; the effect is very sensible between 0 and 0.5, a
low decrease being observed between 0.5 and 1.
For "high" concentrations of polyisoprenyllithium (10^{-2} mole.1^{-1}),
PMDT affects the propagation rate like TMEDA. Respectively a low
and a high increase are observed for r values less than 0.5 and
r values included between 0.5 and 1 (figure 5).
The presence of a plateau for r values $\geqslant$ 1 suggests the existence
of a 1 : 1 active chelate. This complexe justifies the micro-

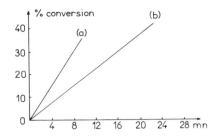

Figure 1. Influence of the chelating agent on the propagation rate in the polymerization of isoprene at 18°C in cyclohexane: (a) r = [TMEDA]/[PILi] = 1.0; *PILi* = 5.4 × 10⁻³ *mol · L⁻¹; (b)* r = [PMDT]/[PILi] = 0.9; *PILi* = 10⁻² *mol · L⁻¹.*

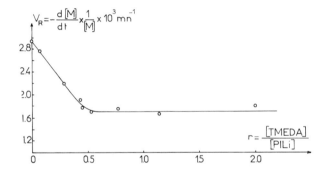

Figure 2.   Influence of TMEDA on the propagation rate in the polymerization of isoprene in cyclohexane at low PILi concentration.

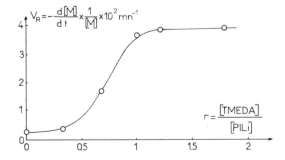

Figure 3.   Influence of TMEDA on the propagation rate in the polymerization of isoprene in cyclohexane at high PILi concentration: [PILi] = 5 × 10⁻³ *mol · L⁻¹; 18°C.*

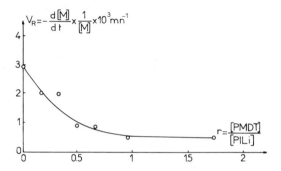

Figure 4. Influence of PMDT on the propagation rate in the polymerization of isoprene in cyclohexane at low PILi concentration: [M] = 0.34 mol⁻¹; [PILi] = 4.5 × 10⁻⁴ mol · L⁻¹; 18°C.

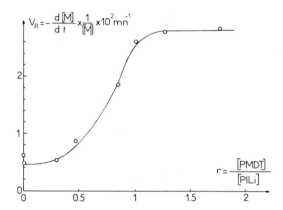

Figure 5. Influence of PMDT on the propagation rate in the polymerization of isoprene in cyclohexane at high PILi concentration: [PILi] = 10⁻² mol · L⁻¹; 18°C.

structure of the polymers obtained by polymerizing anionically isoprene in presence of tertiary polyamines (17, 19). Moreover, ultra violet spectroscopy and ^{1}H NMR studies of the living species take in account the 1 : 1 chelate (17, 18). According to some published results, it seems that the complexe is not aggregated. By determining the orders with respect to the complexes polyisoprenyllithium/TMEDA and polyisoprenyllithium/PMDT some new informations may be available about the aggregation when r is equal to the unity.

Determination of the orders

Order with respect to the polyisoprenyllithium concentration [PILi] . In the propagation reaction we have determined the order with respect to polyisoprenyllithium in absence of any chelating agent at 18°C (figure 6). We found 0.3, value which may be assimilated to 0.25 as given by BYWATER AND WORSFOLD (1). The following mechanism is proposed (1) :

$$(PILi)_4 \rightleftharpoons 4\ PILi$$

$$PILi + monomer \longrightarrow propagation$$

The non aggregated monomeric polyisoprenyllithium is the active specie.

Complexation with TMEDA. The following experimental conditions are required :
temperature : 18°C
r = 1
[PILi] and [TMEDA] varying from 3.10^{-4} to 10^{-2} mole.1^{-1}
The results plotted on figure 6 show that the order with respect to the active species concentration is equal to 1.1 presuming that the complex PILi/TMEDA is monomeric.

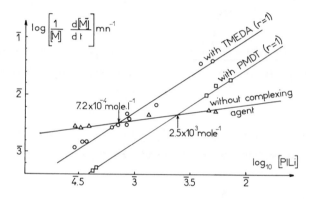

Figure 6. Influence of PILi concentration of the propagation rate in the polymerization of isoprene in cyclohexane without and with complexing agents
(TMEDA and PMDT).

The two straight lines obtained in presence and in absence of TMEDA cross at a point corresponding to a polyisoprenyllithium concentration of 7.2×10^{-4} mole.1^{-1} ; for that value any addition of TMEDA does not affect the propagation rate of the polymerization. Below this point the rate decreases and above this point the rate increases in presence of catalytic amounts of TMEDA.

Complexation with PMDT. The study developed above has been repeated using the tertiary triamine PMDT. The results are analogous to those found in presence of TMEDA (figure 6). A first order reaction with respect to the complexe PILi/PMDT is evidenced : so it seems that the active propagating species is the 1 : 1 chelate.

The two crossing lines obtained in presence and in absence of PMDT (figure 6) show a critical point at 2.5×10^{-3} mole.1^{-1} (7.2×10^{-4} mole.1^{-1} for TMEDA).

Moreover, it is possible to determine from figure 6 the values of v_R corresponding to the complexes PILi/TMEDA and PILi/PMDT. Such investigation evidences the higher reactivity of the 1 : 1 chelate PILi/TMEDA toward the 1 : 1 complexe PILi/PMDT.

Discussion

The results obtained can be interpreted by considering the following equilibria also proposed by HELARY and FONTANILLE in the case of polystyryllithium (16) :

$$(PILi)_4 \quad \rightleftharpoons \quad 4\ PILi \qquad\qquad\qquad\qquad (a)$$

$$(PILi)_4 + 4\ amine \quad \rightleftharpoons \quad 4\ (PILi/amine) \qquad (b)$$

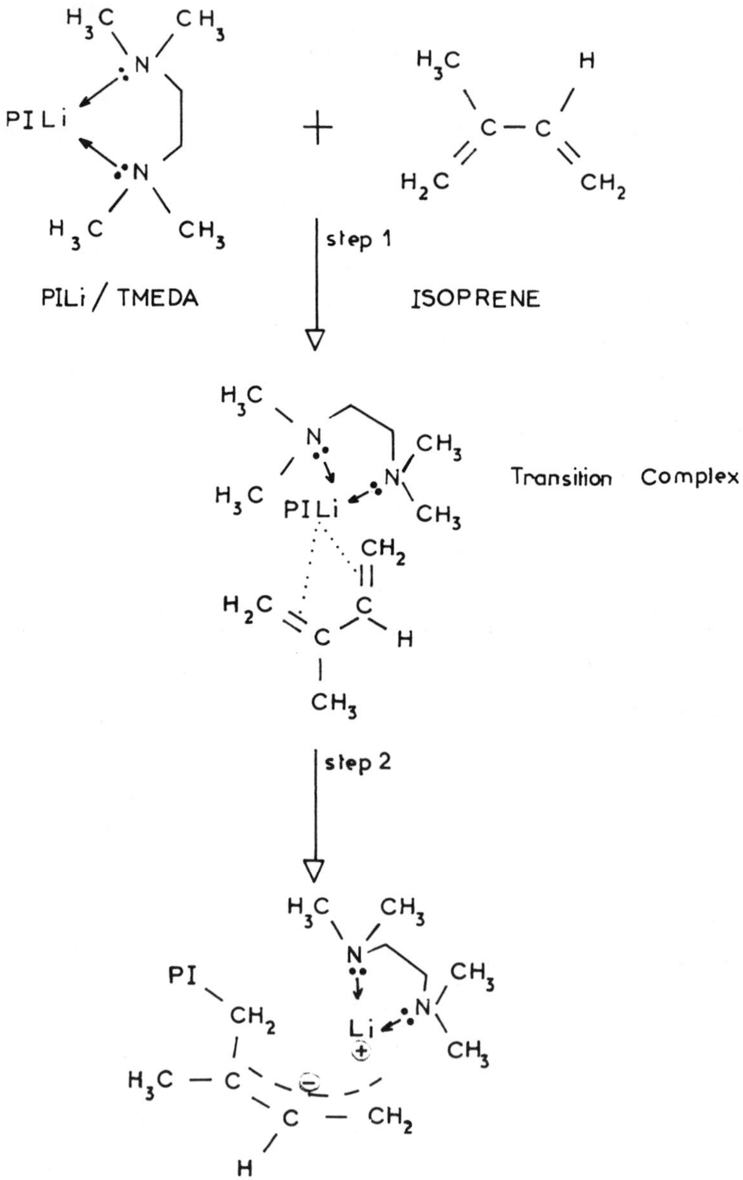

PILi / TMEDA   ISOPRENE

step 1

Transition   Complex

step 2

The equilibrium (a) should be stroungly displaced to the left
and equilibrium (b) to the right.

At low concentrations range the species tend to dissociate,
increasing the concentration of non aggregated PILi, but the
complexation by means of the amine drops the reactivity. At high
concentrations range, the enhanced reactivity is explained by a
high concentration of complexed specie (PILi/amine) despite its
low reactivity compared to the non complexed specie.

The following reactivity order must be respected :

PILi > (PILi/amine)

A two step mechanism can be developed.

Using the same scheme, PMDT leads to a complexe in which three
nitrogen atoms surround the lithium atom. In step 1 occurs a
precomplexation of the monomer which is then inserted in the
propagating chain (step 2).

In both cases, the protonic nmr study of the living species (18)
shows the existence of the 1 : 1 chelate ; the structural (19)
and the kinetic studies confirm that existence.

It seems also reasonable to think that substituting TMEDA by
PMDT provokes a steric hindrance in the monomer insertion, involving a diminution of reactivity ; this point is more developed in a coming paper about the polymerization of isoprene with
the complexe oligoisoprenyllithium/hexamethyltriethylenetetrami-
ne.

Experimental

The purification of chemicals, the synthesis and the study
of the living species using protonic nuclear magnetic resonance
and ultra violet spectroscopy have been intensively described in
other papers (18, 19).

The kinetics have been performed on a CARRY 118 C spectrometer.

Isoprene presents a maximum of absorption at 223 nm but its molar absorptivity is very high ($\varepsilon_{max.}$ = 23,000 $cm^{-1}.1.mole^{-1}$) ; for that reason and in order to follow the propagation rate, the consumption of isoprene has been studied at 255 nm (ε = 39). Oligoisoprenyllithium of polymerization degree equal to 10, uncomplexed or complexed (with TMEDA or PMDT), was the living specie used to "initiate" the propagation of the monomer.
All the experiments were carried out at 18°C under high vacuum (10^{-6} mm of mercury) in sealed glass vessel supplied with an ultra-violet cell. Before polymerization the apparatus is carefully and successively washed with a cyclohexane solution of butyllithium and cyclohexane. The stability of complexed polyisoprenyllithium has been verified by protonic nuclear magnetic resonance and ultra violet spectroscopy (18)

Literature cited

1. Worsfold D.J. and Bywater S.. Can. J. Chem., 1964, 42, 1.

2. Bywater S. and Worsfold D.J.. Adv. Chem. Sci., 1966, 52, 36

3. Roovers J.E.L. and Bywater S.. Macromolecules, 1968, 1, 328.

4. Roovers J.E.L. and Bywater S. Macromolecules, 1973, 8, 251

5. Worsfold D.J. and Bywater S.. Can. J. Chem., 1960, 38, 1891

6. Bywater S. and Worsfold D.J.. J. Organometal. Chem., 1967, 10, 1

7. Kropacheva E.N.. Dokl. Akad. Nauk, SSR, 1960, 130, 1253

8. Bywater S. and Worsfold D.J.. Can.J. Chem., 1962, 40, 1964.

9. Szwarc M. "Carbanions, living polymers and electron transfer processes" ; Interscience : New York, 1968.

10. Bywater S. and Alexander I.J. J. Polym. Sci., part A1, 1968, 6, 3407

11. a) Langer, A.W. Trans N.Y. Acad. Sci., 1965, 27, 742

    b) Langer A.W. Polym. Preprints, 1966, 7, 132.

12. Hay J.N. ; McCABE J.F. and Robb J.C. Faraday Trans. I, 1972, 68, 1.

13. Hay J.N. and McCABE J.F.. J. Polym. Sci., Chem. Ed., 1972 10, 3451

14. Davidjan A. ; Nikolaew N. ; Sgonnik V. ; Belenkii B. ; Nesterow V. and Erussalimsky B.. Makromol. Chem., 1976, 177, 2469.

15. Smirnowa N. ; Sgonnik V. ; Kalninsch K. and Erussalimsky B. Makromol. Chem., 1977, 178, 773.

16. a) Helary G. and Fontanille M. 1st European Discussion meeting on Polymer Science, Strasbourg (FRANCE), Feb. 27-March 2, 1978

    b) Helary G. and Fontanille M. Europ. Polym.J., 1978, 14(5) 345.

17. Dumas S. ; Marti V. ; Sledz J. and Schué F. 1st European Discussion meeting on Polymer Science, Strasbourg (FRANCE), Feb. 27-March 2, 1978.

18. Part II, Collet V. ; Dumas S. ; Sledz J. and Schué F. to be published.

19. Part III, Dumas S. ; Sledz J. and Schué F. to be published.

RECEIVED March 24, 1981.

Preparation of Polymeric Free-Radical Initiators by Anionic Synthesis: Polymeric Azo Derivatives

G. RIESS and R. REEB

Ecole Nationale Supérieure de Chimie de Mulhouse, France

The anionic polymerization is particularly suitable for the preparation of block copolymers ($\underline{1}$, $\underline{2}$). This method, which leads to well defined copolymers of low polydispersity, is however restricted to monomers of low polarity like isoprene, butadiene or styrene.

More polar monomers, as for instance vinyl chloride or maleic anhydride cannot be polymerized by this technique. Thus, in order to obtain block copolymers containing a sequence with such polar monomer units, it is necessary to use a *free radical polymerization step*.

The initiation of this type of polymerization can be achieved with a *polymeric initiator*, e.g. a polymer containing peroxide or azo end groups. As the different types of synthesis which have been described for the preparation of such polymeric initiators, have generally the **disadvantage to need several reaction steps** ($\underline{3}$-$\underline{5}$), it seemed interesting to prepare these initiators in a *one step reaction*, by keeping the different advantages of the anionic polymerization.

Polymeric peroxides have thus been obtained by desactivation of a living anionic polymer either with functionalized peroxide ($\underline{6}$) or directly with oxygen ($\underline{7}$-$\underline{10}$).

In a similar way, polymeric azo initiators have been prepared by desactivation of a living anionic polymer with functionalized azo derivatives as azobisisobutyronitrile ($\underline{11}$, $\underline{12}$). Concerning this reaction with azobisisobutyronitrile (AIBN), no details on the methanism have been given in the patent litterature.

In a preliminary study ($\underline{13}$) of the reaction betwen monofunctional living polystyrene and AIBN, we have shown that a *coupling reaction* occurs leading to the incorporation of azo groups either in the middle of the chain or at the end.

The polymers containing such azo groups can be represented schematically by

$$P - N = N - P \quad \text{and} \quad P - N = N - R$$

0097–6156/81/0166–0477$05.00/0

where P is a polystyrene chain and

$$R = \begin{array}{c} CH_3 \\ | \\ C \\ | \\ CN \end{array} \begin{array}{c} CH_3 \\ \end{array}$$

To complete this previous work, it was interesting to study in detail the reaction between AIBN and a living anionic polystyrene, for monofunctional polystyrene $P^{\ominus} Me^{\oplus}$ as well as for difunctional polystyrene $Me^{\oplus \ominus}P^{\ominus} Me^{\oplus}$.

The efficiency of the coupling reaction will therefore be examined as a function of :
- the type of counter-ion $Me^{\oplus}$ where

$$Me^{\oplus} = Li^{\oplus}, Na^{\oplus}, K^{\oplus}, Rb^{\oplus}, Cs^{\oplus}$$

- the type of carbanion

$$P - CH_2 - \overset{H}{\underset{\text{(phenyl)}}{\overset{|}{C}}}^{\ominus} \qquad P - CH_2 - \overset{CH_3}{\underset{\text{(phenyl)}}{\overset{|}{C}}}^{\ominus} \qquad P - CH_2 - \overset{\text{(phenyl)}}{\underset{\text{(phenyl)}}{C}}^{\ominus}$$

- the temperature.
Furthermore, by using AIBN labeled with ^{14}C on the C ≡ N group, the problem of the reaction mechanism will be approached, which might be of 2 types :

$$P^{\ominus} Me^{\oplus} + \overset{CH_3}{\underset{CH_3}{>}} \overset{}{\underset{C \equiv N}{\overset{|}{C}}} - N = N - \overset{}{\underset{C \equiv N}{\overset{|}{C}}} - \overset{CH_3}{\underset{CH_3}{<}} \longrightarrow$$

(1a) $\overset{}{\underset{P}{>}}C - N = N - \overset{}{\underset{\underset{Me^{\oplus}}{N^{\ominus}}}{\overset{|}{C}}}< \xrightarrow[H_2O]{H^{\oplus}} \overset{}{\underset{P}{>}}C - N = N - \overset{}{\underset{\underset{Me^{\oplus}}{N^{\ominus}}}{\overset{|}{C}}}<$

(1b) $\overset{}{\underset{P}{>}}C - N = N - \overset{}{\underset{C = O}{\overset{|}{C}}}< \qquad \overset{}{\underset{C = O}{}}$

(2) $\overset{}{\underset{P}{>}}C - N = N - \overset{}{\underset{P}{\overset{|}{C}}}< \qquad + \begin{array}{c} 2 CN^{\ominus} \\ 2 Me^{\oplus} \end{array}$

EXPERIMENTAL

The living anionic polystyrene are prepared according to actually well known techniques (14). The initiation is achieved

with the following *monofonctional initiators* : sec BuLi, phenylisopropyl K, phenylisopropyl Rb or Cs, benzyl Na (15). The *difunctional initiators* were naphthalene Li,α methylstyrene tetramer of
Na or K.

The polymerization is carried out at -78°C in THF. In order
to modify the reactivity of the terminal carbanion, a small
excess, with respect to the initial concentration of initiator,
either of α methylstyrene or diphenylethylene, is added to the
living polymer.

At this stage a sample of the living polymer is withdrawn and
desactivated with methanol. The molecular weight of this "precursor PS" is determined by GPC.

The remaining part of living polymer is desactivated with
AIBN. AIBN, which is a C_{14} labeled product in some experiments, is
added dropwise to the living polymer during the temperature rise,
between 0 and 20°C. A series of experiments have also been
achieved at constant temperature : -80°C, -30°C and +20°C.

After precipitation in methanol, the coupled polymer is
analysed by GPC and eventually by its radioactivity.

RESULTS

The coupling efficiency for the reaction of *monofunctional*
PS with AIBN is given by

$$\rho = 2 \left(1 - \frac{\overline{M}_{no}}{\overline{M}_n} \right)$$

where $\overline{M}_{no}$ is the molecular weight (number average) of the precursor PS
$\overline{M}_n$ is the molecular weight of the reaction product containing coupled and non coupled chains.

Table 1 shows the variation of ρ as a function of the terminal carbanion and as a function of the counter-ion.

It appears that the highest coupling efficiency is obtained
for a living polystyrene with a α methylstyrene carbanion end-
group and in presence of potassium as counter-ion.

In a similar way, the amount of coupling in the case of
difunctional PS is given by

$$\rho' = \overline{M}_n / \overline{M}_{no}$$

where $\overline{M}_{no}$ and $\overline{M}_n$ have the same signification as before.

The corresponding results are given in table 2, which shows
also the evolution of $\overline{M}_n$ and $\overline{M}_w$ of the different samples. This
gives a clear indication that a multicoupling (coupling of more
than 2 precursor chains) occurs and that the efficiency is again

TABLE 1 : Coupling efficiency ρ as a function of the counter-ion $Me^{\oplus}$ and the end group carbanion of the living anionic polystyrene. System : monofunctional $PS^{\ominus}$ + AIBN

Counter-ion $Me^{\oplus}$	End group carbanion		
	$-CH_2-\overset{H}{\underset{\varnothing}{C}}{}^{\ominus}$	$-CH_2-\overset{CH_3}{\underset{\varnothing}{C}}{}^{\ominus}$	$CH_2-\overset{\varnothing}{\underset{\varnothing}{C}}{}^{\ominus}$
$Li^{\oplus}$	24 %	41 %	6 %
$Na^{\oplus}$	11 %	53 %	21 %
$K^{\oplus}$	19 %	67 %	54 %
$Rb^{\oplus}$	18 %	37 %	29 %
$Cs^{\oplus}$	12 %	34 %	24 %

TABLE 2 : Amount of coupling $\rho' = M_n/M_{no}$ as a function of the counter-ion and the end group carbanion. System : difunctional $^{\ominus}PS^{\ominus}$ + AIBN

$Me^{\oplus}$		Precursor PS	End group carbanion		
			$-CH_2-\overset{H}{\underset{\varnothing}{C}}{}^{\ominus}$	$-CH_2-\overset{CH_3}{\underset{\varnothing}{C}}{}^{\ominus}$	$CH_2-\overset{\varnothing}{\underset{\varnothing}{C}}{}^{\ominus}$
$Li^{\oplus}$	$\overline{M}_n$	98 800	113 400	197 200	112 200
	$\overline{M}_w$	111 900	144 700	389 600	140 300
	ρ'		1.15	2.00	1.14
$Na^{\oplus}$	$\overline{M}_n$	41 400	47 900	85 500	57 400
	$\overline{M}_w$	55 200	83 600	124 100	93 400
	ρ'		1.16	2.06	1.39
$K^{\oplus}$	$\overline{M}_n$	64 900	75 400	182 700	153 500
	$\overline{M}_w$	70 700	100 900	1 040 700	379 500
	ρ'		1.16	2.81	2.36

the highest with PS having α methylstyrene end groups and potassium as counter-ion.

Table 3 shows the coupling efficiency as a function of the desactivation temperature. It gives also the amount of radioactive groups : C_{14} labeled C = O or C = NH groups derived from CN, as determined by radioactivity measurements. It can be noticed that the polymers present only a low amount of C_{14}, ranging from 0.1 % to about 4 % of the theoretical value by considering the reaction mechanism (1a) or (1b) . The reaction mechanism (2) seems therefore to be the predominant one.

The other proves to support this mechanism are :

- *the formation of CN$^{\ominus}$ ions* ; their formation have been shown qualitatively by analysing the methanolic solution after precipitation of the polymer (analysis by colorimetry and detection of radioactive HCN)

- *the absence of oxygen containing groups* like C = O in the polymer after hydrolysis, as expected according to reaction (1b) ; the microanalysis of oxygen has been achieved on a coupled PS fraction isolated by preparative chromatography

- *the low nitrogen content* determined on the same coupled fraction by Kjeldahl analysis ; by this technique only nitrogen of CN and C = NH groups is determined but not that of - N = N - groups.

According to these characteristics of the coupled polymers, reaction (2) can be considered as the predominant one, especially in the case where the coupling efficiency ρ is high. Reaction (1) might enter to some **extent** into competition with reaction (2) for low ρ values.

Thus, the reaction of living anionic polystyrene with AIBN, leading to chain coupling with elimination of CN$^{\ominus}$, seems to be similar to that proposed by YOSHIMURA (15) for the reaction of carbanions with substituted α aminonitriles.

Furthermore, the presence of - N = N - in the middle of a coupled polymer has been proved by thermal or UV cleavage and by formation of free radicals according to the reaction :

$$P - N = N - P \xrightarrow{k_d} 2\ P^{\cdot\cdot} + N_2 \nearrow$$

where $P^{\cdot\cdot}$ is a polymeric free radical.

The kinetics of this decomposition have been approached by microgasometry and by GPC analysis.

Starting with these polymeric initiators, we have finally shown their possibilities for the preparation of block copolymers.

Acknowledgments

Acknowledgment is made to the "Délégation Générale à la Recherche Scientifique et Technique" (Contract N° 76.7.0250) for **partial** support of this research.

The authors are grateful to Professor J.P. FLEURY and Professor R. QUIRK for helpful discussions.

TABLE 3 : Coupling efficiency ρ as a function of the desactivation temperature.
System : monofunctional $PS^{\ominus}$ + AIBN
         carbanion end group : α methylstyrene
         counter-ion   : $K^{\oplus}$

	Precursor PS	Desactivation temperature		
		$- 80°C$	$- 30°C$	$+ 20°C$
$\overline{M}_n$	13 200	15 100	20 000	17 600
$\overline{M}_w$	14 900	18 300	26 600	22 800
ρ	-	24 %	68 %	49 %
mole AIBN /mole PS	-	$3.4 \ 10^{-2}$	$1.1 \ 10^{-3}$	$3.7 \ 10^{-3}$
mole AIBN /mole precursor PS	-	$3.9 \ 10^{-2}$	$2.1 \ 10^{-3}$	$5.4 \ 10^{-3}$

Literature Cited

1. Szwarc M.   Carbanions, Living Polymers and Electron Transfer Processes.   Interscience, New-York (1968)
2. Rempp P. Bull. Soc. Chim. Fr., 1964, 221
3. Foucault A. Thesis, Mulhouse, 1972
4. Vuillemenot J. ; Barbier B. ; Riess G. ; Banderet A.   J. Polym. Sci., 1965, A 3, 1969
5. Vollmert B. ; Bolte H. Makromol. Chem., 1960, 36, 17
6. Riess G ; Palacin F. Inf. Chim., 1973, 116, 9
7. Sumitomo Chemical Co., Ger. Offen. 1971, 2009066
8. Brossas J. ; Catala J.M. ; Clouet G. ; Gallot Z. C.R. Hebd. Séanc. Acad. Sci. Paris, 1974, 278, 1031
9. Vinchon Y. Thesis, Mulhouse, 1975
10. Catala J.M. Thesis, Strasbourg, 1975
11. Mc Leay R.E. ; Sheppard C.S., French Patent 1970, 1586468
12. Echte A. ; Stein D. ; Fahrbach G. ; Adler H. ; Geberding K. Ger. Offen. 21-4-1977, 2546377
13. Vinchon Y. ; Reeb R. ; Riess G. Europ. Polymer J., 1976, 12, 317
14. Shimomura T. ; Tölle K.J. ; Smid J. ; Szwarc M.   J. Am. Chem. Soc., 1967, 89, 796
15. Yoshimura J. ; Ogho Y. ; Sato T. Bull. Chem. Soc. Japan, 1965 38, 1809

RECEIVED June 5, 1981.

Coupling Reactions of Carbanionic Polymers by Elemental Compounds Such as Oxygen and Sulfur

J. M. CATALA, J. F. BOSCATO, E. FRANTA, and J. BROSSAS

Universite Louis Pasteur, Centre de Recherches sur les Macromolecules (CNRS),
6 rue Boussingault, 67038 Strasbourg Cedex, France

Synthesis of well defined functionalized (- telechelic or multifunctional-) macromolecules is an important task for polymer chemists. The polymers with $PO(OR)_2$, - $Si(OR)_3$, -OH, - OOH ... functional groups[1-8] are produced in limited quantities. The need for polymeric materials possessing specific properties has led to a renewed interest is functional polymers, especially if the initial material is a common hydrocarbon polymer. One of the techniques that we use in our laboratory to prepare these new molecules is based on anionic processes. This anionic technique is best suited to control the length of the chains prepared and to obtain samples with low polydispersity. Although the functionalization of carbanionic sites with various deactivating reagents is easier than with other methods because of the long lived species, it is still necessary to carefully control the deactivation reaction to prevent secondary reactions.

In this paper we describe the reaction between carbanionic ends of oligomers or polymers and elemental compounds such as oxygen and sulfur; we observe two sets of reactions: coupling and functionalization.

If we can control the coupling reaction then one can efficiently functionalize polymers: on the other hand the coupling reaction may often be a novel reaction.

Reactions of Anionic Sites With Oxygen as a Deactivation Reagent

The oxidation of living polymers by oxygen was first reported by SZWARC[2] who observed that the viscosity of the living mixture increased when the oxygen was passed through a polystyrylsodium solution.

Indeed, the ω-anionic polymers[8,10] deactivated with oxygen exhibit a shoulder on the GPC chromatogram. The molecular weight of the macromolecules corresponding to the shoulder, is double that of the initial aliquot sample deactivated with methanol. A

0097–6156/81/0166–0483$05.00/0

coupling reaction has obviously taken place. We have studied the
mechanism of this reaction and have described our results below.

Mechanism of the reaction

The reaction between carbanion and oxygen is not a usual one
because the oxygen molecule is a paramagnetic, with two single
electrons in the π^x levels. When oxygen reacts with a carbanion,
the most probable reaction which is permitted is the electron
transfer reaction from the carbanion to the oxygen according to
the following scheme:

$$R^- + O_2 \longrightarrow R^{\cdot} + O_2^- \qquad (I)$$

This mechanism was suggested for the first time by Russell[11] for
the oxidation of 2-nitropropane in basic medium. Lamb[12] in the
case of Grignard compounds has also shown the existence of a
radical intermediate. The high reactivity of organolithium compounds has probably prevented the early study of this reaction.
When we carry out the oxidation of the dianionic dimers of vinylic monomers, the first generation of compounds which is expected
is the radical anion B:

$$\qquad (II)$$

This species can undergo two reactions:

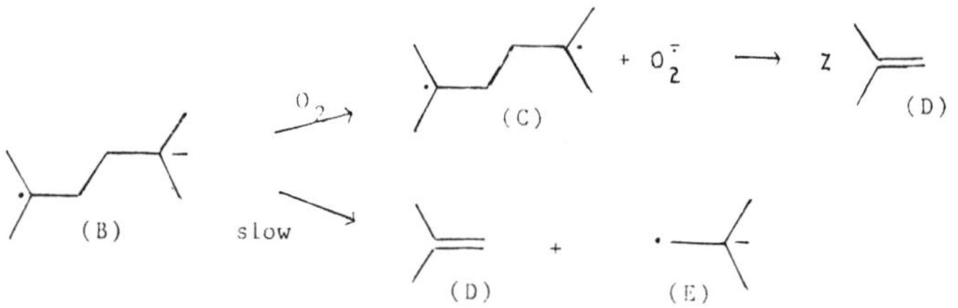

$$\qquad (III)$$

$$\qquad (IV)$$

The second electronic transfer to the oxygen produces the diradi-
cal (C) which evolves into monomer formation. The latter possibility (IV) is a homolytic cleavage giving another anion radical.
If the process follows scheme III or IV, we must obtain monomer
formation after the oxidation reaction in all cases. We have
carried out the oxidation of carbanionic dimers derived from:
isoprene, α-methylstyrene, styrene, 1,1-diphenylethylene.

We have used two oxidation procedures:

- direct oxidation: the oxygen is directly passed through
  the living solution

● inverse oxidation: the anionic solution is deactivated
by adding it dropwise into a solution of THF cooled to
-50°C and saturated with oxygen.
The different results are collected in the Table I.

Table I						
Yield of Monomer Obtained by Action of Oxygen on Dianionic Dimers						
dianionic dimers	$10^{+2}[C^-],M$	τ_a	τ_b	cation	C_a	C_b
α-methylstyrene	2	67	38	Na	0	10
	2	68	37.5	Na	0	10
1,1-diphenylethylene	2	64	66	Na	0	0
isoprene	2	0	0	Na	30	30
	2	0	0	Na	30	30
styrene	2	36	3	Na	10	30
	2	41	3.5	Na	10	30
1,1-diphenylethylene	2	57.5	52	Li	0	0
isoprene	5	0	0	Li	30	30
	5	0	0	Li	30	30
styrene	5	34.5	7	Li	10	10

τ: monomer yield   a) "inverse oxidation"
C: coupling reaction yield   b) "direct oxidation"

We observe the presence of monomer formation for every oxidation
of carbanions in the benzylic position. This first observation
is the proof that a radical mechanism takes place during the
oxidation of organolithium. The second remark, concerns the
coupling reaction: there is none for the diphenylmethyl
carbanion: this can be related to the important steric hindrance
of the latter. The third observation concerns the monomer yield,
which depends on the oxidation mode. With "inverse oxidation" we
obtain a high yield - between 40 and 70% - with styrene,
diphenylethylene and α-methylstyrene. In these cases, the functionalization reaction and the electronic transfer from the
carbanion to the oxygen are competitive, and the coupling reaction (VI) is minimized.

- functionalization
  reaction

$$\text{(structure)} \xrightarrow{O_2} \text{(structure)} \qquad (V)$$

- coupling
  reaction

$$2 \cdot \text{(structure)} \longrightarrow \text{(structure)} \qquad (VI)$$

With "direct oxidation" there is a low amount of oxygen in the mixture and the coupling reaction and the alcoholate formation are favoured. This sequence prevents the monomer formation by the disappearance of the radical.

The last observation concerns the oxidation of the isoprene carbanionic dimers.[4] We do not observe the monomer formation, but the presence of monoalcohols with a high yield. In "inverse oxidation" we obtain 52% of monoalcohol and 80% in "direct oxidation". These monoalcohols have been isolated and characterized (Fig. 1). They come from the transfer reaction from the primary radical to the solvent. The high yield of monoalcohol shows that the hydrogen transfer from the solvent is a competitive reaction towards functionalization. This fact is confirmed by the similar values of the two rate constants
$(k \quad 10^8 \text{ M}^{-1}.\text{s}^{-1})$[13,14].

To summarize we can say that the first step of the reaction of oxygen onto carbanion is of radical nature.

2 - <u>Oxidation of anionic polymers in solution</u>

By extrapolation of these results to the monocarbanionic polymers, it is possible to predict the synthesis of the following compounds:

$$\text{(structure)} Li + O_2 \longrightarrow \text{(structure)} \cdot \qquad (I)$$

$$\text{(structure)} \cdot + O_2 \longrightarrow \text{(structure)} OO \cdot \qquad (V)$$

$$2 \text{(structure)} \cdot \longrightarrow \text{(structure)} \qquad (VI)$$

$$\text{(structure)} OO \cdot + \text{(structure)} Li \longrightarrow \text{(structure)} OOLi + \text{(structure)} \cdot \qquad (VI)$$

$$\text{(structure)} OOLi + Li \text{(structure)} \longrightarrow 2 \text{(structure)} OLi \qquad (VIII)$$

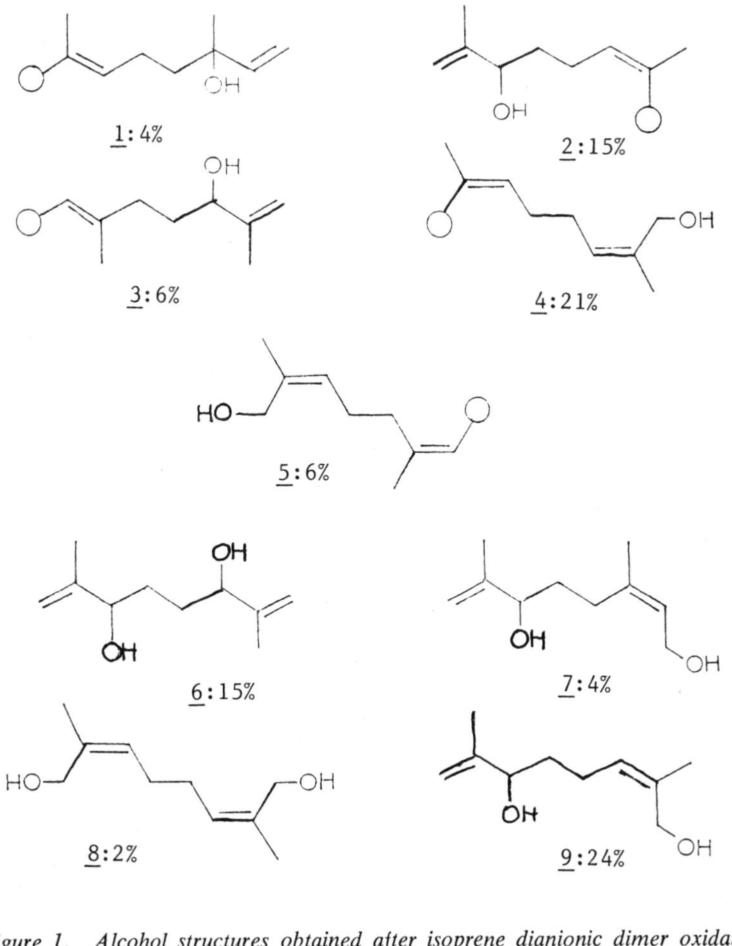

Figure 1. *Alcohol structures obtained after isoprene dianionic dimer oxidation: hydrogen transfer position (○).*

To obtain well defined functionalized macromolecules, it is necessary to check the coupling reactions (VI) which is inherent to the nature of the reaction as we have shown before. We have studied several parameters:

- the oxygen concentration
- the living end concentrations
- the solvent polarity
- the temperature
- the structure of the carbanionic end groups

We have defined the coupling yield as the ratio of the number of living chains ($2N_2$) which have participated to the coupling reactions over the total number of initial living chains ($N_1 + 2N_2$)

$$C = \frac{2N_2}{N_1 + 2N_2}$$

N_1 : non coupled macromolecules

$\overline{M}_{n,o}$: average number molecular weight of non coupled chains

$$\text{and} \quad C = 2\left[1 - \frac{\overline{M}_{n,o}}{\overline{M}_n}\right]$$

$\overline{M}_n$ . . : average number molecular weight of the final polymer

C has been determined by deconvolution of the GPC curves.

We have defined also, another parameter F, which corresponds to a theoretical hydroperoxide yield. F is the ratio of the number of non-coupled macromolecule (N_1) to the total number of macromolecules after reaction:

$$F = \frac{N_1}{N_1 + N_2}$$

$$\text{with} \quad F = 2 - \frac{\overline{M}_n}{\overline{M}_{n,o}} = 2 - \frac{2}{2-C}$$

We have determined the number of hydroperoxide groups per chain, analyzed by leucobase titration. The comparison between the analytical determination of hydroperoxide end groups f and the theoretical values F, allows us to know whether all the non-coupled chains are functionalized by a hydroperoxide group.

2.1 - <u>Influence of the oxidation mode</u>

When the living ends are deactivated by "inverse oxidation" we observe (Table II) a low amount of coupling reaction: 10% as opposed to the "direct oxidation" mode (20%). This difference comes from the oxygen concentration in the medium. In "inverse oxidation" the oxygen concentration is in excess towards the living ends concentration, the the secondary reactions VI and

VIII are minimized. The similar values of F and f confirm this explanation.

Table II

Influence of the Oxygen Concentration (Oxidation Mode)
on the Hydroperoxide Functionality and on the Coupling Yield

	Oxidation Mode	$\bar{M}_{n,o}$	$\bar{M}_n$	$10^3 \cdot [PS^-], M$	C	F	f
PS I		5 400		2.6			
PS I$_1$	Inverse	-	5 800	2.6	0.09	0.96	0.95
PSO I$_2$	Direct	-	5 900	2.6	0.17	0.90	0.40
PS II		5 900		2			
PSO II$_1$	Inverse	-	6 300	2	0.10	0.94	0.86
PSO II$_2$	Direct	-	6 700	2	0.21	0.88	0.20

$[PS^-]$: living ends concentration
PS : anionic chain deactivated by a proton
PSO : polymer chain which have undergone the oxidation

Deactivation temperature : - 65°C
Solvent .... : THF

2.2 - Influence of the Living End Concentration

When the living end concentration increases, the probability of collision between two macroradicals increases and then the coupling reaction is favoured (Table III).

Table III

Influence of the Living Concentrations on the Coupling Reaction

	$\bar{M}_{n,o}$	$\bar{M}_n$	$10^3 \cdot [PS^-], M$	C	F	f
PS I	5 500		2.6			
PSO I$_1$		5 800	2.6	0.08	0.96	0.95
PS II	6 600		4.6			
PSO II$_1$		7 100	4.6	0.15	0.90	0.85
PS III	6 300		6.1			
PSO III$_1$		7 200	6.1	0.23	0.87	0.81
PS IV	5 700		16			
PSO IV$_1$		6 300	7.8	0.20	0.88	0.86
PSO IV$_2$		6 700	16	0.25	0.86	0.85

Deactivation mode: "inverse oxidation"
Deactivation tem-
perature :- 65°C
Solvent .... : THF

When low living end concentration are used, we obtain a good hydroperoxide functionality 0.95.

2.3 – Influence of the Solvent

If we investigate the influence of the solvent, it is necessary to take into account the nature of the species in the medium. In non-polar solvent we have aggregated species[15]; when we add tetramethylenediamine (TMEDA) we destroy the aggregates, and we have an equilibrium between ion pairs and complexed species.[16] In polar solvent, we have another equilibrium between the ion pairs, the solvated ion pairs and loose ion pairs.[17] If we compare the oxidation results (Table IV), we observe that for similar oxygen concentration, the coupling yield and the hydroperoxide functionality increase like the ionicity of the C-Li linkage.

Table IV
Influence of the Solvent on the Coupling Reaction

	Solvent	$T°C$	$\bar{M}_{n,o}$	$\bar{M}_n$	C	F	f
PS I	benzene/heptane 30/70		7 000				
PSO I$_1$	" " "	-18		7 000	0.03	0.98	0.24
PSO I$_2$	benzene/heptane " TMEDA	-18		7 600	0.16	0.91	0.24
PS II	tetrahydrofurane		8 100				
PSO II$_1$	" "	-20		9 100	0.20	0.89	0.50

Living end concentration : 2.10^{-3} M

$[PS^-]$ / $[TMEDA]$ = 0,5

Deactivating mode : "inverse oxidation"

The simultaneous increase of these two ratios: C and f, is due to the increase of the macroradical concentration. Consequently this phenomenon implies that the electronic transfer to the oxygen (reaction I) is easier with ionic species, than non ionic species.

2.4 – Influence of the Temperature

The decrease of the temperature gives a functionality and a low amount of coupling reactions (Table V).

Table V
Influence of the Temperature on the Coupling Reaction

	T°C	$\bar{M}_{n,o}$	$\bar{M}_n$	C	F	f
PS I		8 100				
PSO I_1	-65		8 600	0.10	0.95	0.91
PSO I_2	-20		9 100	0.20	0.89	0.50
PSO I_3	+20		9 100	0.20	0.89	0.44

Living end concentration   : 2.10^{-3} M

Solvent : THF

Deactivatint mode .... : "inverse oxidation"

At low temperature the oxygen concentration in the medium is twice as high as that at room temperature. In addition the low temperature increases the viscosity[18] and the diffusion of oxygen molecules is easier than the macroradical which can explain the results observed.

2.5 – Influence of the Anionic Ends Structures

In adding different monomer units at the end of monocarbanionic polystyrenes, we obtain a set of carbanionic structures which habe been deactivated in the same way. The results (Table VI) show that the terminal unit, which allows the more delocalized anion or radical charge, and presents the more steric hindrance, gives the lower coupling ratio, and the best functionality.

Table VI
Influence of the Anionic Ends Structures on the Coupling
Reaction and on the Functionality

Samples	PS^- .10^3,M	$\bar{M}_{n,o}$	$\bar{M}_n$	C	F	f
PS I	2.5	8 400				
PSO I₁	"		9 100	0.14	0.92	0.89
PS IO I₂	"		9 800	0.29	0.83	0.40
PS II	6	6 300				
PSO II₁	"		7 300	0.24	0.86	0.84
PSDPEO II₂	"		6 700	0.11	0.94	0.92
PSMMAO II₃	"		6 700	0.08	0.94	0.90
PS III	7	6 800				
PSDPEJIII₁	"		7 300	0.13	0.93	0.89
PSMMAO III₂	"		7 300	0.07	0.97	0.90

Solvent : THF
PSI : anionic polystyrene with isoprenyl terminal units
PSDPE : anionic polystyrene with diphenylethyl   terminal unit
PSMMA : anionic polystyrene with methylmethacrylyl terminal unit
Deactivating temperature : - 65°C
Deactivating mode ...... : "inverse oxidation"

3 - Oxidation of Anionic Polymers in the Solid State

The ability of the macroradical and of the macroions to
diffuse in the mixture, and to interreact is responsible for the
secondary products formation: coupling reaction and alcoholate
synthesis. To prevent the diffusion phenomenon, we have carried
out the deactivation in the solid state. The living polymers
have been prepared in benzene, with or without a solvating agent
(THF or TMEDA) and the solution has been freeze dried before the
oxygen introduction. The experimental results are collected in
Table VII.

Table VII
Oxidation of ω-Carbanionic Polymers in Solid State

Samples	$10^3 \cdot [PS^-]$, M	Additives	$\bar{M}_{n,o}$	$\bar{M}_n$	C	F	f
PS I	5	-	10 500				
PSO I$_1$	5	none		12 200	0.27	0.83	0.23
PS II	5	-	12 500				
PSO II$_1$	5	none		14 700	0.32	0.82	0.20
PS III	9.5	-	8 900				
PSO III$_1$	9.5	none		12 700	0.60	0.57	0.20
PSO II$_2$	5	THF/R'=10)		13 300	0.12	0 93	0.87
PSO III$_2$	9.5	THF/R'=20)		9 900	0 20	0.89	0.87
PSO III$_3$	9.5	THF/R'=200)		9 600	0.16	0.92	0.90
PSO III$_4$	9.5	TMEDA		9 300	0.08	0.95	0.96
PSO II$_3$	5	THF Vapor		12 500	≃ 0	1	0.97

$$R' = [THF] / [PS^-]$$

When the oxygen deactivates the freeze dried living polymers without additives, the coupling reaction becomes predominant and can reach up to 60%. We explain this result by the association of the living ends preexisting in the solution, and kept in the solid state. The proximity of these carbanions and consequently of the radical after the electronic transfer favours the coupling reactions.

Alternatively, the disappearance of the aggregates by the addition of a solvating agent, allows us to obtain a good hydroperoxide functionality and yields a low level of coupling.

Finally, is the solvating reagent is added in the vapor phase to the freeze dried polymer, it is possible to prevent the coupling reaction completely. We think that the THF molecules are preferentially located around the living ends and this hindrance prevents the coupling when the macroradicals are formed. In this case, the hydroperoxide functionality is quantitative.

III- REACTION OF ANIONIC SITES WITH SULFUR: S_8, AS A DEACTIVATING REAGENT

In addition to our investigation of oxygen reactions with carbanions, we have also studied the reaction of sulfur on carbanions. Sulfur exists naturally under the form of an eight membered ring : S_8. Different authors have introduced sulfur in organic compounds via various methods.

Schonberg[19] and Gilman[20] have reacted some carbanions with alkylpolysulfides and have obtained various sulfide compounds. Eller[21] has reacted the 2,3,4,5 tetrafluorophenyllithium with sulfur below stoichiometry and obtained thiolates. Hallensleben[22] has added some carbanions onto a cycloalkylpolysulfide which causes the sulfur bridge to open.

We have used monocarbanionic oligomers and characterized the different compounds which are formed after reacting them onto S_8[23,24]. The ratio of carbanion on sulfur is very important as far as the resulting compounds are concerned. We have investigated in detail the reaction on a model molecule s.butyllithium in order to characterize the different compounds obtained.

Table VII
Nature and Proportions of the Compounds Obtained When Sec.BuLi is Reacted on S_8

K	$\bar{x}$	R - Sx - R % molar					% weight RSH / (RSH+RSxR)	S_8 unreacted
		X:1	2	3	4	> 4		
8	1	100					60	No
4	2.3	21	31	48			30	No
2	3.6	5	11	36	46	traces	0	traces
1	3.6						0	some
0,5	3.6						0	some

Solvent : Benzene,   Temperature : 25°C
$$K = \frac{[sec.BuLi]}{[S_8]}$$

- For a ratio K = 8 we obtain only one coupled compound and a high yield of sec.butyl-thiolate (about 60% of the products formed). The coupled compound is well defined and contains only one sulfur atom in the sulfur bridge.

- When K = 4 we obtain always the coupled compounds and only a small amount of thiolate. Let us remark that in both cases K = 8 and K = 4, all the sulfur has been incorporated. The average number of sulfur atoms is about x - 2.3 and we observe sulfur bridges containing 1, 2, and 3 atoms.

- For K = 2 we observe that there are traces of sulfur unreacted; we do not observe any thiolate formation but only coupled compounds with 3.6 average number of sulfur atoms in the bridge of the dialkylpolysulfides. The number of sulfur atoms in the polysulfides varies from 1 to 4, and one observes even traces of a higher polysulfide.

- When K decreases down to 0.5 there is a large excess of sulfur and some is left unreacted. There is no thiolate formed and the average number of sulfur per polysulfide does not change from the value obtained for K = 2. This fact shows that the carbanion reaction on the elemental sulfur is faster than those of the carbanion on the alkylpolysulfides.

All the characterizations have been made by ^{1}H NMR. This is possible because the protons located on the carbon α to the sulfur atom have different chemical shifts, which depend on the number of sulfur atoms in the polysulfide (Table IX).

Table IX
^{1}H NMR Chemical Shifts of Di sec.Butylpolysulfides

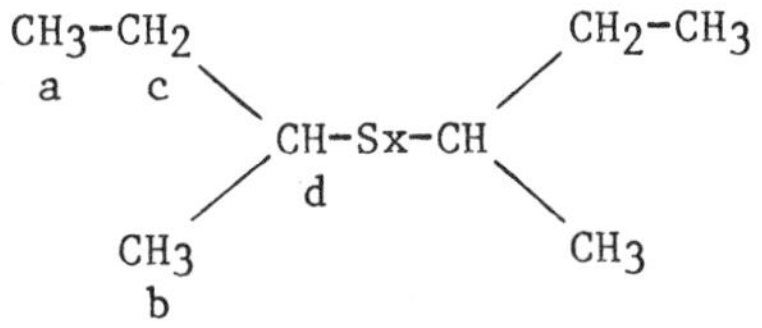

$$CH_3-CH_2 \quad\quad CH_2-CH_3$$

```
CH3-CH2              CH2-CH3
 a    c   \        /
           CH-Sx-CH
          /    d    \
       CH3              CH3
        b
```

- NMR apparatus Cameca 250 MHz

- solvent: CDCl$_3$

- ref. TMS

x	a (t)	b (d)	c (m)	d (m)
1	0.97	1.25	1.60	2.74
2	0.98	1.29	1.60	2.74
3	0.99	1.35	1.60	2.98
4	1.00	1.39	1.60	3.07

The alkylpolysulfides have been separated by vapor phase chromatography. A linear relationship is found between the logarithm
of the retention volume and the molecular weight of the dialkyl-
polysulfide (Figure 2) this linear correlation is similar to
those observed by Hillen with various alkylsulfides.[25] The vapor
phase chromatography was coupled with a mass spectrometer which
enable us to determine the exact molecular weights, and the frag-
mentation of the different dialkylpolysulfides.

We carried out the same investigations using a stablized and
sterically hindered carbanion: the 3—methyl-1,1 diphenylpentyl-
lithium.

Table X

Nature and Properties of the Compounds Obtained When
3—Methyl-1,1-Diphenylpenthlithium is Reacted on S_8

K	$R - S_x - R$ $\overline{x}$	% weight $\dfrac{RSH}{RSH + R-S_x-R}$	S_8 unreacted
8	1.2	66	No
4	2.5	33	No
2	3.4	8	traces
1	4	8	some
0.5	4	< 8	some

We have not separated the individual polysulfides. We can observe as before with the secondary butyllithium the predominant
formation of alkylpolysulfide R-Sx-R with x = 4, when K is below
2. This average number decreases sharply when K reached 8. We
can observe as well that with K increasing, the proportion of
thiolate increases also.

We have not succeeded in separating the different polysulfides with different sulfur atoms in the likage (using vapor
phase chromatography, or liquid chromatography). The NMR spectrum has then been run on the mixture. Because there is no proton on the carbon α to the sulfide, there is not clear separation
of the different compounds. The proton on the β position is not
sensitive enough to achieve this.

The ^{1}H NMR spectrum shows that the methyl groups are strongly shifted upfield. They appear at δ = 0.5 ppm from TMS as reference. We can explain this by the stronly hindered polysulfide,
which causes the methyl groups to be located in the shielding
cone of the phenyl rings.

We have extended this reaction to oligomeric carbanions
(Table XI).

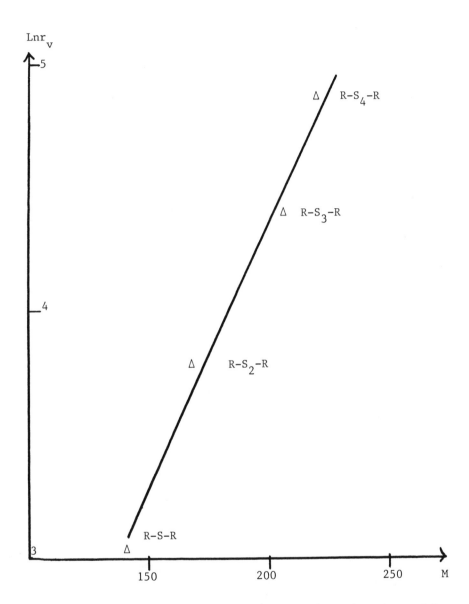

Figure 2. Relationship between the logarithm of 1 VPC retention volume and the molecular weight of the Disec-butylpolysulfides: R = *sec-BuLi;* T^m = *150°C;* r = *retention volume; column = 3% OV 17 1.2m.*

Table XI
Molecular Weights of Oligomers After Reaction of
Carbanions with Elemental Sulfur

Li compound	M_n (H)	M_n (S) calc.	found (a)	Yield %	R-Sx-R $\bar{x}$
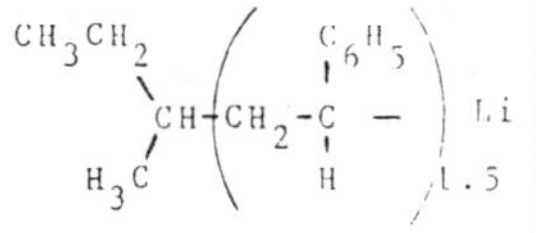 CH_3CH_2 $\quad\diagdown$ $\quad\quad CH\mathord{-}CH_2\mathord{-}C\ (\ C_6H_5) -\ )\ Li$ $H_3C\diagup\quad\quad H\quad\quad 1.5$	214	554	520	79	4
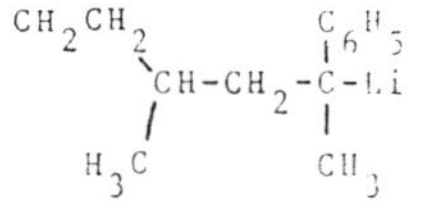 CH_2CH_2 $\quad\diagdown$ $\quad\quad CH\mathord{-}CH_2\mathord{-}C(\ C_6H_5)\mathord{-}Li$ $H_3C\diagup\quad\quad CH_3$	176	446	420	77	3
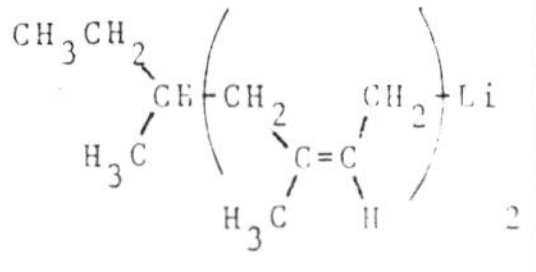 CH_3CH_2 $\quad\diagdown$ $\quad\quad CH\mathord{-}CH_2\quad CH_2\mathord{-}Li$ $H_3C\diagup\quad C=C$ $\quad\quad H_3C\quad H\quad 2$	194	482	450	82	3

(a) : determined by cryometry

K = 2

Solvent : benzene

The results obtained are quite similar to those described above.
For K < 2, we observe high yields of coupled compounds (around
80%) and some thiolate formation.

The number of sulfur atoms in the bridge are also between 3
and 4. When any of these dialkylpolysulfides are reacted with
carbanions such as secondary butyllithium, in excess, one
observes the gradual disappearance of the initial compound and
the formation of the disymetric dialkylthioether resulting from a
scrambling reaction.

Instead of monocarbanionic oligomers, we have used sites
carried by polymers and observed the formation of thiolates and
coupling reactions as described above. When dicarbanionic polymers are used, if conditions are chosen properly, a polycondensation type reaction takes place between the carbanionic polymers
and the sulfur. Polystyryl, polyisoprenyl and poly-α-methylstyryl
dianions have been used. In very good agreement with our
expectation, we have observed an increase of the molecular weight
by a factor of 2 to 5 depending on the conditions. The number of
sulfur atoms in the bridge varies from 1 to 3.

To account for these results we can propose the following mechanism for the reaction of carbanions onto elemental sulfur:

$$R - Li + S_8 \longrightarrow \left[R - S_8 - Li\right] \qquad \text{non isolated}$$

$$\left[R-S_8-Li\right] + RLi \longrightarrow R-S_x - R + Li_2S_y \qquad \left(x + y = 8\right)$$

$$R-S_x-R + RLi \longrightarrow R - S_z - R + R - S_{x-z}Li \qquad x > z$$

IV - EXPERIMENTAL PART
Experimental conditions have been described elsewhere.[2,3,8]

V - CONCLUSION
To conclude, we have been successful in preventing the coupling reactions when oxygen deactivates freeze dried living polymers. Moreover, we have shown that oxidation may be a general method to functionalize living polymers. With a small quantity of deactivating reagent, the functionality is quantitative, and one can prevent completely secondary reactions.

In our investigation of deactivating living polymers with sulfur, we obtain thiolates or polydialkylpolysulfides. For K = 2 the polymers contain about 3 sulfur atoms in the bridge. The technique of the sulfur coupling reaction can be used also to prepare multiblock polymers. The technique of deactivation of carbanionic polymer with oxygen or sulfur is able to yield numerous interesting organic compounds such as novel macromolecular initiators, new macromolecular additives, and telechelic polymers. Finally, the coupling reactions can be used to build block polymers.

Acknowledgments

This work was rendered possible thanks to the support of the Societe National Elf Aquitaine (Polymeres) to one of us, J. F. Boscato.

Literature Cited

1. L. J. Fetters, J. Polym. Sci. 26, 1 (1969).
2. J. Brossas, G. Clouet, C. R. Acad. Sci. C-280, 1459 (1975).
3. J. Brossas, C. P. Pinazzi, G. Clouet, F. Clouet, Makromol. Chem., 170, 105 (1973).
4. R. Rupprecht, J. Brossas, J. Polym. Sci. 52, 67 (1975).
5. J. Brossas, G. Clouet, Makromol. Chem. 175, 3067 (1974).
6. J. M. Catala, J. Brossas, B. Ville, M. Fontanille, C. R. Acad. Sci., C-285, 417 (1977).

7. J. M. Catala, J. Brossas, Polym. Bull. 2, 137 (1980).
8. J. M. Catala, G. Reiss, J. Brossas, Makromol. Chem. 178,
        1249 (1977).
9. M. Szwarc, Nature, 178, 1168 (1956).
10. L. J. Fetters, E. R. Firer, Polym., 18, 306 (1977).
11. G. A. Russel, J. Amer. Chem. Soc., 76, 1595 (1954).
12. R. C. Lamb, P. W. Ayers, M. K. Roney, J. F. Garst,
        J. Amer. Chem. Soc., 88, 4261 (1966).
13. J. K. Kochi, G. A. Olah, Free Radical, Ed. J. Wiley & Sons,
        1, 66.
14. R. H. Gobran, M. B. Berenbaum, A. V. Tobolsky,
        J. Polym. Sci., 46, 431 (1960).
15. M. Morton, L. J. Fetters, J. Polym. Sci., A-2, 2211 (1964).
16. G. Helary, M. Fontanille, 1st Europ. Disc. Meeting on
        Polymer Science, Strasbourg (1978).
17. S. Wenstein, G. C. Robinson, J. Amer. Chem. Soc., 80,
        169 (1958).
18. C. Carbagal, K. J. Tolle, J. Smid, M. Szwarc, J. Amer.
        Chem. Soc., 87, 5548 (1965).
19. A. Schonberg, A. Stephen, H. Kaltschmitt, E. Petersen,
        H. Schulten, Ber. Deutsch. Chem. Gesell. 66 237 (1933).
20. H. Gilman, C. G. Stuckwisch, J. Amer. Chem. Soc. 65,
        1461 (1943).
21. G. Eller, O. W. Meek, J. Org. Metal Chem. 22, 631 (1970).
22. M. Hallensleben, Makromol. Chem. 175, 3315 (1974).
23. J. F. Boscato, J. M. Catala, E. Franta, J. Brossas,
        Makromol. Chem. 180, 1571 (1979).
24. J. F. Boscato, J. M. Catala, E. Franta, J. Brossas,
        French Pat. 78.31.819 (1978).
25. L. W. Hillen, R. L. Werner, J. Chromatogr. 79, 318 (1973).

RECEIVED May 5, 1981.

Cyclic Sulfonium Zwitterions

Novel Anionic Initiators for the Control of Cyclooligomerization of Isocyanates

J. E. KRESTA, C. S. SHEN, K. H. HSIEH, and K. C. FRISCH

Polymer Institute, University of Detroit, Detroit, MI 48221

Isocyanates can be polymerized in the presence of anionic catalysts to form 1-nylons:

$$\underline{n}\ \ R\text{-}N{=}C{=}O \longrightarrow \left[\!\!\!-N{-}\overset{\overset{O}{\|}}{C}-\!\!\!\right]_{\underline{n}}$$

According to Shashoua et al (1), the linear polymerization preferentially proceeds at temperatures below -30°C. At room temperature and at elevated temperatures, no formation of linear polymers was observed, but cyclotrimerization of isocyanates occurred.

The polycyclotrimerization of difunctional isocyanates (or NCO-terminated prepolymers) produces polymer networks containing heterocyclic, thermostable perhydro-1,3,5-triazine-2,4,6-trione (isocyanurate) rings as crosslinks:

$$3 \sim\!\!\sim NCO \longrightarrow$$

Recently polycyclotrimerization of isocyanates has found successful applications in the preparation of high temperature resistant cellular plastics, coatings and elastomers.

0097–6156/81/0166–0501$05.00/0

The volatility of difunctional isocyanates (such as tolylene diisocyanates, hexamethylene diisocyanate, etc.) creates many environmental problems in the urethane industry. These difficulties can be overcome by preparation of NCO-terminated oligomers with low vapor pressure. One approach is the preparation of NCO-terminated oligomers by partial cyclotrimerization of difunctional isocyanates. Usually this is achieved by a multi-step process which includes also deactivation of the catalyst at a certain conversion. During our work on cyclotrimerization of isocyanates we found that cyclic sulfonium zwitterions are very active cyclotrimerization catalysts (2). Recently we found that cyclic sulfonium zwitterions under certain reaction conditions act as anionic initiators. This behavior of cyclic sulfonium zwitterions permits preparation of isocyanate oligomers containing isocyanurate rings by a one-step procedure, eliminating the deactivation step. In this communication this unusual behavior of cyclic sulfonium zwitterions will be discussed.

Experimental

The kinetics of catalysis of cyclotrimerization was studied on the model system phenyl isocyanate/acetonitrile (solvent). Acetonitrile (AN, 99.64%, from Vinstron Corp.) was purified by refluxing with phosphorus pentoxide (5 g/l), then with calcium hydride (2 g/l) followed by distillation under nitrogen. Phenyl isocyanate was obtained from the Upjohn Company with a purity of 99.5%, and was purified by distillation. Tolylene diisocyanates (2,4 and 80/20 2,4/2,6 isomers) were obtained from the Mobay Chemical Co., and were purified by distillation. Cyclic sulfonium zwitterions (SZ) were obtained from the Dow Chemical Co.

Measurement of Kinetics

Polymerizations were carried out in a dried 250 ml three-necked flask equipped with a nitrogen inlet, magnetic stirrer, reflux condenser, dropping funnel, and recording thermocouple. The reaction flask was immersed in a constant temperature bath, 50 ml solution of phenyl isocyanate in acetonitrile (conc.: 1 mol kg^{-1}) was pipetted into the flask and stirred magnetically while the catalyst solution in acetonitrile (50 ml) was placed in the thermostat-controlled dropping funnel. When constant temperature ($\pm$ 0.1°C) was reached in the flask as well as in the funnel, the catalyst solution was added to the flask and the contents were thoroughly mixed. Samples of 3-4 g were taken at regular intervals and the isocyanate content was determined by the dibutylamine method (3). The solvent was then removed in vacuo after the reaction mixture reached 100% conversion, and the resulting polymer was isolated and purified by extraction with methanol. (The trimer is soluble whereas the linear polymer is insoluble in methanol).

Polymer: IR (KBr): 1740 cm^{-1} (ν_{CO}).

Trimer: mp 281°C IR (KBr): 1700 cm^{-1} (ν_{CO}). (Lit. mp 281°C) (3).

Preparation of NCO-Terminated Oligomer

The preparation was carried out in a dried 250 ml three-necked flask equipped with a nitrogen inlet, magnetic stirrer, reflux condenser, micro-dropping funnel and recording thermocouple. The reaction flask was immersed in a constant temperature bath (25°C) and 100 ml of the 2,4-tolylene diisocyanate (TDI) were placed into the flask. When constant temperature was reached in the flask, one ml of a solution of a catalyst in acetonitrile was added dropwise to the reaction mixture and the resultant mixture was intensively mixed for 4 hours. The reaction product was analyzed by GPC and the conversion was determined by the dibutylamine method.

Results and Discussion

The cyclotrimerization of isocyanates is initiated by anionic type of catalysts and proceeds via propagation, transfer and termination steps (4). It was found that in the case where the cyclotrimerization reaction proceeds with a long kinetic chain length, the kinetics of the reaction followed second order with respect to the isocyanate as measured by the disappearance of the isocyanate groups and was first order with respect to the initial concentration of the catalyst (4,5).

During the initial stages of cyclotrimerization of difunctional isocyanates, the formation of polyfunctional NCO-terminated oligomers takes place:

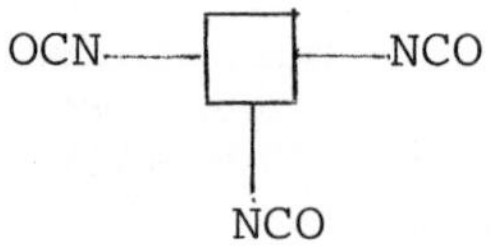

OCN⸺☐⸺NCO

NCO

I

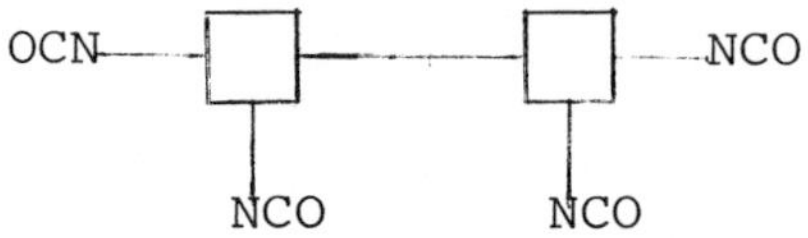

OCN⸺☐⸺☐⸺NCO

NCO   NCO

II

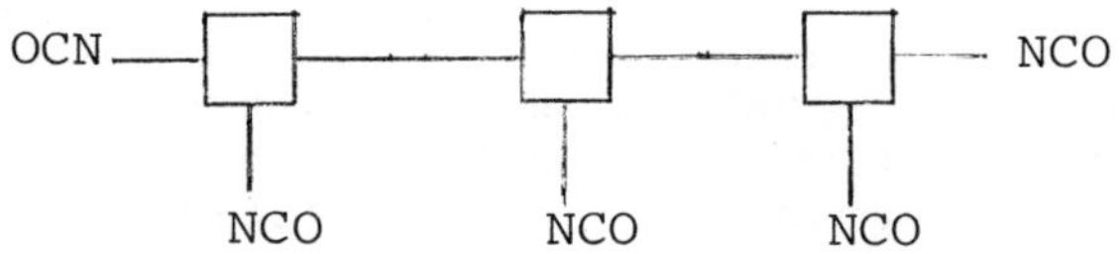

OCN⸺☐⸺☐⸺☐⸺NCO

NCO   NCO   NCO

III

☐ = isocyanurate ring

It was found that the cyclic sulfonium zwitterions of the general formula IV are very active catalysts for cyclotrimerization of isocyanates ($\underline{2}$).

$\underline{IV}$ ☐S$^+$⸺⬡⸺O$^-$ 2 H$_2$O   X$_1$,X$_2$ = H,R

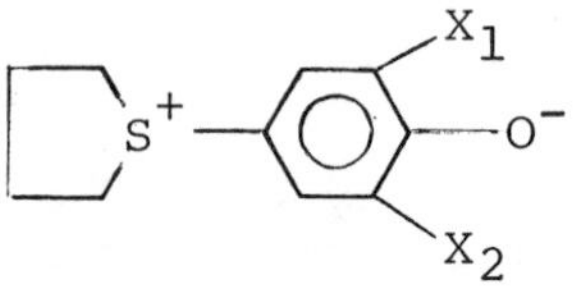

However, cyclic sulfonium zwitterions without two molecules of water are unstable and polymerize due to the nucleophilic attack of the anion (O$^-$) on the tetrahydrothiophenium ring followed by ring opening with consecutive polymerization. Products of polymerization are low molecular weight linear polymers and cyclic dimers and trimers ($\underline{6},\underline{7}$):

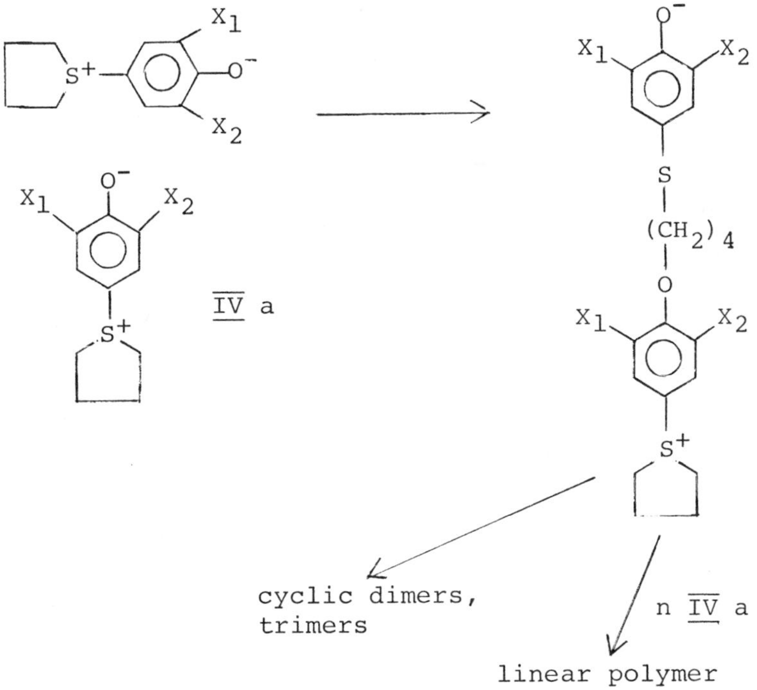

$$\text{IV a}$$

cyclic dimers, trimers

n IV a

linear polymer

Protic dipolar solvents such as water and alcohol stabilize the cyclic sulfonium zwitterions; the decomposition in solution proceeds at a fast rate above 100°C. Cyclic sulfonium zwitterions in stable form contain usually two molecules of water. The water molecules solvate the sulfonium cation forming a protective barrier against nucleophilic attack. The cyclotrimerization activity of the cyclic sulfonium zwitterions is a function of the electron density on the oxygen as can be seen from Table I. The methyl substitution on the phenyl group in the meta position decreased significantly the cyclotrimerization activity. This is probably due to the interference of the methyl group with the solvate molecules around the cation. The "loose" solvate envelope decreases the extent of the charge separation, resulting in lower electron density on the oxygen (shift in the direction to the quinoid resonance formula):

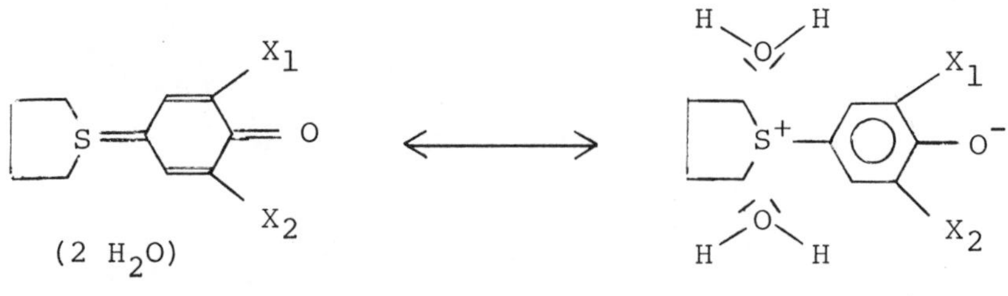

The effect of relative permitivity D on the cyclo-trimerization of phenyl isocyanate was studied in the solvent system acetonitrile (AN) - ethylacetate (EA) using cyclic sulfonium zwitterion VI as catalyst. It was found that the rate constant increased with the increase of relative permitivity D of the solvent system, the experimental data correlate well with the Kirkwood equation:

$$\ln k_{cat} = \ln k_{cat_o} - \frac{1}{kT}\left[\frac{D-1}{2D+1}\right]\left[\frac{\mu^2_C}{r^3_C} + \frac{\mu^2_M}{r^3_M} - \frac{\mu^2_{CM^\ddagger}}{r^3_{CM^\ddagger}}\right]$$

where μ_C, r_C, μ_M, r_M, $\mu_{CM}^\ddagger$, $r_{CM}^\ddagger$ are dipole moments and radii for catalyst, isocyanate and activated complex $_{CM}\ddagger$, respectively.

The observed positive value for the second term (see slope, Fig. 1) indicated that the activated complex (created between monomer and catalyst, or pro-pagating anion) has a larger separation of charges than the reactants in the initial stage.

Similar effects of relative permitivity D on the cyclotrimerization were also observed in the case of substituted ammonium carboxylate catalysts. It was also observed that, besides relative permitivity, the specific solvation of reactants by aprotic dipolar solvents had a considerable effect on the rate constant of cyclotrimerization of isocyanates (see Table II).

In diluted aprotic solutions of isocyanates, the reaction rate between isocyanates and stabilizing water of cyclic sulfonium zwitterions is negligible but be-comes important at high concentrations of isocyanates or high reaction temperatures.

Under these conditions, isocyanate reacts with the stabilizing water according to the following reactions:

TABLE I

EFFECT OF SUBSTITUTION ON CYCLOTRIMERIZATION
ACTIVITY OF CYCLIC SULFONIUM ZWITTERIONS

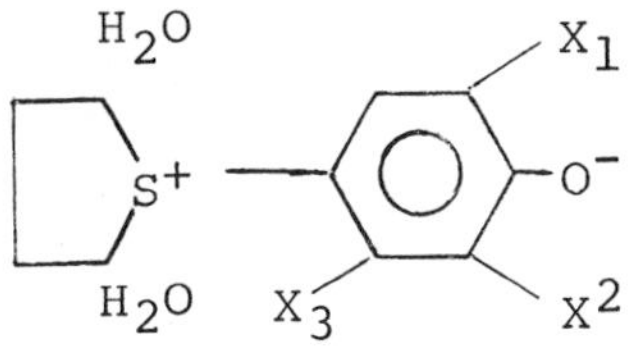

Sulfonium Zwitterion	Substituents			k_{cat} $kg^2 mole^{-2} min^{-1}$
	X_1	X_2	X_3	
V	CH_3	CH_3	H	327.2
VI	CH_3	H	H	283.5
VII	H	H	H	198.8
VIII	H	H	CH_3	155.1

25°C, Solvent AN, PhNCO

TABLE II

SOLVENT EFFECT IN CYCLOTRIMERIZATION OF PhNCO

Catalyst: Cyclic Sulfonium Zwitterion VI
T: 25°C

Solvent System D = 22.54	$k_{cat} (kg^2 mole^{-2} min^{-1})$
AN - CCl_4	139.7
AN - Bu_2O	65.3
AN - EA	57.1
AN - cyclohexanone	35.3
AN - benzene	28.8

polymerization

The deactivation reaction transfers an active catalyst
into the inert (non-reactive) polymer. This phenomenon, when cyclic sulfonium zwitterions act as anionic
initiators, can be utilized for the control of the
cyclotrimerization of difunctional isocyanates.
Therefore the degree of oligomerization of difunctional
isocyanates can be controlled by the concentration of
the initiator, rate of addition of the initiator, as
well as by the temperature of the reaction system.
The effect of the concentration of the initiator and
temperature on the final conversion of cyclotrimerization of 80/20 2,4- and 2,6-tolylene diisocyanate is
depicted in Figs. 2 and 3. As can be seen, the extent
of the reaction increases with an increase of the concentration of initiator and with decrease of temperature. The rate of addition of the initiator also plays
an important role, as can be seen from Fig. 4. A typical distribution of oligomers is shown on the GP
chromatogram shown in Fig. 5. The resulting NCO-terminated oligomers are stable because the polymer formed
from the initiator does not contain ionic or other
catalytically active centers.

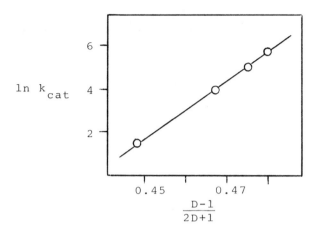

*Figure 1. Effect of relative permitivity D of the solvent system on cyclotrimeriza-
tion of isocyanates: Catalyst; cyclic sulfonium zwitterion VI; 25°C; solvent system,
acetonitrile (D = 37.5) and ethyl acetate (D = 6.02).*

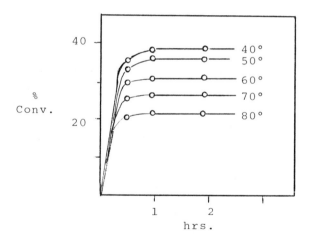

*Figure 2. Trimerization of TDI. Effect of temperature. (TDI) = 100 mL;
[Catalyst VI] = 7.4 mmol; rate of catalyst addition, 20 min.*

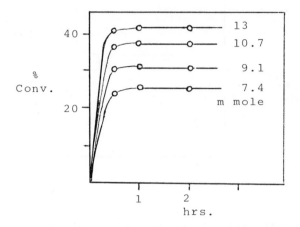

Figure 3. Trimerization of TDI. Effect of concentration of Catalyst (VI): 70°C; rate of catalyst addition, 21 min. (Other conditions same as in Figure 2.)

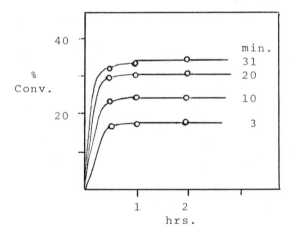

Figure 4. Trimerization of TDI. Effect of rate of catalyst addition: 70°C; (TDI), 100 mL; [Catalyst VI] = 9.1 mmol.

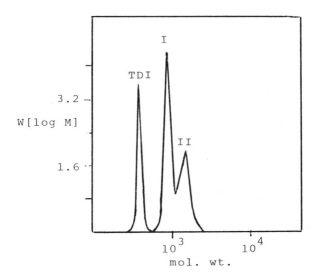

Figure 5. GPC of oligomers: 70°C; (TDI) = 100 mL; [Catalyst VI], 9.1 mmol; rate of catalyst addition, 20 min; conversion, 30.4%.

Acknowledgments

The support of the Dow Chemical Company is gratefully acknowledged. The authors wish also to express
their appreciation to Dr. J. R. Runyon of Dow Chemical
Company for the GPC measurements.

Literature Cited

1. Shashoua, V. E., Sweeny, W., Tietz, R. F.; J. Am.
    Chem. Soc., 82, 866 (1960).

2. Kresta, J. E., Shen, C. S.; U. S. Pat. 4,111,914,
    Sept. 5, 1978.

3. David, D. J., Staley, H. B.; "Analytical Chemistry
    of Polyurethanes," Wiley-Interscience, New York,
    1969.

4. Kresta, J. E., Shen, C. S., Frisch, K. C.; ACS
    Coat. Plast. Preprints, 36, 2, 674 (1976).

5. Kresta, J. E., Shen, C. S., Frisch, K. C., Makrom-
    ol. Chem., 178, 2495 (1977).

6. Hatch, M. J., Yoshimine, M., Schmidt, D. L., Smith,
    H. B; U. S. Pat. 3,636,052, (1972).

7. Schmidt, D. L., Smith, H. B., Yoshimine, M., Hatch,
    M. J; J. Polym. Sci., A-1, 10, 2951 (1972).

RECEIVED March 3, 1981.

Anionic Polymerization. VII

Polymerization and Copolymerization with Lithium–Nitrogen-Bonded Initiator

TAI CHUN CHENG

Central Research Laboratories, The Firestone Tire and Rubber Co.,
Akron, OH 44317

Polymerization of butadiene with lithium morpholinide, an initiator with a built-in microstructure modifier, has been carried out in hexane. In general, the vinyl content of the polymers prepared with this initiator is dependent on the initiator concentrations and on the polymerization temperatures. This dependence is identical to that observed in a THF-modified lithium diethylamide polymerization initiator system. A comparison of these initiator systems for polymerization of butadiene is presented. In addition, a study of the effect of metal alkoxides on the vinyl content of lithium morpholinide initiated butadiene polymerization is included.

Copolymerization of butadiene and styrene in hexane with a number of initiators, such as lithium morpholinide, lithium dialkylamide, lithium piperidinide, etc., has also been examined. In general, the microstructure and styrene content of the polymers are dependent on the type of initiator and the polymerization conditions. Detailed results including a postulated mechanism for these polymerizations are discussed.

Lithium diethylamide has been shown to be an effective initiator for the homopolymerization of dienes and styrene (1,2). It is also known that such a polymerization process is markedly affected by the presence of polar compounds, such as ethers and amines (2,3). However, there has been no report of the use of a lithium amide containing a built-in polar modifier as a diene polymerization initiator. This paper describes the preparation and use of such an initiator, lithium morpholinide. A comparison between polymerization with this initiator and lithium diethyl amide, with and without polar modifiers, is included. Furthermore, we have examined the effects of lithium-nitrogen initiators on the copolymerization of butadiene and styrene.

0097–6156/81/0166–0513$05.00/0

Experimental

A. Preparation of Initiators. The lithium amides were
prepared by the reaction of n-butyllithium with the corresponding amines in hexane with the amounts shown in Table I. In
general, the n-butyllithium in hexane (15%) was added to the
amine-hexane solution slowly because a strong exotherm is
associated with the reaction. The lithium amides precipitated
from the solution in the form of crystalline compounds. The
lithium amides were isolated by means of filtration and were
then washed four times with a total of two liters of hexane.
The materials were dried in a vacuum oven at room temperature
and then suspended in hexane solution for use as polymerization
initiators. A GC-mass spectroscopic method was used to verify
the purity of this material, and a titration (11) was used for
the determination of total alkalinity.

B. Polymerization Procedure. Butadiene was used as a
blend in hexane. Neat butadiene was treated with mercuric
sulfate, washed with caustic and water, distilled and dried by
passage through molecular sieves. At this point, the butadiene
was blended with hexane, and the blend was passed through
molecular sieve drying columns. Precautions were taken to
avoid the presence of water, oxygen and other impurities which
would inhibit the polymerization or react with the initiator.
    The homopolymerizations were all run in 28 oz. beverage
bottles. The bottles were baked for at least 24 hrs. and then
capped with crown, three-hole caps and rubber liners. Cooling
of the bottles was effected while purging with nitrogen. After
cooling, the bottles were charged with the butadiene blend,
the heterogeneous initiator-hexane suspension and modifiers.
The bottles were then placed in a constant temperature bath
equipped with an agitation device for 24 hrs. or more.
    The copolymerizations were performed with a butadiene/
styrene/hexane blend. This blend was prepared by a procedure
identical to the above method for the butadiene blend, with
the exception that distilled styrene was blended with the
butadiene and hexane prior to final drying.
    Polymers were isolated by coagulation in methanol, dis-
solving in toluene, filtering and recoagulation in methanol.
The polymer was then dried in an oven at $40^{\circ}C$ and 60-70 mm
pressure. Purified butadiene polymers and butadiene-styrene
copolymers were analyzed by infrared and nmr spectroscopy(4,5,6).

C. Preparation of the 1:1 Adduct of Lithium Diethylamide
and Butadiene. Butadiene (4.9 g, 9.07 mmole) was dissolved in
400 ml of purified diethylamine. n-Butyllithium (9% mmole) in
hexane was added dropwise over a period of one hour at $0^{\circ}C$.
The reaction mixture was then stirred two hours at room temperature and heated at $50^{\circ}C$ for 48 hours. Upon cooling, a mixture

of cis- and trans–1-diethylamino-2-butene was isolated by means
of distillation (b.p. 134-6°C). The yield of the product based
on butadiene is 65-70% (7.5-8.0 g). The ratio of cis to trans
by G.C. is 1 to 2. In addition, a small amount of 1,2 product
was observed ($\sim$5%).

Results and Discussion

Homopolymerization. Lithium Morpholinide. This compound
was synthesized by the reaction of n-butyllithium and morpholine
in hexane solution. The lithium morpholinide precipitated from
the solution in the form of a crystalline compound. The
reaction was accompanied by the liberation of gas (butane).
The former was detected by gas chromatography and mass spectroscopy. The lithium morpholinide powder was isolated by means
of filtration. In order to be sure that there was no butyllithium or morpholine in the dry powder, G.C.-mass spectroscopic
analysis of D_2O-quenched powder was performed. No deuterated
butane or undeuterated morpholine was found.

The polymerization of butadiene with lithium morpholinide
was carried out in hexane at several different temperatures.
The data presented in Table II show that the 1,2 content of the
polybutadiene is dependent upon initiator level and polymerization temperature. At 30°C, as the concentration of the
initiator increases from 3.25 moles to 13 mmoles, the 1,2 content of the polybutadiene increases from 24.5% to 36.3%,
respectively. However, when the temperature was increased
from 30°C to 80°C, the 1,2 content of the polybutadiene
decreased from 24.5% to 12.8%, respectively at constant initiator charge (3.2 mmoles).

The effect of the addition of alkali metal t-butoxides on
the lithium morpholinide-initiated heterogeneous butadiene
polymerization was also studied at 30°C in hexane. The data in
Table III show that the 1,2 content of the polybutadiene is
sensitive to the type of alkali metal used. For example, by
changing metal alkoxides from Li to Na to K, the 1,2 content
of polybutadiene changes from 46.0% to 58.2% to 55.4%, respectively, at an initiator concentration of about 10 mmoles.
However, the 1,2 content of the polybutadienes is unaffected
by the metal alkoxide concentration (1,2 content, 42-46% for
LiOtBu, 55-58% for NaOtBu, 53-55% for KOtBu) when the alkoxide/
morpholinide ratio is greater than 1:1.

Lithium Diethylamide. This compound has been used as an
initiator for the polymerization of diene by Vinogrador and
Basayeva (1). In order to compare this initiator with lithium
morpholinide (a lithium-nitrogen initiator with a built-in
polar modifier), we have prepared lithium diethylamide according
to the procedure described by Vinogrador and Basayeva (1) and
utilized it as an initiator for THF-modified butadiene
polymerizations.

TABLE I

PREPARATION OF INITIATORS

No.	mmole of nBuLi	Amine Type	mmole
1	349.0	Morpholine	382.0
2	5.0	Dimethylamine	5.0
3	5.0	Diethylamine	5.0
4	5.0	Di-n-butylamine	5.0
5	154.0	Di-i-propylamine	154.0
6	154.0	Piperidine	154.0
7	154.0	2,6-Dimethylpiperidine	154.0
8	154.0	2,2,6,6-Tetramethylpiperidine	154.0
9	112.0 and 224.0	Piperazine	112.0

All the amines were purchased from either MCB or Aldrich
Chemical Co. Lithium t-butoxide and n-butyllithium (15%) in
hexane were purchased from Lithcoa Co. Sodium and potassium
t-butoxides were purchased from MSA Research Corp.

TABLE II

POLYMERIZATION OF BUTADIENE WITH LITHIUM MORPHOLINIDE IN HEXANE[a]

No.	Pzn Temp $^\circ$C	mmole Initiator	Microstructure, mole %[b]			% Yield
			Cis 1.4%	Trans 1.4%	1.2%	
1	30	3.25	27.6	47.9	24.5	71.4
2	30	6.5	25.9	44.6	29.5	78.6
3	30	13.0	22.0	41.6	36.3	75.7
4	50	3.25	28.4	54.7	15.9	76.0
5	70	3.25	35.7	50.3	14.0	75.5
6	80	3.25	41.1	46.0	12.8	70.2

(a) Weight of butadiene 70 g (1.30 moles).
(b) By Infrared method.

The polymerization of butadiene with lithium diethylamide was conducted at several different temperatures. In general, the conversion to polymer was reasonable (75-89%), and the microstructure was independent of polymerization temperatures and initiator levels over the range investigated. These results are shown in Table IV. It is interesting to note that the microstructure of the polybutadiene prepared by us at 30°C is different from the microstructure reported by Vinogrador and Basayeva (1) (1,2%:10-11.9 vs 27-29, cis 1,4%; 30.7-32.3 vs 6-16). Currently, we do not have an explanation for this discrepancy.

The effect of THF on the microstructure of lithium diethylamide initiated polymerization of butadiene-1,3 was studied. The 1,2 content of the polymers is dependent on the ratio of THF to lithium as expected. Table V shows that as the ratio of THF to lithium diethylamide varies from 0/1.62 to 30.5/1.62, the 1,2 content changes from 11% to 51.9%, respectively at 30°C. Thus, the level of THF at constant lithium diethylamide levels controls the overall 1,2 content of the polybutadiene formed.

A possible explanation for the observed dependence of polymer microstructure on the level of lithium morpholinide is that the presence of the oxygen atom in morpholine acts as a modifier. Such a modifier, as shown below, can be depicted in terms of

$$\text{------CH}_2 \xrightarrow{} \overset{+}{\text{Li}} \xleftarrow{} :O\!\!\!\diagdown\!\!\!\diagup\!\!N\text{-R}$$

where R=polymer or Li

complexation between oxygen and polybutadienyl lithium. The result of electron donation by the modifier is an enhancement of ionic character of the polybutadienyl lithium. Thus, increasing the concentration of lithium morpholimide will increase the level of complexed initiator. Consequently, one would expect the increasing of 1,2 content of polybutadiene in the polymerization. This explanation has been supported by the use of THF as a modifier in lithium diethylamide initiated polymerization of butadiene (Table V). Similar results using THF as a modifier in alkyllithium catalyzed butadiene polymerization have been reported by Antkowiak and his associates (7).

The temperature dependency of 1,2 content shown in Table II is also consistent with complex formation between polybutadienyllithium and the oxygen atom in the lithium morpholinide moleculre. One can visualize an equilibrium between noncomplexed and complexed molecules which would be influenced by temperature. Higher temperatures would favor dissociation of the complex; and, therefore, the 1,2 content of the polymer would be lower than that from the low temperature polymerization. This explanation is supported by the polymerization of butadiene with lithium diethylamide, in which the microstructure of the polybutadiene remains constant regardless of the polymerization temperature (Table IV). This is presumably due to the fact that trialkylamines are known to be poor

TABLE III

EFFECT OF METAL T-BUTOXIDE ON LITHIUM MORPHOLINIDE [a]
INITIATED BUTADIENE POLYMERIZATION AT 30°C IN HEXANE

No.	Modifier Type	mmole	Microstructure, mole % [b] Cis 1.4%	Trans 1.4%	1.2%	% Yield
1	-	-	25.9	44.6	29.5	78.6
2	LiO-t-Bu	6.8	15.7	42.2	42.1	75.5
3	"	10.6	13.2	40.8	46.0	73.2
4	"	18.4	15.5	37.8	46.7	72.5
5	NaO-t-Bu	10.0	3.6	34.8	58.2	76.3
6	"	20.0	7.0	38.3	55.5	70.2
7	"	30.0	9.3	36.6	55.2	69.5
8	KO-t-Bu	2.6	12.3	34.6	53.1	75.5
9	"	10.6	8.0	36.6	55.4	72.5
10	"	21.2	9.1	35.7	55.2	67.3

(a)   70 g (1.30 moles) of butadiene and 0.578 g (6.8 mmoles)
      of lithium morpholinide were used.

(b)   By Infrared method.

TABLE IV

POLYMERIZATION OF BUTADIENE WITH LITHIUM
DIETHYLAMIDE IN HEXANE [a]

No.	Pzn Temp °C	mmole Initiator	Microstructure, mole % [b] Cis 1.4%	Trans 1.4%	1.2%	% Yield
1	30	2.70	31.2	57.9	10.9	75.0
2	30	3.78	30.9	58.3	10.8	89.0
3	50	1.62	32.2	58.0	9.8	84.3
4	50	2.70	32.3	57.7	10.1	85.9
5	70	1.62	31.3	57.7	11.0	87.5
6	70	2.70	32.0	57.0	11.0	89.0
7	80	1.62	31.1	57.1	11.7	84.3
8	80	2.70	30.7	57.4	11.9	85.9
9 [c]	30	-	6-16	55-57	27-29	-

(a)   64 g (1.19 moles) of butadiene monomer was used.

(b)   By Infrared method.

(c)   See reference 1.

modifiers for alkyllithium system (7). A similar complexed
species can also be postulated in the butadiene polymerization
with metal alkoxides modified lithium morpholinide system.
Such complexation involving alkyl metals (Li, Na and K) and
metal alkoxides (Li, Na and K) has been previously reported
from our laboratories (8,9,10).

The initiation step for the lithium diethylamide catalyzed
butadiene polymerization can be postulated in terms of the
following equations:

$$
\begin{array}{c}
\mathrm{CH_3CH_2} \\
\diagdown \\
\qquad \mathrm{N-Li} + \mathrm{CH_2 = CH-CH = CH_2} \longrightarrow \\
\diagup \\
\mathrm{CH_3CH_2}
\end{array}
$$

$$
\begin{array}{cc}
\mathrm{CH_3CH_2} & \mathrm{CH_3CH_2} \\
\diagdown & \diagdown \\
\qquad \mathrm{N-CH_2-CH = CH-CH_2Li} + & \qquad \mathrm{N-CH_2-CH-Li} \\
\diagup & \diagup \qquad \mathrm{CH} \\
\mathrm{CH_3CH_2} & \mathrm{CH_3CH_2} \quad \| \\
 & \qquad\qquad \mathrm{CH_2} \\
\quad (\mathrm{I}) & \qquad\qquad (\mathrm{II})
\end{array}
$$

The compound (I) then reacts with additional butadiene monomer
to form a polymer.

In order to verify this hypothesis, we have isolated a
mixture of 1-diethylamino-2-butene in 65-70% of yield by the
reaction of equimolar amounts of lithium diethylamide and
butadiene in diethylamine. This product consists of a mixture
of the cis and trans olefins in a ratio of 1 to 2. In addition,
we have also observed about 5% of 1-diethylamino-3-butene by
G.C. analysis. All the samples have been identified by NMR,
mass spectroscopy, I.R. and G.C. With this information in
mind, it is believed that other systems, such as lithium
morpholinide initiator, lithium morpholinide-metal alkoxide,
etc., follow the same polymerization mechanism.

Copolymerization. The copolymerization of butadiene-
styrene with alkyllithium initiator has drawn considerable
attention in the last decade because of the inversion phenomenon (12) and commercial importance (13). It has been known
that the rate of styrene homopolymerization with alkyllithium
is more rapid than butadiene homopolymerization in hydrocarbon
solvent. However, the story is different when a mixture of
butadiene and styrene is used. The propagating polymer chains
are rich in butadiene until late in reaction when styrene
content suddenly increases. This phenomenon is called inversion because of the rate of butadiene polymerization is now
faster than the styrene. As a result, a block copolymer is
obtained in this system. However, the copolymerization
characteristic is changed if a small amount of polar solvent

is used as a modifier. The resulting copolymer is essentially randomized (14,15,17). We felt it would be interesting to study initiators with built-in modifier for randomizing the styrene.

Lithium Morpholinide. Lithium morpholinide was the first one chosen as an initiator for butadiene-styrene copolymerization because of the built-in oxygen. It was felt that the presence of oxygen in the initiator may lead to a copolymer with unusual properties.

The copolymerization of butadiene-styrene with this initiator was carried out heterogeneously in hexane. The data in Table VI show that the styrene content is highly dependent on the percent of conversion. It appears that a block styrene would be obtained if 100% conversion were reached. A detailed discussion regarding this subject will be mentioned later.

From thé Table IV, it also shows that the low styrene content in the copolymer may relate to the polymerization temperature. As the polymerization temperature was increased from 5° to 70°C, the styrene content of the butadiene-styrene copolymer decreased from 21.7% to 9.1%, respectively. The decreasing in styrene content at higher temperature is consistent with the paper reported by Adams and his associates (16) for thermal stability of "living" polymer-lithium system. In Adams' paper, it was concluded that the formation of lithium hydride from polystyryllithium and polybutadienyllithium did occur at high temperature in hydrocarbon solvent. The thermal stability of polystyryllithium in cyclohexane is poorer than polybutadienyllithium. From these results, it appears that the decreasing in styrene content in lithium morpholinide initiated copolymerization at higher temperature is believed to be associated with the formation of lithium hydride.

Since the reversal of activity of butadiene with respect to styrene in alkyllithium system has been observed (12), it would be of interest to find out whether the inversion phenomenon still holds in the case of the lithium morpholinide system. Four temperatures, namely 30°, 40°, 50° and 60°C were chosen for this study. At 30°C polymerization temperature the curve is characteristic of block copolymerization when one plots percent bound styrene vs percent conversion (Fig. 1). Initially, a small amount ($\sim$3%) of styrene is polymerized. This is followed by a block of butadiene. The remaining styrene is then polymerized after all the butadiene is consumed. This result is identical to the alkyllithium initiated copolymerization. Thus it appears that the order of reactivity for lithium morpholinide system again is: butadiene $>$ styrene

Identical results have also been observed for 40°, 50° and 60°C polymerization temperature. This is also shown in Fig. 1. It appears that the polymerization temperature has no effect on the rate of polymerization even with the presence of built-in oxygen as a modifier in the initiator. However, it does change

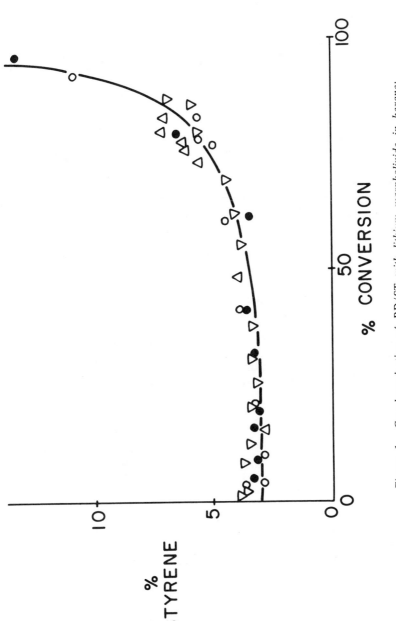

Figure 1.   Copolymerization of BD/ST with lithium morpholinide in hexane:
60°C (△); 50°C (○); 40°C (●); 30°C (▽).

the styrene content in the copolymer as mentioned previously. Nevertheless, the result did suggest the formation of block copolymer in lithium morpholinide initiated copolymerization.

In addition, we have also found that the styrene content of butadiene-styrene copolymer is independent of the catalyst concentration over the range studied. For example, at 30°C, the styrene content of the butadiene-styrene copolymer is virtually unchanged between 6.5 mmoles and 9.75 mmoles of catalyst concentration (19.2-21.6%). Similar results were also obtained at 5° and 50°C (Table VI). It is believed that such phenomenon is due to the heterogeneous nature of the catalyst.

Lithium dialkylamide. Four catalysts, namely lithium dimethylamide, lithium diethylamide, lithium di-n-butylamide and lithium di-i-propylamide, have been prepared by the reaction of n-butyllithium and the corresponding amines. Characterization of these amides was performed by G.C.-Mass Spectroscopy after D_2O quenching of the amides. Further analysis for the total alkalinity was performed by a titration (11). After quenching with D_2O, no deuterobutane or undeuteriated amines were observed.

The copolymerizations of butadiene and styrene with these four amides were carried out in hexane with the insoluble initiators at several different temperatures. In general, the styrene content in the lithium diethylamide or lithium di-n-butylamide initiated butadiene-styrene polymerizations is higher than in the lithium dimethylamide or lithium di-i-propylamide system (Table VII). A high styrene content (23.4% to 25.3%) in the lithium diethylamide and the lithium di-n-butylamide systems was obtained at 50°C polymerization temperature. From this result, it appears that the best co-polymerization temperature for these systems is 50°C.

In lithium dimethylamide and lithium di-i-propylamide systems, the styrene content in the copolymer is generally low (7-15%) even though the initial change of styrene is 25%. This phenomenon is more clearly in case of lithium di-i-propylamide system. One explanation for the low styrene content could be attributed to the poor solubility of the initiators in solution. As a result, it takes a long time to initiate the polymerization. Such hypothesis has also been suggested by Angood and his associates when they polymerized styrene with lithium diethylamide initiator (2). Another explanation could be the slow acting impurities terminated the few active chains before reaching the quantitative conversion. As a result, the rate of copolymerization will be slow. In other words, the induction period for this initiator will be longer due to the presence of less active initiators. Nevertheless, all the initiators are classified to be an active catalyst for the polymerization of butadiene-styrene (Table VII).

TABLE V

EFFECT OF THF ON LITHIUM DIETHYLAMIDE INITIATED BUTADIENE
POLYMERIZATION AT 30°C IN HEXANE [a]

No.	mmole THF	Microstructure, mole % [b]			% Yield
		Cis 1.4%	Trans 1.4%	1.2%	
1	–	31.7	57.7	11.0	87
2	6.17	20.8	49.8	29.4	85
3	12.3	17.4	45.3	37.3	91
4	30.5	13.5	34.5	51.9	88

(a) 70 g (1.3 moles) of butadiene monomer and 1.62 mmole of
lithium diethylamide were used.

(b) By Infrared method.

TABLE VI

COPOLYMERIZATION OF BUTADIENE/STYRENE WITH LITHIUM
MORPHOLINIDE IN HEXANE [a]

No.	Pzn Temp °C	mmole Initiator	Microstructure, mole % [b]				% Yield
			Cis 1.4%	Trans 1.4%	1.2%	% Sty	
1	5	13.6	14.1	39.4	46.6	18.7	78
2	5	20.4	12.1	38.3	49.5	21.7	79
3	30	6.5	26.0	49.7	19.6	19.2	79
4	30	9.75	19.2	52.5	22.2	21.6	84
5	50	3.25	34.7	49.4	16.0	12.8	76
6	50	6.5	33.8	47.0	19.2	13.9	71
7	70	3.25	41.4	45.2	13.3	9.1	63

(a) 64.2 g (1.2 moles) of butadiene and 16.8 g (0.162 moles)
of styrene were used.

(b) By Infrared method.

TABLE VII

EFFECT OF ALKYL GROUP OF LITHIUM DIALKYLAMIDE ON MICROSTRUCTURE
AND STYRENE CONTENT OF BUTADIENE/STYRENE COPOLYMER [a]

No.	Pzn Temp °C	Initiator Alkyl	mmole	Microstructure, mole % [b] Cis 1.4%	Trans 1.4%	1.2%	% Sty	% Yield
1	30	Methyl	9.6	21.5	56.2	22.2	14.8	75
2	30	"	19.2	18.9	53.7	27.4	14.6	70
3	50	"	9.6	20.4	62.3	17.3	13.6	73
4	50	"	19.2	22.3	57.7	19.9	12.4	70
5	80	"	9.6	22.2	63.2	14.6	10.9	69
6	30	Ethyl	2.7	29.1	60.2	10.7	19.0	90
7	30	"	9.6	28.0	61.1	10.8	11.8	78
8	30	"	19.2	29.3	59.2	11.6	6.5	70
9	50	"	1.62	28.5	61.5	10.0	24.4	92
10	50	"	9.6	27.1	62.4	10.5	24.6	93
11	50	"	19.2	29.3	60.1	10.6	23.4	95
12	80	"	1.62	27.6	60.1	12.3	11.3	80
13	30	n-Butyl	0.9	30.8	60.4	8.8	10.7	83
14	30	"	1.5	34.4	57.5	8.1	17.3	80
15	50	"	0.9	28.9	61.6	9.5	25.3	94
16	50	"	1.5	30.0	60.6	9.5	24.1	93
17	70	"	0.6	29.5	59.5	11.0	21.3	92
18	70	"	0.9	27.7	61.0	11.3	21.6	93
19	50	i-propyl	2.8	30.9	57.7	11.4	7.7	68
20	50	"	5.6	28.4	59.2	12.4	8.1	67
21	50	"	11.2	26.7	58.6	14.8	8.7	71
22	80	"	2.8	27.3	60.4	12.4	8.7	70
23	80	"	5.6	30.8	57.3	11.8	6.9	70

(a) 64.2 g (1.2 moles) of butadiene and 16.8 g (0.162 moles)
     of styrene were used (21% styrene content).

(b) By Infrared method.

At this stage, we do not know whether the styrene incorporated in polymer chain is a block styrene or not. However, it is speculated that the block styrene will be resulted based on the lithium morpholinide system. More work along this line is in need in order to qualify this statement.

Lithium Piperidinide. Since the butadiene-styrene copolymer prepared from lithium di-i-propylamide contained low styrene content (∼7%) regardless of the polymerization temperature and initiator concentration, it would be interesting to know if there is a good reason to explain it. One possible explanation could be based on the steric effects of the i-propyl group besides the problem of insolubility. To test this theory, we prepared initiators, namely lithium piperidinide, lithium 2,6-dimethylpiperidinide and lithium 2,2,6,6-tetramethylpiperidinide, in hexane. The copolymerization of butadiene and styrene was carried out in hexane at 30°C at several different initiator concentrations using these insoluble initiators. The data presented in Table VIII show that the styrene content of the copolymers is generally high (20-24%) in either the lithium piperidinide or the lithium 2,6-dimethylpiperidinide systems regardless of the initiator level. The conversion of the polymer based on initial monomer change is also high (80-90%) over the range 3 mmole to 10.4 mmole of initiator. However, the styrene content (∼9.6%) and conversion (<20%) of the polymer are low when one uses lithium 2,2,6,6-tetramethylpiperidinide as an initiator for the copolymerization of butadiene and styrene (Table VIII). Based on this finding, it appears that the steric effects do have some influences on the rate of polymerization. However, it is also possible that the reduced rate of polymerization may be due to the poor solubility of the initiator in monomer or solvent. More works along this line are needed in order to understand the mechanism of this system.

Mono and Dilithium Piperazinide. Both initiators were prepared by the reaction of n-butyllithium with piperazine. Monolithium piperazinide was prepared by the use of a 1:1 ratio of n-butyllithium to piperazine, while dilithium piperazinide was prepared with a 2:1 ratio of n-butyllithium to piperazine. Both salts are crystalline and insoluble in hexane.

The copolymerization of butadiene and styrene with these two initiators was performed at several different temperatures and initiator concentrations. In general, a low styrene content (10-12%) was observed regardless of the polymerization temperature and initiator concentrations (Tables IX and X). It is believed that the low styrene content is due to the insolubility of the initiators.

The 1,2 content of this butadiene/styrene copolymer is also found to be low (9-11%). It behaves like an alkyllithium system. This result is expected since the structure of the

TABLE VIII

COPOLYMERIZATION OF BUTADIENE AND STYRENE WITH
LITHIUM PIPERIDINIDE IN HEXANE[a]

No.	Initiator Type	Initiator mmole	Cis 1.4%	Microstructure, mole %[b] Trans 1.4%	1.2%	% Sty	% Yield
1	A	3.1	38.9	60.5	10.6	23.4	98.7
2	A	6.2	27.4	61.1	11.5	24.5	92.6
3	A	10.4	28.6	58.9	12.6	23.2	98.7
4	B	2.6	24.0	60.8	15.2	22.0	96.2
5	B	5.2	23.1	57.1	19.8	22.8	98.0
6	B	10.4	22.3	51.1	26.6	20.3	80.2
7	C	5.4	19.7	56.8	23.6	9.6	<20
8	C	10.4	20.4	53.4	26.2	9.4	<20

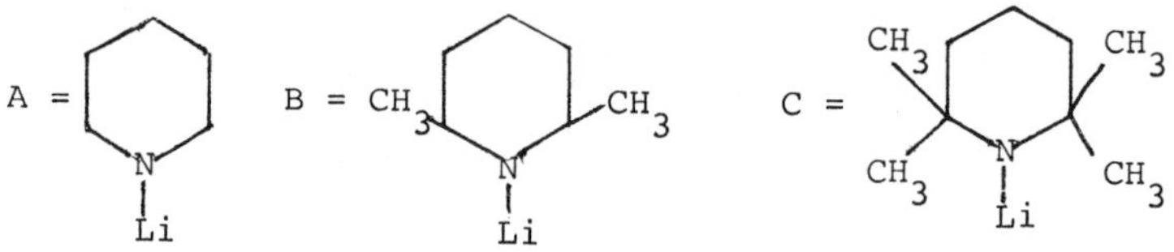

(a) 64.2 g (1.2 moles) of butadiene and 16.8 g (0.162 moles)
of styrene were used.

(b) By Infrared method.

TABLE IX

COPOLYMERIZATION OF BUTADIENE/STYRENE WITH MONOLITHIUM
PIPERAZINIDE IN HEXANE[a]

No.	Pzn Temp °C	mmole Initiator	Cis 1.4%	Microstructure, mole %[b] Trans 1.4%	1.2%	% Sty	% Yield
1	50	2.1	33.8	55.5	10.7	11.3	83.7
2	50	4.3	31.5	56.8	11.8	13.3	87.5
3	70	2.1	40.4	49.4	10.2	10.0	75.0
4	70	4.3	36.5	52.6	10.9	9.1	78.7
5	80	2.1	44.3	45.7	10.2	10.2	78.0
6	80	4.3	42.3	46.9	10.7	10.7	77.5

(a) 64.2 g (1.2 moles) of butadiene and 16.8 g (0.162 moles)
of styrene monomer was used.

(b) By Infrared analysis.

TABLE X

COPOLYMERIZATION OF BUTADIENE/STYRENE WITH DILITHIUM
PIPERAZINIDE IN HEXANE[a]

No.	Pzn Temp °C	mmole Initiator	Microstructure, mole %[b]				
			Cis 1.4%	Trans 1.4%	1.2%	% Sty	% Yield
1	30	6.4	30.6	58.0	11.4	12.1	81.0
2	50	3.2	47.9	44.3	7.8	11.4	78.7
3	50	6.4	41.8	49.4	8.8	11.8	83.7
4	50	1.28	46.0	45.0	8.9	11.7	81.3
5	70	1.9	47.7	43.2	8.9	10.5	86.0
6	70	3.2	49.5	41.7	8.8	12.2	85.0
7	70	6.4	44.2	46.7	9.0	12.7	82.5
8	80	1.28	48.2	42.1	9.6	12.3	85.0
9	80	1.9	53.1	38.3	8.7	10.0	85.0
10	80	3.2	50.6	40.3	9.1	12.4	85.0
11	80	6.4	44.8	45.3	9.9	12.5	87.5

(a) 64.2 g (1.2 moles) of butadiene and 16.8 g (0.162 moles) of styrene were used.

(b) By Infrared analysis.

lithium or dilithium piperazinide is similar to the structure of the lithium diethylamide.

From the above studies, it is clear that a copolymer of butadiene and styrene can be prepared from lithium–nitrogen bond initiators. The styrene content of the copolymer is highly dependent on the type of initiator and the polymerization conversion. These lithium–nitrogen bond initiators do not yield a randomized copolymer even with the presence of built-in polar modifier. This may be due to the heterogeneous nature of the initiator. In order to understand the mechanism of copolymerization with lithium–nitrogen bonded initiator. More work along these lines is needed.

Anionic Polymerization VI appeared in *J. Polymer Science*: *Polymer Chem. Ed.* 17, 1771 (1979). Preliminary communication of this paper has appeared in *J. Polymer Sci., Polym. Chem. Ed.,* 17, 1847 (1979).

Acknowledgement

The author wishes to thank The Firestone Tire and Rubber Company for permission to publish these results.

Literature Cited

1. Vinogrador, P. A. and Basayeva, N. N.; _Polymer Sci._, USSR
    (Eng. Ed.) <u>4</u>, 1568, 1963.
2. Angood, A. C.: Hurley, S. A.; Tait, P. J. T.; _J. Polymer
    Sci._; _Polym. Chem. Ed._, <u>11</u>, 2777, 1973.
3. Angood, A. C.; Hurley, S. A.; Tait, P. J. T.; _J. Polymer
    Sci._; _Polym. Chem. Ed._, <u>13</u>, 2437, 1975.
4. Binder, J. L.; _Anal. Chem._, <u>26</u>, 1877, 1954.
5. Binder, J. L.; _J. Polym. Sci._, A, <u>1</u>, 47, 1963.
6. Mochel, V. D.; _Rubber Chem. Technol._, <u>40</u>, 1200, 1967.
7. Antkowiak, T. A.; Oberster, A. E.; Halasa, A. F.; Tate,
    D. P.; _J. Polym. Sci._, A, <u>10</u>, 1319, 1972.
8. Cheng, T. C.; Halasa, A. F.; Tate, D. P.; _J. Polym. Sci._,
    A-1, <u>9</u>, 2493, 1971.
9. Cheng, T. C.; Halasa, A. F.; Tate, D. P.; _J. Polym. Sci._,
    Polym. Ed., <u>2</u>, 253, 1973.
10. Cheng, T. C. and Halasa, A. F.; _J. Polym. Sci._; _Polym. Ed._,
    <u>14</u>, 573, 1976.
11. Turner, R. R.; Altenau, A. G.; Cheng, T. C.; _Analytical
    Chem._, <u>42</u>, 1835, 1970.
12. Korothor, A. A. and Chesnokova, N. N.; _Vysokovmol Soedin_,
    <u>2</u>, 365, 1960.
13. Zelinski, R. P. and Childers, C. W.; _Rubber Chem. Technol._,
    <u>41</u>, 161, 1968.
14. Zelinski, R. P.; U. S. Pat. 2,975,160, 1961.
15. Kuntz, I.; _J. Polym. Sci._, <u>54</u>, 569, 1961.
16. Kern, W. J.; Anderson, J. N.; Adams, H. E.; Bouton, T. C.;
    and Bethea, J. W.; _J. Applied Polym. Science_, <u>16</u>, 3123,
    1972.
17. Hsieh, H. L. and Glaze, W. H.; _Rubber Chem. Technol._, <u>43</u>,
    22, 1970.

RECEIVED February 12, 1981.

Anionic Copolymerization of Butadiene and Isoprene with Organolithium Initiators in Hexane

I.–C. WANG, Y. MOHAJER, T. C. WARD, G. L. WILKES, and JAMES E. McGRATH

Chemistry and Chemical Engineering Departments, Polymer Materials and Interfaces Laboratory, Virginia Polytechnic Institute and State University, Blacksburg, VA 24061

In homopolymerization initiated by sec–butyl–lithium in hexane, isoprene is a more active monomer than butadiene (with $k_1 = 5.53 \times 10^{-5}$ sec^{-1} vs. $k_1 = 0.98 \times 10^{-5}$ sec^{-1} at 20°C). This is also true for reactions at 30° and 40°C. The apparent activation energy for both monomers has been found to be roughly the same, i.e., 19.4 kcal/mole. In the case of copolymerization, butadiene reacts preferentially, but nonexclusively. Significant amounts of isoprene units are also incorporated in a rather random fashion during the early stage of copolymerization. At 50 mole percent isoprene or higher, one observes a second stage of the polymerization that is faster and essentially identical to that observed for isoprene homopolymerization under similar conditions. Various methods have been used to estimate reactivity ratios. The average values at 20°C are $r_B = 2.64$ and $r_I = 0.40$. Preliminary evidence suggests that the copolymerization becomes more selective at lower temperature where the inversion phenomenon is more significant.

Copolymerization, of course, involves the simultaneous polymerization of a mixture of two (or more) monomers.

$$\sim\!\!\sim\!\!\sim\!\!M_1^* + M_1 \xrightarrow{k_{11}} \sim\!\!\sim\!\!\sim\!\!M_1\!-\!M_1^* \qquad (1)$$

$$\sim\!\!\sim\!\!\sim\!\!M_1^* + M_2 \xrightarrow{k_{12}} \sim\!\!\sim\!\!\sim\!\!M_1\!-\!M_2^* \qquad (2)$$

$$\sim\!\!\sim\!\!\sim\!\!M_2^* + M_2 \xrightarrow{k_{22}} \sim\!\!\sim\!\!\sim\!\!M_2\!-\!M_2^* \qquad (3)$$

$$\sim\!\!\sim\!\!\sim\!\!M_2^* + M_1 \xrightarrow{k_{21}} \sim\!\!\sim\!\!\sim\!\!M_2\!-\!M_1^* \qquad (4)$$

In its most simplified form, one assumes a steady state and thus deals with the probabilities of an activated macromolecular chain end either adding another chemically identical unit, e.g., its own monomer, or "cross–initiating" the second monomer. The latter situation amounts to a copolymerization and is illustrated schematically in Equations (2) and (4) above. Typically one defines reactivity ratios r_1 and r_2 on the basis that:

$$r_1 = \frac{k_{11}}{k_{12}} \quad \text{and} \quad r_2 = \frac{k_{22}}{k_{21}}$$

The most studied activated chain ends are the macromolecular free radicals. Vast numbers of monomers have been investigated (1) in hopes of identifying new random (statistical) copolymers which might show a favorable balance of averaged physical properties. In the case of free radical intermediates, one must be very much concerned with both the radical lifetime and the need for a facile cross-initiation. Otherwise, premature termination can lead to generation of (usually) incompatible homopolymers which in turn possess rather unattractive physical properties. (2) In the case of anionic copolymerization, the "living end" must still make a choice between reacting with its own monomer or the second monomer. Relative basicities are an important consideration. (3,4) However, for organolithium initiated polymerizations of butadiene, isoprene or styrene in hexane, cyclohexane or benzene, it is possible to study homogeneous termination free systems. The reactivity of the carbanion end would be expected to be dependent on several parameters such as the counterion, solvent, temperature, etc. It is known for example, that carbanions can exist in "tight", "loose", or even "free" structures as a function of counterion and solvent. (5,6) In hydrocarbons, one should expect organolithiums to exist in more or less tight ion-pairs. Perhaps the most investigated anionic system is the organolithium initiated copolymerization of the monomer pairs, styrene-butadiene and styrene-isoprene. (7) In hydrocarbon solvents, it is the diene which dominates the initial copolymerization to the virtual exclusion of styrene, and one observes a rate nearly identical to that of the diene alone. (9,10) Only when the diene supply is nearly depleted does styrene begin to be incorporated in the polymer chain. Interestingly, a faster polymerization rate is then observed for the styrene segment. Thus in these systems, there is an apparent "reversal of reactivity" of styrene and the diene, since for the homopolymerization situation styrene is a much more reactive monomer than either diene.

Relatively little information is available for the copolymerization of butadiene with isoprene. In an early paper by Rakova and Korotkov (8), it was concluded that in n-hexane with n-butyllithium as the initiator, the reactivity ratios for butadiene and isoprene were r_B = 3.38 and r_I = 0.47, respectively. In other words, the butadiene monomer reactivity significantly differs from isoprene but by a smaller factor than for a styrene-diene system. Nevertheless, the phenomenon of "reversal of reactivity" still reportedly occured in the copolymerization between these two dienes. We felt that if these values were essentially correct that relatively pure blocks of butadiene might well be formed at the early stage of reaction by directly copolymerizing butadiene and isoprene simultaneously. We are quite interested in synthesizing block-type thermoplastic elastomers of this type. The butadiene-isoprene-butadiene copolymer per se may not be influenced by architecture arrangement. However, in the totally or selectively hydrogenated derivatives, one might expect major differences if there are crystallizable polyethylene blocks derived from the 1,4 butadiene units in polyethylene-poly(ethylene-co-propylene) or polyethylene-poly-isoprene block copolymer. By analogy with the styrene-diene (7) systems, the crystalline polyethylene end blocks in the triblock should associate to form crystalline tie-down points which will desirably reinforce the soft interior block of polyethylene-co-propylene or polyisoprene in the resulting two hydrogenated copolymers. Our kinetic investigation was thus principally motivated by the desire to learn whether the possible "reversal of reactivity" between butadiene and isoprene via anionic co-polymerization would be of sufficient magnitude to produce crystalization sequences in the hydrogenated derivatives. We were also interested in investigating possible subtle temperature or solvent effects. High vacuum techniques were employed (12) with the view that this work might improve on the research published two decades ago. (8) Our new results on the reactivity ratios for butadiene and isoprene have also been analyzed via both conventional (13,14) and the more recent statistical methods. (15-18)

Experimental

Copolymerization Apparatus and Techniques

The high reactivity of alkyllithium compounds requires that these polymerizations be performed under extremely high purity conditions. In order to achieve this we have utilized a high vacuum system. The basic design of a high vacuum apparatus and purification procedures has been described in detail elsewhere. (12) Accordingly, the required techniques used for the purification of n-hexane, the monomers and the sec-butyllithium initiator have thus been performed. A typical glassware reactor which also permits purging of the reaction vessels is shown in Figure 1. The reactors were always flamed at a constant pressure of 10^{-5}mm Hg until the flame took on the characteristic sodium color. It was then sealed off the vacuum system, purged with n-butyllithium/hexane solution and rinsed with hexane by back distillation at least four times. The purging section was then sealed off the reactor. Initiator which has previously been vacuum distilled and diluted with pure hexane was used. A

measured volume and known concentration was introduced from the
attached ampoule into the reactor, followed by careful rinsing
again with hexane. The purged reactor was then sealed onto the
vacuum line through one of the break seals. The additional
solvent and monomers were then quantitatively distilled in.
Volumes were recorded at suitable low temperatures (e.g. $-78°C$)
where density values were available. The values of 0.73 gm/cc
for butadiene at $-78°C$ and 0.68 gm/cc for isoprene at $20°C$ were
used. It should be stressed that the sealing of all constric-
tions or dilatometers was performed with the solution frozen or
cooled down at $-196°C$ or $-78°C$, respectively.

Kinetic Study

Four to six dilatometers (2 mm inner bore) of 3-5 ml each
were attached to one of the side arms of the reactor shown in
Figure 1. The actual solution volume depended on the monomer
concentration and was usually adjusted to allow about a 10-15
cm drop at 100% conversions. After the contents of the reaction
were equilibrated near room temperature, the dilatometer volume
was usually adjusted such that the stems were about half-full.
To prevent any undesirable distillation and/or bumping which
would cause concentration fluctuations, the procedures outlined
by Juliano (19) were carefully followed. Next, each detached
dilatometer was securely clamped in a constant temperature bath
which was maintained at 20, 30, or $40°C$ as desired. Readings
were taken with a cathetometer after the initial thermal expansion to the bath temperature. The data were treated as described
by Pett. (20) The dilatometric treatment was also used for those
reactions where a conversion versus time curve was required for
later estimation of extent of conversion.

The extent of conversion was cross-checked by precipitation
of the polymer into methanol, followed by filtration and drying
at room temperature under mechanical pump vacuum ($\sim 10^{-2}$ Torr)
until constant weight was attained.

The composition of the copolymer was determined by either
NMR analysis at 90 MHz according to the equations derived by
Mochel (21) or by infrared. (22) The agreement of these methods
was $\pm$ 2% when applied to copolymer taken to 100% conversion. The
reactivity ratios were calculated according to the Mayo-Lewis
Plot (13,15), the Fineman-Ross Method (14), or by the Kelen-Tudos
equation. (16,17,18) The statistical variations recently noted
by O'Driscoll (23), were also considered.

The total hydrogenation of the copolymers utilized the
diimide method (24), generated in situ from p-toluenesulfonylhy-
drazide (TSH) in xylene for six hours at reflux temperature
($132-134°C$). In general, 5.0 moles of TSH per 100 grams of polymer were used. We have found (25) that the addition of a
phenolic antioxidant such as Irganox 1010 effectively decreases
the minor, but detectable side reactions. The saturation of the

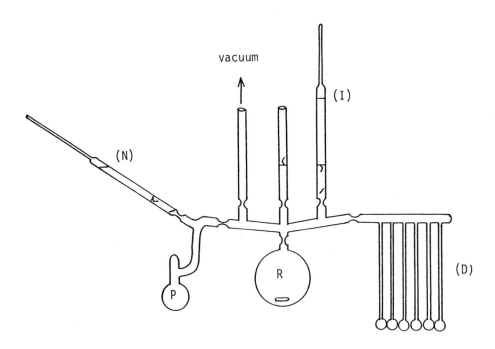

vacuum

(I)

(N)

R

(D)

P

Figure 1. Copolymerization purging apparatus and reactor with attached dilatometers: initiator ampule (I); dilatometers (D); reactor (R); purging solution collector (P); n-butyllithium in hexane solution (N).

polydiene was determined to be complete either by high temperature H-NMR analysis in hexachloro-1,4-butadiene using hexamethy-
disiloxane as an internal standard, or by infrared spectra of
the polymer film cast on a KBr plate. Thickness of between 1/2
to 1 mil was required for a good spectra. The hydrogenated
copolymers were further checked for crystallinity content on a
Model 2 Perkin-Elmer differential scanning calorimeter (DSC) at
a heating rate of 20°C/minute.

Results And Discussion

Homopolymerization of Butadiene and Isoprene in n-Hexane

Homopolymerizations were first conducted to establish a
basis for the copolymerization study. The polymerization of the
diene monomers in n-hexane has been investigated in the literature via two approaches: (a) by mixing all of the monomers and
solvent directly with initiator, and (b) by premixing the initiator with a slight molar excess of the monomer to form a
living "seeded" polymer. The later method was particularly
employed with n-butyllithium initiator (26) where the initiation
rate was slow enough that the sample organolithium persisted
during a substantial period of the polymerization process. In
order to enhance the rate of initiation relative to that of
propagation, sec-butyllithium has evolved (27) as the preferred
initiator, especially for kinetic studies. Branched alkyllithiums have been reported to increase the initiation rate of
dienes in hydrocarbon solvents. (28) When the initiation rate
is of the same order of magnitude as the rate of propagation,
the homogeneous anionic polymerizations allows the synthesis of
polymers possessing a very narrow molecular weight distribution.
Relatively linear polymerization curves can be obtained for both
butadiene and isoprene as shown for example in Figure 2. This
was true at three different temperatures, namely, 20°, 30°, and
40°C.

The kinetic results obtained by dilatometry for the polymerization of butadiene and isoprene are shown in Figures 3 and 4.
As usual, ΔH_t refers to the change in height at a given time.
It can be seen that the kinetic studies on dienes confirm that
the propagation reaction has a first-order dependence on the
monomer concentration. These observations are to be expected
since in the absence of adventitious impurities the number of
growing chains should remain constant, only the monomer concentration decreases.

The observed apparent first-order rate constants are listed
in Table I. From the constants at different temperatures but
at the same initiator concentration, it is possible to calculate
an apparent activation energy for the propagation reaction in
each case. This has been done, and the Arrhenius plot is shown
in Figure 5. It is interesting to note the similarity of these

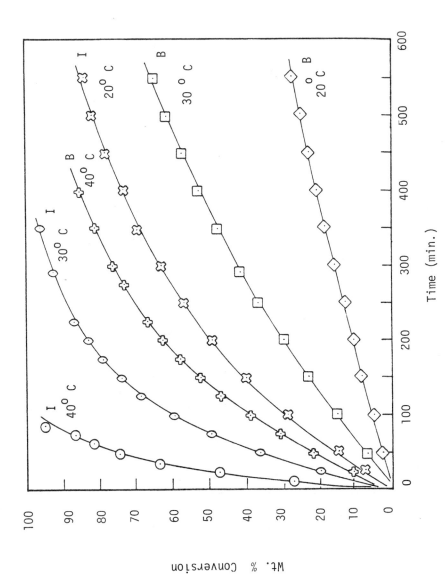

Figure 2. Homopolymerization rates of butadiene and isoprene in hexane.

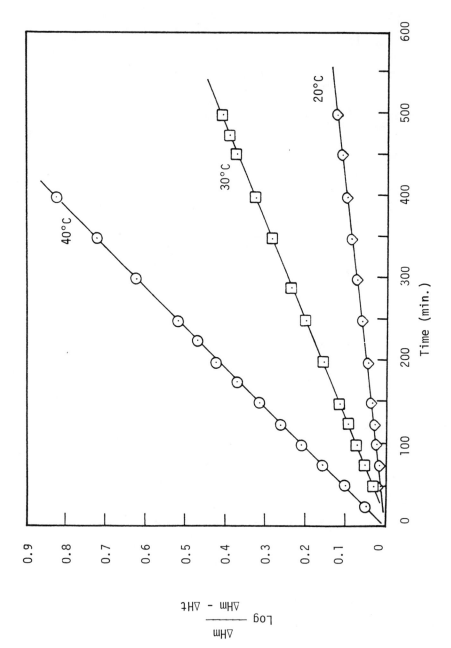

Figure 3. First-order plots for butadiene in hexane.

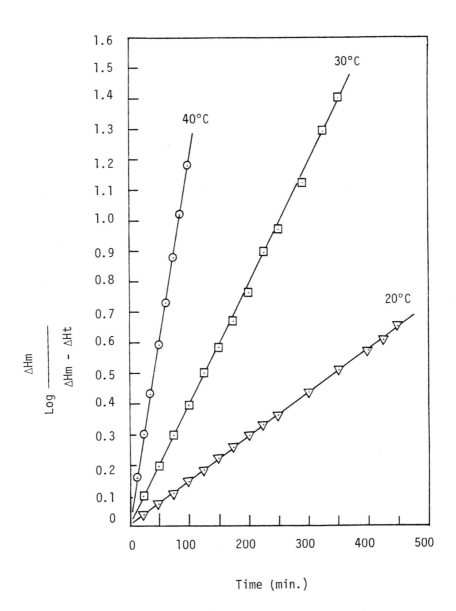

Figure 4.   First-order plots for isoprene in hexane.

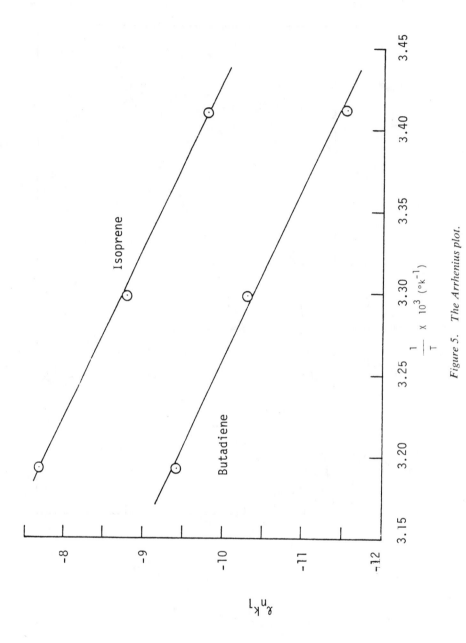

Figure 5.   The Arrhenius plot.

Table I

First-Order Propagation Rate Constants (k_1) of the
Homopolymerizations of Butadiene and Isoprene

System	Temp. $^{\circ}C$	$\dfrac{\text{sec-BuLi}}{\text{mole/1.}}_{\circ}$ x 10^3	Monomer $_{\circ}$ mole/1.	k_1, sec^{-1} x 10^4
Butadiene –Hexane	20	1.27	1.00	0.0978
	30	1.27	1.00	0.327
	40	1.27	1.00	0.796
Isoprene –Hexane	20	1.26	1.00	0.553
	30	1.26	1.00	1.48
	40	1.26	1.00	4.56

activation energies, 19.2 kcal/mol, for polymerization of the
two monomers in n-hexane. The value obtained is much higher
that what has been reported for the case where THF was the
solvent. (26) Our value is also somewhat higher than the 14.3
k-cal/mole published by Worsfold and Bywater. (29) From the
viscosity studies of Morton, et. al., (30,31) one may assume
that the active polydienyllithium is associated in pairs in
n-hexane, but no such association is observed in THF. On this
basis, the high activation energy found for the propagation
reactions in n-hexane should include the heat of dissociation of
the associated ion pairs if the dimer species do not react
directly with the incoming monomer. (32)

The dependence of the propagation rate on the
concentration of growing chains is illustrated in Figures 6
and 7, and is listed in Table II. The first-order rate constant
from Table II are plotted as a function of the initiator concentration. Although the kinetics of organolithium polymerization
in nonpolar solvents have been subjected for intensive studies,
the results were still somewhat controversial. In view of the
strong experimental evidence for association between the organolithium species, the kinetic order ascribed to this phenomenon
was postulated (30,31) as shown in Equations (5) and (6).

$$(RMj^-Li^+)_n \rightleftharpoons nRMj^-Li^+ \qquad (5)$$

$$RMj^-Li^+ + M \longrightarrow RM^-_{j+1}Li^+ \qquad (6)$$

This assumes that only the dissociated chain ends are active.
Actually, measurements on the state of association, i.e., the
value of n in Equation (5), have been carried out for styrene,

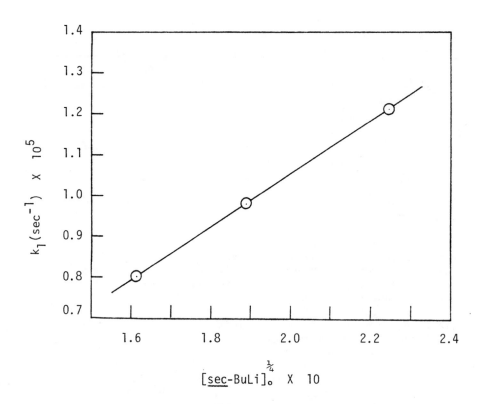

Figure 6.   Dependence of butadiene rate on initiator concentration at 20°C.

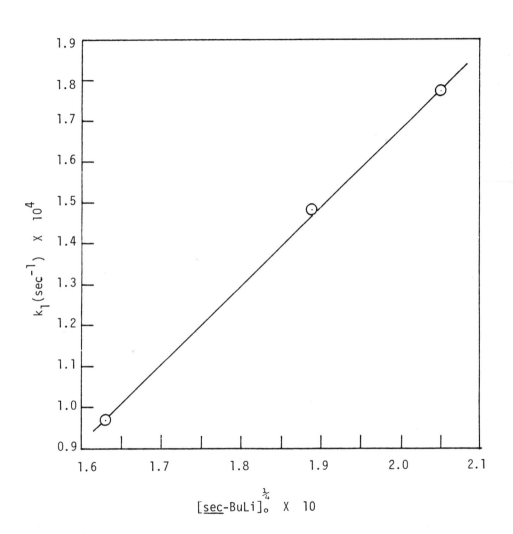

Figure 7. Dependence of isoprene rate on initiator concentration at 30°C.

Table II

Effect of Initiator Concentration of the Propagation
Rate of Butadiene and Isoprene in Hexane

System	Temp. $^{\circ}$C	sec-BuLi mole/1. $^{\circ}$ x 10^3	Monomer mole/1. $^{\circ}$	k_1, sec^{-1} x 10
Butadiene –Hexane	20	0.683 1.27 2.54	1.00	0.0802 0.0978 0.121
Isoprene –Hexane	30	0.704 1.26 2.51	1.00	0.967 1.482 1.769

butadiene and isoprene. The n values were measured to be consistently very close to 2. While this could explain the half-order kinetics demonstrated by most investigators (12) in the case of styrene, it could not account for the lower orders (1/4 to 1/6) found for butadiene and isoprene in various hydrocarbon concentrations. Our data (Fig6) fit a 1/4-order type kinetics, especially for the case for butadiene at 20°C. The same kinetics data for isoprene at 30°C shown in Figure 7 are not so precise on the basis of a 1/4-order dependence. Nevertheless, our results confirm the idea that the propagation rates of the diene monomers in hydrocarbon are certainly a fractional order type dependence. Whether 1/4-order or something different order probably needs to be further defined. It appears that the propagation reaction of these associated growing chains in nonpolar media may be more complicated than what is proposed in Equations (5) and (6), and probably involves a direct interaction between the monomer and the associated complex. (32)

Kinetic Study on the Copolymerization of Butadiene and Isoprene in Hexane

It has been emphasized in the copolymerization of styrene with butadiene or isoprene in hydrocarbon media, that the diene is preferentially incorporated. (7,9,10) The rate of copolymerization is initially slow, being comparable to the homopolymerization of the diene. After the diene is consumed, the rate increases to that of the homopolymerization of styrene. Analogously our current investigation of the copolymerization of butadiene with isoprene shows similar behavior. However, the

reversal of reactivity is much less. The degree of reversal
phenomenon among these two dienes can be correlated with the
reaction temperature for a fixed monomer feed ratio. The dilatometric data are plotted in Figures 8-10 for three different
temperatures, and are summarized in Table III for the purposes of
clarity and comparison. Curves 1 and 5 in these figures represent the homopolymerization rates of isoprene and butadiene,
respectively. In addition, data for three different initial
monomer feed ratios are expressed by B/I's with 25/75, 50/50,
and 75/25. The numbers denote the monomer ratio of butadiene
content to isoprene for curve 2, 3, and 4, respectively. The
initial propagation rate constants were measured and are listed
in Table III, along with the homopolymerization rates of butadiene and isoprene under the same conditions. For higher content
of butadiene (i.e. with a mole fraction of 0.75), the overall
<u>copolymerization</u> rate is almost identical to the <u>homopolymeriza-</u>
<u>tion</u> rate of butadiene. This behavior was observed at three
different temperatures, 20°, 30°, and 40°C. Curve 4 shown in
Figures 8-10 is almost parallel to Curve 5, and shows no sign of
the inversion phenomenon. Accordingly, during the copolymerization with a butadiene-to-isoprene molar ratio of 75 to 25, there
is a considerable amount of isoprene incorporated in a rather
random fashion with the butadiene. The overall rate is nevertheless controlled by the slower rate-determining step of butadiene
polymerization. The copolymer composition at low conversion has
been determined to be rich in butadiene, for example 83.5 to
88.5 mole % were found compared to 75 mole % in charge (cf also
Table IV). However, when a mixture of butadiene and isoprene
with 25/75 or 50/50 molar ratio is polymerized, the initial
propagating rate is enhanced slightly due to the isoprene. All
the prematurely terminated copolymers with butadiene mole fraction from 0.20 to 0.80 in the feed are nevertheless found to be
rich in butadiene, as seen by the copolymerization data listed
in Table IV. Moreover, there is another interesting point observed with the molar ratios of butadiene to isoprene of 25/75
or 50/50. Here one can observe the occurance of the "inversion"
phenomenon which is manifested by the inflection point of the
kinetic Curves 2 and 3 of Figures 8-10 at all three temperatures.
The lower the copolymerization temperature is, for example 20°C,
the sharper the inflection appears to become. This behavior
leads to the conclusion that the copolymerization is more
selective at lower temperatures. Conversely, when temperature
is increased selectivity is noticeably decreased. The derived
reactivity ratios reflect this trend and are shown in Table V.

By measuring the kinetic rate of second stage reaction
after inflection, one can observe that rate is very analogous to
the homopolymerization rate of isoprene. The data are listed in
Table III, and can also be detected by the straight portion of
Curves 2 and 3 after inflection. The "inversion" phenomenon can
be easily explained by the fact that, although the isoprene is

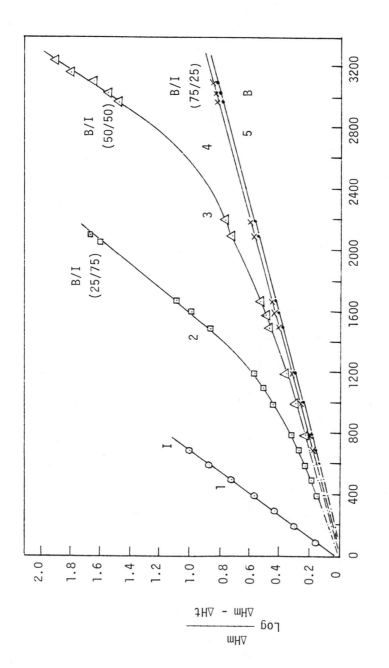

Figure 8. Effect of comonomer feed ratio on the rate of polymerization in hexane at 20°C.

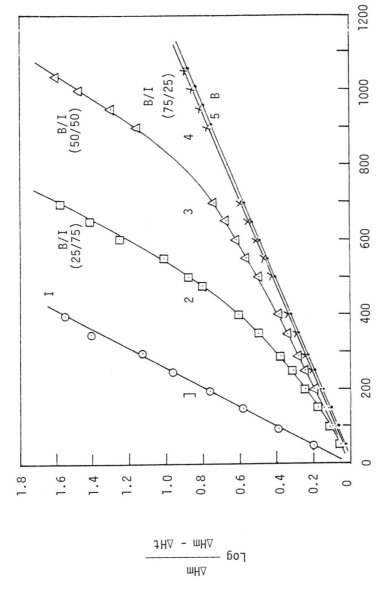

Figure 9. Effect of comonomer feed ratio on the rate of polymerization in hexane at 30°C.

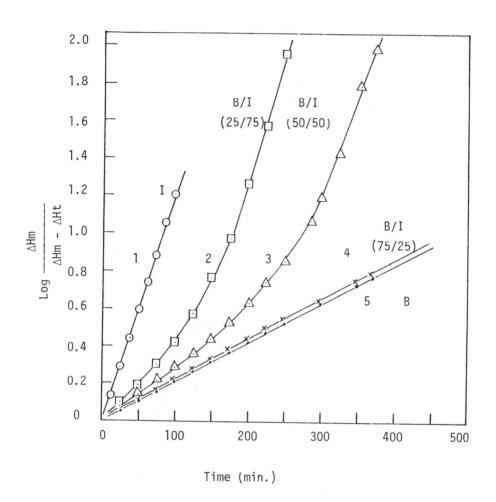

Figure 10. Effect of comonomer feed ratio on the rate of polymerization in hexane at 40°C.

Table III

Summary of Kinetic Results for the Homo- and Co-Polymerization of Butadiene and Isoprene in Hexane

Curve #	Monomer Feed Ratio B/I	Temp. °C	sec-BuLi $\frac{mole/l.}{}$ x 10^3	Propagation Rate Constant k_1 (sec^{-1}) x 10^4	
				Initial	After Inversion
1	0/100	20 30 40	1.26	0.553 1.48 4.56	
2	25/75	20 30 40	1.22	0.126 0.450 1.382	0.514 1.324 4.95
3	50/50	20 30 40	1.26	0.105 0.359 1.076	0.710 1.244 4.52
4	75/25	20 30 40	1.31	0.0979 0.320 0.864	no inversions
5	100/0	20 30 40	1.27	0.0978 0.327 0.796	

Table IV

Anionic Copolymerization Data

System	Feed Mole % Butadiene	Temp °C	Copolymer Composition mole % Butadiene	Conversion Wt %
Butadiene/Isoprene/Sec-BuLi/Hexane	20	20	37.3	2.63
		30	37.2	4.40
		40	38.9	10.37
	30	20	54.2	3.47
		30	46.8	6.08
		40	48.2	9.06
	40	20	62.3	5.52
		30	58.0	4.08
		40	61.7	7.56
	50	20	72.7	4.58
		30	66.3	5.10
		40	69.7	10.62
	60	20	80.0	2.66
		30	74.3	8.33
		40	77.7	14.75
	75	20	88.5	10.00
		30	83.5	38.00
		40	--	--
	80	20	91.3	4.94
		30	87.1	9.67
		40	88.1	11.82

Table V

Copolymerization Reactivity Ratios
of Butadiene (M_1) and Isoprene (M_2) initiated by Sec-BuLi in Hexane

Temperature °C	Reactivity Ratios				
	FR(a)	Inverted FR(b)	ML–JJ(c)	KT(d)	YBR(e)
20	r_1 = 2.62	2.71	2.62 ± 0.06	2.62	2.64 ± 0.04
	r_2 = 0.39	0.43	0.39 ± 0.07	0.40	0.41 ± 0.02
30	r_1 = 1.57	1.60	1.68 ± 0.09	1.60	1.59 ± 0.03
	r_2 – 0.33	0.35	0.36 ± 0.08	0.35	0.35 ± 0.02
40	r_1 = 1.74	1.97	1.82 ± 0.07	1.89	1.85 ± 0.11
	r_2 – 0.24	0.34	0.28 ± 0.09	0.32	0.32 ± 0.04

(a) FR = Fineman-Ross Method

(b) Inverted FR = Inverted Fineman-Ross Method

(c) ML-JJ - Graphical Mayo-Lewis Method modified by Joshi-Joshi estimation

(d) KT = Kelen-Tudos Method

(e) YBR = Yezrielev-Brokhina-Roskin Method (Ref. 36)

more reactive than butadiene in homopolymerization, in the mixture butadiene tends to preferentially (not exclusively) polymerize first with some random incorporation of isoprene units. After the butadiene is depleted at the early stage of copolymerization, the isoprene rate does eventually take over.

One can try to explain the reversal of reactivity of butadiene and isoprene during copolymerization in hexane solvent. A number of different effects may be important. Major parameters include the electronic and steric character of the monomers and how they may relate to the slow step of the polymerization. One might argue that the electron-donating methyl group on isoprene could decrease the electrophilic character of isoprene and render it less reactive toward a living carbanion end. However, one must also recall that the homopolymerization rate is higher! The copolymerization of butadiene, isoprene, and 2,3-dimethyl-butadiene with polystyryl sodium in tetrahydrofuran solution showed a retarding effect of the methyl group on the addition of diene. (33) Again, though these results were, in THF, not hexane where association effects are important, steric effects, possibly related to the 4,1 nature of the active chain, seems to be a more likely possibility in nonpolar medium.

The NMR analysis (21) of the chemical composition for copolymers from various monomer feed ratios at fairly low conversion are shown in Table IV. The results were then used to estimate the reactivity ratios for the diene monomers under the conditions employed. Various published methods of calculating monomer reactivity ratios have been examined. These include the once popular but now somewhat out of favor Fineman-Ross method (14), the graphical Mayo-Lewis solution (13) modified by Joshi-Joshi estimation (15), and the recent Kelen-Tudos method. (16, 17,18) The results are all summarized in Table V. Although the determination of reactivity ratios based on the usual relation between copolymer composition and monomer ratio can only be considered as an approximation, the values recorded in Table V do indicate a trend, that the ratios differ by a slightly greater factor at 20°C. (r_B=2.69, r_I=0.40) than at 30° or 40° C. This observation is corroborated with what has been found in Figures 8-10. There is more of an inversion phenomenon occurance at 20°C. However, the difference between 30°C and 40°C is small and apparently similar, within experimental error. Nevertheless, the new established reactivity ratios of butadiene and isoprene at all three temperatures differ by a smaller factor than what were reported by the work of Korotkov (8) (e.g. r_1 = 3.38 and r_2 = 0.47). Moreover, butadiene is more reactive and initial copolymer contains a larger proportion of butadiene randomly placed along with some incorporation of isoprene units. The randomness of the copolymer via direct copolymerization has been confirmed by the comparison with pure diblock copolymer produced by sequential monomer addition. Both copolymers have similar chemical composition (50/50) and molecular weight. Their characterization data are presented in Table VI. The near "mono-dispersities" were also confirmed by the GPC measurements. After hydrogenation via diimide in situ the block copolymer of poly-butadiene-polyisoprene produces a polyethylene-poly(ethylene-co-propylene) system. The partially crystalline blocks of "poly-ethylene" show melting point about $100°C$ by DSC as shown in Figure 11. Parallel behavior has also been reported for partially hydrogenated polybutadiene in the literature. (34) The properties of our hydrogenated block polymers are interesting and will be re-ported elsewhere. (35) For the copolymer made via direct co-polymerization with butadiene-to-isoprene weight ratio of 50 to 50, its totally hydrogenated derivative shows different pattern. Only a very small melting transition at $92°C$ can be seen in the DSC thermogram. This indicated that only an insignificant por-tion of polyethylene has been induced through hydrogenating the copolymer block rich in polybutadiene. The sequence length of the polybutadiene is shortened by the incorporation of some isoprene. The isoprene units thus eliminate the crystallinity in the subsequently hydrogenated material.

Conclusions

In the current study of the homopolymerization and copoly-merization of butadiene and isoprene by secondary-butyllithium in hexane the following conclusions can be made.

(1) Isoprene is a more active monomer than butadiene in homopolymerization, but the apparent activation energy of the propagation reaction is 19.2 kcal/mole for both monomers.

(2) In copolymerization butadiene reacts preferentially, however, significant concentration of isoprene units are also incorporated in a random manner during the early stage of reaction. Butadiene has higher reactivity ratio than isoprene. The reversal of reactivity is caused by the less steric and polar factors in the interaction of butadiene molecules with an active anionic center.

(3) Interestingly, kinetic results show that essentially pure isoprene blocks are formed in the latter stage of reaction after "inversion" point. The phenomenon is somewhat dependent on the relative molar concentration and temperature.

(4) Various methods have been applied to estimate their reactivity ratios which are tabulated in Table V. Typical values at $20°C$ are $r_B = 2.64$ and $r_I = 0.404$. Preliminary evidence was reported that suggests the copolymerization is more selec-tive at lower temperatures.

Table VI

Comparison of Copolymers via Sequential and Direct Monomer Addition

Curve #	Total Hydrogenation of Polymer Code #	Monomers Addition	wt % PB/PI	M_n (a) (x10^{-3})	M_w (b) (x10^{-3})	25°C η THF
1	9-ICW-29	Sequential	49/51	190	208	1.80
2	9-ICW-88	Direct	48.8/51.2	183	213	2.24

(a) Determined via membrane osmometer in chlorobenzene at 25°C.

(b) Obtained from gel permeation chromatography calibration curve.

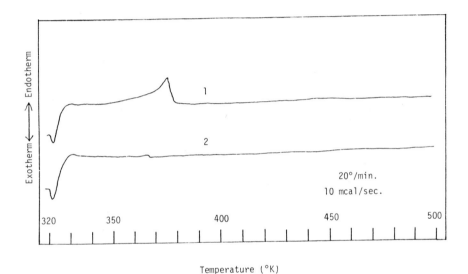

Figure 11. DSC diagrams of totally hydrogenated derivatives of copolymers: diblock copolymer, 9-ICW-29, by sequential addition of monomers with 49/51 wt % of PB/PI (1); random copolymer, 9-ICW-88, by direct addition of monomers with 48.8/51.2 wt % of PB/PI (2).

Acknowledgement

The authors would like to thank the Army Research Office for their support of this work. They would like to thank the NSF Polymers Program for purchase of some of the equipment used in this research.

LITERATURE CITED

1. Greenley, R. Z.; J. Macromol. Sci.-Chem. 1980, A14 (4), 427, 445.
2. Olabisi, O.; Robeson, L. M.; Shaw, M. T. "Polymer-Polymer Viscibility"; Academic Press: New York, 1979.
3. Fetters, L. J.; J. Polymer Sci. 1969, C26, 1.
4. Dart, E. C.; "Reactivity, Mechanism and Structure in Polymer Chemistry"; Jenkins, A. D.; Ledwith, A.; Fds.; Interscience, New York, 1974; Chap. 10.
5. Szwarc, M.; Acct. Chem. Res. 1969, 2, 87.
6. Szwarc, M.; "Ions and Ion Pairs in Organic Reactions"; Wiley-Interscience, New York, 1972; Vol. 1, Chap. 1.
7. Noshay, A.; McGrath, J. E.; "Block Copolymer: Overview and Critical Survey"; Academic Press, New York, 1977.
8. Kakova, G. V.; Korotokov, A. A.; Dokl. Akad. Nauk SSSR 1958, 119, 982; Chem. Abstr. 1959, 53, 4089i; Rubber Chem. Tech. 1960, 33, 623.
9. Korotokov, A. A.; Rakova, R. V.; Vysokomol. Soedin. 1961, 3, 1482; Poly. Sci. USSR 1962, 3, 990.
10. Korotkov, A. A.; Vysokomol. Soedin. 1960, 2, 365; Poly. Sci. USSR 1962, 2, 284.
11. Kuntz, I.; J. Poly. Sci. 1961, 54, 569.
12. Morton, M.; Fetters, L. J.; Rubber Chem. Tech. 1975, 48, 359.
13. Mayo, F. R.; Lewis, F. M.; J. Am. Chem. Soc. 1944, 66, 1594.
14. Fineman, M.; Ross, S. D.; J. Poly. Sci. 1950, 5, 269.
15. Joshi, R. M.; Joshi, S. G.; J. Macromol. Sci.-Chem. 1971, A5 (8), 1329.
16. Kelen, T.; Tudos, F.; ibid. 1975, A9(1), 1.
17. Tudos, F.; Kelen, T.; Foldes-Berezsnich, T.; TurcsANYi, B.; ibid. 1976, A10 (8), 1513.
18. Kennedy, J. P.; Kelen, T.; Tudos, F.; J. Poly. Sci. Chem. Ed. 1975, 13, 2277.
19. Juliano, P.; Ph.D. Thesis, Univ. of Akron, 1968.
20. Pett, R. A.; Ph.D. Thesis, Univ. of Akron, 1965.
21. Mochel, V.D.; Rubber Chem. Tech. 1968, 41, 1200.
22. Haslen J.; Willis, H.A.; "Identification and Analysis of Plastics"; Iliffe Book Ltd: London, 1965; Chapters 1, 5 and 6.

23. McFarlane, R. C.; Reilly, P. M.; O'Driscoll,K. F.; J. Poly. Sci. Chem. Ed. 1980, 18, 251.
24. Mango, L. A.; Lenz, R. W.; Makromol. Chem. 1973, 163, 13; Harwood, H. J.; Russell, D. B.; Verthe, J.J.A.; Zymonas, J. ibid. 1973, 163, 1.
25. Wang, I.-C.; Mohajer, Y.; Wilkes, G. L.; McGrath, J.E.; Polymer Bulletin, 1981, in preparation.
26. Morton, M.; Bostick, E. E.; Livigni, R. A.; J. Poly. Sci. 1963, Part A, 1, 1735.
27. Milkovich, R.; S. African Pat. No. 280,712 (1963).
28. Hsieh, H. L.; J. Poly. Sci. 1965, Part A, 3, 153, 163, 173.
29. Worsfold, D. J.; Bywater, S.; Can. J. Chem. 1960, 38, 1891.
30. Morton, M.; Fetters, L. J.; Pett, R. A.; Meier . F.; Macromol. 1970, 3, 327.
31. Fetters, L. J.; Morton, M.; Macromol. 1974, 7, 552.
32. Smart, J. B.; Hogan, R.; Scherr, P. A.; Emerson, M. T.; Oliver, J. P. J. Organomet. Chem. 1974, 64, 1.
33. Shima, M.; Smid, J.; Szwarc, M.; Poly. Letters 1964, 2, 735.
34. Cowie, J. M.G.; McEven, I. J. Macromol. 1977, 11, 1124.
35. Mohajer, Y.; Wilkes, G. L.; Wang, I.-C.; McGrath, J. E. Polymer 1981, in press.
36. Yezrielev, A. I.; Brokhina, E. L.; Roskin, Y. S.; Vysokomol. Soedin. 1969, A11(8), 1670.

RECEIVED February 17, 1981.

The Reactivity of Polydiene Anions with Divinylbenzene

M. K. MARTIN[*], T. C. WARD, and JAMES E. McGRATH

Chemistry Department and Polymer Materials and Interfaces Laboratory,
Virginia Polytechnic Institute and State University, Blacksburg, VA 24061

Star-branched homopolymers of butadiene and isoprene have been synthesized utilizing anionic polymerization techniques. The commercial mixture of divinylbenzene was employed as the star linking agent for the "living" polydienyllithium anions. The nature of the star-branching reaction was studied with respect to: the molar ratio of divinylbenzene (DVB)/alkyllithium (RLi), reaction temperature, reaction time and the nature of the polydienyllithium chain end. In general, a higher number of arms is achieved via increased DVB/RLi ratio, reaction temperature and through the choice of polydienyllithium chain ends. The polymers were characterized by gel permeation chromatography (G.P.C.), membrane osmometry, intrinsic viscosity, and H-NMR spectroscopy. A Chromatix low-angle light scattering G.P.C. detector was also employed for the determination of the weight average molecular weight ($\overline{Mw}$), as well as a sensitive detector, in order to assay absolutely the $\overline{Mw}$ versus elution volume profile for a series of star-branched polyisoprenes and polybutadienes. The results indicate that under optimum conditions a relatively well defined number of arms can be achieved with DVB linking.

In 1965, Milkovich (1) reported that divinylbenzene could be utilized for the formation of star-branched macromolecules. Later, Rempp and coworkers (2, 3, 4) successfully applied this method for the synthesis of star-branched polystyrenes. Moreover, Fetters and coworkers (5-8) used this procedure for the synthesis of multi-arm star-branched polyisoprene homopolymers and poly-

[*] Current Address: Shell Development Company, P.O. Box 1380, Houston, TX 77001.

0097–6156/81/0166–0557$06.00/0

styrene-polydiene block polymers. This method of star-branched polymer formation involves the sequential polymerization of monomer giving the "arm", followed by the addition of divinylbenzene. The polymerization of divinylbenzene then results in a "microgel" nucleus containing pendant vinyl groups which serve as branch points for the star-shaped polymer. This procedure has led to the formation of star-branched polyisoprenes containing up to 56 weight-average numbers of arms (5, 6). The arms of these star macromolecules possess narrow molecular weight distributions due to the termination free aspect of diene polymerizations initiated by organolithium species in hydrocarbon solvents (9).

Star-branched block and homopolymers are of interest both from a theoretical and practical viewpoint. Studies on the melt rheology have shown the viscosity to be independent of the extent of branching, yet dependent upon the arm molecular weight (6, 7, 8). Star-branched styrene-butadiene block polymers appear to have improved mechanical properties and processability over the linear styrene-butadiene-styrene analogs (7, 8). A high styrene content star-shaped block copolymer introduced by Phillips (K-resin) offers the rigidity, transparency, and good impact strength necessary for packaging and thermoforming applications (10). High butadiene content star block polymers have been used in hot melt and pressure sensitive adhesives (11, 12). Due to this growing interest in star-branched polymer behavior, it became of interest to study the nature of the star-branching reaction between polydienyllithium anions and the commercial mixture of divinylbenzene. The subject of our research discussed herein was to investigate the variables that influence the star formation process. These certainly include the molar ratio of DVB/RLi, the linking reaction temperature, reaction time, and the nature of the "living" diene chain end.

Experimental

Star-polymer synthesis was achieved by utilizing an all glass high vacuum design as shown in Figure 1. Monomers and solvents were purified by drying over calcium hydride; with periodic degassing, followed by distillation under vacuum over a succession of sodium mirrors. The commercial DVB monomer was further purged over dibutylmagnesium (Lithium Corporation of America), which produced a bright-yellow complex that was indicative of dryness. Solvents and diene monomer were further stored over an organometallic purge until needed. The DVB was split down into ampoules in benzene or hexane to form 0.2N-0.8N solutions. This was necessary due to the high reactivity of DVB.

A typical polymerization reactor is shown in Figure 2. The reactor contains a side-ampoule used to isolate a portion of the "living" linear precursor before star-linking by the addition of DVB. The reactor was purified by flaming with a hand torch while continuously pumping until a pressure of 10^{-5} was reached. The

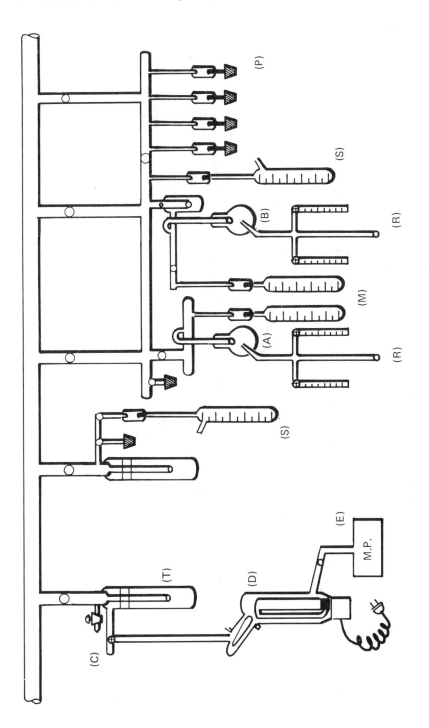

Figure 1. High-vacuum apparatus.

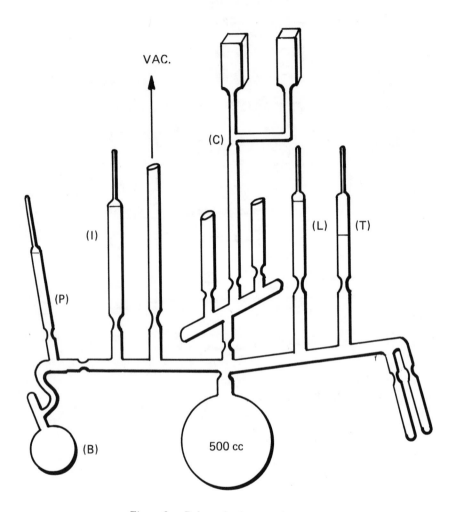

VAC.

(C)

(I)

(P)

(L)   (T)

(B)

500 cc

Figure 2. Polymerization reactor.

apparatus was then heat-sealed from the vacuum line via the constriction and the interior walls of the reactor rinsed with n-butyllithium-hexane solution in order to remove any adsorbed impurities on the interior walls of the glassware.

sec-Butyllithium was used as the initiator to prepare the polydienes. The sec-butyllithium was purified by high vacuum distillation as described in previous work (13). The pure sec-butyllithium was then diluted with hexane and split-down into ampoules containing breakseals. Each ampoule could be used for subsequent polymerization. A typical split-down apparatus is shown in Figure 3. Initiator concentrations were determined by titration with standard HCl to a phenolphthalein endpoint.

The diene polymerizations were generally allowed to proceed for 48 hours, which ensured quantitative conversion. In the case of butadiene, the reactor was cooled periodically to draw residual butadiene monomer into the solution from the void volume of the reactor.

The star-shaped polymers were prepared by the addition of the DVB solution to the "living" polydienyllithium hexane solution. Prior to the addition of divinylbenzene (DVB), a portion of the linear polymer solution was isolated via the sidearm, which enabled the arm molecular weight to be determined.

Upon addition of divinylbenzene the characteristic red-orange color of the styryllithium type anion resulted. After the star-polymer formation was complete, the solutions were terminated with degassed methanol. The polymer was precipitated in methanol and dried to a constant weight under vacuum. Characterization techniques included membrane osmometry, intrinsic viscosity, GPC and spectroscopic methods. A Wescan recording osmometer was used at 43°C with chlorobenzene as the solvent. The accuracy of the instrument was checked by the use of several solutions of known molecular weight polystyrene standards. The values of $\bar{M}n$ were determined from measurements made on solutions of at least four concentrations, followed by the typical $(\pi/c)^{1/2}$ versus concentration plot procedure. The gel permeation chromatography analysis was achieved using a Waters HPLC equipped with both refractive index and U.V. detectors. The carrier solvent was tetrahydrofuran at 25°C, with a flow rate of 1.0 ml/min. The column arrangement consisted of four one-ft. columns of micro-styragel with a continuous porosity range of 500 Å, 10^3 Å, 10^4 Å, and 10^5 Å.

The weight average molecular weight $(\bar{M}w)$ measurements were determined by utilizing the Chromatix low angle laser light scattering G.P.C. detector (14). The G.P.C. column set was similar to those mentioned above and the carrier solvent was tetrahydrofuran at 25°C with a flow rate of 1.5 ml/min. The (dn/dc) values of both polybutadiene and polyisoprene were determined in THF with the Chromatix KMX-16 differential refractometer.

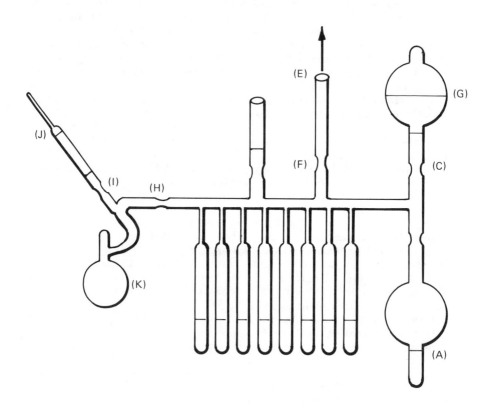

Figure 3.   The sec-*butyllithium split-down apparatus.*

Results and Discussion

One would anticipate the linking of polyisoprenyllithium as well as polybutadienyllithium anions with commercial DVB to be of a complex nature. The control of several reaction parameters could be expected to influence the efficiency of the linking reaction. Young and Fetters (15) have studied the nature of the linking reaction by observing the change in the ultraviolet and visible spectrum throughout the course of the reaction. It was concluded that meta-divinylbenzene was more efficient than para-DVB with respect to star-formation. These authors also observed that significant differences existed between polybutadienyllithium and polyisoprenyllithium anions with respect to their crossover reaction with the DVB isomers.

Qualitatively, we have observed that the initial crossover reaction from the polybutadienyllithium anion to that of the vinylbenzyl anion (II) (Reaction 1) was much slower than the corresponding reaction involving polyisoprenyllithium anions. This observation was in agreement with the Young and Fetters U.V. - visible analysis (15).

$$\text{~~CH=CH-CH}_2^{\ominus}\text{Li}^{\oplus} + \quad \xrightarrow{K_1} \quad \text{~CH=CH-CH}_2\text{-CH}_2\text{-----CH}^{\ominus}\text{Li}^{\oplus} \quad (1)$$

(I)   (II)

Once the vinylbenzylanion (II) forms, it may add to another divinylbenzene molecule resulting in DVB "homopolymerization" (Reaction 2). Likewise, the vinylbenzylanion (II) may attach a pendant vinyl group of another polymer chain (Reaction 3) to give the alkylbenzylanion (III).

$$\text{~~CH}_2\text{CH}^{\ominus}\text{Li}^{\oplus} + \text{CH}_2\text{=CH} \xrightarrow{k_2} \text{~~CH}_2\text{CH-CH}_2\text{CH}^{\ominus}\text{Li}^{\oplus} \quad (2)$$

(II)

$$\text{~~CH}_2\text{CH}^{\ominus}\text{Li}^{\oplus} + \text{~CH}_2\text{CH-CH}_2\text{CH}^{\ominus}\text{Li}^{\oplus} \xrightarrow{k_3} \text{~~CH}_2\text{CHCH}_2\text{CH}^{\ominus}\text{Li}^{\oplus} \quad (3)$$

$$\text{~~CH}_2\text{-CHCH}_2\text{CH}^{\ominus}\text{Li}^{\oplus}$$

(III)

In analogous fashion, an uncapped polybutadienyllithium or poly-isoprenyllithium anion may attack an unreacted pendant vinyl group of another polymer chain (Reaction 4) to give the alkylbenzyl anion (III).

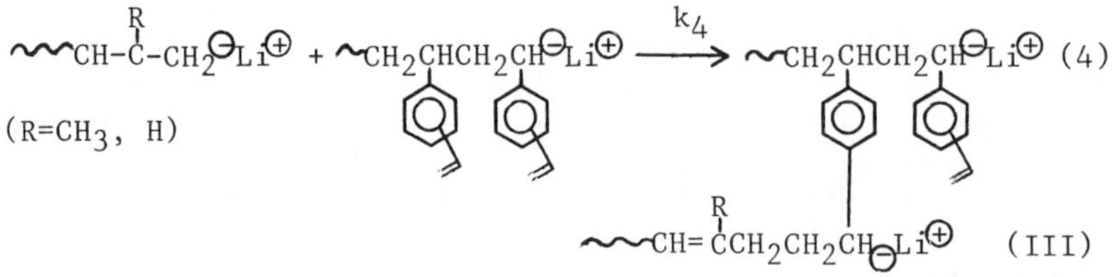

Reactions (3) and (4) described above result in the beginning of star-branched polymer formation. Once the alkylbenzylanion (III) has formed it may further react with any available DVB monomer as illustrated in reaction sequence (5).

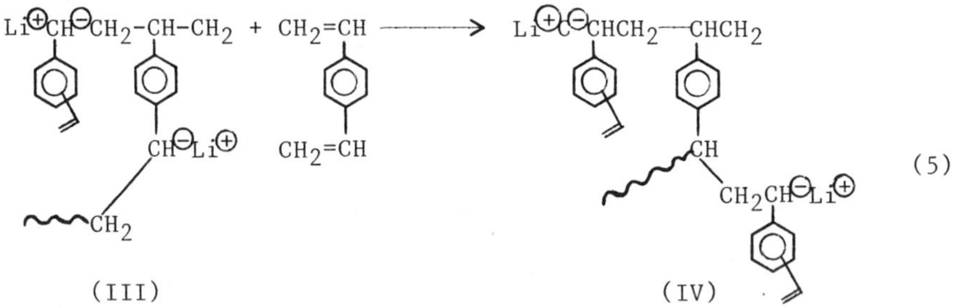

Reaction 5 above would be more probable at the higher molar ratios of DVB/RLi. Finally, in the latter stages of the star formation process, the remaining residual vinyl groups would react as illustrated in (Reaction 6).

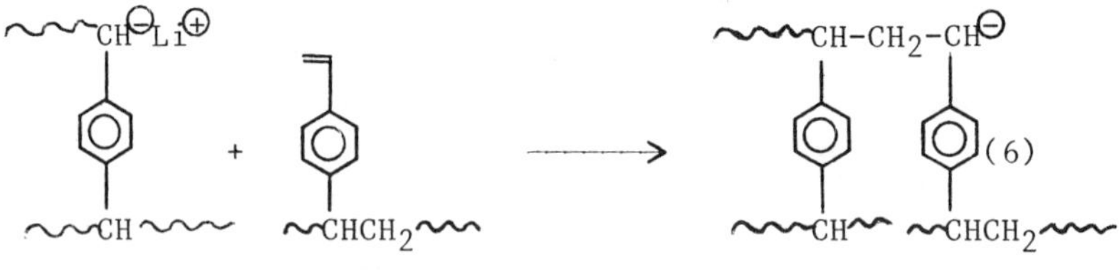

The reaction sequences described above are simplified, since "living" chain end self-association and cross-association may further complicate the reaction.

As seen from the above reaction steps, the DVB procedures for star-branched polymer formation consists of several competitive and consecutive reactions. It was the aim of our work to study various reaction variables which effect this process. As men-tioned earlier, these include: DVB/RLi molar ratio, reaction temperature, reaction time, and the nature of the "living" diene chain end. In the ensuing discussion, we will describe how these above mentioned factors influence the efficiency of the star-formation process.

Influence of the DVB/RLi Molar Ratio

By using various molar concentrations of sec-butyllithium and DVB, the effect of the DVB/RLi ratio was investigated on these star-branched polymerizations. First, butadiene was distilled into a known concentration of a sec-butyllithium hexane solution. The homopolymerization of butadiene was carried out for 48 hours and the polymer analyzed by G.P.C.; a typical G.P.C. chromatogram is shown in Figure 4.   One can observe the narrow distribution ($\overline{Mw}/\overline{Mn}$ = 1.07) and the peak height elution volume at 29 ml. The DVB solution was then added via the break-seal into the poly-butadienyllithium solution at 25°C.   The yellow-orange color developed, indicative of the cross-over reaction.   The star-branching was allowed to proceed for 72 hours after which the polymer was isolated. The G.P.C. chromatogram of the resulting polymer is shown in Figure 5. One notices a considerable portion of "unlinked" material with an elution volume at 27 ml. The linear precursor, however, had a peak elution volume at 29 ml. From the G.P.C. analysis it was concluded the polymer fraction eluting at 27 ml. was a coupled dimer or two-arm star. The DVB/RLi ratio in this case was 3.0 (corrected for 44% EVB). From this observation it became of interest to study the influence of DVB/RLi ratio on the efficiency of star formation. The reaction time, temperature, and arm molecular weight were held constant while the DVB/RLi ratio varied.

Figure 6 illustrates the effect of the DVB/RLi ratio. It was observed that as the DVB/RLi ratio increased, the percentage of unlinked linear material decreased. At the lower DVB/RLi ratios, linear coupled two arm stars were formed (see Ref. 15), while at the higher ratios, the linear material was not coupled. Thus, there is a strong influence on the efficiency of the star-branched polymer formation as the DVB/RLi ratio is varied for the case of poly(butadienyllithium). At the low molar ratios of DVB/RLi and two arm coupling reaction competes with the star-formation process.

Figure 7 illustrates the influence of the increase in the DVB/RLi ratio as well as reaction time for polybutadienyllithium anions.   The efficiency is plotted as the ratio of linked to unlinked chains.   At the very high ratios (11.9-12.9) nearly quantitative linking is observed as seen from the G.P.C. analysis in Figure 8.   However, the moelcular weight distribution is broadened at the higher ratios, possibly indicating intermolecular or inter-nodule star coupling between two different star macro-molecules. As mentioned previously, Reaction 5 would be more likely to occur at the higher ratios; thus the overlap of the

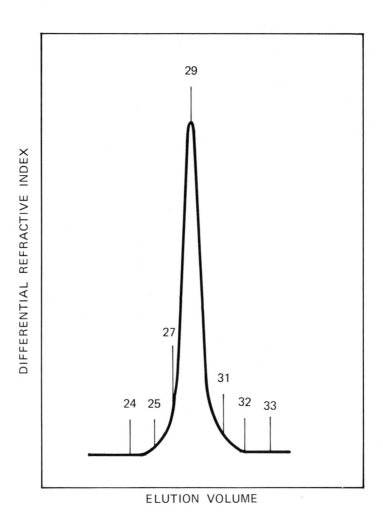

Figure 4.   GPC of linear precursor, Sample 2-MM-S-1.

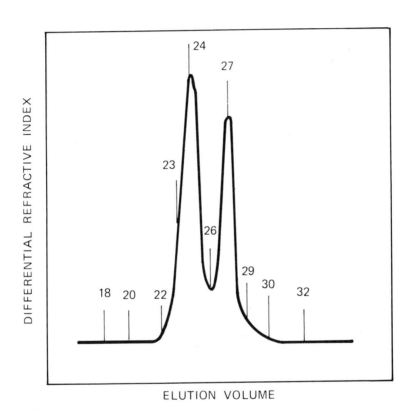

Figure 5. GPC of star-branch polybutadiene linked for 72 h at 25°C, Sample 2-MM-S-1.

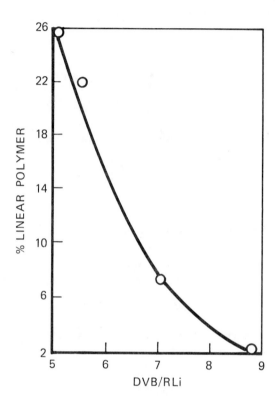

Figure 6.   Percent linear polymer vs. DVB/RLi ratio for polybutadiene anions linked with DVB at 45°C for 48 h ($\overline{M}_n$ arm = 35,000).

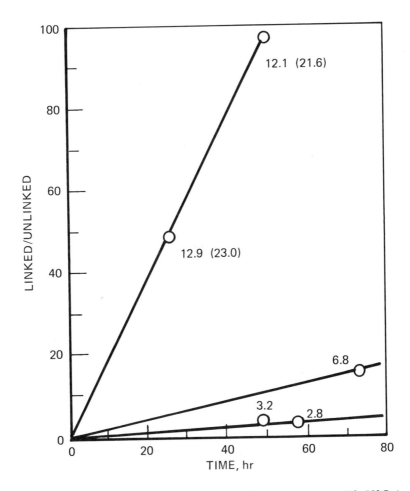

Figure 7. Linking efficiency of polybutadienyllithium anions at 45°–50°C for various DVB/RLi molar ratios.

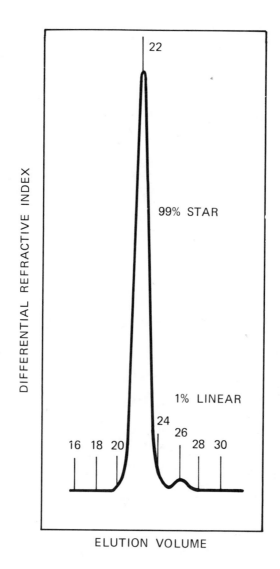

Figure 8.   GPC trace for a star-branched polybutadiene linked for 48 h at 50°C
with DVB/RLi = 12.1, Sample 3-MM-S-12.

vinylbenzyl anion of the polymer chain to that of adjacent "living" star nodule would result in inter-star coupling. A broadened G.P.C. curve may be observed by using the Chromatix low angle laser light scattering detector (LALLS). One advantage this detector offers is its high sensitivity to high molecular weight species. As observed in Figure 9, the weight average molecular weight elution volume profile shows a shoulder at the high molecular weight end of the chromatogram. This shoulder reflects the percentage of internodule coupling observed at the higher DVB/RLi ratio. This shoulder also reflects an inconsistency in the relationship of molecular weight to hydrodynamic volume. In other words, an increase in the eluant molecular weight is observed with increased molecular branching or it may be related to the column exclusion limit. Inter-nodule coupling gives increased branching along with increased molecular weight as evidenced by the Chromatix LALLS analysis.

For polyisoprenyllithium anions, it was also observed that increased linking efficiency was achieved at higher DVB/RLi ratios, Figure 10. In comparison, the polyisoprenyllithium anions link more efficiently than polybutadienyllithium. Also, in contrast to polybutadienyllithium anions, polyisoprene star formation does not result in coupled dimer at the lower DVB/RLi ratios. This is perhaps reflected in the faster cross-over rate from the polydienyllithium anion to the vinylbenzylanion (reaction 1) for polyisoprene.

The reaction temperature was also found to influence the efficiency of the linking reaction. As summarized in Table I, the increase in reaction temperature from 25°C to 45°C resulted in an increase in star-branch polymer formation. Qualitatively, it was also observed that the formation of vinylbenzylanion occurred more rapidly at the elevated temperatures. Apparently, the increased reaction temperatures render the intermolecular attack on the pendant vinyl groups by the polydienyllithium anions more favorable in comparison to the intramolecular intra-nodule alkylbenzyl anion-vinyl group reaction (Table I on the following page).

From the summary in Table I, it can be seen that increased reaction temperatures lead to more rapid and efficient linking, when compared to the reactions carried out at 25°C.

Comparing samples in Table I (4-MM-S-17, a-e), one observes that increasing the reacting time also results in more quantitative linking. However, the percent of further increased linking is smaller at longer reaction times. As the "living" star nodule continues of react, increased steric hindrance slows down the ability of a macromolecular carbanion to enter into the "microgel" nodule.

From the data in Tables I and II, one observes an increase in the DVB/RLi ratio does not necessarily result in more highly branched star-shaped polydienes. In other words, an increase in the DVB/RLi ratio results in more quantitative star formation but not always increased branch functionality.

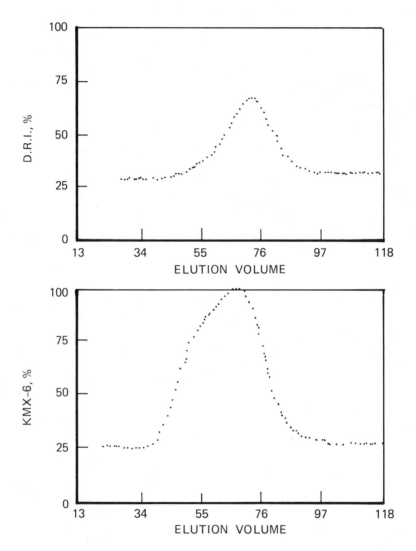

*Figure 9.   KMK-6/DP GPC detector response for Sample PI-S-16, a star-branched
polymer linked with DVB/RLi = 11.9, Sample 3-MM-S-16.*

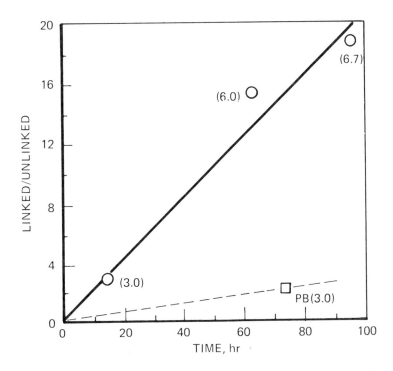

Figure 10.   Linking efficiency of polyisoprenyllithium anions at 25°C as a function of time and DVB/RLi ratio.

TABLE I

POLYBUTADIENE STAR POLYMERS

Sample	DVB/RLi[a]	$\overline{N}a$[b]	% Linear[c]	Reaction Time (HRS)	T°C	Solvent
2-MM-S-1	3.0	18	36D[d]	72	25°C	Hexane
2-MM-S-5	5.9	--	29 [e]	24	45°C	Hexane
2-MM-S-6	4.9	15	6	36	45°C	Hexane
2-MM-S-7	3.2	23.4	21.9D	48	45°C	Hexane
2-MM-S-8	2.8	13.4	25.7D	56	45°C	Hexane
3-MM-S-12	12.1	16	1	48	50°C	Benzene
3-MM-S-13	12.9	--	2	24	50°C	Benzene
3-MM-S-15	6.8	23.8	5.9	72	50°C	Hexane
4-MM-S-17a	4.0	--	33.1	6	25°C	Benzene
4-MM-S-17b	4.0	--	13.8	24	25°C	Benzene
4-MM-S-17c	4.0	--	4	152	25°C	Benzene
4-MM-S-17d	4.0	--	2.4	163	25°C	Benzene
4-MM-S-17e	4.0	--	1.7	200	25°C	Benzene

(a) DVB/RLi ratio has been corrected for the 44% ethylvinylbenzene present in the commercial DVB.

(b) $\overline{N}a = \dfrac{\overline{M}n \text{ Star}}{\overline{M}n \text{ Linear}}$

(c) % linear from G.P.C.

(d) D represents coupled dimer.

(e) It is suspected some premature termination resulted for 2-MM-S-5.

TABLE II

POLYISOPRENE STAR POLYMERS

Sample	DVB/RLi(a)	$\overline{N}a$(b)	% Linear(c)	Reaction Time (hrs.)	Temp. (°C)	Solvent
2-MM-S-2	3.0	7.7	23%	14	25°C	Hexane
2-MM-S-3	6.1	13.1	6%	62	25°C	Hexane
2-MM-S-4	6.7	9.8	5%	96	25°C	Hexane
2-MM-S-9	5.3	12.7	3%	62	45°C	Hexane
2-MM-S-10	3.2	9.8	8.5%	62	45°C	Hexane
2-MM-S-16	11.9	12.1	9.8%	48	50°C	Benzene

(a) DVB/RLi has been corrected for 44% Ethylvinylbenzene present in commercial DVB.

(b) $\overline{N}a = \dfrac{Mn\ star}{Mn\ arm}$

(c) Determined from G.P.C.

In comparison to the polybutadiene stars under similar reaction conditions, the polyisoprene stars showed slightly lower degrees of branching. The added steric hindrance from the methyl group on the polyisoprene anion perhaps makes entry into the DVB "microgel" nodule difficult.

LALLS Analysis

The Chromatix ($\underline{14}$) low angle laser light scattering (LALLS) G.P.C. detector was employed for further analysis of a few star-shaped polydienes. The unique feature of the (LALLS) detector is its high sensitivity to the high molecular weight fractions within the molecular weight distribution. The LALLS detector is also able to measure the weight average molecular weight $\overline{M}w$ as a function of elution volume from the G.P.C. Thus, the "universal" calibration procedure is not required. Table III summarized the LALLS results. From these results, it can be concluded that higher DVB/RLi ratios result in broader molecular weight distributions. As described previously, this is due to an increased amount of inter-nodule linking between two growing star-shaped molecules. The light-scattering $\overline{M}n$ values agree with those determined independently from membrane osmometry measurements.

Star-Polymer Characterization

Characterization was achieved by using a Waters Gel Permeation chromatograph. The $\overline{M}n$ molecular weight characterization was achieved by membrane osmometry. Table IV summarizes the G.P.C. as well as the membrane osmometry results. The elution volume behavior for the star-branched materials appear insensitive to the overall star-branched polymer molecular weight, while more dependent upon arm molecular weight. This is what one might expect as the G.P.C. separation process occurs by differences in hydrodynamic volume not actual molecular weight. As the star-branched arm molecular weight increases so does the hydrodynamic volume, hence earlier elution volumes would be expected with increasing arm molecular weight.

Conclusion

Star-branched poly(butadienes) and poly(isoprenes) have been synthesized by linking the "living" chain ends with commercial DVB. Higher DVB/RLi molar ratios result in a more efficient star-branching process. DVB/RLi molar ratios greater than 10-11 give nearly quantitative linking; however, these star-branched polymers have higher polydispersity ratios (e.g. $\overline{M}w/\overline{M}n = 1.3$). This can be rationalized by an increased probability of intermolecular star-nodule coupling result in a high molecular weight fraction within the molecular weight distribution. Increased reaction

TABLE III

CHROMATIX LASER LIGHT SCATTERING GPC SUMMARY

Sample	Light-Scattering $\overline{M}_n$(a) x10⁻⁵	$\overline{M}_w$(b) x10⁻⁵	Osmotic $\overline{M}_n$ x10⁻⁵	$(\overline{M}_w/\overline{M}_n)$ l.s.	$(\overline{M}_w/\overline{M}_n)$ o.p.	DVB/RLi(c)
PI-S-16	4.23	5.63	4.58	1.33	1.23	21.2 (11.9)
PI-S-16	4.46	6.10	4.58	1.37	1.33	21.2 (11.9)
PI-S-9	5.21	5.75	5.22	1.10	1.10	9.5 (5.3)
PB-S-15	8.34	8.65	--	1.04	--	12.2 (6.8)
PB-S-8	4.68	4.86	5.49	1.04	--	5.0 (2.8)

$\overline{(DN/DC)}$ = 0.1407 PI/THF 25°C $\overline{(DN/DC)}$ = 0.1398 PI/THF 25°C

(a) $\overline{M}_n = \dfrac{\Sigma\, c_I}{\Sigma\, \dfrac{c_I}{M_I}}$

(b) $\overline{M}_w = \dfrac{\Sigma\, c_I M_I}{\Sigma\, c_I}$

(c) Values in parentheses are corrected for 44% ethylvinyl-benzene.

TABLE IV

Sample	$\overline{M}n$ Linear Precursor	$\overline{M}n$ Star	Elution Volume	
			Star	Linear
Polybutadiene	10^{-3}	10^{-4}	ml	
2-MM-S-1	32.0	58.9	24.2[a]	29
2-MM-S-6	64.0	95.9	22.7	27
2-MM-S-7	35.2	82.4	22.1	26.2
2-MM-S-8	34.8	61.4	22.1	26.0
3-MM-S-12	35.0	56.2	22.1	26.2
3-MM-S-15	35.0	83.4	22.0	26.2
Polyisoprene	$\overline{M}n$ Linear Precursor 10^{-3}	$\overline{M}n$ Star 10^{-3}	Star	Linear
2-MM-S-2	27.6	21.2	24.5	28.3
2-MM-S-3	51.0	67.3	23.6	27.5
2-MM-S-4	55.7	56.2	23.5	27.6
2-MM-S-9	41.2	52.2	23.8	27.9
2-MM-S-10	39.8	38.9	24	28.0
2-MM-S-16	37.7	45.8	22[b]	27.1

(a) GPC column set different for sample S-1.
(b) GPC column set different for sample S-16.

times and temperatures likewise give more quantitative linking for a given DVB/RLi molar ratio.

The star-formation process was also found to be sensitive to the nature of the "living" polydienyllithium chain end. Poly-butadienyllithium anions cross-over to give the vinylbenzylani-ons, upon reaction with DVB, at a slower rate than the cor-responding polyisoprenyllithium anions. Furthermore, it was observed that when low DVB/RLi ratios were employed to link polybutadienyllithium anions, that a two-arm coupling reaction resulted, in effective competition with star-branched polymer formation. However, in the case of polyisoprenyllithium anions no such dimerization was observed under similar linking reaction conditions. This can perhaps be reflected in the overall faster cross-over reaction of polyisoprenyllithium chain ends with DVB as compared to the corresponding reaction for polybutadienyllithium.

Low Angle Laser Light Scattering (LALLS) results indicate, that at lower DVB/RLi molar ratios, a relatively narrow molecular weight distribution (1.04-1.1) of star-branched polymers can be synthesized. In contrast, DVB/RLi ratios greater than 11 can lead to rather broad molecular weight distributions ($\overline{Mw}/\overline{Mn}$ = 1.3).

It is not certain how the ethylvinylbenzene isomers influence the star-formation process and further experiments are necessary. The influence of reaction temperature on the stability of the vinylbenzyl and polydienyllithium chain ends is presently under investigation utilizing U.V.-visible spectroscopic techniques. Future experiments are in progress (16) to further advance the existing knowledge of this interesting polymerization process.

Acknowledgments

The authors wish to thank the Army Research office for financial support of this work. We are also grateful to Dr. Jordan and Phil Christ of Chromatix for the opportunity to obtain the LALLS G.P.C. results.

Literature Cited

1.   R. Milkovich, Canadian Patent 712245 (Aug. 24, 1965).

2.   D. Decker and P. Rempp, C. R. Hebd. Sceances Acad. Sci., 261, 1977 (1965).

3.   J. G. Zilliox, D. Decker, and P. Rempp, C. R. Hebd. Sceances Acad. Sci., Ser. C, 262, 726 (1966).

4.   D. J. Worsfold, J. G. Zillox, and P. Rempp, Can. J. Chem., 47, 3379, (1969).

5.   G. Quack, L. J. Fetters, N. Hadjichristidis, and R. N. Young, Ind. Eng. Chem. Prod. Res. Dev., 19, 587 (1980).

6.   G. Quack, L. J. Fetters, Polym. Preprints, Am. Chem. Soc., Div. Polym. Chem., 18(2), 558, (1977).

7.   Li-K. Bi, and L. J. Fetters, Macromolecules, 8, 90, (1975).

8.   Li-K. Bi, and L. J. Fetters, Macromolecules, 9, 732, (1976).

9.   K. Ziegler, Angew. Chem., 49, 499, (1936).

10. L. M. Foclur, A. G. Kitchen, C. C. Baird, Am. Chem. Soc. Div. Org. Coat. Plast. Chem. Pap., 34(1), 130, (1974).

11. O. L. Marrs, F. E. Naylor, and L. O. Edmonds, J. Adhes., 4, 211, (1972).

12. O. L. Marrs and L. O. Edmonds, Adhesives Age, 14(12), 15, (1971).

13. M. Morton and L. J. Fetters, Rubber Reviews, 48, 359, (1975).

14. M. L. McConnell, American Laboratory, May, 1978.

15. R. N. Young, and L. J. Fetters, Macromolecules, 11, 899, (1978).

16. M. Martin, and J. E. McGrath, Polymer Preprints, 23(1), (Atlanta ACS Preprints), 000, (1981).

RECEIVED March 17, 1981.

INDEX

Jacket design by Carol Conway.

Production by V. J. DeVeaux.

Elements typeset by Service Composition Co., Baltimore, MD.

Printed and bound by The Maple Press Co., York, PA.